The Chemokine Receptors

THE RECEPTORS

KIM A. NEVE, SERIES EDITOR

The Chemokine Receptors, EDITED BY Jeffrey K. Harrison and Nicholas W. Lukacs, 2007

The Serotonin Receptors: From Molecular Pharmacology To Human Therapeutics, EDITED BY Bryan L. Roth, 2006

The Adrenergic Receptors: In the 21st Century, EDITED BY Dianne M. Perez, 2005

The Melanocortin Receptors, EDITED BY Roger D. Cone, 2000

The GABA Receptors, Second Editon, EDITED BY S.J. Enna and Norman G. Bowery, 1997

The Ionotropic Glutamate Receptors, EDITED BY Daniel T. Monaghan and Robert Wenthold, 1997

The Dopamine Receptors, EDITED BY Kim A. Neve and Rachael L. Neve, 1997

The Metabotropic Glutamate Receptors, EDITED BY P. Jeffrey Conn and Jitendra Patel, 1994

The Tachykinin Receptors, EDITED BY Stephen H. Buck, 1994

The Beta-Adrenergic Receptors, EDITED BY John P. Perkins, 1991

Adenosine and Adenosine Receptors, EDITED BY Michael Williams, 1990

The Muscarinic Receptors, EDITED BY Joan Heller Brown, 1989

The Serotonin Receptors, EDITED BY Elaine Sanders-Bush, 1988

The Chemokine Receptors

Edited by

Jeffrey K. Harrison, PhD

Department of Pharmacology and Therapeutics
University of Florida College of Medicine
Gainesville, FL

and

Nicholas W. Lukacs, PhD

Department of Pathology and Graduate Program in Immunology
University of Michigan Medical School
Ann Arbor, MI

HUMANA PRESS ✷ TOTOWA, NEW JERSEY

999 Riverview Drive, Suite 208
Totowa, New Jersey 07512

www.humanapress.com

For additional copies, pricing for bulk purchases, and/or information about other Humana titles, contact Humana at the above address or at any of the following numbers: Tel.: 973-256-1699; Fax: 973-256-8341, E-mail: order@humanapr.com; or visit our Website: http://www.humanapress.com.

This publication is printed on acid-free paper. ∞
ANSI Z39.48-1984 (American National Standards Institute) Permanence of Paper for Printed Library Materials

10 9 8 7 6 5 4 3 2 1

Library of Congress Control Number: 2006939954

ISBN: 978-1-58829-773-0 e-ISBN: 978-1-59745-020-1

Preface

G-protein coupled receptors (GPCRs) constitute a large superfamily of proteins that have been historically well represented as drug targets. As such, many of the volumes within *The Receptors* series have focused their attention on receptor subfamilies in the GPCR superfamily (e.g., adrenergic, serotonergic, dopaminergic). While chemokine receptors are relatively young, in terms of the knowledge that has accumulated regarding their roles in physiology and pathology, it is clear that this GPCR subfamily is dwarfing all others with respect to the total number of receptors and their functional diversity. The breadth of the chemokine field was not fully appreciated in the early days of chemokine discovery, when the characterization of these molecules was largely a curiosity of a small set of biologists primarily interested in inflammation. Presently, the field of *chemokinology* is quite diverse, and it is nearly impossible to stay informed on all aspects of chemokine biology. Nonetheless, this book seeks to distill information regarding the molecular and cell biology, physiology, and pathology of chemokines and their receptors into a single volume, providing an excellent resource for those entering the field or the most experienced chemokinologist seeking new insights. In the chapters, leading authorities consider structural and cellular aspects of chemokines and chemokine receptors, and more notably, discuss the roles of these molecules in a variety of disease states that transcend their classically defined role(s) in inflammation. The reader will become acquainted with the functional diversity of these gene families and more importantly, we hope, will grasp the significance of these receptors to researchers in academia and the pharmaceutical industry.

We are grateful to the many experts who have contributed their time and energy to this book. These authors have played significant roles in the early discoveries and continue to expand the knowledge base related to their subfield specialties. Finally, we thank Dr. Kim Neve for the kind invitation to direct this project and the editorial staff for its help in bringing this book to production.

***Jeffrey K. Harrison,* PhD**
***Nicholas W. Lukacs,* PhD**

Contents

Contributors

SEEMA S. AHUJA • *South Texas Veterans Health Care System, Audie L. Murphy Division, San Antonio, Veterans Administration Center for Research on AIDS and HIV-1 Infection, San Antonio, TX; and Department of Medicine, University of Texas Health Science Center at San Antonio, San Antonio, TX, USA*

JOHN A. BELPERIO • *Department of Medicine, Division of Pulmonary and Critical Care Medicine, David Geffen School of Medicine at University of California at Los Angeles, Los Angeles, CA, USA*

CHRISTOPHER C. BRODER • *Department of Microbiology and Immunology, Uniformed Services University, Bethesda, MD, USA*

ASTRID E. CARDONA • *Neuroinflammation Research Center, Department of Neurosciences, Lerner Research Institute, Cleveland Clinic, Cleveland, OH, USA*

CARLOS A. ESTRADA • *South Texas Veterans Health Care System, Audie L. Murphy Division, San Antonio, Veterans Administration Center for Research on AIDS and HIV-1 Infection, San Antonio, TX; and Department of Medicine, University of Texas Health Science Center at San Antonio, San Antonio, TX, USA*

JOSHUA M. FARBER • *Inflammation Biology Section, Laboratory of Molecular Immunology, National Institute of Allergy and Infectious Diseases, National Institutes of Health, Bethesda, MD, USA*

AMY M. FULTON • *Department of Pathology, School of Medicine, and Marlene and Stewart Greenebaum Cancer Center, University of Maryland, Baltimore, MD, USA*

GERARD J. GRAHAM • *Division of Immunology, Infection and Inflammation, University of Glasgow, Glasgow G12 8TA, United Kingdom*

JEFFREY K. HARRISON • *Department of Pharmacology and Therapeutics, University of Florida College of Medicine, Gainesville, FL, USA*

MICHAEL N. HEDRICK • *Inflammation Biology Section, Laboratory of Molecular Immunology, National Institute of Allergy and Infectious Diseases, National Institutes of Health, Bethesda, MD, USA*

CORY M. HOGABOAM • *Department of Pathology, University of Michigan Medical School, Ann Arbor, MI, USA*

RICHARD HORUK • *Departments of Molecular Pharmacology and Immunology, Berlex Biosciences, Richmond, CA, USA*

MAYA R. JERATH • *Department of Medicine, Thurston Arthritis Research Center, University of North Carolina at Chapel Hill, Chapel Hill, NC, USA*

FABIO JIMENEZ • *South Texas Veterans Health Care System, Audie L. Murphy Division, San Antonio, Veterans Administration Center for Research on AIDS and HIV-1 Infection, San Antonio, TX; and Department of Medicine, University of Texas Health Science Center at San Antonio, San Antonio, TX, USA*

MICHAEL P. KEANE • *Department of Medicine, Division of Pulmonary and Critical Care Medicine, David Geffen School of Medicine at University of California at Los Angeles, Los Angeles, CA, USA*

MILDRED KWAN • *Department of Medicine, Thurston Arthritis Research Center, University of North Carolina at Chapel Hill, Chapel Hill, NC, USA*

CHANG H. KIM • *Laboratory of Immunology and Hematopoiesis, Department of Pathobiology, Purdue Cancer Center, Purdue University, West Lafayette, IN, USA*

DENNIS M. LINDELL • *Department of Pathology, University of Michigan Medical School, Ann Arbor, MI, USA*

PENG LIU • *Department of Medicine, Thurston Arthritis Research Center, University of North Carolina at Chapel Hill, Chapel Hill, NC, USA*

ELIAS LOLIS • *Department of Pharmacology, Yale University School of Medicine, New Haven, CT, USA*

NICHOLAS W. LUKACS • *Department of Pathology and Graduate Program in Immunology, University of Michigan Medical School, Ann Arbor, MI, USA*

HERNAN G. MARTINIEZ, • *South Texas Veterans Health Care System, Audie L. Murphy Division, San Antonio, Veterans Administration Center for Research on AIDS and HIV-1 Infection, San Antonio, TX; and Department of Medicine, University of Texas Health Science Center at San Antonio, San Antonio, TX, USA*

KENJIRO MATSUNO • *Department of Anatomy (Macro), Dokkyo University School of Medicine, Tochigi 321-0293, Japan*

KOUJI MATSUSHIMA • *Department of Molecular Preventive Medicine, Graduate School of Medicine, The University of Tokyo, Bunkyo-city, Tokyo 113-0033, Japan*

James W. Murphy • *Department of Pharmacology, Yale University School of Medicine, New Haven, CT, USA*

Peter Jon Nelson • *Medizinische Poliklinik-Innenstadt, Ludwig-Maximilians-University Munich, Schillerstrasse 42, 80336 Munich, Germany*

Robert J. Nibbs • *Division of Immunology, Infection and Inflammation, University of Glasgow, Glasgow G12 8TA, United Kingdom*

Dhavalkumar D. Patel • *Department of Medicine, Thurston Arthritis Research Center, University of North Carolina at Chapel Hill, Chapel Hill, NC, USA; and Novartis Institutes of Biomedical Research, CH-4002, Basel, Switzerland*

Marlon P. Quinones • *South Texas Veterans Health Care System, Audie L. Murphy Division, San Antonio, Veterans Administration Center for Research on AIDS and HIV-1 Infection, San Antonio, TX; and Department of Medicine, University of Texas Health Science Center at San Antonio, San Antonio, TX, USA*

Richard M. Ransohoff • *Neuroinflammation Research Center, Department of Neurosciences, Lerner Research Institute, Cleveland Clinic, Cleveland, OH, USA*

Sofia Ribeiro • *Departments of Molecular Pharmacology and Immunology, Berlex Biosciences, Richmond, CA, USA*

Detlef Schlondorff • *Medizinische Poliklinik-Innenstadt, Ludwig-Maximilians-University Munich, Schillerstrasse 42, 80336 Munich, Germany*

Stephan Segerer • *Medizinische Poliklinik-Innenstadt, Ludwig-Maximilians-University Munich, Schillerstrasse 42, 80336 Munich, Germany*

Laura Smith • *Department of Pharmacy and Pharmacology, University of Bath, Bath BA2 7AY, United Kingdom*

Tzanko S. Stantchev • *Department of Microbiology and Immunology, Uniformed Services University, Bethesda, MD, USA*

Robert M. Strieter • *Department of Internal Medicine, University of Virginia School of Medicine, Charlottesville, VA, USA*

Glenda Trujillo • *Department of Pathology, University of Michigan Medical School, Ann Arbor, MI, USA*

Tonya C. Walser • *Department of Pathology, School of Medicine, University of Maryland, Baltimore, MD, USA*

Stephen G. Ward • *Department of Pharmacy and Pharmacology, University of Bath, Bath BA2 7AY, United Kingdom*

ADAM WEBB, • *Department of Pharmacy and Pharmacology, University of Bath, Bath BA2 7AY, United Kingdom*

HIROYUKI YONEYAMA • *Laboratory of Stem Cell Dynamism, Stelic Institute of Regenerative Medicine, 1-9-15 Higashi-Azabu, Minato-city, Tokyo 106-0044; and Department of Molecular Preventive Medicine, Graduate School of Medicine, The University of Tokyo, Bunkyo-city, Tokyo 113-0033, Japan*

TEIZO YOSHIMURA • *Laboratory of Molecular Immunoregulation, National Cancer Institute at Frederick, Frederick, MD, USA*

Color Plate

The following illustrations appear in color in the insert that follows page 178.

1

The Birth and Maturation of Chemokines and Their Receptors

Nicholas W. Lukacs and Jeffrey K. Harrison

Summary

Chemokines were initially discovered in the context of inflammatory pathologies and were the curiosity of a limited number of researchers. Receptors for these cytokine molecules were identified shortly thereafter and determined to be members of the large G protein–coupled receptor (GPCR) superfamily. Collectively, these basic observations provided a framework to understand mechanisms by which leukocyte subsets could migrate into tissues in a specific manner. Nonetheless, as the field evolved, so came a realization of the broader impact of this family on diverse biological processes, which include the regulation of leukocyte trafficking in hematopoiesis, innate and adaptive immunity, angiogenesis, cancer, and viral pathogenesis. We begin this volume with an overview of research on chemokines and their receptors and then subsequently migrate into a brief discussion of key findings that defined the maturation of this field. As members of the GPCR superfamily have historically dominated drug development programs, so too has interest in chemokine receptors as therapeutic targets.

Key Words: Chemotactic cytokine; GPCR; monocyte; neutrophil; proinflammatory.

1.1. Birth of Chemotactic Cytokines, or Chemokines

Not more than 20 years ago, a limited number of laboratories around the world were attempting to investigate a small peptide mediator that seemed to have neutrophil chemotactic properties and was thought to be involved in regulation of inflammatory responses. This molecule was identified and

From: *The Receptors: The Chemokine Receptors*
Edited by: J. K. Harrison and N. W. Lukacs © Humana Press Inc., Totowa, NJ

initially termed *monocyte-derived neutrophil chemotactic factor* (MDNCF)/ *neutrophil-activating factor* (NAF) and was subsequently termed *interleukin-8* (IL-8). At this same time, Anthony Cerami and coworkers had identified a series of macrophage-expressed molecules with functional similarities that they termed *macrophage inflammatory proteins* (MIPs), which had either monocyte (MIP-1) or neutrophil (MIP-2) chemotactic activity *(1)*. These MIP molecules were later determined to have a number of important properties aside from their chemotactic activity, including acting as endogenous pyrogens and having effects on myelopoiesis. These heparin-binding molecules were subsequently found to have significant sequence homology with a number of other small protein cytokines (8 to 10 kd) including the MIP-1–like molecules MIP-1β, TCA-3/I-309, RANTES, and JE/monocyte/macrophage chemotactic protein (MCP)-1 and the MIP-2–similar proteins PF4, IP-10, and MGSA/GRO. At an international meeting on chemokines in 1992 in Vienna, Austria, these proteins were collectively termed *chemokines*, short for chemotactic cytokines. Initially, the few chemokines identified were characterized by the robust induction and high expression levels during inflammatory conditions, both in vitro and in vivo. These newly discovered molecules soon became intensely researched, and interest in them as viable targets for therapeutic intervention rapidly grew. This seemed quite plausible and initially straightforward. It was envisioned that IL-8 would be targeted to block neutrophil recruitment, whereas MCP and/or MIP-1 would be targeted to block monocyte recruitment. However, these early investigations and discoveries clearly failed to comprehend the magnitude and impact the chemokines would have on the biology related to inflammation/immunology and every aspect of research from development to wound healing, as well as in diseases ranging from bacterial infections to cancer.

Since those initial days of discovery, the chemokine family has grown to more than 40 ligands. Chemokines were originally divided based on the position of the first 2 cysteine residues in their amino acid sequence (i.e., CXC or CC) but can now be further subdivided in a number of ways based on differing functional parameters. These include their ability to initiate inflammatory versus homeostatic migration and/or recirculation of lymphocytes, granulocytes, and mononuclear cells (homeostatic vs. inflammatory/inducible). Other delineations include, for instance, an ability to promote or inhibit angiogenesis (ELR vs. non-ELR CXC chemokines). The use of broad-based programs that searched out new chemokine molecules by sequence homology as opposed to biological function allowed an efficient identification of the various chemokine ligands. However, in doing so the biology of the individual ligands has not always been thoroughly examined. In addition to the two main families, CXC and CC, two other subfamilies have also been identified: the C and CX3C families. These two subfamilies have a single member, and each appears to have important functions in immune responses with unique niches. The impact of chemokines on the research community can clearly be appreciated by the accelerated pace of publications on these molecules over the past two decades. In 1990, there were less than 500 searchable references (via PubMed) on chemokines. By mid-2006, there were more than 31,000 papers published related to chemokines. Many of the recent publications still focus on some of these initially characterized chemokines, further demonstrating the continuous efforts to understand and clarify a still evolving biology of the various chemokines.

By 1999, it became clear that with simultaneous discovery of multiple ligands with different names, a more simplified nomenclature for this now large family of ligands and receptors was needed. An international group of investigators, headed by Osamu Yoshie and Albert Zlotnik, worked together and established a logical and numerical nomenclature that used order of discovery in the two largest families, CXC and CC, for identifying the different ligands *(2)*. Although initially confusing for those already working in the field, it led to an easier identification and categorization of ligands for investigators who were beginning research on chemokine biology. A table highlighting the nomenclature of the chemokine family is presented in Chapter 2. Internet access to this information is available at http://cytokine.medic.kumamoto-u.ac.jp/CFC/CK/Chemokine.html. This site provides a wealth of information related to chemokines and affords researchers the opportunity for updates as new ligands are discovered. The use of this standardized nomenclature is strongly recommended, although concurrent use of new and old names is acceptable as new publications arise, particularly as it will be necessary to bridge the old with the new literature.

The ability of chemokines to bind to glycosaminoglycans (GAGs) was initially a biochemical tool, but soon it was realized that this function allowed chemokines to be deposited onto cells and into tissues establishing solid-phase gradients. This had multiple effects on the biology of chemokines. First, one can quickly imagine how a solid-phase gradient would be a more stable platform than trying to maintain a soluble gradient in a dynamic environment. Second, it was soon realized that chemokine signals were required to change the affinity of β-integrins on leukocytes to transition from rolling along the activated endothelium to firmly adhering and spreading on adhesion molecules prior to emigration to inflamed tissue. Third, the ability of chemokines to bind to GAGs at a site of inflammation or nidus of recruitment may also allow a concentrating effect that could intensify the signal for leukocyte stasis or activation of effector function, such as degranulation, once the cell has been recruited to the site. Finally, the ability to bind to GAGs may also serve as an efficient clearance mechanism in circulation and tissue compartments. Although the full functionality of GAGs on chemokine biology is only beginning to be explored more thoroughly, its importance is clear from data already generated.

1.2. Identification of Chemokine Receptors as G Protein–Coupled Receptors

Some of the early excitement of targeting chemokines came from the identification of receptors for these cytokines as members of the family of G protein–coupled receptors (GPCRs), molecules that have traditionally been the most successful for target compound development. It was imagined that by blocking a single receptor, inflammation in numerous diseases could be attenuated, although this was clearly an oversimplification of the biology. In the early 1990s, the first chemokine receptors were identified with the discovery of two CXC (IL-8) receptors *(3,4)*. By 1993, the first CC chemokine receptor was reported *(5,6)*, and thus the quest began to identify the multiple GPCRs that were important for cellular responses to these chemotactic factors. There are now 19 known chemokine receptors, binding different subfamilies of chemokines, although this is by no means complete. Similar to the chemokine ligands, receptors for chemokines are so named based on a standardized nomenclature system *(7)*. Despite identification of chemokine ligands prior to the discovery of the first chemokine receptor, rules for receptor nomenclature were adopted prior to the establishment of the chemokine nomenclature. Receptor nomenclature has its basis in the human genes, and formal "blessing" is dependent on not only establishing binding characteristics of the receptor but also determining some signaling properties. For instance, some of these receptors clearly bind chemokines (e.g., D6 and duffy antigen receptor for chemokines (DARC)) but lack an ability to stimulate any known signaling pathways. Given that signaling associated with these chemokine-binding proteins is established in the future, these will be accepted with a standardized nomenclature. A complete list of receptors is found in Chapter 3 and can also be accessed on the Web at the International Union of Pharmacology (IUPHAR)/GPCR database: http://www.iuphar-db.org/GPCR/ReceptorFamiliesForward.

Laboratories continue to identify potentially new chemokine binding receptors from a vast array of orphan receptors. The newest receptor, tentatively called CXCR7, has now been characterized *(8,9)* and binds CXCL11 and CXCL12, chemokines formally thought to bind exclusively to CXCR3 and CXCR4, respectively. Thus, this area of research continuously seeks to identify new receptor members while at the same time attempts to better understand the biology of known receptors and the signaling events associated with ligand interactions.

The initially identified receptors were generally ones that bound the inflammatory chemokines. In most instances, these receptors bind chemokine ligands in a very promiscuous manner. Thus, the redundant nature of these interactions coupled with broad cellular expression patterns made the task of identifying disease-associated targets more difficult. The chemokine receptor family also contains several examples of highly selective receptors. The number and diversity of chemokines and their receptors also begs the question: Which of these genes represents the primordial ligand-receptor pair? Chemokines have been characterized in many species and appear to extend back to early vertebrates. Phylogenetic analysis suggests that stromal cell-derived factor-1 (SDF-1)/CXCL12 and CXCR4 likely represent the earliest ancestral chemokine-chemokine receptor pair and may have initially evolved within the context of the central nervous system and not, surprisingly, within the immune system *(10)*. This suggestion is rather ironic given, once again, that chemokine research had its beginnings in the field of immunology.

The biological usefulness of these genes is also highlighted by the discovery of virally encoded chemokine and receptor homologues. These include chemokine homologues expressed by members of the herpesvirus and poxvirus families as well as receptors found in certain herpesviruses. Although a full appreciation of the function of these virally pirated genes is far from clear, these chemokine homologues provide a convenient mechanism for the infectious agent to escape the wrath of the host immune system. As such, they will likely have value in the drug discovery process. Virally encoded chemokine peptides and receptors are discussed further in Chapters 2 and 15.

1.3. Chemokine Receptors: Beyond Migration of Leukocytes

Although the initial focus on this family of molecules concerned their effects on inflammatory leukocyte populations, it nonetheless became clear that chemokines and their receptors functioned in arenas beyond this aspect of the immune system. One of the prominent early findings that catapulted chemokines into the limelight was the discovery that chemokine receptors acted as coreceptors for HIV-1 entry into susceptible cells (discussed in depth in Chapter 13). These included compelling experiments that identified the coreceptors, CCR5 and CXCR4, as well as studies that showed that chemokine ligands could block HIV-1 entry *(11–17)*. Moreover, a polymorphism in CCR5 (delta-32) was discovered and shown to confer resistance to HIV infection. Although the lack of CCR5 in individuals harboring the CCR5 delta-32 mutation did not alter function from the chemokine biology point of view, it introduced investigators from multiple fields to chemokines biology. The significance of chemokine receptor usage by HIV became more striking when it was realized that the virus mutated toward utilization of CXCR4, correlating with the infection of T cells and subsequent development of AIDS. This set the stage for chemokines to become one of the most commonly investigated cytokine families. The next few years would establish an intense period of investigation and discovery in chemokine receptor biology, and the search for specific chemokine receptor blocking agents intensified.

Beyond roles of chemokine receptors in hematopoiesis and innate immunity, roles for chemokines in adaptive immunity emerged. Moreover, other non–leukocyte migration properties of chemokine receptors have been identified. These include roles in the biology of endothelial cells (Chapter 15), cancer (Chapter 16), smooth muscle (Chapter 11), fibroblasts (Chapter 14), stem cells (Chapter 8), and all cell types associated with nervous system tissues (Chapter 17). In many instances, broad functional overlap is evident as chemokines can direct the migration of these cells just as they do with leukocytes. In certain instances, the ability of chemokines to retain cell populations within a specific microenvironment is as important as their migration-promoting properties. However, it is also clear that migration and retention are not the sole end points.

1.4. The Future of Chemokine Receptors: Are They Valid Drug Targets?

As will be obvious from reading subsequent chapters, chemokine receptors have provided the pharmaceutical industry a rich list of new drug targets. This contention is supported by the clear roles of receptors in a variety of diseases. Technological advances in recombinant DNA methods and the ability to manipulate the genetics of laboratory mice have enhanced our understanding of the roles of specific genes in physiology and pathology. Chemokines and their receptors have not been immune to these approaches. Mice engineered to lack specific receptors and chemokine ligands have yielded insights into the functions of chemokine systems, and these are highlighted throughout this volume. Phenotypes associated with some of these gene-disrupted animals make perfect sense given what was known about the function of these receptors from other approaches. On the other hand, a complete understanding of the functions of many chemokines and chemokine receptors have not been clarified through characterization of gene-deleted animals. Continued efforts are needed to characterize chemokine functions through these approaches.

Functional roles for chemokines and receptors have also been obtained through identification and analysis of polymorphisms present in a handful of these genes. As discussed previously, individuals harboring the delta-32 mutation of CCR5 are significantly resistant to HIV-1 infection. Polymorphisms in other receptors and chemokines have also been identified. Phenotypes associated with these variants are summarized in the chemokine and chemokine receptor databases; some are given consideration in other chapters of this book.

The first small-molecular-weight antagonists targeting chemokine receptors have begun to surface. As outlined in the final chapter (Chapter 18), the path to the development of chemokine receptor antagonists has not been particularly ripe with success. These efforts have largely been hampered by species specificity and the inability to use the antagonists in mouse models. Nonetheless, there is still a paucity of data related to the biology of most of the chemokines, and it is likely that many of these targets were established prior to fully understanding the biology of the systems and/or diseases that were being examined. Despite any skepticism by pharmaceutical companies to identify chemokine receptor antagonists, these efforts should and will continue. As is evident from the chapters that follow, chemokine receptors are clearly involved in a myriad of physiologic and pathologic processes and, as such, deserve considerable attention in the arena of therapeutic development.

References

1. Wolpe SD, Cerami A. Macrophage inflammatory proteins 1 and 2: members of a novel superfamily of cytokines. FASEB J 1989;3:2565–2573.
2. Zlotnick A, Yoshie O. Chemokines: a new classification system and their role in immunity. Immunity 2000;12:121–127.
3. Holmes WE, Lee J, Kuang WJ, Rice GC, Wood WI. Structure and functional expression of a human interleukin-8 receptor. Science 1991;253:1278–1280.
4. Murphy PM, Tiffany HL. Cloning of complementary DNA encoding a functional human interleukin-8 receptor. Science 1991;253:1280–1283.
5. Neote K, DiGregorio D, Mak JY, Horuk R, Schall TJ. Molecular cloning, functional expression, and signaling characteristics of a C-C chemokine receptor. Cell 1993;72:415–425.
6. Gao JL, Kuhns DB, Tiffany HL, et al. Structure and functional expression of the human macrophage inflammatory protein 1 alpha/RANTES receptor. J Exp Med 1993;177;1421–1427.
7. Murphy PM, Baggiolini M, Charo IF, et al. International union of pharmacology. XXII. Nomenclature for chemokine receptors. Pharmacol Rev 2000;52:145–176.
8. Balabanian K, Lagane B, Infantino S, et al. The chemokine SDF-1/CXCL12 binds to and signals through the orphan receptor RDC1 in T lymphocytes. J Biol Chem 2005;280:35760–35766.
9. Burns JM, Summers BC, Wang Y, et al. A novel chemokine receptor for SDF-1 and I-TAC involved in cell survival, cell adhesion, and tumor development. J Exp Med 2006;203:2201–2213.

10. Huising MO, Stet RJM, Kruiswijk CP, Savelkoul HFJ, Verburg-van Kemenade BML. Molecular evolution of CXC chemokines: extant CXC chemokines orginate from the CNS. Trends Immunol 2003;24:306–312.
11. Cocchi F, DeVico AL, Garzino-Demo A, Arya SK, Gallo RC, Lusso P. Identification of RANTES, MIP-1 alpha, and MIP-1 beta as the major HIV-suppressive factors produced by CD8+ T cells. Science 1995;270:1811–1815.
12. Feng Y, Broder CC, Kennedy PE, Berger EA. HIV-1 entry cofactor: functional cDNA cloning of a seven-transmembrane, G protein-coupled receptor. Science 1996;272:872–877.
13. Alkhatib G, Combadiere C, Broder CC, et al. CC CKR5: a RANTES, MIP-1alpha, MIP-1beta receptor as a fusion cofactor for macrophage-tropic HIV-1. Science 1996;272:1955–1958.
14. Choe H, Farzan M, Sun Y, et al. The beta-chemokine receptors CCR3 and CCR5 facilitate infection by primary HIV-1 isolates. Cell 1996;85:1135–1148.
15. Deng H, Liu R, Ellmeier W, et al. Identification of a major co-receptor for primary isolates of HIV-1. Nature 1996;381:661–666.
16. Doranz BJ, Rucker J, Yi Y, et al. A dual-tropic primary HIV-1 isolate that uses fusin and the beta-chemokine receptors CKR-5, CKR-3, and CKR-2b as fusion cofactors. Cell 1996;85:1149–1158.
17. Dragic T, Litwin V, Allaway GP, et al. HIV-1 entry into CD4+ cells is mediated by the chemokine receptor CC-CKR-5. Nature 1996;381:667–673.

2

The Structural Biology of Chemokines

Elias Lolis and James W. Murphy

Summary

This chapter provides an overview of the literature in the field of the structural biology of chemokines. The secondary, tertiary, and quaternary structures as determined by x-ray crystallography and nuclear magnetic resonance are compared among the four chemokine families. The biological significance of chemokine structures is explored through a discussion of additional molecules that interact with the chemokines. Specific interactions of chemokines and their receptors are discussed as are interactions between chemokines and glycosaminoglycans. Additionally, a set of tables and figures summarizes the structural information available in the databases.

Key Words: Chemokines; oligomerization; GPCR; GAGs; heparin; crystallography; NMR; structure.

2.1. Introduction

Infection and tissue injury induces the migration of cells that results in the innate and/or adaptive response. The migration of these cells—macrophages, neutrophils, mast cells, basophils, natural killer (NK) cells, and lymphocytes—is initiated by a superfamily of secreted proteins known as chemokines. In addition to the proinflammatory response first described for chemokines, some chemokines and their receptors serve other important physiologic functions. These include, but are not limited to, roles in embryonic development during organ vascularization *(1–3)*, T- and B-cell development *(4,5)*, neuronal communication *(6)*, and CD34 stem cell mobilization *(7)*. Chemokines function by activating specific G protein–coupled receptors (GPCRs). In certain

From: *The Receptors: The Chemokine Receptors*
Edited by: J. K. Harrison and N. W. Lukacs © Humana Press Inc., Totowa, NJ

circumstances infectious agents have usurped chemokine receptors for their own benefit. Additionally, chemokines and chemokine receptors are known to be responsible for pathophysiologic or genetic diseases. A primary example of the chemokine receptors involved in pathologic roles are CCR5 and CXCR4, which are the two major HIV-1 coreceptors that allow entry of the virus into cells. Some forms of the genetic disease warts-hypogammaglobulinema-infections-myelkathexis (WHIM) syndrome are due to mutants at the cytosolic C-terminal end of CXCR4 *(8)*. Among many other diseases that involve chemokines and their receptors are atherosclerosis *(9,10)* and cancer proliferation *(11–13)* and metastasis *(14)*. The physiologic and pathologic roles of chemokines and their receptors are described in greater detail in other chapters in this book.

Chemokines were initially discovered as secreted proinflammatory proteins of about 8 kD defined by four conserved cysteine residues that are disulfide bonded. The chemokine superfamily has now been extended to proteins that contain one, two, or three disulfides. The chemokine superfamily is divided into two major families containing multiple proteins and two minor families, each with only one chemokine (Table 1). The nomenclature of chemokines and their receptors was standardized in the year 2000 based on the sequence of the cysteine residues in chemokines *(15)*. One of the two major families is defined by an intervening amino acid between the first two cysteines and is known as the CXC (or α-) chemokine family. In the CC (β-) chemokine family, the first two cysteines are adjacent to each other. The protein in the CX3C family has three amino acids between the first two cysteines. In these three families, the first cysteine forms a disulfide bond with the third conserved cysteine in the family, and the second cysteine forms a disulfide with the fourth cysteine residue. Some chemokines have six cysteines, with the two extra nonconserved cysteines forming a third disulfide. The protein in the XC family only has one disulfide. The designation of each receptor is based on the family of the chemokine(s) that activates it. For example, the receptor CXCR1 is activated by CXCL8, CCR1 is activated by numerous CCL agonists (Table 1), and CX3CR1 and XCR1 are activated by CX3CL1 and XCL, respectively. It should also be noted that two chemokines, CX3CL1 and CXCL16, are domains of single transmembrane proteins that are important for the physiologic function of these proteins. CX3CL1 is the N-terminal domain of fractalkine, a heavily glycosylated protein with a mucin stalk domain of 227 amino acids and a cytoplasmic domain of 37 amino acids *(16)*. CXCL16 is the N-terminal domain of a 176-residue ectodomain of another protein with a 27-amino-acid cytoplasmic sequence *(17)*.

The organization of chemokine families based on the cysteine sequence has functional significance. Some human chemokines can compete for binding and activation of receptors with other intrafamily chemokines. This raised the possibility that significant structural differences in chemokine-receptor interactions exist among proteins in the four families that result in activation of receptors exclusively within a family. The first three-dimensional structures of CXC and CC chemokines showed remarkable differences in the quaternary arrangement of dimers that could explain the source of the family-specific binding and activation of receptors. However, subsequent experiments indicated that the interactions between chemokines and their receptors are not as straightforward as originally thought. For example, the three-dimensional structure of the CC chemokine CCL20 *(18)* displays the quaternary structure of a CXC chemokine. In addition, the CXC agonists of CXCR3 also function as antagonists of the CCR3 receptor *(19,20)*, violating the hypothesis regarding competition for binding to receptors with only intrafamily chemokines. However, *activation* of receptors across chemokine families has not been detected. Interestingly, a herpesvirus-8 CC chemokine, viral macrophage inflammatory protein (vMIP)-II, has evolved to bind to receptors from all four families. In one case, vMIP-II functions as a CC receptor agonist *(21)*, but in most cases it is a chemokine receptor antagonist *(22–24)*. These observations complicate the development of a general mechanism to explain the structural basis of functional activity (agonism or antagonism) for chemokines.

2.2. Chemokine Structures

Tables 2 and 3 present lists of the known three-dimensional structures of human and viral chemokines. Virtually all chemokine structures, regardless of family, have the same monomeric structure. A flexible N-terminal region precedes the first cysteine and is involved in receptor activation. After the N-terminal region is the 10- to 20-residue N-terminal loop that is generally involved in receptor specificity, a short 3_{10} helix, a β-sheet composed of three antiparallel β-strands, and a C-terminal α-helix that packs against the β-sheet (Figure 1A). The loops connecting the first and second β-strand and the second and third β-strand are known as the 30s and 40s loops, respectively. One generic difference between monomers from each major family has to do with the chirality of the disulfide bonds. The chirality of the first disulfide is predominately right-handed in CXC chemokines and left-handed in the CC chemokines, whereas the second disulfide is left-handed in both families. However, both disulfides for CXCL12 stromal cel-derived factor (SDF-1α) are left-handed. This is consistent with a hypothesis that CXCL12 may be most closely related to the evolutionary ancestor of the chemokine superfamily based on low sequence homology to both CXC and CC chemokines *(25)*.

The oligomeric structures of chemokines vary. CXCL8 (IL-8), the prototypical CXC chemokine, is dimeric in solution *(26)* and in its crystalline form (due to twofold crystallographic symmetry) *(27)*. The three-stranded β-sheet of each subunit joins to form a six-stranded β-sheet (Figure 1B). The two C-terminal

Table 1
Families of the Known Chemokines

CXCL family		CCL family	
Ligand	Corresponding CXCR	Ligand	Corresponding CCR
CXCL1 (GRO-α)	CXCR2	CCL1 (I309)	CCR8
CXCL2 (GRO-β)	CXCR2	CCL2 (MCP-1)	CCR2, CCR11
CXCL3 (GRO-γ)	CXCR2	CCL3 (MIP-1α)	CCR1, CCR5
CXCL4 (PF-4)	CXCR3B	CCL4 (MIP-1β)	CCR5, CCR8
CXCL5 (ENA-78)	CXCR2	CCL5 (RANTES)	CCR1, CCR3, CCR4, CCR5
CXCL6 (GCP2)	CXCR1, CXCR2	CCL6 (MuC10)	CCR1
CXCL7 (NAP-2)	CXCR2	CCL7 (MCP-3)	CCR1, CCR2, CCR3
CXCL8 (IL-8)	CXCR1, CXCR2	CCL8 (MCP-2)	CCR1, CCR2, CCR3, CCR11
CXCL9 (Mig)	CXCR3	CCL9 (MIP-1γ)	CCR1
CXCL10 (IP-10)	CXCR3	CCL10	Unknown
CXCL11 (ITAC)	CXCR3, CXCR7	CCL11 (Eotaxin)	CCR3
CXCL12 (SDF-1)	CXCR4, CXCR7	CCL12 (MCP-5)	CCR2
CXCL13 (BLC)	CXCR5	CCL13 (MCP-4)	CCR1, CCR2, CCR3, CCR11

CXCL14 (BRAK)	Unknown	CCL14 (HCC1)	CCR1
CXCL15 (Lungkine)	Unknown	CCL15 (HCC2)	CCR1
CXCL16 (SRPSOX)	CXCR6	CCL16 (HCC4)	CCR1
CXCL17 (VCC)	Unknown	CCL17 (TARC)	CCR4, CCR8
		CCL18 (PARC)	CCR3
		CCL19 (ELC)	CCR7
		CCL20 (LARC)	CCR6
		CCL21 (SLC)	CCR7
CX3CL1 (fractalkine)	CX3CR1	CCL22 (MDC)	CCR4
		CCL23 (MPIF1)	CCR1
		CCL24 (Eotaxin-2)	CCR3
		CCL25 (TECK)	CCR9
XCL family		CCL26 (Eotaxin-3)	CCR3
Ligand	Corresponding XCR	CCL27 (Eskine)	CCR10
XCL1 (lymphotactin)	XCR1	CCL28 (MEC)	CCR10

Table 2
Three-Dimensional Structures of Human Chemokines

Name	Common name	PDB code	Method	Quaternary structure	Notes
CXCL1	Gro-α	1MGS	NMR	Dimer	
CXCL2	Gro-β	1QNK	NMR	Dimer	
CXCL4	Platelet factor-4	1RHP	X-ray (2.4Å)	Tetramer	
CXCL4	CTAP-III	1F9P	X-ray (1.93Å)	Monomer	Complex with heparin analogue
CXCL7	NAP-2	1NAP	X-ray (1.9Å)	Tetramer	
CXCL8	IL-8	1ICW	X-ray (2.01Å)	Dimer	E38C/C50A
CXCL8	IL-8	1QE6	X-ray (2.35Å)	Dimer	L5C/H33C
CXCL8	IL-8	1ILQ	NMR	Dimer	With $CXCR1^{9-29}$
CXCL8	IL-8	3IL8	X-ray (2.0Å)	Dimer	
CXCL8	IL-8	1ILP	NMR	Dimer	With $CXCR1^{1-40}$
CXCL10	IP-10	1LV9	NMR	Monomer	
CXCL10	IP-10 A,B,C,D	1O7Y	X-ray (3.0Å)	Tetramer	M-form
CXCL10	IP-10	1O7Z	X-ray (1.92Å)	Dimer	T-form
CXCL10	IP-10	1O80	X-ray (2.0Å)	Dimer	H-form
CXCL11	ITAC	1RJT	NMR	Monomer	
CXCL12	SDF-1α	1SDF	NMR	Monomer	
CXCL12	SDF-1α	1A15	X-ray (2.2Å)	Dimer	N33A mutant
CXCL12	SDF-1α	1QG7	X-ray (2.0Å)	Dimer	
CXCL12	SDF-1α	1VMC	NMR	Monomer	With $CXCR4^{1-27}$
CCL1	I-309	1EL0	NMR	Monomer	Three disulfides
CCL2	MCP-1	1DOK	X-ray (1.85Å)	Dimer	P-form
CCL2	MCP-1	1DOL	X-ray (2.4Å)	Monomer	I-form, tetramer in crystal

CCL2	MCP-1	1DOM	NMR	Dimer	
CCL3	MIP-1α mutant	1B53	NMR	Dimer	D27A (10 structures)
CCL4	MIP-1β mutant	1JE4	NMR	Monomer	Monomeric variant
CCL4	MIP-1β	1HUM	NMR	Dimer	
CCL5	AOP-RANTES	1B3A	X-ray (1.6Å)	Dimer	
CCL5	RANTES	1HRJ	NMR	Dimer	
CCL5	RANTES	1RTO	NMR	Dimer	
CCL5	RANTES	1U4L	X-ray (2.0Å)	Dimer	Complex with heparin I-S
CCL5	RANTES	1U4M	X-ray (2.0Å)	Dimer	Complex with heparin III-S
CCL7	MCP-3	1BO0	NMR	Monomer	
CCL8	MCP-2	1ESR	X-ray (2.0Å)	Monomer	
CCL15	HCC-2	2HCC	NMR	Monomer	
CCL17	TARC	1NR2	X-ray (2.18Å)	Monomer	
CCL17	TARC	1NR4	X-ray (1.72Å)	Dimer	
CCL20	MIP-3α	1HA6	NMR	Monomer	
CCL20	MIP-3α	1M8A	X-ray (1.7Å)	Dimer	
CCL23	MPIF	1G91	NMR	Monomer	Three disulfides
CCL24	Eotaxin-2	1EIG	NMR	Monomer	$CCR3^{1-35}$
CCL26	Eotaxin-3	1G2S	NMR	Monomer	$CCR3^{1-35}$
CX3CL	Fractalkine	1B2T	NMR	Monomer	With $CX3CR1^{2-19}$
CX3CL	Fractalkine	1F2L	X-ray (2.0Å)	Tetramer	Novel 4° structure
XCL1	Lymphotactine	1J9O	NMR	Monomer	

Table 3
Three-Dimensional Structures of Viral Chemokines

Common name	PDB code	Method	Quaternary structure
vMIP-I	1ZXT	X-ray (1.7Å)	Dimer
vMIP-II	1CM9	X-ray (2.1Å)	Dimer
vMIP-II	1HF6	NMR	Dimer
vMIP-II	1HHV	NMR	Monomer
vMIP-II and vMIP-II (1–10)	1HFG	NMR	Monomer
vMIP-II	1VMP	NMR	Monomer

α-helices are packed against the β-sheet and provide interactions that stabilize the dimeric structure. This structure resembles the major histocompatibility complex (MHC) with the exception that CXCL8 is much smaller and compact and does not contain a major groove as is present between the MHC α-helices, which provides the antigen binding site. The oligomeric structures of specific CXC chemokines, as well as CC chemokines, are not always the same in nuclear magnetic resonance (NMR) and crystal structures. The solution structure of CXCL10 interferon inducible protein (IP-10) is monomeric *(28)*, and CXCL12 exists as a monomer or dimer depending on solution conditions *(29,30)*, yet the oligomeric states in the crystal structures are different: CXCL12 is a dimer *(31)*, whereas CXCL10 was crystallized as both a dimer and tetramer *(32)*. The tetrameric form of CXCL10 is a dimer of dimers formed by β-sheet sandwiching of two AB-type dimers (Figure 1C) similar to CXCL4 (platelet factor 4) *(33)*, the first chemokine structure to be determined.

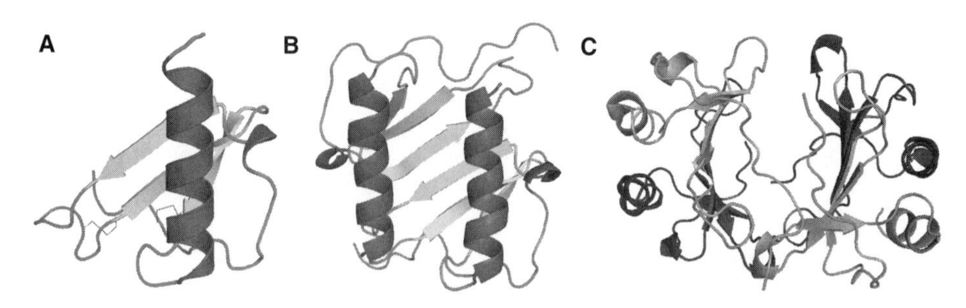

Fig. 1. Oligomerization of CXC chemokines. (A) Monomeric form of CXCL8 (IL-8) with disulfide bridges shown as black sticks. (B) Dimeric form of CXCL8 showing six-stranded β-sheet. (C) Tetrameric form of CXCL10 (IP-10).

The structure of CC chemokines also comes in a variety of oligomeric forms that include monomers, dimers, and tetramers. The monomeric structure is similar to that of CXC chemokines (Figure 1A). However, the dimeric and tetrameric chemokine structures are significantly different. The first identified structure of a CC chemokine was that of CCL4 (MIP-1β), which, like CXCL8, was also dimeric. The difference in the dimeric structure of CCL4 relative to CXCL8 was unexpectedly striking (Figure 2A). The subunit interface of CCL4 involves mostly the N-terminal region and parts of a loop connecting β-strands 1 and 2. In contrast with most CXC chemokines, the N-terminal regions of some CC chemokines are not as flexible due to their role in forming the dimeric interface. These interactions lead to an elongated protein with each helix at opposite ends of the molecule. The structure of CCL2 monocyte chemoattractant protein (MCP-1) determined by x-ray crystallography was tetrameric *(34)*. In this case, one of the dimers within the tetramer has similarities with CXC dimers due to crystallographic symmetry (Figure 2B). Whether this tetramer is due only to the symmetry present in crystals or is physiologically important remains to be determined. It is important to note that in the presence of a heparin octasaccharide, a tetrameric CCL2 is observed in sedimentation equilibrium studies, but the three-dimensional configuration of this tetramer is unknown *(35)*. Another CC chemokine, CCL20, crystallizes as a CXC-type dimer (Figure 2C) *(18)*, although an equilibrium between monomer and dimer was observed by NMR *(36)* and analytical ultracentrifugation *(18)*.

The structure of fractalkine, the CX3C chemokine consisting of residues 1 to 76 of a full-length single transmembrane protein, was determined by both NMR *(37)* and x-ray crystallography *(38)*. The solution form of the chemokine domain of fractalkine is monomeric (Figure 3A), but the crystal form is tetrameric (Figure 3B)—a dimer of dimers where the interactions between the dimers are mediated by water molecules. Although this tetrameric structure is unique, it is likely to be an artifact of crystallization. The in vivo oligomerization of fractalkine, either as part of, or proteolytically cleaved from, the full-length protein is unknown. Glycosaminoglycan (GAG) binding to the fractalkine chemokine is weak and therefore improbable in inducing dimerization. However, electrostatic analysis of the fractalkine chemokine domain indicates the dimeric interface is relatively uncharged (in contrast with the opposite site, which is fully charged) and could form an interface as part of the full-length protein *(38)*.

The dimeric form of fractalkine (Figure 3C) resembles a compact CC chemokine compared with the elongated form seen in Figure 2A. The first disulfide of fractalkine forces the N-terminal region to remain close to the core of the molecule *(38)*. The interface between subunits is asymmetrical and involves a β-strand from residues Cys-8 to Thr-11 from one monomer and residues Thr-11

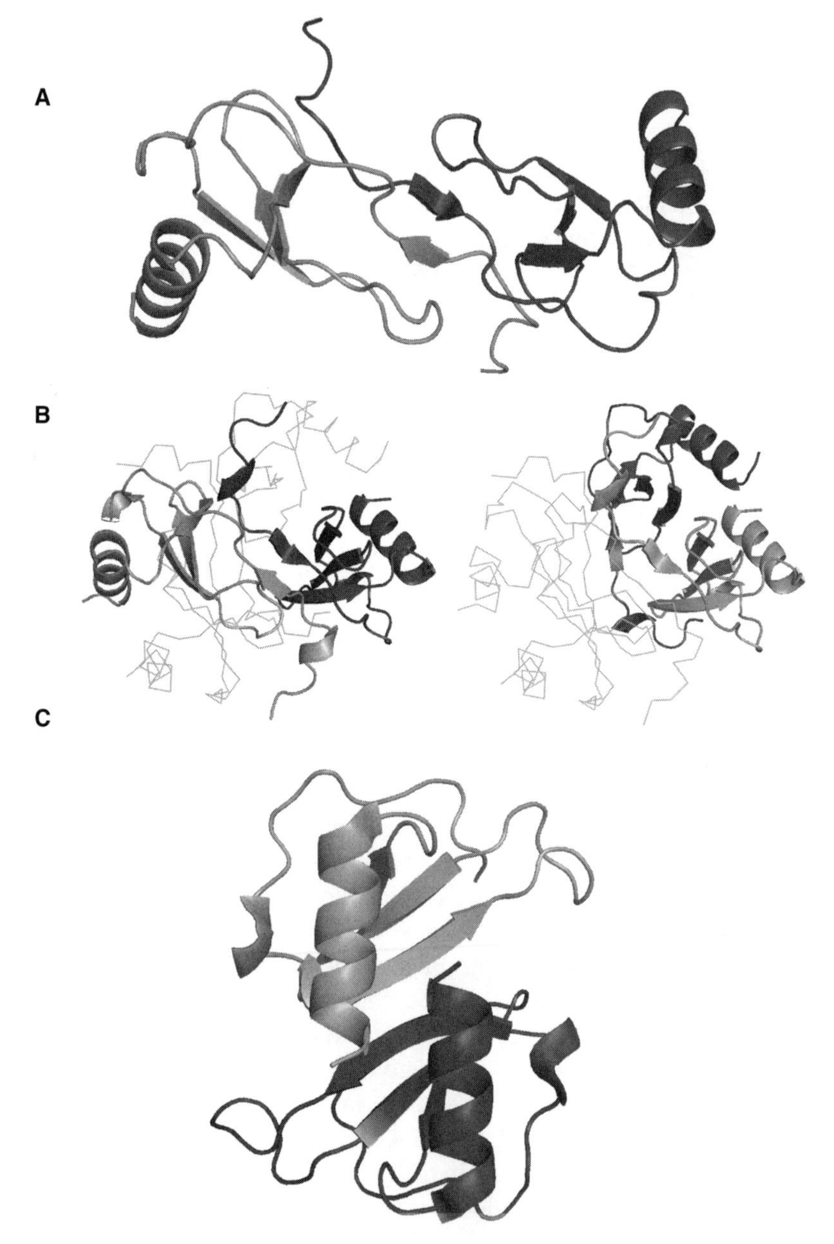

Fig. 2. Oligomerization of CC chemokines. (A) Dimeric form of CCL4 (MIP-1β). (B) Tetrameric form of CCL2 (MCP-1) highlighting both CC dimerization (left) and CXC dimerization (right). (C) CCL20 forms liver and activation regulated chemokine (LARC) a CXC dimer.

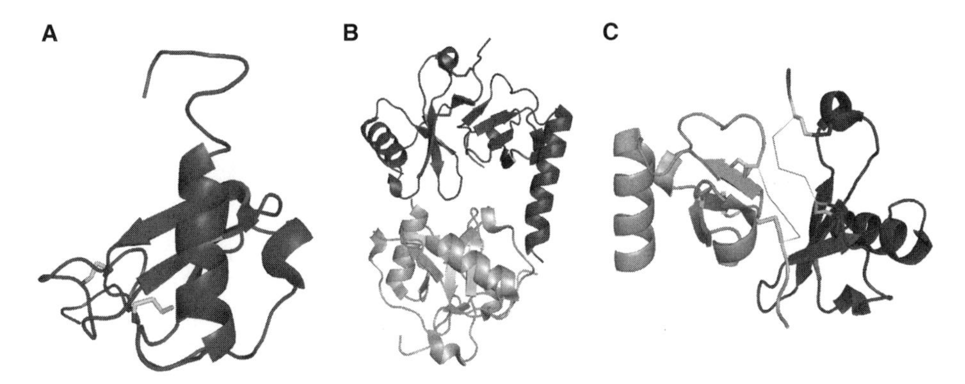

Fig. 3. Oligomerization of fractalkine. (A) Monomeric form of CX3CL1 showing disulfide bonds as sticks. (B) Tetrameric form of CX3CL1 in which a dimer of dimers is formed through water contacts. (C) Dimer of CX3CL1 with the three residues between the first two conserved cysteines shown as a backbone trace.

to Lys-14 from the other monomer to form a two-stranded β-sheet. It is not surprising that fractalkine forms a CC-type dimer given that the sequence homology of fractalkine is closest to CC chemokines.

Lymphotactin (XCL1) also displays unique structural properties. It lacks one of the two disulfides that is present in all other chemokines and has an extended C-terminal sequence that is required for biological activity *(39,40)*. At low temperature (10°C) in the presence of 200 mM NaCl and 20 mM phosphate buffer (pH 6.0), lymphotactin possesses the typical monomeric chemokine fold with a disordered C-terminal extension (that follows the C-terminal α-helix) *(41)*. The NMR chemical shifts are dramatically different at higher temperature and lower ionic strength. The experimental data indicate that lymphotactin undergoes a structural transition leading to an equilibrium of tertiary and quaternary structures that are not observed with any other chemokine *(42)*. The three-stranded β-sheet is refolded into a four-stranded β-sheet, with the loss of the C-terminal helix. Analytical ultracentrifugation indicates that at 40°C and 100 mM NaCl, the K_d decreases from 1.2 mM (at low temperature and 200 mM NaCl) to ~40 μM, significantly increasing the amount of dimeric lymphotactin present in solution. Although this K_d may appear too high given that nanomolar levels of chemokines are detected in the plasma, the actual concentration in the presence of glycosaminoglycans at the surface of endothelial cells or in tissues where cells react to a haptotactic gradient is probably large enough for the formation of dimers or higher molecular weight complexes. The presence of a single disulfide, formation of a four-stranded β-sheet in the monomer without an α-helix, and the possibility that the four-stranded β-sheet forms a dimer makes lymphotactin unique among all chemokines.

2.3. Chemokine-Receptor Interactions

The identification of GPCRs as chemokine receptors led to the expectation that small-molecule antagonists (or agonists) would be easier to identify and develop into drugs for diseases caused by dysregulation of these proteins *(43)*. This expectation was complicated by the realization that chemokine receptors display a great deal of redundancy, with different receptors having overlapping activities. Moreover, multiple chemokines can activate the same receptor. The possibility was suggested that chemokines bind to distinct but overlapping regions of GPCRs *(44)*, which may add to the complexity of identifying a single receptor antagonist that blocks activation by all activating chemokines. The cocrystallization of a chemokine receptor bound to its ligand as a means for drug design or for understanding the properties necessary to induce or inhibit activation also presents a serious obstacle. To date, the only experimentally determined structure of a GPCR is that of bovine rhodopsin *(45)*. Rhodopsin is unique among GPCRs as it constitutes greater than 70% of the proteins in the membranes of the rod outer segments of the retina allowing milligram quantities to be purified from a single bovine eye. Recombinant GPCRs expressed in various cells may produce higher expression levels of proteins than in endogenous cells, but they are not as stable during purification and are difficult to crystallize. Nonetheless, the determination of crystal structures of other GPCRs is inevitable. For the moment, alternative approaches have been used to study the structural basis of chemokine-receptor interactions.

The N-terminal region of chemokine receptors is known to interact with specific chemokines *(37,46–49)*. One of the approaches to study chemokine-receptor interactions involves NMR studies of chemokines with peptides derived from the N-terminal region of chemokine receptors. These peptides have dissociation constants for their respective chemokines that are in the micromolar to millimolar range compared with the picomolar to nanomolar range for the full-length receptor. Such studies have been carried out for two CXC chemokines, two CC chemokines, and the CX3CL chemokine fractalkine. One of the limiting factors in these studies is the inability to discern whether these peptides interact with the monomeric or dimeric forms of the chemokines. Another interesting approach that is still being optimized for structural studies involves using a soluble protein scaffold to present multiple extracellular regions from receptors to chemokines *(50,51)*. Although the studies with receptor peptides or protein scaffolds are important starting points to increase our understanding of how chemokines interact with receptors, it is important to interpret the initial results with caution. For example, it is not known whether the peptides or protein scaffolds correctly mimic the interactions that occur in vivo. Roles for receptor glycosylation or sulfation *(52–55)* as well as functions of other molecules such as GAGs, which are known to be important for chemokine interactions, are ignored in these studies. The results from these studies will provide models for interactions that need to be verified by other methods.

CXCL8 has been a major focus in structural studies of chemokine-receptor interactions *(46,47)*. An initial study used the peptide consisting of CXCR1 residues 1 to 40 ($CXCR1^{1-40}$). Measuring chemical shift changes as $CXCR1^{1-40}$ was titrated into a solution containing ^{15}N-labeled CXCL8, a dissociation constant between 170 and 340 μM (depending on whether the CXCL8 dimer was interacting with one or two peptides) was determined *(46,47)*. The residues with the greatest ^{1}H and ^{15}N chemical shifts for CXCL8 were the side chain of Gln-8, and the backbone atoms of Thr-12, Lys-15, Phe-17, His-18, Lys-20, Phe-21, Ser-44, Glu-48, Leu-49, Cys-50, and Val-61. Assuming the largest chemical shifts are associated with interacting amino acids, the receptor peptide forms interactions with CXCL8 residues from (i) the N-terminal loop, (ii) the 40s loop, (iii) β-strand 3, and (iv) the α-helix. Of interest is that there is no evidence that Glu-4, Leu-5, Arg-6 (the ELR motif) of CXCL8 interacts with the N-terminus of the CXCR1 peptide. The ELR sequence had been identified in earlier studies as important for CXCR1 and CXCR2 receptor activation *(56,57)*. In a more detailed crystallographic characterization of the interactions between CXCL8 and a receptor-based peptide, the CXCR1 N-terminal region was reduced to 21 amino acids corresponding with residues Met-9 to Pro-29 and was further modified by replacing residues Leu-15 to Gly-19 with a 6-amino hexanoic acid linker *(47)*. CXCL8 residues most affected by addition of the peptide containing the linker were similar to those described in the NMR study above. These residues form the boundary of a cleft containing solvent-exposed hydrophobic groups that binds only a small segment of the $CXCR1^{4-29}$ peptide, as the cleft is not large enough to accommodate either of the $CXCR1^{1-40}$ or the modified $CXCR1^{9-29}$ peptides (Figure 4A and B, see color plate). These studies partly agree with a mutagenesis study of 27 residues of CXCL8 that identified a hydrophobic groove composed of residues Phe-17, Phe-21, Ile-22, and Leu-43 *(58)*. This groove is more than 20 Å from the ELR motif in either subunit of the CXCL8 dimer, suggesting that the CXCL8 hydrophobic groove binds to the CXCR1 N-terminus and the CXCL8 ELR motif binds to an uncharacterized site of the receptor.

CXCL12 interactions with CXCR4 have been studied by a similar NMR-based approach *(49)*. CXCR4 N-terminal peptides 1–27, 1–17, 7–27, and 18–27 were initially screened for binding to ^{15}N-labeled CXCL12. The dissociation constants determined by NMR titration experiments for $CXCR4^{1-27}$ and $CXCR4^{1-17}$ were 45 μM and 133 μM, respectively. The magnitude of the change in chemical shifts of ^{15}N-CXCL12 upon addition of $CXCR4^{7-27}$ was too small to determine a dissociation constant and $CXCR4^{18-27}$ provided no evidence for binding. The largest chemical shifts were produced by the presence of the $CXCR4^{1-27}$ peptide for CXCL12 residues Val-23, His-25, and Ala-40. Additional changes in chemical shifts are present for (i) the N-terminal loop residue Phe-13, (ii) Lys-24 of the first β-strand, (iii) Gln-37, Val-39, Arg-41, and Leu-42 of the second β-strand, (iv) 40s loop residue Asn-45, (v) Gln-48 and Val-49 of the third β-strand, and (vi) Ile-58 and Tyr-61 of the α-helix. Many of these residues contribute to a shallow groove in the top half of the β-sheet (Figure 4C). Again, it is interesting to note that there are no interactions between the N-terminal–based peptide of CXCR4 and the flexible N-terminus of CXCL12, which is responsible for receptor activation, providing evidence for the two-stage binding and activation model of chemokine-receptor interaction *(59)*.

Interactions between a receptor-based peptide and CC chemokines are available for CCR3 with CCL11 and CCL24 *(48,60)*. The study with CCR3 underscores the complexity of chemokine-receptor interactions. The two chemokines have a sequence identity of about 35% and exclusively bind to CCR3. CCL2 shares 65% identity with CCL11, functions as an agonist for CCR2 and CCR4, but has no affinity for CCR3. These observations highlight that the interactions between chemokines and receptors are subtle and require high-resolution information concerning chemokines, receptors, and other molecules that participate in these interactions.

In the CCL11 study, a variety of peptides based on the predicted extracellular domains of CCR3 were screened for binding to ^{15}N-labeled CCL11. These peptides included the linear sequence of the N-terminal region ($CCR3^{1-35}$) and each of the three extracellular loops connected to the transmembrane domains. The three extracellular loops were also synthesized with a Cys-Gly-Gly at the N-terminus and a Gly-Gly-Cys at the C-terminus. In an attempt to mimic constraints for each loop linked to the transmembrane helices, each peptide was oxidized to form an intramolecular disulfide bond between the cysteine at the termini. Of the seven CCR3 peptides screened by NMR, only $CCR3^{1-35}$ induced concentration-dependent changes in the chemical shifts of CCL11. The linear and cyclic peptides from loops 2 and 3 did not induce any changes in the heteronuclear single quantum correlation (HSQC) spectrum of CCL11, whereas the linear and cyclic peptides of loop 1 caused precipitation and could not be used for further experiments. Titration of $CCR3^{1-35}$ affected the chemical shift of residues mainly in the N-loop (residues 13 to 18), the 3_{10} helix (residues 20 to 22), and β-strand 3 (residues 46 to 49, 51). Other residues with significant changes in chemical shift included Arg-28 (β-strand 1), Gln-36 and Ala-38 (30s loop), and Lys-55 and Trp-57 (residues at the N-terminus of the α-helix) (Figure 4D). CCL24 interactions with CCR3 were studied with the same receptor-based peptide approach *(60)*. In addition to $CCR3^{1-35}$ concentration-dependent chemical shift changes, the line shape of other residues—Ser-5, Cys-7, Cys-8, Phe-11, Ile-16, Lys-34, Ala-35, and Val-37—was altered by broadening and/or distortion. The peaks of Arg-15 and Glu-64 disappeared from the spectrum in the presence of CCR3 peptide. Despite the low similarity in primary sequence between CCL11 and CCL24, the major changes occurred in a conserved area with CCL11 consisting of a hydrophobic base surrounded by charged or hydrophilic residues (Figure 4E). The electrostatic potential of CCL2, CCL11, and CCL24 was calculated and analyzed in an attempt to explain the difference between CCR3-binding properties of CCL11 and CCL24 and the lack of binding by CCL2. Significant differences were found in the $CCR3^{1-35}$ binding region of CCL11 and CCL24 with respect to the same location on CCL2.

The interaction of the chemokine domain of fractalkine (CX3CL1) with a CX3CR1 peptide was also characterized. Titration of the $CX3CR1^{2-19}$ peptide into a sample of ^{15}N-labeled fractalkine resulted in changes in chemical shifts for residues 6 to 17 (N-terminal region and N-loop), Gln-31 and Gly-35 (the 30s loop), and residues Ile-40, Gln-45, and Leu-48 (Figure 4F) *(37)*. This is the only chemokine that has a change in chemical shift for the N-terminal region, but it is important to reemphasize that these changes may have nothing to do with direct interactions.

An examination of the panels in Figure 4 (see color plate) indicates that the N-terminal peptides from CXCR1, CXCR4, CCR3, and CX3CR1 all interact with the similar regions of their respective agonists. The similarity in the N-terminal receptor binding region for chemokines from three different families strongly suggests a common mechanism of initial binding. Site-directed mutagenesis has identified some of the subtleties of chemokine-receptor interactions *(56,61)*. The absence of any direct interactions between the receptor peptides and the activating N-terminal sequence of chemokines from these structural studies (with the exception of CX3CL1) support the proposed two-step model of receptor binding and activation for most chemokines. The initial binding will certainly involve the chemokine regions shown in Figure 2–4 with the N-terminal region *of the receptor* followed by activation due to interactions of the N-terminal sequence *of the chemokine* with the receptor.

2.4. Chemokine-Glycosaminoglycan Interactions

The activation of a chemokine receptor is more complex than the traditional agonist-receptor paradigm. For example, chemokine activity is mediated by GAGs (heparin, heparan, and heparin sulfate; chondroitin sulfate; and dermatan sulfate) at various sites during the chemotactic process. Chemokines released by tissue injury, infection, or inflammation activate adjacent endothelial cells and induce rolling and extravasation of leukocytes. These interactions between endothelial cells and leukocytes involve selectins, integrins, and intracellular adhesion molecules (ICAMs). Chemokines are also involved in the interactions between the two cell types. Endothelial cells are active in transcytosing chemokines to their luminal surface, where they interact with GAGs, are presented for interaction to chemokine receptors on leukocytes, are involved in integrin activation, and induce firm attachment to the endothelium prior to extravasation *(62)*. The GAG-chemokine interaction may also be important in the connective tissue and extracellular matrix. GAGs induce the oligomerization of chemokines *(63)* and may be responsible for creating haptotactic gradients through the release of chemokines from the GAG-polychemokine complex necessary for the migration of leukocytes *(64)*. On the cell surface, the GAG-induced oligomerized chemokines increase the local concentration of chemokines and promote interactions with the cell surface receptor *(63,65)*.

The importance of GAGs in the biological function of chemokines is highlighted by a number of studies. Mutants of CCL2, CCL4, and CCL5 deficient in GAG binding retain chemotactic activity in vitro but are unable to induce migration of cells when injected intraperitoneally in mice compared with wild-type controls *(66)*. In the same study, mutant monomers of CCL2, CCL4, and CCL5 were chemotactic in vitro but lost this activity in vivo, suggesting that the oligomerization of chemokines in vivo by GAGs is necessary for mediating a chemotactic response. A variant of CCL5 (with a sequence from residues 44 to 47 of AANA) that does not bind heparin inhibits the in vivo activity of endogenous CCL5, presumably by the formation of inactive heterodimers with endogenous wild-type CCL5 *(67)*. These important regulatory roles for GAGs in various aspects of chemokine activities suggest that the GAG-binding site of a chemokine might be a more specific drug target than the receptor binding pocket *(67)*.

Three-dimensional structures of chemokine-GAG complexes are difficult to determine because of the propensity of GAGs to induce the precipitation of chemokines at the high concentrations required for structural work. Nuclear magnetic resonance studies have been used to study the affinity, oligomerization, and/or GAG binding site of CXCL4 *(68)*, CXCL8 *(69)*, CXCL12 (Murphy et al., unpublished results), and CCL5 *(65,70)*. The crystal structures of CCL5 and two heparin-derived disaccharides have been determined by x-ray crystallography (Figure 5A; see color plate) *(71)*. The interaction between chemokines and the disaccharides are largely electrostatic, yet specific. However, these studies may reveal a potential template for structure-based drug optimization that increases both the affinity and selectivity for disaccharide-based derivatives *(67)*. Interestingly, there is variation in the GAG interaction site within a chemokine family. Nuclear magnetic resonance studies of CXCL4 *(68)* indicate the GAG binding site is located at two loops and the C-terminal helix (Figure

5B; see color plate) that in the tetramer forms a ring of positive charges (not shown) as predicted by Stuckey et al. *(72)*. Nuclear magnetic resonance studies of CXCL8 *(69)* (Figure 5C; see color plate) indicate that only the C-terminal helix is involved in binding to GAGs. In contrast, structural data on CXCL12 (Murphy et al., unpublished results) verify the results of mutagenesis experiments that heparin disaccharide I-S binds to basic residues within the subunit interface at the first and second β-strands (Figure 5D; see color plate) *(73)*. Mutagenesis of murine CXCL10 reveals a heparin binding site that is similar to that of CXCL4 (Figure 5E compared with B; see color plate) *(74)*. This location involves N-terminal loop residues Arg-20, Arg-22, and Ile-24, β-strand 1 residue Lys-26, and 40s loop residues Lys-46 and Lys-47 (Figure 5E; see color plate).

2.5. Future Directions

The three-dimensional structures of chemokines have been straightforward to determine. This is due to a number of factors. The small size of these proteins allows for either chemical synthesis or production by recombinant methods. The physiochemical properties of these proteins are generally well-suited for NMR or crystallization. As a consequence, we have learned a lot from these structures. It is much more challenging to structurally characterize complexes of chemokines with GAGs or chemokine receptors. The future of structural biology for chemokines now depends on developing methods to overcome the difficulties associated with determining structures of chemokine-GAG and chemokine-receptor complexes. Once these structures are determined, our knowledge of how the chemokine activity is regulated and how receptors are activated will increase dramatically.

References

1. Nagasawa T, Hirota S, Tachibana K, et al. Defects of B-cell lymphopoiesis and bone-marrow myelopoiesis in mice lacking the CXC chemokine PBSF/SDF-1. Nature 1996;382:635–8.
2. Tachibana K, Hirota S, Iizasa H, et al. The chemokine receptor CXCR4 is essential for vascularization of the gastrointestinal tract. Nature 1998;393:591–4.
3. Zou YR, Kottmann AH, Kuroda M, Taniuchi I, Littman DR. Function of the chemokine receptor CXCR4 in haematopoiesis and in cerebellar development. Nature 1998;393:595–9.
4. Forster R, Emrich T, Kremmer E, Lipp M. Expression of the G-protein–coupled receptor BLR1 defines mature, recirculating B cells and a subset of T-helper memory cells. Blood 1994;84:830–40.
5. Vicari AP, Figueroa DJ, Hedrick JA, et al. TECK: a novel CC chemokine specifically expressed by thymic dendritic cells and potentially involved in T cell development. Immunity 1997;7:291–301.

6. Meucci O, Fatatis A, Simen AA, Bushell TJ, Gray PW, Miller RJ. Chemokines regulate hippocampal neuronal signaling and gp120 neurotoxicity. Proc Natl Acad Sci U S A 1998;95:14500–5.
7. Flomenberg N, Devine SM, DiPersio JF, et al. The use of AMD3100 plus G-CSF for autologous hematopoietic progenitor cell mobilization is superior to G-CSF alone. Blood 2005;106:1867–74.
8. Hernandez PA, Gorlin RJ, Lukens JN, et al. Mutations in the chemokine receptor gene CXCR4 are associated with WHIM syndrome, a combined immunodeficiency disease. Nat Genet 2003;34:70–4.
9. Boring L, Gosling J, Cleary M, Charo IF. Decreased lesion formation in $CCR2^{-/-}$ mice reveals a role for chemokines in the initiation of atherosclerosis. Nature 1998;394:894–7.
10. Gu L, Okada Y, Clinton SK, et al. Absence of monocyte chemoattractant protein-1 reduces atherosclerosis in low density lipoprotein receptor-deficient mice. Mol Cell 1998;2:275–81.
11. Zhou Y, Larsen PH, Hao C, Yong VW. CXCR4 is a major chemokine receptor on glioma cells and mediates their survival. J Biol Chem 2002;277:49481–7.
12. Rubin JB, Kung AL, Klein RS, et al. A small-molecule antagonist of CXCR4 inhibits intracranial growth of primary brain tumors. Proc Natl Acad Sci U S A 2003;100:13513–8.
13. Zhang J, Sarkar S, Yong VW. The chemokine stromal cell derived factor-1 (CXCL12) promotes glioma invasiveness through MT2-matrix metalloproteinase. Carcinogenesis 2005;26:2069–77.
14. Muller A, Homey B, Soto H, et al. Involvement of chemokine receptors in breast cancer metastasis. Nature 2001;410:50–6.
15. Murphy PM, Baggiolini M, Charo IF, et al. International union of pharmacology. XXII. Nomenclature for chemokine receptors. Pharmacol Rev 2000;52:145–76.
16. Bazan JF, Bacon KB, Hardiman G, et al. A new class of membrane-bound chemokine with a CX3C motif. Nature 1997;385:640–4.
17. Matloubian M, David A, Engel S, Ryan JE, Cyster JG. A transmembrane CXC chemokine is a ligand for HIV-coreceptor Bonzo. Nat Immunol 2000;1:298–304.
18. Hoover DM, Boulegue C, Yang D, et al. The structure of human macrophage inflammatory protein-3alpha/CCL20. Linking antimicrobial and CC chemokine receptor-6-binding activities with human beta-defensins. J Biol Chem 2002;277:37647–54.
19. Loetscher P, Pellegrino A, Gong J-H, et al. The ligands of CXC chemokine receptor 3, I-TAC, Mig, and IP10, are natural antagonists for CCR3. J Biol Chem 2001;276:2986–91.
20. Xanthou G, Duchesnes CE, Williams TJ, Pease JE. CCR3 functional responses are regulated by both CXCR3 and its ligands CXCL9, CXCL10 and CXCL11. Eur J Immunol 2003;33:2241–50.
21. Sozzani S, Luini W, Bianchi G, et al. The viral chemokine macrophage inflammatory protein-II is a selective Th2 chemoattractant. Blood 1998;92:4036–9.

22. Kledal TN, Rosenkilde MM, Coulin F, et al. A broad-spectrum chemokine antagonist encoded by Kaposi's sarcoma-associated herpesvirus. Science 1997;277: 1656–9.
23. Chen S, Bacon KB, Li L, et al. In vivo inhibition of CC and CX3C chemokine-induced leukocyte infiltration and attenuation of glomerulonephritis in Wistar-Kyoto (WKY) rats by vMIP-II. J Exp Med 1998;188:193–8.
24. Luttichau HR, Stine J, Boesen TP, et al. A highly selective CC chemokine receptor (CCR)8 antagonist encoded by the poxvirus molluscum contagiosum. J Exp Med 2000;191:171–80.
25. Bleul CC, Fuhlbrigge RC, Casasnovas JM, Aiuti A, Springer TA. A highly efficacious lymphocyte chemoattractant, stromal cell-derived factor 1 (SDF-1). J Exp Med 1996;184:1101–9.
26. Clore GM, Appella E, Yamada M, Matsushima K, Gronenborn AM. Three-dimensional structure of interleukin 8 in solution. Biochemistry 1990;29:1689–96.
27. Baldwin ET, Weber IT, St Charles R, et al. Crystal structure of interleukin 8: symbiosis of NMR and crystallography. Proc Natl Acad Sci U S A 1991; 88:502–6.
28. Booth V, Keizer DW, Kamphuis MB, Clark-Lewis I, Sykes BD. The CXCR3 binding chemokine IP-10/CXCL10: structure and receptor interactions. Biochemistry 2002;41:10418–25.
29. Crump MP, Gong JH, Loetscher P, et al. Solution structure and basis for functional activity of stromal cell-derived factor-1; dissociation of CXCR4 activation from binding and inhibition of HIV-1. EMBO J 1997;16:6996–7007.
30. Veldkamp CT, Peterson FC, Pelzek AJ, Volkman BF. The monomer-dimer equilibrium of stromal cell-derived factor-1 (CXCL 12) is altered by pH, phosphate, sulfate, and heparin. Protein Sci 2005;14:1071–81.
31. Dealwis C, Fernandez EJ, Thompson DA, Simon RJ, Siani MA, Lolis E. Crystal structure of chemically synthesized [N33A] stromal cell-derived factor 1alpha, a potent ligand for the HIV-1 "fusin" coreceptor. Proc Natl Acad Sci U S A 1998;95:6941–6.
32. Swaminathan GJ, Holloway DE, Colvin RA, et al. Crystal structures of oligomeric forms of the IP-10/CXCL10 chemokine. Structure 2003;11:521–32.
33. St Charles R, Walz DA, Edwards BF. The three-dimensional structure of bovine platelet factor 4 at 3.0-A resolution. J Biol Chem 1989;264:2092–9.
34. Lubkowski J, Bujacz G, Boque L, Domaille PJ, Handel TM, Wlodawer A. The structure of MCP-1 in two crystal forms provides a rare example of variable quaternary interactions. Nat Struct Biol 1997;4:64–9.
35. Lau EK, Paavola CD, Johnson Z, et al. Identification of the glycosaminoglycan binding site of the CC chemokine, MCP-1: implications for structure and function in vivo. J Biol Chem 2004;279:22294–305.
36. Perez-Canadillas JM, Zaballos A, Gutierrez J, et al. NMR solution structure of murine CCL20/MIP-3alpha, a chemokine that specifically chemoattracts immature dendritic cells and lymphocytes through its highly specific interaction with the beta-chemokine receptor CCR6. J Biol Chem 2001;276:28372–9.

37. Mizoue LS, Bazan JF, Johnson EC, Handel TM. Solution structure and dynamics of the CX3C chemokine domain of fractalkine and its interaction with an N-terminal fragment of CX3CR1. Biochemistry 1999;38:1402–14.
38. Hoover DM, Mizoue LS, Handel TM, Lubkowski J. The crystal structure of the chemokine domain of fractalkine shows a novel quaternary arrangement. J Biol Chem 2000;275:23187–93.
39. Hedrick JA, Saylor V, Figueroa D, et al. Lymphotactin is produced by NK cells and attracts both NK cells and T cells in vivo. J Immunol 1997;158:1533–40.
40. Marcaurelle LA, Mizoue LS, Wilken J, et al. Chemical synthesis of lymphotactin: a glycosylated chemokine with a C-terminal mucin-like domain. Chemistry 2001;7:1129–32.
41. Kuloglu ES, McCaslin DR, Kitabwalla M, Pauza CD, Markley JL, Volkman BF. Monomeric solution structure of the prototypical "C" chemokine lymphotactin. Biochemistry 2001;40:12486–96.
42. Kuloglu ES, McCaslin DR, Markley JL, Volkman BF. Structural rearrangement of human lymphotactin, a C chemokine, under physiological solution conditions. J Biol Chem 2002;277:17863–70.
43. Fernandez EJ, Lolis E. Structure, function, and inhibition of chemokines. Annu Rev Pharmacol Toxicol 2002;42:469–99.
44. Fernandez EJ, Wilken J, Thompson DA, Peiper SC, Lolis E. Comparison of the structure of vMIP-II with eotaxin-1, RANTES, and MCP-3 suggests a unique mechanism for CCR3 activation. Biochemistry 2000;39:12837–44.
45. Palczewski K, Kumasaka T, Hori T, et al. Crystal structure of rhodopsin: A G protein-coupled receptor. Science 2000;289:739–45.
46. Clubb RT, Omichinski JG, Clore GM, Gronenborn AM. Mapping the binding surface of interleukin-8 complexes with an N-terminal fragment of the type 1 human interleukin-8 receptor. FEBS Lett 1994;338:93–7.
47. Skelton NJ, Quan C, Reilly D, Lowman H. Structure of a CXC chemokine-receptor fragment in complex with interleukin-8. Structure 1999;7:157–68.
48. Ye J, Kohli LL, Stone MJ. Characterization of binding between the chemokine eotaxin and peptides derived from the chemokine receptor CCR3. J Biol Chem 2000;275:27250–7.
49. Gozansky E, Louis J, Caffrey M, Clore G. Mapping the binding of the N-terminal extracellular tail of the CXCR4 receptor to stromal cell-derived factor-1. J Mol Biol 2005;345:651–8.
50. Datta A, Stone MJ. Soluble mimics of a chemokine receptor: chemokine binding by receptor elements juxtaposed on a soluble scaffold. Protein Sci 2003;12: 2482–91.
51. Datta-Mannan A, Stone MJ. Chemokine-binding specificity of soluble chemokine-receptor analogues: identification of interacting elements by chimera complementation. Biochemistry 2004;43:14602–11.
52. Farzan M, Mirzabekov T, Kolchinsky P, et al. Tyrosine sulfation of the amino terminus of CCR5 facilitates HIV-1 entry. Cell 1999;96:667–76.

53. Farzan M, Babcock GJ, Vasilieva N, et al. The role of post-translational modifications of the CXCR4 amino terminus in stromal-derived factor 1 alpha association and HIV-1 entry. J Biol Chem 2002;277:29484–9.
54. Fong AM, Alam SM, Imai T, Haribabu B, Patel DD. CX3CR1 tyrosine sulfation enhances fractalkine-induced cell adhesion. J Biol Chem 2002;277: 19418–23.
55. Preobrazhensky AA, Dragan S, Kawano T, et al. Monocyte chemotactic protein-1 receptor CCR2B is a glycoprotein that has tyrosine sulfation in a conserved extracellular N-terminal region. J Immunol 2000;165:5295–303.
56. Hebert CA, Vitangcol RV, Baker JB. Scanning mutagenesis of interleukin-8 identifies a cluster of residues required for receptor binding. J Biol Chem 1991;266: 18989–94.
57. Clark-Lewis I, Schumacher C, Baggiolini M, Moser B. Structure-activity relationships of interleukin-8 determined using chemically synthesized analogs. Critical role of NH2-terminal residues and evidence for uncoupling of neutrophil chemotaxis, exocytosis, and receptor binding activities. J Biol Chem 1991;266: 23128–34.
58. Williams G, Borkakoti N, Bottomley GA, et al. Mutagenesis studies of interleukin-8. Identification of a second epitope involved in receptor binding. J Biol Chem 1996;271:9579–86.
59. Monteclaro FS, Charo IF. The amino-terminal extracellular domain of the MCP-1 receptor, but not the RANTES/MIP-1alpha receptor, confers chemokine selectivity. Evidence for a two-step mechanism for MCP-1 receptor activation. J Biol Chem 1996;271:19084–92.
60. Mayer KL, Stone MJ. NMR solution structure and receptor peptide binding of the CC chemokine eotaxin-2. Biochemistry 2000;39:8382–95.
61. Pakianathan DR, Kuta EG, Artis DR, Skelton NJ, Hebert CA. Distinct but overlapping epitopes for the interaction of a CC-chemokine with CCR1, CCR3 and CCR5. Biochemistry 1997;36:9642–8.
62. Middleton J, Patterson AM, Gardner L, Schmutz C, Ashton BA. Leukocyte extravasation: chemokine transport and presentation by the endothelium. Blood 2002;100: 3853–60.
63. Hoogewerf AJ, Kuschert GS, Proudfoot AE, et al. Glycosaminoglycans mediate cell surface oligomerization of chemokines. Biochemistry 1997;36:13570–8.
64. Tanaka Y, Adams DH, Hubscher S, Hirano H, Siebenlist U, Shaw S. T-cell adhesion induced by proteoglycan-immobilized cytokine MIP-1 beta. Nature 1993; 361:79–82.
65. McCornack MA, Boren DM, LiWang PJ. Glycosaminoglycan disaccharide alters the dimer dissociation constant of the chemokine MIP-1 beta. Biochemistry 2004; 43:10090–101.
66. Proudfoot AE, Handel TM, Johnson Z, et al. Glycosaminoglycan binding and oligomerization are essential for the in vivo activity of certain chemokines. Proc Natl Acad Sci U S A 2003;100:1885–90.

67. Johnson Z, Kosco-Vilbois MH, Herren S, et al. Interference with heparin binding and oligomerization creates a novel anti-inflammatory strategy targeting the chemokine system. J Immunol 2004;173:5776–85.
68. Mayo KH, Ilyina E, Roongta V, et al. Heparin binding to platelet factor-4. An NMR and site-directed mutagenesis study: arginine residues are crucial for binding. Biochem J 1995;312:357–65.
69. Kuschert GS, Hoogewerf AJ, Proudfoot AE, et al. Identification of a glycosaminoglycan binding surface on human interleukin-8. Biochemistry 1998;37: 11193–201.
70. McCornack MA, Cassidy CK, LiWang PJ. The binding surface and affinity of monomeric and dimeric chemokine macrophage inflammatory protein 1 beta for various glycosaminoglycan disaccharides. J Biol Chem 2003;278:1946–56.
71. Shaw JP, Johnson Z, Borlat F, et al. The X-ray structure of RANTES: heparin-derived disaccharides allows the rational design of chemokine inhibitors. Structure (Cambridge) 2004;12:2081–93.
72. Stuckey JA, St Charles R, Edwards BF. A model of the platelet factor 4 complex with heparin. Proteins 1992;14:277–87.
73. Amara A, Lorthioir O, Valenzuela A, et al. Stromal cell-derived factor-1alpha associates with heparan sulfates through the first beta-strand of the chemokine. J Biol Chem 1999;274:23916–25.
74. Campanella GS, Lee EM, Sun J, Luster AD. CXCR3 and heparin binding sites of the chemokine IP-10 (CXCL10). J Biol Chem 2003;278:17066–74.

3

Chemokine Receptors: A Structural Overview

Gerard J. Graham and Robert J. Nibbs

Summary

To date, all mammalian chemokine receptors and chemokine-binding proteins belong to the seven-transmembrane (7TM)-spanning family of receptors, which are typically, although not exclusively, coupled to G proteins *(1)*. The large G protein–coupled receptor (GPCR) family comprises almost 1000 members in the human genome *(2,3)* accounting for in excess of 3% of transcribed sequences. This family is subdivided into the class A, B, C, and F/S families on the basis of shared motifs, and chemokine receptors specifically belong to the class A GPCR family. This review summarizes the current state of knowledge of chemokine receptor structure and discusses recent developments toward the determination of the full three-dimensional structure of these important receptors.

Key Words: Chemokines; receptors; GPCR; structure; inflammation; immunology; biochemistry.

3.1. Background of the Chemokine Receptors

3.1.1. Discovery of the Chemokine Receptors

It had been suspected for some time that chemokine receptors belonged to the G protein–coupled receptor (GPCR) family. Evidence for this included the previous identification of GPCRs as chemotactic receptors for ligands, such as formal-methionine, leucine, phenylalanine (fMLP), and the sensitivity of chemokine responses to inhibition by pertussis toxin and induction of calcium ion mobilization in response to ligand binding. In 1991, two groups reported the cloning of receptors for the CXC chemokine IL8 (CXCL8) *(4,5)*, which were initially referred to as IL8RA and IL8RB (now referred to as CXCR1 and CXCR2). These receptors, as expected, turned out to be seven-transmembrane

From: *The Receptors: The Chemokine Receptors*
Edited by: J. K. Harrison and N. W. Lukacs © Humana Press Inc., Totowa, NJ

(7TM) GPCRs, and the identification of these two molecules facilitated the discovery of many of the other chemokine receptors. The identification of the two IL8 receptors was followed, shortly after, by the cloning of a receptor for CC chemokines. This receptor was initially called CC-CKR1 (now CCR1) and was shown to promiscuously bind a number of members of the CC chemokine family *(6)*. Subsequent to the identification of these three chemokine receptors, two principal methods have been used to clone the cDNAs for the other currently identified receptors:

1. Degenerate-PCR cloning: The availability of three chemokine receptor sequences allowed the identification of conserved motifs spanning nonconserved sequences within the receptor cDNAs. These motifs have been used by a number of laboratories to facilitate the design of "degenerate" oligonucleotides capable of amplifying similar, but not necessarily identical, regions from other chemokine receptors. A major advantage of the chemokine receptors in this regard is that, in general, almost the entire coding sequence is represented in a single genomic exon. Thus, a number of the known chemokine receptors have been characterized through such degenerate-PCR cloning methodology using genomic DNA as a template.
2. Orphan GPCR screening: Another successful approach to identifying novel chemokine receptors has involved the screening of previously identified 7TM receptors. Such receptors, if they have no identified ligands, are referred to as *orphan* receptors and the assignment of ligands to such receptors as *de-orphanization*. The de-orphanization of many orphan GPCRs as chemokine receptors has again benefited from the identification of conserved motifs (see below) in the chemokine receptors, which have enabled researchers to focus de-orphanizing efforts at orphan GPCRs bearing structural similarities to other chemokine receptors.

These varied approaches have contributed to the characterization of 19 chemokine receptors and 3 chemokine-binding proteins that have now been identified.

3.1.2. Chemokine Receptor Nomenclature

A systematic nomenclature has been developed for the chemokine receptors (see Table 1). Thus, receptors for CC chemokines are referred to as CCR, receptors for CXC chemokines as CXCR, and the receptors for the XC and CX3C chemokines as XCR and CX3CR, respectively. To date, there are 10 CCRs (CCRs 1 to 10), 7 CXCRs, a single XCR, and a single CX3CR. The numbering is based on the date of deposition of the chemokine receptor sequence within the nucleic acid databases. For orphan receptors, this date refers to the point of identification of the orphan receptors as chemokine receptors and not to the date of initial deposition in the cDNA databases.

Table 1
Basic Properties of the Chemokine Receptors

Name	Alternative names	Accession number	Chromosomal location	Ligands
CCR1	CD191	NM_001295	3p21	CCLs 3, 5, 7, 8, 13, 14a, 15, 23
CCR2	CD912	NM_000647	3p21	CCLs 2, 7, 8, 13, 16
CCR3	CD193	NM_178329	3p21	CCLs 3, 5, 7, 8, 11, 13, 15, 24, 26, 28
CCR4		NM_005508	3p24	CCLs 17, 22
CCR5	CD195	NM_000579	3p21	CCLs 3, 4, 5, 8, 11, 14a, 16
CCR6	CD196, STRL22	NM_004367	6q27	CCL20
CCR7	CD197, EBI1, BLR2	NM_001838	17q12-21.1	CCLs 19, 21
CCR8	CD198, TER1	NM_005201	3p22	CCL1
CCR9	CD199, GPR-9-6	NM_006641	3p21	CCL25
CCR10	GPR2	NM_016602	17q21.1-q21.3	CCLs 27, 28
CXCR1	CD181, IL8R1	NM_000634	2q35	CXCLs 6, 8
CXCR2	CD182, IL8R2	NM_001557	2q35	CXCLs 1, 2, 3, 5, 6, 7, 8
CXCR3	CD183, GPR9	NM_001504	Xq13	CXCLs 9, 10, 11
CXCR4	CD184, LESTR	NM_001008540	2q21	CXCL12
CXCR5	CD185	NM_001716	11q23	CXCL13
CXCR6	CD186, BONZO	NM_006564	3p21	CXCL16
CXCR7	RDC1	NM_020311	2q37	CXCLs 11, 12
XCR1		NM_005283	3p21	XCL1
CX3CR1		NM_001337	3p21	CX3CL1
D6	CCBP2	NM_001296	3p21	CCLs 2, 3, 4, 5, 7, 8, 11, 13, 17, 23, 24
DARC	CD234, Duffy antigen	NM_002036	1q21-q22	CCLs 2, 5, 7, 8, 17 CXCLs 1, 5
CCR11*	CCX-CKR	AJ344142	3q22	CCLs 19, 21, 25

*Denotes the disqualification of CCX-CKR as CCR11.

In addition to the classical chemokine receptors, there exist three other chemokine-binding molecules that are also 7TM molecules. These are the Duffy antigen, also called DARC (Duffy antigen receptor for chemokines), the receptor formerly known as CCR11 and now more commonly referred to as CCX-CKR, and the D6 molecule *(7)*. These three molecules are characterized by an inability to signal in response to ligand binding and thus are not covered by the accepted definition of a receptor *(8)*. These are therefore more accurately referred to as chemokine-binding proteins and are not included within the systematic nomenclature system. Interestingly, CCX-CKR and D6 are clearly homologous to the other chemokine receptors and, as shown in Table 1, are encoded by genes that are located within the dominant chemokine receptor cluster on human chromosome 3. Both chemokine-binding proteins have therefore evolved from within the larger chemokine receptor family. In contrast, DARC is structurally quite distinct from the other chemokine receptors and is found at a chromosomal locus not otherwise associated with chemokine receptors. It may be, therefore, that DARC has entered the field of chemokine biology through a "convergent" evolutionary process rather than from a divergent mechanism within the chemokine receptor family.

3.2. Ligand Binding by Chemokine Receptors

3.2.1. Chemokine Receptor Ligand Binding Profiles

On one level, chemokine receptor biology and biochemistry is easy to understand in that CCRs bind CC chemokines, CXCRs bind CXC chemokines, and the XCR and CX3CR receptors bind their respective ligand partners. There are a few reported exceptions to this general rule *(8)*, but the only mammalian chemokine receptor that comprehensively binds ligands from more than one subfamily is the DARC receptor, which binds a number of inflammatory CC and CXC chemokines with high affinity. In many ways, however, the biology and biochemistry of chemokine receptors is extremely complicated. This is, in the main, a consequence of the fact that many chemokine receptors display marked promiscuity of ligand binding. In addition, many ligands are unfaithful to their receptors and can bind multiple chemokine receptors with high affinity. Thus, many chemokine-receptor relationships are characterized by promiscuity and unfaithfulness. In general, the receptors involved in inflammatory leukocyte migration display the most overt promiscuity (see Table 1). For example, CCR3 is known to bind at least 10 separate ligands. An additional level of complexity is that a number of studies have indicated that different ligands binding to the same receptor can do so in slightly different ways and can even induce distinct downstream responses within target cells *(9–11)*. Thus, this promiscuity of ligand binding potentially allows for extreme subtlety of regulation of cell function by the different ligands. On the other hand, the ability of multiple ligands to bind to receptors involved in inflammatory leukocyte migration may be a protective mechanism that ensures both redundancy and subtlety in the orchestration of an in vivo inflammatory response. This is borne out by the observations of individual ligand-null mice in which it is often difficult to find discrete pathologies in which the ligands play an indispensable and unique role.

In contrast, the receptors involved in homeostatic (constitutive) chemokine function display much more limited promiscuity of ligand binding and in some cases enjoy a monogamous relationship with their ligands. For example, CCR9 binds only CCL25, and CXCR5 binds only CXCL13. Curiously, three of the homeostatic CCRs (CCRs 4, 7, and 10) each bind two ligands. It has been shown for CCRs 4 and 7 that the ligands interact with the receptors in quite specific ways. Thus with CCR7, CCL21 binds but does not induce internalization, whereas CCL19 binds and induces internalization in the expected manner *(12–14)*. Similarly, for CCR4, whereas CCL22 induces efficient receptor internalization, CCL17 does not *(15)*. It remains to be determined whether CCLs 27 and 28 also display differences in terms of their interactions with CCR10, but the apparent coevolution of CCR7 and CCR10 and their ligands might suggest that this is likely to be the case.

In addition to chemokines, there is abundant evidence of other types of ligands binding to chemokine receptors. For example, β-defensins bind to CCR6 *(16)*, and a number of pathogen-derived peptides *(17)* and even autoantigens *(18,19)* have been shown to bind to chemokine receptors. In addition, the glycoprotein (GP120) peptide of M-tropic strains of the human immunodeficiency virus (HIV) binds to CCR5, and the GP120 molecule of T-tropic HIV strains binds to CXCR4 *(20)*. Abundant evidence has been published demonstrating the importance of these interactions in HIV pathogenesis.

3.2.2. The Structural Basis for Ligand Recognition by Chemokine Receptors

The primary structure of a typical chemokine receptor (CCR7) is represented diagrammatically in Figure 1, and the primary structures of all the known CC chemokine receptors are aligned in Table 2; CXCRs, XCR1, and CX3CR1 are aligned in Table 3. There have been many studies designed to determine the structural domains of the chemokine receptors that are essential for ligand binding. Many of these studies have relied on the ligand binding selectivity of receptors to generate chimeric molecules with which to investigate the structural basis for ligand binding. Some of the earliest studies in this context involved analysis of the differential ligand binding by the two receptors for CXCL8. CXCR1 is relatively specific for CXCL8, whereas CXCR2 is much more promiscuous and binds a number of other CXC chemokines in addition to CXCL8.

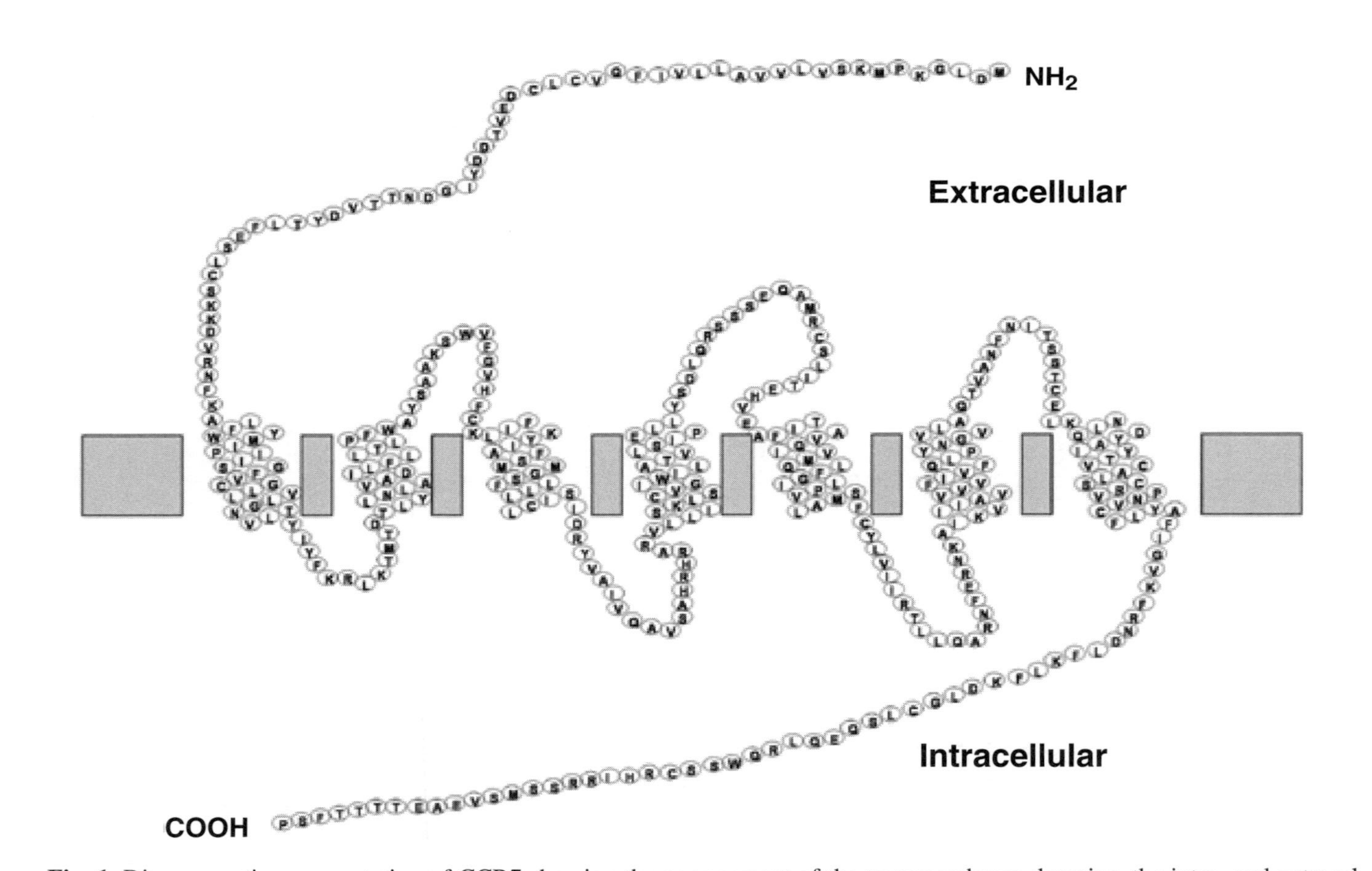

Fig. 1. Diagrammatic representation of CCR7 showing the arrangement of the transmembrane domains, the intra- and extracellular loops, and the amino-terminus and carboxy-terminus.

Chimeric receptor studies demonstrated the amino-terminus of CXCR2 to be the primary determinant of ligand binding and specificity *(21)*. Later studies have also implicated the first extracellular loop in ligand binding by CXCR2 and have further demonstrated the subtly different mechanisms of ligation of different ligands to CXCR2 *(10)*. A number of other domain-swap studies have demonstrated the variability in the domains responsible for ligand binding in other chemokine receptors. Thus, for CCR1 and CCR3, a multisite binding model has been proposed on the basis of the requirements for multiple extracellular receptor domains for effective ligand binding *(22)*. The use of cross-linkable macrophage inflammatory protein (MIP)-1α has further revealed the importance of the second extracellular loop for ligand binding to CCR1 *(23)*. Studies on CCR5 have strongly implicated the second extracellular loop as a primary determinant of ligand binding *(24,25)*, although the involvement of the amino-terminus again has also been demonstrated in subsequent studies *(26)*. Detailed analyses of ligand binding have highlighted different modes of interaction for regulated on activation normal T-cell expressed and secreted (RANTES) and MIP-1α with CCR5 again demonstrating the complexity of the interactions between chemokines and their receptors *(26)*. For CCR2, the amino-terminus has been shown to be both essential and sufficient for ligand binding *(27,28)*, although extracellular loops, specifically the first extracellular loop, have also been implicated in high-affinity binding and signaling *(24,29)*. As well as the extracellular regions of the receptors, other studies have indicated a requirement for other receptor regions for ligand interaction. It is likely therefore that the contribution of transmembrane domains, and even intracellular domains, to receptor structure are important in the proper orientation of the extracellular domains for ligand binding. To summarize, the majority of data generated support a two-stage model of ligand binding involving the amino-terminal region of the chemokine receptors along with one or more of the other extracellular domains.

3.3. Primary Structural Determinants of Chemokine Receptor Function

3.3.1. Cysteines in Chemokine Receptor Structure and Function

Data from studies with other GPCRs have highlighted the importance of extracellular cysteines in ligand binding and the maintenance of the conformational integrity of the receptors. There are typically four conserved cysteine residues found on extracellular domains of chemokine receptors (see Figure 1 and Tables 2 and 3): one on the amino-terminus and one on each of the three extracellular loops. It is clear that the cysteines on extracellular loops 1 and 2 form a disulfide bond that is essential for the proper trafficking of the receptors

Table 2
Alignment of the CC Chemokine Receptors with CCX-CKR (Formerly Called CCR11) and D6

```
CCR1      ------------------METPNTTEDYDTT-------TEFDYGDATPCQKVNERAFGAQ 35
CCR2      -----------------MLSTSRSRFIRNTNESGEEVTTFFDYDYGAPCHKFDVKQIGAQ 43
CCR3      -----------------MTTSLDTVETFGTT-------SYYDD-VGLLCEKADTRALMAQ 35
CCR4      -----------------MNPTDIADTTLDESIYSN---YYLYESIPKPCTKEGIKAFGEL 40
CCR5      ------------------MDYQVSSPIYDIN-----------YYTSEPCQKINVKQIAAR 31
CCR6      -MSGESMN------------FSDVFDSSEDYFVSVNTSYYSVDSEMLLCSLQEVRQFSRL 47
CCR7      MDLGKPMKSVLVVALLVIFQVCLCQDEVTDDYIGDNT-TVDYTLFESLCSKKDVRNFKAW 59
CCR8      -----------------MDYTLDLSVTTVTD-------YYYPDIFSSPCDAELIQTNGKL 36
CCR9      ----------------------MADDYGSESTSSMED-YVNFNFTDFYCEKNNVRQFASH 37
CCR10     ------------------MGTEATEQVSWGHYSGDEEDAYSAEPLPELCYKADVQAFSRA 42
CCX-CKR   ------------------MALEQNQSTDYYYEENEMNGTYDYSQYELICIKEDVREFAKV 42
D6        -------------MAATASPQPLATEDADSENSSFYYYDYLDEVAFMLCRKDAVVSFGKV 47
                                                          C

CCR1      LLPPLYSLVFVIGLVGNILVVLVLVQYK-RLKNMTSIYLLNLAISDLLFLFTLPFWIDYK 94
CCR2      LLPPLYSLVFIFGFVGNMLVVLILINCK-KLKCLTDIYLLNLAISDLLFLITLPLWAHSA 102
CCR3      FVPPLYSLVFTVGLLGNVVVVMILIKYR-RLRIMTNIYLLNLAISDLLFLVTLPFWIHYV 94
CCR4      FLPPLYSLVFVFGLLGNSVVVLVLFKYK-RLRSMTDVYLLNLAISDLLFVFSLPFWGYYA 99
CCR5      LLPPLYSLVFIFGFVGNMLVILILINCK-RLKSMTDIYLLNLAISDLFFLLTVPFWAHYA 90
CCR6      FVPIAYSLICVFGLLGNILVVITFAFYK-KARSMTDVYLLNMAIADILFVLTLPFWAVSH 106
CCR7      FLPIMYSIICFVGLLGNGLVVLTYIYFK-RLKTMTDTYLLNLAVADILFLLTLPFWAYSA 118
CCR8      LLAVFYCLLFVFSLLGNSLVILVLVVCK-KLRSITDVYLLNLALSDLLFVFSFPFQTYYL 95
CCR9      FLPPLYWLVFIVGALGNSLVILVYWYCT-RVKTMTDMFLLNLAIADLLFLVTLPFWAIAA 96
CCR10     FQPSVSLTVAALGLAGNGLVLATHLAARRAARSPTSAHLLQLALADLLLALTLPFAAAGA 102
CCX-CKR   FLPVFLTIVFVIGLAGNSMVVAIYAYYK-KQRTKTDVYILNLAVADLLLLFTLPFWAVNA 101
D6        FLPVFYSLIFVLGLSGNLLLLMVLLRYV-PRRRMVEIYLLNLAISNLLFLVTLPFWG-IS 105
          : .     :  ..  GN :::          :  .. .:L::A:::::  .:.P:

CCR1      LKDDWVFGDAMCKILSGFYYTGLYSEIFFIILLTIDRYLAIVHAVFALRARTVTFG--VI 152
CCR2      AN-EWVFGNAMCKLFTGLYHIGYFGGIFFIILLTIDRYLAIVHAVFALKARTVTFG--VV 159
CCR3      RGHNWVFGHGMCKLLSGFYHTGLYSEIFFIILLTIDRYLAIVHAVFALRARTVTFG--VI 152
CCR4      AD-QWVFGLGLCKMISWMYLVGFYSGIFFVMLMSIDRYLAIVHAVFSLRARTLTYG--VI 156
CCR5      AA-QWDFGNTMCQLLTGLYFIGFFSGIFFIILLTIDRYLAVVHAVFALKARTVTFG--VV 147
CCR6      ATGAWVFSNATCKLLKGIYAINFNCGMLLLTCISMDRYIAIVQATKSFRLRSRTLPRSKI 166
CCR7      AK-SWVFGVHFCKLIFAIYKMSFFSGMLLLLCISIDRYVAIVQAVSAHRHRARVLLISKL 177
CCR8      LD-QWVFGTVMCKVVSGFYYIGFYSSMFFITLMSVDRYLAVVHAVYALKVRTIRMG--TT 152
CCR9      AD-QWKFQTFMCKVVNSMYKMNFYSCVLLIMCISVDRYIAIAQAMRAHTWREKRLLYSKM 155
CCR10     LQ-GWSLGSATCRTISGLYSASFHAGFLFLACISADRYVAIARALPAG-PRPSTPGRAHL 160
CCX-CKR   VH-GWVLGKIMCKITSALYTLNFVSGMQFLACISIDRYVAVTKVP----SQSGVGKPCWI 156
D6        VAWHWVFGSFLCKMVSTLYTINFYSGIFFISCMSLDKYLEIVHAQPYHRLRTRAKS--LL 163
              W :    C:    :Y  .    . ::  :: D:Y: :.:.      :

CCR1      TSIIIWALAILASMPGLYFSKTQWE--FTHHTCSLHFPHESLREWKLFQALKLNLFGLVL 210
CCR2      TSVITWLVAVFASVPGIIFTKCQKE--DSVYVCGPYFPRG----WNNFHTIMRNILGLVL 213
CCR3      TSIVTWGLAVLAALPEFIFYETEEL--FEETLCSALYPEDTVYSWRHFHTLRMTIFCLVL 210
CCR4      TSLATWSVAVFASLPGFLFSTCYTE--RNHTYCKTKYSLNS-TTWKVLSSLEINILGLVI 213
CCR5      TSVITWVVAVFASLPGIIFTRSQKE--GLHYTCSSHFPYSQYQFWKNFQTLKIVILGLVL 205
CCR6      ICLVVWGLSVIISSSTFVFNQKYNTQGSDVCEPKYQTVS-EPIRWKLLMLGLELLFGFFI 225
CCR7      SCVGIWILATVLSIPELLYSDLQRSSSEQAMRCSLIT---EHVEAFITIQVAQMVIGFLV 234
CCR8      LCLAVWLTAIMATIPLLVFYQVASE--DGVLQCYSFYNQQT-LKWKIFTNFKMNILGLLI 209
CCR9      VCFTIWVLAAALCIPEILYSQIKEESG--IAICTMVYPSDESTKLKSAVLTLKVILGFFL 213
CCR10     VSVIVWLLSLLLALPALLFSQDGQREG--QRRCRLIFPEGLTQTVKGASAVAQVALGFAL 218
CCX-CKR   ICFCVWMAAILLSIPQLVFYTVNDN-----ARCIPIFPRYLGTSMKALIQMLEICIGFVV 211
D6        LATIVWAVSLAVSIPDMVFVQTHENP-KGVWNCHADFGGHG-TIWKLFLRFQQNLLGFLL 221
           .   W  :     . : :                                    : : :

CCR1      PLLVMIICYTGIIKILLRRPNEKK-SKAVRLIFVIMIIFFLFWTPYNLTILISVFQDFLF 269
CCR2      PLLIMVICYSGILKTLLRCRNEKKRHRAVRVIFTIMIVYFLFWTPYNIVILLNTFQEFFG 273
CCR3      PLLVMAICYTGIIKTLLRCPSKKK-YKAIRLIFVIMAVFFIFWTPYNVAILLSSYQSILF 269
CCR4      PLGIMLFCYSMIIRTLQHCKNEKK-NKAVKMIFAVVVLFLGFWTPYNIVLFLETLVELEV 272
CCR5      PLLVMVICYSGILKTLLRCRNEKKRHRAVRLIFTIMIVYFLFWAPYNIVLLLNTFQEFFG 265
CCR6      PLMFMIFCYTFIVKTLVQAQNSKR-HKAIRVIIAVVLVFLACQIPHN-MVLLVTAANLGK 283
CCR7      PLLAMSFCYLVIIRTLLQARNFER-NKAIKVIIAVVVVFIVFQLPYNGVVLAQTVANFNI 293
```

Table 2 *(Continued)*

```
CCR8            PFTIFMFCYIKILHQLKRCQNHNK-TKAIRLVLIVVIASLLFWVPFNVVLFLTSLHSMHI 268
CCR9            PFVVMACCYTIIIHTLIQAKKSSK-HKALKVTITVLTVFVLSQFPYNCILLVQTIDAYAM 272
CCR10           PLGVMVACYALLGRTLLAARGPER-RRALRVVVALVAAFVVLQLPYSLALLLDTADLLAA 277
CCX-CKR         PFLIMGVCYFITARTLMKMPNIKI-SRPLKVLLTVVIVFIVTQLPYNIVKFCRAIDIIYS 270
D6              PLLAMIFFYSRIGCVLVRLRPAGQ-GRALKIAAALVVAFFVLWFPYNLTLFLHTLLDLQV 280
                P:  :   Y      L          :.:::   ::   .    P..   :

CCR1            TH-ECEQSRHLDLAVQVTEVIAYTHCCVNPVIYAFVGERFRKYLRQLFHR--RVAVHLVK 326
CCR2            LS-NCESTSQLDQATQVTETLGMTHCCINPIIYAFVGEKFRRYLSVFFRK--HITKRFCK 330
CCR3            GN-DCERSKHLDLVMLVTEVIAYSHCCMNPVIYAFVGERFRKYLRHFFHR--HLLMHLGR 326
CCR4            LQ-DCTFERYLDYAIQATETLAFVHCCLNPIIYFFLGEKFRKYILQLFKTC-RGLFVLCQ 330
CCR5            LN-NCSSSNRLDQAMQVTETLGMTHCCINPIIYAFVGEKFRNYLLVFFQK--HIAKRFCK 322
CCR6            MNRSCQSEKLIGYTKTVTEVLAFLHCCLNPVLYAFIGQKFRNYFLKILKDLWCVRRKYKS 343
CCR7            TSSTCELSKQLNIAYDVTYSLACVRCCVNPFLYAFIGVKFRNDLFKLFKDLGCLSQEQLR 353
CCR8            LD-GCSISQQLTYATHVTEIISFTHCCVNPVIYAFVGEKFKKHLSEIFQK--SCSQIFNY 325
CCR9            FISNCAVSTNIDICFQVTQTIAFFHSCLNPVLYVFVGERFRRDLVKTLKNLGCISQAQWV 332
CCR10           RERSCPASKRKDVALLVTSGLALARCGLNPVLYAFLGLRFRQDLRRLLRGGSSPSGPQPR 337
CCX-CKR         LITSCNMSKRMDIAIQVTESIALFHSCLNPILYVFMGASFKNYVMKVAKKYG--SWRRQR 328
D6              FG-NCEVSQHLDYALQVTESIAFLHCCFSPILYAFSSHRFRQYLKAFLAAVLGWHLAPGT 339
                    C           .T  :.  :. ..P.:Y F .  F:. .

CCR1            WLPFLSV---DRLERVSST-SPSTGEHELSAGF--------------- 355
CCR2            QCPVFYR---ETVDGVTSTNTPSTGEQEVSAGL--------------- 360
CCR3            YIPFLPS---EKLERTSSV-SPSTAEPELSIVF--------------- 355
CCR4            YCGLLQI---YSADTPSSSYTQSTMDHDLHDAL--------------- 360
CCR5            CCSIFQQ---EAPERASSVYTRSTGEQEISVGL--------------- 352
CCR6            SGFSCAGRYSENISRQTSETADNDNASSFTM----------------- 374
CCR7            QWSSCR------HIRRSSMSVEAETTTTFSP----------------- 378
CCR8            LGRQMPR---ESCEKSSSCQQHSSRSSSVDYIL--------------- 355
CCR9            SFTRREG-----SLKLSSMLLETTSG-ALSL----------------- 357
CCR10           RGCPRRP-------RLS-SCSAPTETHSLSWDN--------------- 362
CCX-CKR         QSVEEFP-------FDSEGPTEPTSTFSI------------------- 350
D6              AQASLSS---CSESSILTAQEEMTGMNDLGERQSENYPNKEDVGNKSA 384
```

Note: Fully conserved amino acids are indicated below the alignments.

to the cell membrane and for receptor signaling. Studies on CCR6 have shown that the amino-terminal and extracellular loop 3 cysteines do not form a disulfide bond *(30),* although further studies on CCR5 indicate that they do *(31).* For CCR5, all four cysteines are required for full surface expression, ligand binding, and receptor signaling. Interestingly, despite severely impaired ligand internalization, monocyte tropic strains of HIV-1 bind to CCR5 mutants lacking all four cysteine residues indicating differing receptor requirements for ligand binding compared with HIV GP120 binding *(31).*

3.3.2. Cysteines and Receptor Palmitoylation

A further importance of cysteines lies in the palmitoylation of chemokine receptors. Many chemokine receptors have cysteine residues in their carboxy-terminal regions. In other GPCRs, these have been implicated in palmitoylation and in the anchoring of the carboxy-terminus to the plasma membrane. This effectively generates a fourth intracellular loop in the receptors. Studies on CCR5 have identified a three-cysteine cluster in the carboxy-terminus that is

Table 3
Alignment of the CXC, XC, and CX3C Chemokine Receptors

```
CXCR1     ----------MSNITDPQMWDFDDLN----FTGMPPADEDYSPCMLETETLNKYVVIIAY 46
CXCR2     -----MEDFNMESDSFEDFWKGEDLSNYSYSSTLPPFLLDAAPCEPESLEINKYFVVIIY 55
CXCR3     MVLEVSDHQVLNDAEVAALLENFSSSYDYGENESDSCCTSPPCPQDFSLNFDRAFLPALY 60
CXCR4     -------MEGISSIPLPLLQIYTSDNYTEEMGSGDYDSMKEPCFREENANFNKIFLPTIY 53
CXCR5     --MNYPLTLEMDLENLEDLFWELDRLDNYNDTSLVENHLCPATEGPLMASFKAVFVPVAY 58
CXCR6     ----------------------MAEHDYHEDYGFSSFNDSSQEEHQDFLQFSKVFLPCMY 38
CXCR7     --------MDLHLFDYSEPGNFSDISWPCNSSDCIVVDTVMCPNMPN-KSVLLYTLSFIY 51
XCR1      ----------------------MESSGNPESTTFFYYDLQSQPCENQAWVFATLATTVLY 38
CX3CR1    ----------------------MDQFPESVTENFEYDDLAEACYIGDIVVFGTVFLSIFY 38
                                                              .       Y

CXCR1     ALVFLLSLLGNSLVMLVILYSRVGRSVTDVYLLNLALADLLFALTLPIWAASKVNG--WI 104
CXCR2     ALVFLLSLLGNSLVMLVILYSRVGRSVTDVYLLNLALADLLFALTLPIWAASKVNG--WI 113
CXCR3     SLLFLLGLLGNGAVAAVLLSRRTALSSTDTFLLHLAVADTLLVLTLPLWAVDAAVQ--WV 118
CXCR4     SIIFLTGIVGNGLVILVMGYQKKLRSMTDKYRLHLSVADLLFVITLPFWAVDAVAN--WY 111
CXCR5     SLIFLLGVIGNVLVLVILERHRQTRSSTETFLFHLAVADLLLVFILPFAVAEGSVG--WV 116
CXCR6     LVVFVCGLVGNSLVLVISIFYHKLQSLTDVFLVNLPLADLVFVCTLPFWAYAGIHE--WV 96
CXCR7     IFIFVIGMIANSVVVWVNIQAKTTGYDTHCYILNLAIADLWVVLTIPVWVVSLVQHNQWP 111
XCR1      CLVFLLSLVGNSLVLWVLVKYESLESLTNIFILNLCLSDLVFACLLPVWISPYHWG--WV 96
CX3CR1    SVIFAIGLVGNLLVVFALTNSKKPKSVTDIYLLNLALSDLLFVATLPFWTHYLINE--KG 96
           .:F  .::.N  V        .     T. :  .:L ::D  ..  :P.

CXCR1     FGTFLCKVVSLLKEVNFYSGILLLACISVDRYLAIVHATRTL--TQKR-HLVKFVCLGCW 161
CXCR2     FGTFLCKVVSLLKEVNFYSGILLLACISVDRYLAIVHATRTL--TQKR-YLVKFICLSIW 170
CXCR3     FGSGLCKVAGALFNINFYAGALLLACISFDRYLNIVHATQLY--RRGPPARVTLTCLAVW 176
CXCR4     FGNFLCKAVHVIYTVNLYSSVLILAFISLDRYLAIVHATNSQ--RPRKLLAEKVVYVGVW 169
CXCR5     LGTFLCKTVIALHKVNFYCSSLLLACIAVDRYLAIVHAVHAY--RHRRLLSIHITCGTIW 174
CXCR6     FGQVMCKSLLGIYTINFYTSMLILTCITVDRFIVVVKATKAYNQQAKRMTWGKVTSLLIW 156
CXCR7     MGELTCKVTHLIFSINLFGSIFFLTCMSVDRYLSITYFTNTP--SSRKKMVRRVVCILVW 169
XCR1      LGDFLCKLLNMIFSISLYSSIFFLTIMTIHRYLSVVSPLSTL--RVPTLRCRVLVTMAVW 154
CX3CR1    LHNAMCKFTTAFFFIGFFGSIFFITVISIDRYLAIVLAANSM--NNRTVQHGVTISLGVW 154
          :    CK     :  :.:: . :::: ::..R:: :.                      W

CXCR1     GLSMNLSLPFFLFRQA---YHPNNSSPVCYEVLGNDTAKWRMVLRILPHTFGFIVPLFVM 218
CXCR2     GLSLLLALPVLLFRRT---VYSSNVSPACYEDMGNNTANWRMLLRILPQSFGFIVPLLIM 227
CXCR3     GLCLLFALPDFIFLSA---HHDERLN-ATHCQYNFPQVG-RTALRVLQLVAGFLLPLLVM 231
CXCR4     IPALLLTIPDFIFANVS------EADDRYICDRFYPNDLWVVVFQFQHIMVGLILPGIVI 223
CXCR5     LVGFLLALPEILFAKVSQGHHNNSLPRCTFSQENQAETHAWFTSRFLYHVAGFLLPMLVM 234
CXCR6     VISLLVSLPQIIYGNVFN---------LDKLICGYHDEAISTVVLATQMTLGFFLPLLTM 207
CXCR7     LLAFCVSLPDTYYLKTVT-SASNNETYCRSFYPEHSIKEWLIGMELVSVVLGFAVPFSII 228
XCR1      VASILSSILDTIFHKV-------LSSGCD------YSELTWYLTSVYQHNLFFLLSLGII 201
CX3CR1    AAAILVAAPQFMFTKQ-------KENECLGDYPEVLQEIWPVLRNVETNFLGFLLPLLIM 207
             :  :     :                                       :  :.   :

CXCR1     LFCYGFTLRTLFKAHMGQKH-RAMRVIFAVVLIFLLCWLPYNLVLLADTLMRTQVIQEXC 277
CXCR2     LFCYGFTLRTLFKAHMGQKH-RAMRVIFAVVLIFLLCWLPYNLVLLADTLMRTQVIQETC 286
CXCR3     AYCYAHILAVLLVS-RGQRRLRAMRLVVVVVVAFALCWTPYHLVVLVDILMDLGALARNC 290
CXCR4     LSCYCIIISKLSHSKGHQKR-KALKTTVILILAFFACWLPYYIGISIDSFILLEIIKQGC 282
CXCR5     GWCYVGVVHRLRQAQRRPQRQKAVRVAILVTSIFFLCWSPYHIVIFLDTLARLKAVDNTC 294
CXCR6     IVCYSVIIKTLLHAGGFQKH-RSLKIIFLVMAVFLLTQMPFNLMKFIRSTHWEYYAMT-- 264
CXCR7     AVFYFLLARAISASSDQEKH-SSRKIIFSYVVVFLVCWLPYHVAVLLDIFSILHYIPFTC 287
XCR1      LFCYVEILRTLFRSRSKRRH-RTVKLIFAIVVAYFLSWGPYNFTLFLQTLFRTQIIRS-C 259
CX3CR1    SYCYFRIIQTLFSCKNHKKA-KAIKLILLVVIVFFLFWTPYNVMIFLETLKLYDFFPS-C 265
             Y      :  .    :   : :    .      :     Y: .

CXCR1     ERRNNIGRALDATEILGFLHSCLNPIIYAFIGQNFRHGFLKILAMHGLVSKEFLAR---- 333
CXCR2     ERRNHIDRALDATEILGILHSCLNPLIYAFIGQKFRHGLLKILAIHGLISKDSLPK---- 342
CXCR3     GRESRVDVAKSVTSGLGYMHCCLNPLLYAFVGVKFRERMWMLLLRLGCPNQRGLQRQPSS 350
CXCR4     EFENTVHKWISITEALAFFHCCLNPILYAFLGAKFKTSAQHALTSVSRGSSLKILSKGKR 342
CXCR5     KLNGSLPVAITMCEFLGLAHCCLNPMLYTFAGVKFRSDLSRLLTKLGCTGPASLCQLFPS 354
CXCR6     ----SFHYTIMVTEAIAYLRACLNPVLYAFVSLKFRKNFWKLVKDIGCLPYLGVSHQWKS 320
CXCR7     RLEHALFTALHVTQCLSLVHCCVNPVLYSFINRNYRYELMKAFIFKYSAKTGLTKLIDAS 347
XCR1      EAKQQLEYALLICRNLAFSHCCFNPVLYVFVGVKFRTHLKHVLRQFWFCRLQAPSPASIP 319
CX3CR1    DMRKDLRLALSVTETVAFSHCCLNPLIYAFAGEKFRRYLYHLYGKCLAVLCGRSVHVDFS 325
               .         :.   :.C.NP::Y F . :::

CXCR1     -HRVTSYT-SSSVNVSSNL----------- 350
CXCR2     -DSRPSFVGSSSGHTSTTL----------- 360
CXCR3     SRRDSSWSETSEASYSGL------------ 368
CXCR4     GGHSSVSTESESSSFHSS------------ 360
CXCR5     WRR-SSLSESENATSLTTF----------- 372
CXCR6     SEDNSKTFSASHNVEATSMFQL-------- 342
CXCR7     RVSETEYSALEQSTK--------------- 362
XCR1      HSPGAFAYEGASFY---------------- 333
CX3CR1    SSESQRSRHGSVLSSNFTYHTSDGDALLLL 355
```

Note: Fully conserved amino acids are indicated below the alignments.

essential for palmitoylation *(32,33)*. Mutation of individual cysteines within this cluster does not block palmitoylation but mutation of all three does. One of the most striking consequences of the mutation of the cysteine residues is a severely impaired trafficking of CCR5 to the cell membrane *(33)*. Despite this, the receptor molecules that do make it to the cell surface display wild-type ligand binding affinity, although the compound mutants display significantly impaired ligand-induced signaling and HIV binding. Notably, ligand binding has no effect on the palmitoylation of the receptors. Further studies have shown a reduced half-life and rapid turnover of the nonpalmitoylated CCR5 *(33)*. Interestingly, not all chemokine receptors are palmitoylated, and among the CC chemokine receptors (see Table 2), CCRs 1, 3, 11, and D6 have no carboxy-terminal cysteines and thus are not palmitoylated.

3.3.3. The Second Intracellular Loop and the DRY Box

Prominent among the conserved motifs in the chemokine receptors and in other GPCRs is the single letter amino acid code for aspartate, arginine, tyrosine (DRY) box, which is located on the second intracellular loop. In chemokine receptors, this is typically found within a larger single letter amino acid code for aspartate, arginine, tyrosine, leucine, alanine, isoleucine, valine (DRYLAIV) motif, although variations in this motif are found in the different receptors (see Tables 2 and 3). The strong conservation of the DRY box implicates it in important receptor function, and indeed recent studies have shown this to be the case. The DRY box is the main site for G protein coupling to the chemokine receptors *(34)* and may also be involved in β-arrestin binding *(35)* and in the orchestration of the ligand-dependent internalization of the receptors. Numerous mutagenesis studies have indicated the importance of the DRY box for ligand-induced signaling in a number of chemokine receptors. Interestingly, the DRY box mutations do not appear to impact ligand binding nor do they block the ability of CCR5 or CXCR4 to support HIV entry into cells *(36,37)*. Thus, these data again indicate that the receptor requirements for supporting HIV entry are different from those required for ligand interaction and signaling, and this ties in with the ability of a number of the cysteine mutant receptors described above to interact with HIV in the absence of an ability to interact with ligand.

Notably the constitutively active GPCR, found in Kaposis' Sarcoma Herpes Virus (KSHV), has a mutated DRY box. In this receptor, the DRY motif is altered to single letter amino acid code for valine, arginine, tryrosine (VRY). Introduction of a similar mutation into CXCR2 also induces constitutive activation of the receptor, although the specific nature of the amino acid substitution appears to be important here as alteration to single letter amino acid code for glutamine, arginine, tyrosine (QRY) does not induce constitutive CXCR2 activation *(38)*. Similar VRY substitutions in CXCR1 and CCR5 do not induce constitutive activation suggesting some receptor specificity in this effect *(38,39)*. Alteration of the DRY residues in CCR5 to single letter amino acid code for aspartate, asparagine, tyrosine (DNY) has no effect on ligand binding but blocks signaling responses and is associated with increased basal phosphorylation *(39)*.

A feature of the nonsignaling chemokine receptors such as D6 and DARC is the absence of a DRY box in their second intracellular loop *(7)*. In D6, this is altered to single letter amino acid code for aspartate, lysine, tyrosine (DKY) and in DARC is unrecognizable. Our data on D6 indicate that alteration of the DKY motif to DRY allows this otherwise silent receptor to signal (G.J. Graham and R.J. Nibbs, unpublished data), again highlighting the importance of the DRY box for ligand-induced receptor signaling. Interestingly, in D6, the altered DRY box single letter amino acid code for aspartate, lysine, tyrosine, leucine, glutamate, isoleucine, valine (DKYLEIV) is conserved throughout mammals and thus appears not to be a simple loss-of-function alteration that might not have been conserved. Thus, it is possible that the DRY box, or the DKY box in D6, may have undiscovered importance for other receptor features such as structure or receptor trafficking.

3.3.4. Sulfated Tyrosines and Ligand Binding

Another prominent conserved feature of the chemokine receptors is the presence of clusters of tyrosine residues in the amino-terminal region. Tyrosines, especially those directly following an acidic or neutral amino acid, are targets for posttranslational sulfation by tyrosyl-protein sulfotransferases in the Golgi apparatus. Much evidence has now been published regarding the sulfation of amino-terminal tyrosines in chemokine receptors. In the context of CCR5, the sulfated tyrosines are essential for HIV and ligand binding, and nonsulfated CCR5 is incompetent in supporting HIV entry or ligand-induced responses *(40–42)*. Indeed, sulfated tyrosines not only are essential for CCR5 function but are also, in many respects, sufficient at least for interactions with HIV. This has been demonstrated by the ability of amino-terminal peptides from CCR5, bearing sulfated tyrosines, to interact directly with GP120/CD4 complexes in vitro *(41)*. Nonsulfated amino-terminal peptides are unable to interact with the GP120/CD4 complexes. Curiously, the other major HIV coreceptor, CXCR4, has a lesser requirement for sulfated tyrosines for supporting HIV entry into cells *(43)*. CXCR4 has a cluster of three amino-terminal tyrosines, and alteration of all three of these to phenylalanine does not fully inhibit HIV entry through the mutant receptor. This mutation does however have a marked effect on ligand binding, again highlighting the differences in the nature of the interactions of HIV and ligands with the chemokine receptors. Analyses of CX3CR1, CCR2, and CCR8 have revealed the importance of sulfated tyrosines for ligand binding by these receptors as well *(44–46)*.

3.4. Signaling by Chemokine Receptors

3.4.1. Receptor Structure and Signal Transduction

Chemokine receptors and other GPCRs are thought to undergo a conformational change upon ligand binding that drives intracellular signaling *(47)*. The agonist is envisaged to stabilize the "active" conformation, a modified receptor structure that contains critical alterations in the nature of its interaction with second messenger systems, such as heterotrimeric G-protein complexes. The DRY box in the second intracellular loop, along with the single letter amino acid code for asparagine, proline X-X tyrosine (NPXXY) sequence found in transmembrane region 7 (TM7), are believed to be critical determinants in the activation of class A GPCRs, including chemokine receptors *(36,39,48,49)*. It is thought that the charged Asp-Arg residues in the DRY motif form key ionic bonds (an *ionic lock*) with residues in transmembrane region 6 (TM6) in the inactive state and that these bonds become disrupted in the active receptor to allow G protein access *(48)*. It is unclear how this is achieved during chemokine receptor activation, but it is known that agonist binding to the β_2-adrenergic GPCR changes the precise orientation of transmembrane region 3 (TM3) and TM6 to disrupt the ionic lock and create a binding pocket for heterotrimeric G proteins *(50,51)*. The precise conformational change induced in a given GPCR is likely to be dependent on the nature of the ligand that must be accommodated on the extracellular surface of the receptor. As chemokines are large molecules, TM domain shifting caused by their binding may need to be quite different from that which occurs in the β_2-adrenergic receptor *(50,51)*, other GPCRs for small agonists, or rhodopsin *(52)*. Indeed, it has been noted that the TM domains of chemokine receptors contain proline residues that are not found in rhodopsin and other GPCRs. Prolines introduce kinks and bends into helices so are likely to significantly affect the precise folding of the TM helices in chemokine receptors and the orientation of the extracellular and intracellular loops. Several studies have identified a threonine-x-proline (TxP) motif in transmembrane region 2 (TM2) of CCR5 and CCR2 as a critical determinant in chemokine binding and signaling *(53,54)*. The proline in this motif is proposed to kink TM2 such that its extracellular end is brought in close proximity to TM3 *(53,54)*. This may alter the nature of TM domain reorientation needed to open chemokine receptors for G protein access. Mutagenesis studies have identified many other critical amino acid residues in chemokine receptors, particularly in CCR5, which are required for optimal activation, and models have been constructed using the rhodopsin structure to accommodate these observations. However, only when active chemokine receptor structures are determined by crystallography will these issues be convincingly resolved.

3.4.2. *Covalent Modification of Chemokine Receptors After Chemokine Binding*

Signaling through chemokine receptors initiates intracellular signals that drive biological changes in cell behavior. These signals also feed back onto the receptors themselves introducing covalent modifications that alter receptor structure and activity. For example, the cytoplasmic C-terminal region of many chemokine receptors becomes rapidly phosphorylated, and this change, along with the DRY motif, encourages the recruitment of β-arrestins *(39,49,55–61)*. This is of critical importance in chemokine receptor function. β-Arrestins are key GPCR regulators able to (i) direct receptors to clathrin-coated pits for internalization, (ii) inhibit coupling to G proteins and other signaling molecules, (iii) act as scaffolds for the formation of alternative signaling complexes, and (iv) assist in directing the subsequent intracellular trafficking of GPCRs *(62,63)*. Without β-arrestin recruitment, chemokine receptor function is severely compromised. Other more extensive agonist-induced covalent changes have also been reported to modify chemokine receptor structure and function *(64–66)*. CXCR4, after activation by CXCL12, undergoes ubiquitination whereby the 76-amino-acid ubiquitin protein becomes covalently attached to lysine residues in the intracellular domains of the receptor *(65,66)*. This agonist-induced modification is of critical importance in targeting CXCR4 and many other GPCRs to lysosomes for degradation, leading to receptor downregulation in the continuous presence of agonist *(67)*. Moreover, in breast cancer cells, human epidermal growth factor receptor 2 (HER-2)–mediated inhibition of CXCR4 ubiquitination is thought to contribute to the efficiency of CXCR4-mediated tumor metastasis by maintaining CXCL12 responsiveness *(64)*.

3.5. Chemokine Receptor Dimerization

It has become apparent in recent years that most GPCRs are able to form dimers and other higher order oligomers *(68)*. This conclusion has been reached primarily using heterologous expression systems and by exploiting an array of different methodologies from coimmunoprecipitation and chemical cross-linking to, more recently, biophysical techniques such as bioluminescence resonance energy transfer (BRET) and fluorescence resonance energy transfer (FRET). These techniques have also been used to provide considerable evidence of chemokine receptor homo- and heterodimerization in transfected cells in vitro. Thus, CCR2, CCR5, CXCR1, CXCR2, and CXCR4 are capable of forming homodimers; CCR2 heterodimerizes with CXCR4 and CCR5; CXCR1 and CXCR2 may exist as heterodimers *(68,69)*; and it seems likely that as further studies are undertaken, other chemokine receptors will be added to this list. Formation of these higher order chemokine receptor complexes occurs in the absence of chemokine ligands, and indeed, there is a prevailing view that GPCR homo- and heterodimerization is ligand-independent. Thus, what role does dimerization play in chemokine receptor biology? With some GPCRs, dimerization is involved in efficient trafficking of receptors from the endoplasmic reticulum to the cell surface. For example, the γ-aminobutyric acid (GABA) binding receptor $GABA_{B1}$ is retained in the endoplasmic reticulum (ER), unless it is coexpressed with $GABA_{B2}$, which ensures trafficking to the cell surface *(70–72)*. $GABA_{B2}$ will traffic to the cell surface on its own but cannot bind GABA, thus the heterodimer is essential for surface presentation and function *(70–72)*. Chemokine receptor homodimerization may likewise help ensure appropriate cell surface expression. It is possible that only in the context of a dimer will each monomer assume the appropriate structure to enable trafficking to the surface, though this has yet to receive much experimental support. Moreover, the precise structural characteristics of GPCR dimers are uncertain, but two models have been proposed *(68)*. In the contact dimerization model, each receptor molecule is principally an independent entity, contacting the other member of the dimer by noncovalent interactions between transmembrane helices. In the domain swapping model, the two components of the dimer are more intimately intertwined, sharing helices that form covalent links between the two polypeptide chains. Whereas these models provide a useful theoretical framework in which to envisage chemokine receptor dimerization, it should be emphasized that there is no direct evidence to support either model with regard to chemokine receptor dimerization. Few studies have attempted to define the molecular nature of chemokine receptor dimerization, though one high-profile study by Hernanz-Falcon and colleagues suggested that TM domains 1 and 4 are involved in CCR5 dimerization *(73)*. In this work, single amino acid changes were introduced into these domains of CCR5 to create proteins that apparently retained chemokine binding activity but lost the ability to form homodimers and couple to intracellular signal transduction machinery. Moreover, peptides derived from these domains were reported to block signaling through wild-type CCR5 by perturbing homodimerization. Although intriguing, these observations should be approached with some caution: Others have reported that the same CCR5 mutants are functionally identical to wild-type CCR5 *(74)*. Further investigations are clearly required.

Apart from possible roles in receptor trafficking and folding, it is emerging that dimerization may be important in determining the nature of ligand interaction and defining subsequent downstream intracellular signals *(68)*. However, this aspect of chemokine receptor biology is notable for the remarkably different results obtained by separate groups of researchers. For instance, one group provides compelling evidence that chemokine-induced heterodimerization of CCR2 and CCR5 induces biological responses that are remarkably different from those induced through homodimers of either receptor *(75,76)*. Thus, simultaneous stimulation of cells expressing CCR2 and CCR5 with ligands for both receptors triggers signaling at concentrations up to 100-fold lower than that seen with either chemokine alone, with heterodimers recruiting G proteins distinct from those used by the homodimers *(75,76)*. The same group has also reported on many occasions that chemokine binding is required to drive homo- and heterodimerization of chemokine receptors on the cell surface. However, others have been unable to repeat these observations, reporting instead constitutive receptor dimerization and competition between receptor-specific ligands for binding to the CCR2/CCR5 heterodimer rather than an augmentation of activity *(77)*. From these data, it was hypothesized that pre-formed chemokine receptor dimers are able to bind to only a single chemokine molecule at any one time *(68)*. This was proposed to induce a conformational change in the occupied receptor that is transmitted in some way to its dimeric partner, preventing it from binding chemokine but encouraging it to couple to the signal transduction machinery of the cell.

In summary, whereas physical interactions do undoubtedly occur between chemokine receptor molecules in heterologous expression systems, considerable clarification is required to dissect the functional implications of these interactions. This is worthy of further investigation because the disruption of chemokine receptor dimerization could theoretically represent a novel approach to therapeutically target these molecules in the array of inflammatory and immune pathologies in which they play a key role. Conversely, as it has been suggested that CCR5 dimerization may prevent HIV entry through this receptor *(78)* and that the CCR2 may prevent HIV entry through CXCR4 and CCR5 via heterodimerization *(79)*, enhancing chemokine receptor dimerization could conceivably act as a suitable anti-HIV therapy. However, rather like the rest of the chemokine receptor dimerization literature, there is still a considerable amount of controversy surrounding these observations.

3.6. Determination of the Three-Dimensional Structure of the Chemokine Receptors

3.6.1. The Need for a Structural Model

Approximately 30% to 40% of currently licensed drugs are targeted at GPCRs and thus represent a highly significant source of income for many pharmaceutical companies *(80)*. In the specific context of chemokine receptors, it is clear that they are involved in a number of prominent pathologies and thus represent an important therapeutic target *(81)*. For example, chemokines and their receptors lie at the center of all immune and inflammatory disorders and are responsible for the aberrant accumulation of leukocytes at inflamed sites in autoimmune conditions. In addition, over the past 12 years, it has become clear that the CCR5 and CXCR4 receptors represent the major coreceptors, along with CD4, that are essential for supporting HIV entry into susceptible cells *(82)*. More recent data have indicated an important role for chemokines and their receptors in facilitating the growth and metastasis of many cancers *(83)*. Thus chemokine receptors present themselves as an attractive therapeutic target.

Unfortunately, despite the prevalence of anti-GPCR drugs, the field of rational anti-GPCR drug discovery suffers from the lack of an accurate three-dimensional structure of these receptors. The only 7TM structure that is known is that of rhodopsin *(84)*, which is many hundreds of millions of years evolutionarily distant from other mammalian GPCRs such as chemokine receptors. Thus, there is a pressing need for a full three-dimensional structure of one of the other mammalian GPCRs. In this regard, much work has been done in attempting to purify and determine the structure of mammalian chemokine receptors.

3.6.2. Advances to Date

One of the disadvantages of chemokine receptors and other GPCRs compared with rhodopsin is the relatively low-level expression of these GPCRs. The structural determination of rhodopsin was facilitated by the extremely high levels of expression found in native mammalian sources. Thus, a key feature of many of the initial studies in chemokine receptor purification and structural analysis is that they have involved efforts at increasing expression levels. This has been achieved with CCR5 and CXCR4 by codon optimization using codons typically found in the highly expressed opsin family of GPCRs *(85,86)*. With CCR5, this increased expression two- to fivefold compared with the wild-type receptor. In addition, expression of the receptor in heterologously transfected mammalian cells can be increased by treating cells with the histone-deacetylase inhibitor sodium butyrate. With CCR5, this increases expression by up to threefold. Both CCR5 and CXCR4 have now been purified to homogeneity in an active conformation. Detergent screens have identified cymal-5 as being the optimal detergent for CCR5 solubilization, although other detergents such as dodecyl maltoside and cymal-6 were also effective *(86)*. With CXCR4, 3-[(3-cholamidopropyl) dimethylammonio]-2-hydroxy-1-propanesulfonate (CHAPSO) was found by one study to be optimum *(85)*, although another study found the combination of dodecyl maltoside and cholesterol to be best *(87)*. Thus far, no structural coordinates have been reported for either of these receptors.

We have been interested in studying the structure of the D6 chemokine receptor *(7)*. This receptor has not, so far, been implicated in pathologies as prominent as those involving CCR5 and CXCR4. Nevertheless, it is structurally very similar to the other chemokine receptors and thus a three-dimensional structure of D6 would be an invaluable template that would facilitate the determination of the structure of other chemokine receptors. A major advantage of D6 over the other chemokine receptors is that it can be expressed to extremely high levels in heterologous transfectants. Thus, in human embryonic kidney (HEK) cells and in L1.2 cells, up to 5×10^6 D6 molecules can be expressed per cell. We have been able to purify D6 from large-scale cultures of L1.2 cells using dodecyl maltoside as a detergent *(88)*. All the purified receptor molecules are active and display a binding affinity for ligand that is similar to that seen on whole cells. The ability to generate large quantities of purified D6 on a routine basis makes us hopeful that we may soon be able to determine the full three-dimensional structure of this member of the chemokine receptor family.

3.7. Conclusions

Chemokine receptors represent attractive therapeutic targets and are therefore the focus of intensive investigation. Thus far, much data have been generated regarding the modes of interaction of ligands with the chemokine receptors and on the primary structural basis for receptor function. The next crucial step will involve the determination of a full three-dimensional structure for a chemokine receptor. Once this goal has been achieved, a more complete understanding of the nature of the interactions between chemokine receptors, ligands, and pathogens such as HIV will be available. The three-dimensional structure will also facilitate rational anti–chemokine receptor drug design, which will have an impact on our ability to effectively treat a wide range of prominent pathologies.

References

1. Murphy PM, Baggiolini M, Charo IF, et al. International union of pharmacology. XXII. Nomenclature for chemokine receptors. Pharmacol Rev 2000;52(1):145–176.
2. Vassilatis DK, Hohmann JG, Zeng H, et al. The G protein-coupled receptor repertoires of human and mouse. Proc Natl Acad Sci U S A 2003;100(8):4903–4908.
3. Ono Y, Fujibuchi W, Suwa M. Automatic gene collection system for genome-scale overview of G-protein coupled receptors in eukaryotes. Gene 2005;364:63–73.
4. Holmes WE, Lee J, Kuang WJ, Rice GC, Wood WI. Structure and functional expression of a human interleukin-8 receptor. Science 1991;253(5025):1278–1280.
5. Murphy PM, Tiffany HL. Cloning of complementary DNA encoding a functional human interleukin-8 receptor. Science 1991;253(5025):1280–1283.
6. Neote K, DiGregorio D, Mak JY, Horuk R, Schall TJ. Molecular cloning, functional expression, and signaling characteristics of a C-C chemokine receptor. Cell 1993;72(3):415–425.
7. Nibbs R, Graham G, Rot A. Chemokines on the move: control by the chemokine "interceptors" Duffy blood group antigen and D6. Semin Immunol 2003;15(5):287–294.

8. Alexander SP, Mathie A, Peters JA. Guide to receptors and channels, 2nd edition. Br J Pharmacol 2006;147(Suppl 3):S1–168.
9. Duchesnes CE, Murphy PM, Williams TJ, Pease JE. Alanine scanning mutagenesis of the chemokine receptor CCR3 reveals distinct extracellular residues involved in recognition of the eotaxin family of chemokines. Mol Immunol 2006; 43(8):1221–1231.
10. Katancik JA, Sharma A, de Nardin E. Interleukin 8, neutrophil-activating peptide-2 and GRO-alpha bind to and elicit cell activation via specific and different amino acid residues of CXCR2. Cytokine 2000;12(10):1480–1488.
11. Govaerts C, Bondue A, Springael JY, et al. Activation of CCR5 by chemokines involves an aromatic cluster between transmembrane helices 2 and 3. J Biol Chem 2003;278(3):1892–1903.
12. Bardi G, Lipp M, Baggiolini M, Loetscher P. The T cell chemokine receptor CCR7 is internalized on stimulation with ELC, but not with SLC. Eur J Immunol 2001;31(11):3291–3297.
13. Kohout TA, Nicholas SL, Perry SJ, Reinhart G, Junger S, Struthers RS. Differential desensitization, receptor phosphorylation, beta-arrestin recruitment, and ERK1/2 activation by the two endogenous ligands for the CC chemokine receptor 7. J Biol Chem 2004;279(22):23214–23222.
14. Ott TR, Pahuja A, Nickolls SA, Alleva DG, Struthers RS. Identification of CC chemokine receptor 7 residues important for receptor activation. J Biol Chem 2004;279(41):42383–42392.
15. Mariani M, Lang R, Binda E, Panina-Bordignon P, D'Ambrosio D. Dominance of CCL22 over CCL17 in induction of chemokine receptor CCR4 desensitization and internalization on human Th2 cells. Eur J Immunol 2004;34(1): 231–240.
16. Yang D, Chertov O, Bykovskaia SN, et al. Beta-defensins: linking innate and adaptive immunity through dendritic and T cell CCR6. Science 1999;286(5439): 525–528.
17. Aliberti J, Valenzuela JG, Carruthers VB, et al. Molecular mimicry of a CCR5 binding-domain in the microbial activation of dendritic cells. Nat Immunol 2003;4(5):485–490.
18. Oppenheim JJ, Dong HF, Plotz P, et al. Autoantigens act as tissue-specific chemoattractants. J Leukoc Biol 2005;77(6):854–861.
19. Howard OM, Dong HF, Su SB, et al. Autoantigens signal through chemokine receptors: uveitis antigens induce CXCR3- and CXCR5-expressing lymphocytes and immature dendritic cells to migrate. Blood 2005;105(11):4207–4214.
20. Simmons G, Reeves JD, Hibbitts S, et al. Co-receptor use by HIV and inhibition of HIV infection by chemokine receptor ligands. Immunol Rev 2000;177:112–126.
21. Suzuki H, Prado GN, Wilkinson N, Navarro J. The N terminus of interleukin-8 (IL-8) receptor confers high affinity binding to human IL-8. J Biol Chem 1994; 269(28):18263–18266.
22. Pease JE, Wang J, Ponath PD, Murphy PM. The N-terminal extracellular segments of the chemokine receptors CCR1 and CCR3 are determinants for MIP-1alpha and

eotaxin binding, respectively, but a second domain is essential for efficient receptor activation. J Biol Chem 1998;273(32):19972–19976.
23. Zoffmann S, Chollet A, Galzi JL. Identification of the extracellular loop 2 as the point of interaction between the N terminus of the chemokine MIP-1alpha and its CCR1 receptor. Mol Pharmacol 2002;62(3):729–736.
24. Samson M, LaRosa G, Libert F, et al. The second extracellular loop of CCR5 is the major determinant of ligand specificity. J Biol Chem 1997;272(40):24934–24941.
25. Wu L, LaRosa G, Kassam N, et al. Interaction of chemokine receptor CCR5 with its ligands: multiple domains for HIV-1 gp120 binding and a single domain for chemokine binding. J Exp Med 1997;186(8):1373–1381.
26. Blanpain C, Doranz BJ, Bondue A, et al. The core domain of chemokines binds CCR5 extracellular domains while their amino terminus interacts with the transmembrane helix bundle. J Biol Chem 2003;278(7):5179–5187.
27. Monteclaro FS, Charo IF. The amino-terminal extracellular domain of the MCP-1 receptor, but not the RANTES/MIP-1alpha receptor, confers chemokine selectivity. Evidence for a two-step mechanism for MCP-1 receptor activation. J Biol Chem 1996;271(32):19084–19092.
28. Monteclaro FS, Charo IF. The amino-terminal domain of CCR2 is both necessary and sufficient for high affinity binding of monocyte chemoattractant protein 1. Receptor activation by a pseudo-tethered ligand. J Biol Chem 1997;272(37): 23186–23190.
29. Han KH, Green SR, Tangirala RK, Tanaka S, Quehenberger O. Role of the first extracellular loop in the functional activation of CCR2. The first extracellular loop contains distinct domains necessary for both agonist binding and transmembrane signaling. J Biol Chem 1999;274(45):32055–32062.
30. Ai LS, Liao F. Mutating the four extracellular cysteines in the chemokine receptor CCR6 reveals their differing roles in receptor trafficking, ligand binding, and signaling. Biochemistry 2002;41(26):8332–8341.
31. Blanpain C, Lee B, Vakili J, et al. Extracellular cysteines of CCR5 are required for chemokine binding, but dispensable for HIV-1 coreceptor activity. J Biol Chem 1999;274(27):18902–18908.
32. Blanpain C, Wittamer V, Vanderwinden JM, et al. Palmitoylation of CCR5 is critical for receptor trafficking and efficient activation of intracellular signaling pathways. J Biol Chem 2001;276(26):23795–23804.
33. Percherancier Y, Planchenault T, Valenzuela-Fernandez A, Virelizier JL, Arenzana-Seisdedos F, Bachelerie F. Palmitoylation-dependent control of degradation, life span, and membrane expression of the CCR5 receptor. J Biol Chem 2001;276(34): 31936–31944.
34. Damaj BB, McColl SR, Neote K, et al. Identification of G-protein binding sites of the human interleukin-8 receptors by functional mapping of the intracellular loops. FASEB J 1996;10(12):1426–1434.
35. Kraft K, Olbrich H, Majoul I, Mack M, Proudfoot A, Oppermann M. Characterization of sequence determinants within the carboxyl-terminal domain of chemokine

receptor CCR5 that regulate signaling and receptor internalization. J Biol Chem 2001;276(37):34408–34418.
36. Gosling J, Monteclaro FS, Atchison RE, et al. Molecular uncoupling of C-C chemokine receptor 5-induced chemotaxis and signal transduction from HIV-1 coreceptor activity. Proc Natl Acad Sci U S A 1997;94(10):5061–5066.
37. Doranz BJ, Orsini MJ, Turner JD, et al. Identification of CXCR4 domains that support coreceptor and chemokine receptor functions. J Virol 1999;73(4): 2752–2761.
38. Burger M, Burger JA, Hoch RC, Oades Z, Takamori H, Schraufstatter IU. Point mutation causing constitutive signaling of CXCR2 leads to transforming activity similar to Kaposi's sarcoma herpesvirus-G protein-coupled receptor. J Immunol 1999;163(4):2017–2022.
39. Lagane B, Ballet S, Planchenault T, et al. Mutation of the DRY motif reveals different structural requirements for the CC chemokine receptor 5-mediated signaling and receptor endocytosis. Mol Pharmacol 2005;67(6):1966–1976.
40. Seibert C, Cadene M, Sanfiz A, Chait BT, Sakmar TP. Tyrosine sulfation of CCR5 N-terminal peptide by tyrosylprotein sulfotransferases 1 and 2 follows a discrete pattern and temporal sequence. Proc Natl Acad Sci U S A 2002;99(17):11031–11036.
41. Cormier EG, Persuh M, Thompson DA, et al. Specific interaction of CCR5 amino-terminal domain peptides containing sulfotyrosines with HIV-1 envelope glycoprotein gp120. Proc Natl Acad Sci U S A 2000;97(11):5762–5767.
42. Farzan M, Mirzabekov T, Kolchinsky P, et al. Tyrosine sulfation of the amino terminus of CCR5 facilitates HIV-1 entry. Cell 1999;96(5):667–676.
43. Farzan M, Babcock GJ, Vasilieva N, et al. The role of post-translational modifications of the CXCR4 amino terminus in stromal-derived factor 1 alpha association and HIV-1 entry. J Biol Chem 2002;277(33):29484–29489.
44. Fong AM, Alam SM, Imai T, Haribabu B, Patel DD. CX3CR1 tyrosine sulfation enhances fractalkine-induced cell adhesion. J Biol Chem 2002;277(22): 19418–19423.
45. Gutierrez J, Kremer L, Zaballos A, Goya I, Martinez AC, Marquez G. Analysis of post-translational CCR8 modifications and their influence on receptor activity. J Biol Chem 2004;279(15):14726–14733.
46. Preobrazhensky AA, Dragan S, Kawano T, et al. Monocyte chemotactic protein-1 receptor CCR2B is a glycoprotein that has tyrosine sulfation in a conserved extracellular N-terminal region. J Immunol 2000;165(9):5295–5303.
47. Bissantz C. Conformational changes of G protein-coupled receptors during their activation by agonist binding. J Recept Signal Transduct Res 2003;23(2–3):123–153.
48. Ballesteros JA, Jensen AD, Liapakis G, et al. Activation of the beta 2-adrenergic receptor involves disruption of an ionic lock between the cytoplasmic ends of transmembrane segments 3 and 6. J Biol Chem 2001;276(31):29171–29177.
49. Aramori I, Ferguson SS, Bieniasz PD, Zhang J, Cullen B, Cullen MG. Molecular mechanism of desensitization of the chemokine receptor CCR-5: receptor signaling

and internalization are dissociable from its role as an HIV-1 co-receptor. EMBO J 1997;16(15):4606–4616.
50. Gether U, Lin S, Ghanouni P, Ballesteros JA, Weinstein H, Kobilka BK. Agonists induce conformational changes in transmembrane domains III and VI of the beta2 adrenoceptor. EMBO J 1997;16(22):6737–6747.
51. Ghanouni P, Steenhuis JJ, Farrens DL, Kobilka BK. Agonist-induced conformational changes in the G-protein-coupling domain of the beta 2 adrenergic receptor. Proc Natl Acad Sci U S A 2001;98(11):5997–6002.
52. Farrens DL, Altenbach C, Yang K, Hubbell WL, Khorana HG. Requirement of rigid-body motion of transmembrane helices for light activation of rhodopsin. Science 1996;274(5288):768–770.
53. Arias DA, Navenot JM, Zhang WB, Broach J, Peiper SC. Constitutive activation of CCR5 and CCR2 induced by conformational changes in the conserved TXP motif in transmembrane helix 2. J Biol Chem 2003;278(38): 36513–36521.
54. Govaerts C, Blanpain C, Deupi X, et al. The TXP motif in the second transmembrane helix of CCR5. A structural determinant of chemokine-induced activation. J Biol Chem 2001;276(16):13217–13225.
55. Oppermann M. Chemokine receptor CCR5: insights into structure, function, and regulation. Cell Signal 2004;16(11):1201–1210.
56. Galliera E, Jala VR, Trent JO, et al. beta-Arrestin-dependent constitutive internalization of the human chemokine decoy receptor D6. J Biol Chem 2004; 279(24):25590–25597.
57. Richardson RM, Marjoram RJ, Barak LS, Snyderman R. Role of the cytoplasmic tails of CXCR1 and CXCR2 in mediating leukocyte migration, activation, and regulation. J Immunol 2003;170(6):2904–2911.
58. Sun Y, Cheng Z, Ma L, Pei G. Beta-arrestin2 is critically involved in CXCR4-mediated chemotaxis, and this is mediated by its enhancement of p38 MAPK activation. J Biol Chem 2002;277(51):49212–49219.
59. Barlic J, Andrews JD, Kelvin AA, et al. Regulation of tyrosine kinase activation and granule release through beta-arrestin by CXCRI. Nat Immunol 2000;1(3): 227–233.
60. Orsini MJ, Parent JL, Mundell SJ, Marchese A, Benovic JL. Trafficking of the HIV coreceptor CXCR4: role of arrestins and identification of residues in the C-terminal tail that mediate receptor internalization. J Biol Chem 1999;274(43): 31076–31086.
61. Barlic J, Khandaker MH, Mahon E, et al. Beta-arrestins regulate interleukin-8-induced CXCR1 internalization. J Biol Chem 1999;274(23):16287–16294.
62. Lefkowitz RJ, Shenoy SK. Transduction of receptor signals by beta-arrestins. Science 2005;308(5721):512–517.
63. Perry SJ, Lefkowitz RJ. Arresting developments in heptahelical receptor signaling and regulation. Trends Cell Biol 2002;12(3):130–138.
64. Li YM, Pan Y, Wei Y, et al. Upregulation of CXCR4 is essential for HER2-mediated tumor metastasis. Cancer Cell 2004;6(5):459–469.

65. Marchese A, Benovic JL. Agonist-promoted ubiquitination of the G protein-coupled receptor CXCR4 mediates lysosomal sorting. J Biol Chem 2001;276(49): 45509–45512.
66. Marchese A, Raiborg C, Santini F, Keen JH, Stenmark H, Benovic JL. The E3 ubiquitin ligase AIP4 mediates ubiquitination and sorting of the G protein-coupled receptor CXCR4. Dev Cell 2003;5(5):709–722.
67. Marchese A, Chen C, Kim YM, Benovic JL. The ins and outs of G protein-coupled receptor trafficking. Trends Biochem Sci 2003;28(7):369–376.
68. Springael JY, Urizar E, Parmentier M. Dimerization of chemokine receptors and its functional consequences. Cytokine Growth Factor Rev 2005;16(6):611–623.
69. Wilson S, Wilkinson G, Milligan G. The CXCR1 and CXCR2 receptors form constitutive homo- and heterodimers selectively and with equal apparent affinities. J Biol Chem 2005;280(31):28663–28674.
70. Kuner R, Kohr G, Grunewald S, Eisenhardt G, Bach A, Kornau HC. Role of heteromer formation in GABAB receptor function. Science 1999;283(5398): 74–77.
71. Kaupmann K, Malitschek B, Schuler V, et al. GABA(B)-receptor subtypes assemble into functional heteromeric complexes. Nature 1998;396(6712):683–687.
72. White JH, Wise A, Main MJ, et al. Heterodimerization is required for the formation of a functional GABA(B) receptor. Nature 1998;396(6712):679–682.
73. Hernanz-Falcon P, Rodriguez-Frade JM, Serrano A, et al. Identification of amino acid residues crucial for chemokine receptor dimerization. Nat Immunol 2004; 5(2):216–223.
74. Lemay J, Marullo S, Jockers R, Alizon M, Brelot A. On the dimerization of CCR5. Nat Immunol 2005;6(6):535; author reply 536.
75. Mellado M, Rodriguez-Frade JM, Vila-Coro AJ, et al. Chemokine receptor homo- or heterodimerization activates distinct signaling pathways. EMBO J 2001; 20(10):2497–2507.
76. Mellado M, Rodriguez-Frade JM, Manes S, Martinez AC. Chemokine signaling and functional responses: the role of receptor dimerization and TK pathway activation. Annu Rev Immunol 2001;19:397–421.
77. El-Asmar L, Springael JY, Ballet S, Andrieu EU, Vassart G, Parmentier M. Evidence for negative binding cooperativity within CCR5-CCR2b heterodimers. Mol Pharmacol 2005;67(2):460–469.
78. Vila-Coro AJ, Mellado M, Martin de Ana A, et al. HIV-1 infection through the CCR5 receptor is blocked by receptor dimerization. Proc Natl Acad Sci U S A 2000;97(7):3388–3393.
79. Rodriguez-Frade JM, del Real G, Serrano A, et al. Blocking HIV-1 infection via CCR5 and CXCR4 receptors by acting in trans on the CCR2 chemokine receptor. EMBO J 2004;23(1):66–76.
80. Brink CB, Harvey BH, Bodenstein J, Venter DP, Oliver DW. Recent advances in drug action and therapeutics: relevance of novel concepts in G-protein-coupled receptor and signal transduction pharmacology. Br J Clin Pharmacol 2004; 57(4):373–387.

81. Johnson Z, Schwarz M, Power CA, Wells TN, Proudfoot AE. Multi-faceted strategies to combat disease by interference with the chemokine system. Trends Immunol 2005;26(5):268–274.
82. Doms RW. Chemokine receptors and HIV entry. AIDS 2001;15(Suppl 1): S34-S35.
83. Balkwill F. Chemokine biology in cancer. Semin Immunol 2003;15(1):49–55.
84. Albert AD, Yeagle PL. Structural studies on rhodopsin. Biochim Biophys Acta 2002;1565(2):183–195.
85. Babcock GJ, Mirzabekov T, Wojtowicz W, Sodroski J. Ligand binding characteristics of CXCR4 incorporated into paramagnetic proteoliposomes. J Biol Chem 2001;276(42):38433–38440.
86. Mirzabekov T, Bannert N, Farzan M, et al. Enhanced expression, native purification, and characterization of CCR5, a principal HIV-1 coreceptor. J Biol Chem 1999;274(40):28745–28750.
87. Staudinger R, Bandres JC. Solubilization of the chemokine receptor CXCR4. Biochem Biophys Res Commun 2000;274(1):153–156.
88. Blackburn PE, Simpson CV, Nibbs RJ, et al. Purification and biochemical characterization of the D6 chemokine receptor. Biochem J 2004;379(Pt 2):263–272.

4

Chemokine Signaling in T-Lymphocyte Migration: The Role of Phosphoinositide 3-kinase

Laura Smith, Adam Webb, and Stephen G. Ward

Summary

The biochemical events that are elicited upon chemokine engagement have been a major focus of interest in many cell types responding to a plethora of different chemokines. We now appreciate that collectively, chemokines can couple to a wide range of biochemical signals including phosphoinositide lipid metabolism, elevation of intracellular calcium levels, and activation of a wide array of protein and lipid kinases as well as small GTPases. Chemokine signaling events are particularly well studied in T lymphocytes where the ordered directional migration of T lymphocytes is a key process in development, immune surveillance, and immune responses. These cells therefore offer a splendid model system in which to understand the array of signals activated by chemokines and their functional importance. One of the most robust biochemical signals elicited by chemokines is the activation of several members of the phosphoinositide 3-kinase (PI3K) family. In many cell systems, PI3Ks are known to contribute to several aspects of the migratory machinery including gradient sensing, signal amplification, actin reorganization and hence cell motility. This chapter will therefore focus on the role of PI3K-dependent signaling events in T-lymphocyte migration and points at which these events may integrate with other effectors and signaling cascades.

Key Words: Chemokines; GTPases; phosphoinositide 3-kinase; protein kinase C; T lymphocytes; signaling; tyrosine kinases.

From: *The Receptors: The Chemokine Receptors*
Edited by: J. K. Harrison and N. W. Lukacs © Humana Press Inc., Totowa, NJ

4.1. Introduction

A broad array of biochemical events including cell polarization and movement, immune and inflammatory responses, and prevention of HIV-1 infection are triggered by chemokines, a family of chemoattractant proteins that bind to specific seven-transmembrane-spanning, pertussis toxin–sensitive, G protein–coupled receptors (GPCRs). These chemokine GPCRs have a number of conserved motifs including the Asp-Arg-Tyr-Leu-Ala-Ile-Val (DRYLAIV) motif in the second intracellular loop domain, which appears to be a crucial determinant of GPCR signaling and function *(1)*. The biochemical events that are elicited upon chemokine engagement have been a major focus of interest in many cell types responding to a range of different chemokines. We now appreciate that collectively, chemokines can couple to a wide range of biochemical signals including phosphoinositide lipid metabolism, elevation of intracellular calcium levels, and activation of protein and lipid kinases as well as small GTPases. It is also worth mentioning that there are at least two nonsignaling or "silent" chemokine receptors. First, the Duffy antigen receptor for chemokines (DARC) can bind both CXC and CC chemokines but has less than 20% amino acid identity with other CXC and CC chemokine receptors, lacks the DRYLAIV motif, is not coupled to G proteins, and has not been reported to elicit any detectable signal transduction events. Second, D6, which is known to bind a number of proinflammatory CC chemokines, does not mediate typical chemokine-induced intracellular signals after ligand binding, possibly as a consequence of the classic DRYLAIV motif being altered to Asp-Lys-Tyr-Leu-Glu-Ile-Val (DKYLEIV) in this receptor *(2)*.

The exact biochemical signal profile elicited by chemokines is likely to vary according to individual chemokine-receptor interactions as well as cell type and its activation status. In order to fully appreciate the range of signals activated by chemokines, this chapter will focus on the biochemical events that have been identified in T lymphocytes and assess their relative importance and contribution to the ensuing migratory response.

4.2. The Role of Phosphoinositide 3-kinase(s) in T-Lymphocyte Migration

The coordinated and directional trafficking of T lymphocytes in lymphoid and peripheral tissues is a key process in immunosurveillance and immune responses. This involves a multistep adhesion cascade of selectin- or integrin- (and their corresponding vascular ligands) supported rolling, chemokine-mediated integrin activation, and firm integrin-mediated adhesion to the microvasculature endothelium followed by transmigration through the vessel wall and further migration in extravascular tissue *(3,4)*. For a cell to migrate to a chemoattractant source, it must be polarized, which means that the molecular processes at the front (leading edge) and the back (uropod) of a moving cell are different. Establishing and maintaining cell polarity in response to extracellular stimuli appear to be mediated by a set of interlinked positive-feedback loops involving phosphoinositide 3-kinase (PI3K), Rho family GTPases, integrins, microtubules, and vesicular transport. The relative contributions of these various signals depend on the cell type and the specific stimulus (reviewed in Ref. 5). These intracellular signals result in reorganization of the cytoskeleton and changes in cell adhesion causing the cells to send out pseudopodia and crawl up the chemoattractant gradient. PI3K-dependent signaling events have previously been demonstrated in several cell systems to contribute to several aspects of the migratory machinery including gradient sensing, signal amplification, actin reorganization, and hence cell motility *(5–7)*. We will now focus on the role of PI3K in T-lymphocyte migration and the points at which it may integrate with and/or influence other effectors and signaling cascades.

4.2.1. The Phosphoinositide 3-kinase Family

The major 3′-phosphoinositide products of class I PI3Ks are phosphatidylinositol 3,4,5-trisphosphate [PI(3,4,5)P_3, which is formed primarily from phosphorylation of PI(4,5)P_2) and its metabolite phosphatidylinositol 3,4-bisphosphate, PI(3,4)P_2]. The basal levels of PI(3,4)P_2 and PI(3,4,5)P_3 in cells are usually in low abundance but can rise sharply after cell stimulation to interact with an array of protein effectors via pleckstrin homology (PH) domains, modular segments of about 100 amino acids found in many signaling proteins. It is these PH-domain-containing proteins that are able to propagate and drive downstream signaling events.

The prototypical class I group of PI3Ks consists of two subgroups (Fig. 1). The class IA PI3K consists of an 85-kDa regulatory subunit (responsible for protein-protein interactions via SH2 domain interaction with phosphotyrosine residues of other proteins) and a catalytic 110-kd subunit *(8)*. There are three catalytic isoforms (p110α, p110β, and p110δ) and five regulatory isoforms (p85α, p85β, and p55γ encoded by specific genes and p55α and p50α that are produced by alternate splicing of the p85α gene). A distinct lipid kinase termed class IB PI3Kγ (or p110γ) is activated by G protein βγ subunits from GPCRs and associates with a unique p101 adaptor molecule. Nevertheless, GPCRs such as chemokine receptors are also able to activate the p85/p110 heterodimeric PI3Ks. Expression of p110δ and PI3Kγ is largely restricted to leukocytes, and there is increasing evidence that these isoforms play key roles in immunity *(8)*. The class II PI3Ks are structurally distinct and are thought to use only PI and PI(4)P as substrates (Fig. 1). Mammalian class II PI3Ks predominately include

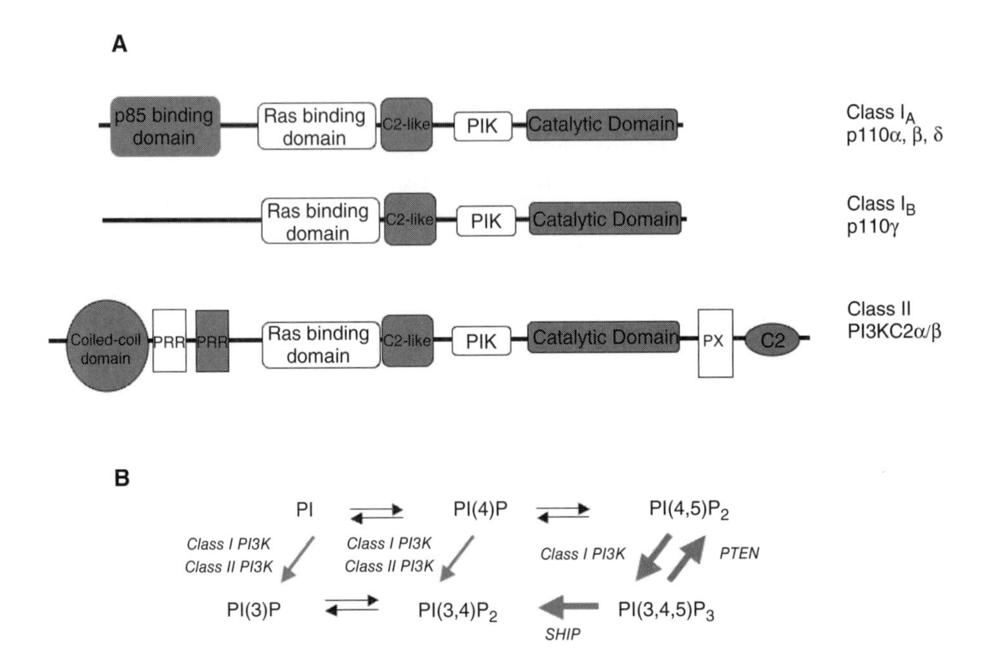

Fig. 1. Structure of class I and class II PI3Ks and their substrate specificity. PRR, proline-rich regions; PX, phox homology domain.

the ubiquitously expressed PI3KC2α and PI3KC2β. Whereas class I PI3Ks reside mainly in the cytoplasm until recruited to active signaling complexes, the class II PI3Ks are mainly constitutively associated with membrane structures (including plasma and intracellular membranes) and with nuclei *(9)*.

The termination of PI3K signaling by degradation of PI(3,4,5)P_3 can be mediated by at least two different types of phosphatases, namely SH2-containing inositol 5-phosphatase (SHIP) and the 3′-phosphatase termed *phosphatase and tensin homology deleted on chromosome ten protein* (PTEN). Dephosphorylation of PI(3,4,5)P_3 by SHIP impairs some downstream effects of PI3K, although the metabolic product of SHIP phosphatase activity is PI(3,4)P_2, which can also mediate PI3K-dependent responses *(8)*.

4.2.2. Class I Phosphoinositide 3-kinases and T-Lymphocyte Migration

Most of what we understand about biochemical events in T cells during migration toward a chemoattractant gradient has been derived from study of chemokine signal transduction. PI3K activation seems to be a signaling event shared by most chemokine receptors expressed on T cells (Table 1). Signaling via CXCR4 has been studied in most detail, and stromal cell-derived factor-1

(SDF-1)/CXCL12 triggers a transient pertussis toxin (PTX)-sensitive accumulation of $PI(3,4,5)P_3$ in the Jurkat leukemic cell line *(10)*. Similar lipid accumulation in murine and human T lymphocytes stimulated with EBL-1-ligand chemokine (ELC)/CCL19 and secondary lymphoid chemokine (SLC)/CCL21 has also been reported *(11)*. The use of PI3K inhibitors has also revealed that in vitro chemotaxis across synthetic membranes of peripheral blood-derived lymphocytes (PBLs) and natural killer (NK) cells in response to CXCL12, and several other chemokines such as regulated upon activation normal T-cell expressed and secreted (RANTES)/CCL5, can be severely attenuated by PI3K inhibitors *(10)*. During migration and chemotaxis, there are at least two chemokine-mediated processes operating, namely chemokine-mediated integrin activation to mediate cell adhesion and chemokine-induced chemotaxis (that is likely to be integrin independent). There is some evidence that enhanced integrin mobility in chemokine-stimulated lymphocytes is dependent on PI3K activation *(11)*. Generally, however, rapid integrin activation on rolling lymphocytes does not depend on PI3K activation processes *(12)*.

Chemokine interaction with GPCRs on lymphocytes has been shown to depend predominately on G_i proteins *(13)*. This has led to the assumption that these receptors are coupled to the βγ-dependent p110γ isoform. Indeed, this does appear to be the case, although several chemokine receptors can additionally activate other PI3K isoforms *(10,14)*. Significant progress has been made in resolving the confusion concerning the specific role of PI3K in lymphocyte migration. This has been made possible by availability of mice deficient in the 110γ catalytic isoform as well as analysis of mice expressing a mutant, catalytically inactive class 1A p110δ isoform. The in vitro migration of p110γ-deficient $CD4^+$ and $CD8^+$ T cells to CCL19, CXCL12 and CCL21 is significantly decreased compared with cells from wild-type mice. In contrast, T cell responses were largely unaffected by p110δ deficiency *(15)*. Hence, in settings where T-cell migration required PI3K activation, the p110γ isoform appears to be the predominant isoform required. This correlates with observations that p110γ selective inhibitors reduce numbers of $CD4^+$ memory T cells in models of systemic lupus *(16)*. Interestingly, B-cell migration to chemokines was not significantly affected by p110γ deficiency, thus implicating involvement of other PI3K isoforms or signaling pathways in B-cell migration *(15,17)*. In this regard, analysis of p110δ-deficient B cells showed a defect in B-cell chemotaxis to CXCL13, whereas responses to CCR7 and CXCR4 ligands were less affected. Adoptive transfer experiments with B cells expressing inactive p110δ revealed diminished CXCR5-mediated homing to Peyer patches and splenic white pulp cords. The ability of p110δ to function downstream of chemokine receptors in lymphocyte chemotactic responses is consistent with the finding that a broad spectrum loss-of-function mutant that disrupts all class 1A

Table 1
Activation of PI3K by Lymphoid Chemokine Receptors

Receptor	Expression	Ligand	PI3K activation	Effect of inhibitors on T-cell migration
CXCR3	Effector T cells	I-TAC/CXCL11	Yes	No effect
		MIG/CXCL9	Yes	
		IP-10/CXCL10	Yes	
CXCR4	Naïve, memory T cells, B cells, thymocytes	SDF-1/CXCL12	Yes	Inhibition
CXCR5	Follicular B helper T cells, B cells	BLC/CXCL13	Yes	ND
CXCR6	Effector T cells	CXCL16	Yes	ND
CCR1	Effector T cells	RANTES/CCL5	Yes	Inhibition
		MIP-1α/CCL3		
		MCP-2/CCL8		
		MCP-3/CCL7		
CCR2	Effector T cells	MCP-1/CCL2	Yes	ND
		MCP-2/CCL8		
		MCP-3/CCL7		
		MCP-4/CCL13		
CCR3	Effector T cells (Th2)	Eotaxin-1/CCL11	ND	ND
		Eotaxin-2/CCL24		
		Eotaxin-3/CCL26		
		RANTES/CCL5		
		MCP-2/CCL8		
		MCP-3/CCL7		
		MCP-4/CCL13		
CCR4	Effector T cells (Th2)	MDC/CCL22	Yes	No effect
		TARC/CCL17	Yes	No effect
CCR5	Effector T cells (Th1)	RANTES/CCL5	Yes	Inhibition
		MIP-1α/CCL3		
		MIP-1β/CCL4	Yes	ND
		MCP-2/CCL8		
		MCP-3/CCL7		
CCR6	Effector T cells, B cells, memory T cells	LARC/CCL20	No	ND

Table 1 *(Continued)*

Receptor	Expression	Ligand	PI3K activation	Effect of inhibitors on T-cell migration
CCR7	Naïve, central memory B cells, mature medullary thymocytes	SLC/CCL21 ELC/CCL19	Yes Yes	No effect No effect
CCR8	Effector T cells (Th2)	I-309/CCL1	Yes	Partial inhibition
CCR9	Memory T cells, B cells, immature thymocytes	TECK/CCL25	Yes	ND

Note: PI3K assays will vary according to the type of T-cell model employed and concentration of chemokine assayed. For clarity, activation of PI3K by chemokines that show promiscuity toward receptors is attributed to signaling via all the receptors bound by these ligands, even though such details have not necessarily been investigated.
ND, not determined.

catalytic isoforms reduced chemotactic responses of leukemic T cells to CXCL12 *(18)*. Together, these data indicate that individual lymphoid chemokine receptors have differing dependence on PI3K-dependent signals for achieving ordered migration.

4.2.3. The Role of Class II Phosphoinositide 3-kinases in Cell Migration

Class II PI3KC2β and its primary product PI-3-P have been implicated in the regulation of cell adhesion, actin reorganization, and migration in response to lysophosphatidic acid–dependent cell migration and wound healing in non-immune cell systems *(19,20)*. Class II PI3K isoforms are activated in leukemic T-cell lines by several chemokines as well as in monocytic cell lines by CCL2/CCR2 *(10)*, although it is unclear how GPCRs couple to class II PI3Ks. In this regard, a unique CCR2-interacting cytoplasmic protein termed *FROUNT* has been identified, which plays a key role in coupling the chemokine receptor CCR2 (but not other chemokine receptors) to the PI3K-dependent signaling cascade and subsequent monocyte chemotactic responses *(21)*. FROUNT shows similarity to clathrin, which is known to co-associate with class II PI3K *(9)*, so it may be involved in coupling CCR2 to class II PI3Ks. FROUNT-like GPCR-interacting proteins may help provide tailored signaling responses for individual chemoattractant receptors.

4.3. What Is the Significance of Rho GTPases in T-Lymphocyte Migration?

Although PI3K activation seems to be a conserved biochemical response common to most chemokines, it is now clear that this can be a dispensable signal for directional migration of T cells *(14,22–24)*. In many cell types (e.g., endothelial cells, monocytes), the small GTPases Rho, Rac, and cdc42 play key roles in regulating the morphology of migrating cells via effects on the actin cytoskeleton *(5)*. The effects of Rac and cdc42 are in turn mediated by a family of regulatory proteins including Wiskott-Aldrich syndrome protein (WASP) and WASP-family verprolin-homologous (WAVE) proteins, which interact with a complex of seven other proteins termed the *actin-related protein* (Arp)2/3 complex, which catalyzes actin polymerization at the front of growing membrane protrusions such as lamellipodia *(5)*. Overexpression of gain-of/loss-of function mutants in T cells has shown that Rho, Rac, cdc42, and the cdc42 target WASP are all variously required for chemotaxis of T cells toward CXCL12 *(25,26)*.

In fibroblast models, the Rho family GTPases RhoA, Rac, and cdc42 have well understood roles in stress fiber and focal adhesion formation, lamellipodal extension, and filopodal extensions, respectively *(5,27)*. T cells do not form easily identifiable adhesion structures, but both cdc42 and Rap1 have been implicated in polarization, integrin activation, and motility *(7)*. Similarly, Rho and its effector Rho-associated kinase (ROCK) signaling are key effectors of T-cell migration and adhesion in response to several chemokines in mature T cells and thymocytes *(22,28,29)*. RhoA activation also appears to be necessary for integrin activation induced by Rap1 and Rac in thymocytes *(29)*. RhoA appears to control both lymphocyte function-associated protein-1 (LFA-1) high-affinity-state triggering by chemokines as well as the lateral mobility induced by chemokines. Two Rho effector regions control these changes in LFA-1 activation in response to CCR7 and CXCR4 ligation *(30)*. The 23–40 effector region controls changes in high affinity, while induction of LFA-1 lateral mobility involves protein kinase C (PKCζ) and further signals generated by the 92–119 effector region of RhoA. The ability of PKCζ to control LFA-1 lateral mobility depends on kinase activity and translocation to the plasma membrane, which in turn depends on PI3K and on the 23–40 effector region of RhoA *(30)*.

Rap1 has been implicated in integrin activation, cell polarization, and motility across vascular endothelium under flow *(31)*, and Rap1 activation occurs in response to CXCL12 stimulation of B cells *(32)*. The biochemical pathways leading to Rap activation by chemokine receptors remain obscure. A distinct family of Ras exchange factors is regulated not only by Ca^{2+} but also by membrane diacylglycerol that is generated along with Ins(1,4,5)P_3 during activation of phospholipase C. These are termed *calcium and diacylglycerol-regulated guanine-nucleotide exchange factors* (CalDAG-GEFs) *(33)*, one of which functions as an exchange factor for Rap. One possibility is that Rap activation by chemokines may involve the phospholipase C (PLC) pathway, because the PI3K-independent CCR4- and CXCR3-mediated chemotaxis of PBLs has been reported to be regulated by a PLC-dependent pathway *(23,34)*.

DOCK-2 Mediates a T-Lymphocyte–Specific Migration Mechanism

There is now a growing appreciation of PI3K-*independent* routes to Rac activation in T cells that involve downstream of Crk-180 homolog-2 (DOCK-2), which is highly expressed in leukocytes. Characterization of mice lacking the gene encoding the Rac-specific guanine nucleotide exchange factor DOCK-2 revealed a striking deficit in lymphocyte migration in response to the lymphoid chemokines SDF-1/CXCL12, B lymphocyte chemoattractant (BLC)/CXCL13, ELC/CCL19, and SLC/CCL21, whereas PI3K activation appeared unaffected *(35,36)*. In contrast, monocyte migration toward chemokines was unaffected. This was the first evidence for the possible existence of cell-specific migration mechanisms.

Closer analysis of the DOCK-$2^{-/-}$–deficient mice revealed that in fact, optimal T-cell migration in vitro and in vivo in response to CCL21, CCL19, or CXCL12 is dependent on expression of both DOCK-2 and p110γ *(17)*. In other words, DOCK-2–deficient T lymphocytes can mount a modest but nonetheless significant 110γ-dependent migratory response. Another unexpected observation was a defect in integrin activation in response to CCL21 that was observed in DOCK-$2^{-/-}$ B cells but not in T cells. DOCK-2 and p110γ thus play distinct roles during T- and B-cell integrin activation and migration *(17)*. It is unclear how DOCK-2 becomes recruited after engagement of surface receptors and whether there is polarized relocalization during migration. However, the scaffolding proteins Elmo and CrkL interact with DOCK-2 in human leukemic cell lines and are required for DOCK-2–mediated Rac activation and cytoskeletal organization *(36,37)*. Elmo contains a putative phosphoinositide lipid binding PH domain *(36,38)*, and a discrete structural domain of the DOCK-2 homologue DOCK-180 has been shown to bind PI(3,4,5)P_3 *(39)*. This domain has been shown to be required for PI(3,4,5)P_3 to induce Elmo-DOCK-180 complexes to translocate from the cytoplasm to the plasma membrane in fibroblast migration models, where they are found in membrane ruffles that are rich in PI(3,4,5)P_3 and polymerized actin *(39)*.

4.4. Activation of Protein Tyrosine Kinases by Chemokines: Relevance to T-Lymphocyte Migration

Several chemokines have been shown to stimulate tyrosine phosphorylation with evidence for activation of both Src family kinases *(40)* as well as Jak2/3 *(41)*. However, recent lines of evidence provide compelling evidence that contrary to previous reports, the Jak family kinases are unlikely to play an essential role in chemokine signaling *(42)*. Studies with the zeta-chain associated protein kinase 70kDa (ZAP-70)–deficient Jurkat leukemic cell line have identified a role for ZAP-70 in CXCR4 signaling and migration where ZAP-70 deficiency results in decreased migration to CXCL12 *(43)*. ZAP-70 deficiency also results in lack of phosphorylation of the adaptor SH2 domain containing leukocyte protein of 76kDa (SLP-76) in response to CXCR4 signaling. Through its interaction with adaptor proteins such as Vav, Nck, and adhesion and deregulation promoting adaptor proteins (ADAP), SLP-76 has been implicated in the regulation of cytoskeletal changes as well as increased integrin adhesion in activated T cells *(44)*. In addition, SLP-76 is a prominent substrate and interaction partner for Itk, a member of the Tec family of tyrosine kinases, most members of which possess PH domains that bind lipid products of PI3K *(45)*. Tec kinases lie downstream of Lck and ZAP-70 and PI3K in the T-cell antigen (TCR) receptor signaling pathway and have been implicated in the regulation of TCR -induced actin polarization and activation of the Rho family GTPase cdc42 *(45)*. In T lymphocytes, CXCL12 stimulates phosphorylation and activation of the Tec kinases Itk and Rlk in a manner dependent on Src tyrosine kinases and PI3Ks *(46,47)*. Expression of a loss-of-function Itk mutant impaired CXCL12-induced migration, cell polarization, and activation of Rac and cdc42 *(46)*. T cells purified from $Rlk^{-/-}Itk^{-/-}$ mice exhibited impaired migration to multiple chemokines in vitro *and* decreased homing to lymph nodes upon transfer to wild-type mice *(46,47)*. These findings raise the possibility that Tec kinases may be one of the key effectors activated by PI3Ks that contribute to cell polarization and migration downstream of chemokine receptors.

4.5. Evidence of a Role for Protein Kinase C Activation in T-Lymphocyte Migration

Activation of phospholipase C (PLC), calcium mobilization, and activation of diacylglycerol (DAG)-dependent PKCs by chemokines have been proposed as regulators of cell adhesion and migration *(1,48)*. However, studies with mice deficient in PLCβ1 and β-3 indicated that the PLC pathway is not required for chemotaxis in neutrophils, although its role in T lymphocytes was not investigated *(49)*. Certainly, during LFA-1–mediated locomotion of activated T cells, PKCβ1 and PKCδ associate with microtubules in the uropod, the trailing extension of the migrating T cells. Curiously, whereas PKCδ is associated with the microtubule organizing center (MTOC) but not the microtubules, PKCβ1 is located at the MTOC and along the microtubules in the trailing cell extensions *(50)*. This may be indicative of discrete functions for individual PKC isoforms in T-cell migration. Activation of both PKCβ and PKCδ is known to involve the master kinase 3′-phosphoinositide–dependent kinase-1 (PDK-1), which phosphorylates several PKC isoforms in their activation loop, which is necessary for their catalytic activation *(51)*.

PKCδ has been implicated in cell motility and migratory responses in many nonimmune cells *(52)*. In T-cell models where migration to CCR4 stimulation can occur independently of PI3K *(22)*, use of pharmacologic tools has indicated that PKCδ is required for chemotactic responses to CCR4 ligands *(34)*. Although this does not fit well with the notion that PKCδ is a substrate for PDK-1, it is worth noting that PKCδ is often a functional enzyme in the absence of phosphorylation in the activation loop *(51)*. It is also interesting to note that PKC contains multiple sites for tyrosine phosphorylation, some of which have been demonstrated to influence its activation *(53)*. Moreover, the conserved domain 2 of PKCδ is a phosphotyrosine-binding domain, which allows for greater adaptability and diversity for integrating with other signaling events *(54)*. It is not well understood how PKC isoforms regulate cell motility/migration, although it is likely that PKC isoforms exert effects on actin reorganization/ polymerization, as well as changes on integrin affinity.

4.6. Importance of Tailoring the Migratory Response

It is important to remember that migrating leukocytes must navigate through complex chemoattractant fields and must migrate from one chemoattractant source to another. Migrating neutrophils have been reported to display "memory" of their recent environment, such that cells' perception of the relative strength of orienting signals is influenced by their history *(55)*. This allows combinations of chemoattractants to guide leukocytes in a step-by-step fashion to their destinations within tissues, and tailored migratory responses involving multiple signaling events would favor this process. Moreover, recent evidence has suggested that nonagonist chemokines are capable of associating with known chemokine agonists resulting in a stronger cellular response, although the molecular basis for this phenomenon has not been determined *(56)*. As a consequence, inflamed and other chemokine-rich tissues would create an environment that renders many leukocyte types more competent to respond to migratory cues.

We are beginning to understand how a relatively small number of chemokine receptors signal under in vitro settings (summarized in Fig. 2), and the challenge now is to understand how these pathways are integrated, tailored, and fine-tuned for each receptor and how they are regulated in inflammatory situations, where there is a vast abundance of chemotactic factors, as well as receptors that can couple to additional G proteins other than $G\alpha_i$ *(57)*. The spatiotemporal regulation of the key components of the PI3K and the Rho GTPases pathway in T cells migrating across endothelial barriers under physiologically relevant

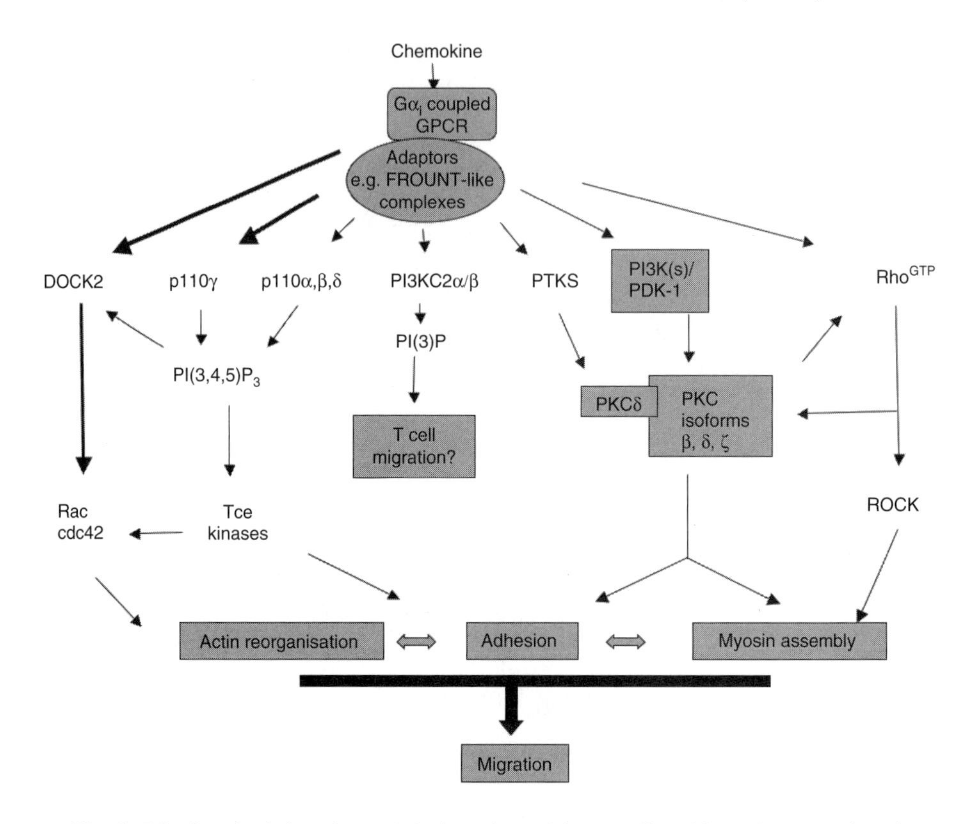

Fig. 2. Biochemical signals used during chemokine-mediated lymphocyte migration. Engagement of a typical $G\alpha_i$-coupled chemokine receptor leads primarily to activation of the Rac-specific guanine nucleotide exchange factor (GEF) DOCK-2. The mechanism by which GPCRs couple to DOCK-2 is unclear, but Rho GTPases are implicated in other systems. In addition, receptor engagement also triggers activation of PI3Ks. The γ isoform of p110 is the major PI3K isoform activated, but other class 1A p110 isoforms can contribute to PI(3,4,5)P_3 accumulation. The relative importance of DOCK-2 and PI3K are indicated by the thickness of the arrows. The activation of p110α, β, γ may involve $G\alpha_i$-activated Src family protein kinases (PTKs) or via βγ-subunit/ PI(3,4,5)P_3-dependent GEF-mediated activation of Rac, which is known to interact with the class 1A p85 regulatory subunit (not depicted). See text for further details.

settings is unexplored. The relative involvement of individual pathways in chemokine-stimulated migration of T cells is likely to be context-dependent, being determined by individual receptor ligands and cell status (e.g., naïve or activated). The putative existence of tailored signaling responses to chemokines is underlined by the observations that CXCL12 stimulates prolonged PKB activation in T cells, whereas other chemokines (e.g., RANTES/CCL5, ELC/CCL19, interferon-inducible protein-10 (IP-10)/CXCL10, and MCP-1/CCL2) stimulate only transient phosphorylation *(58)*. The existence of GPCR-interacting proteins such as FROUNT that can discriminate between chemokine receptors likely contributes to the mechanisms by which biochemical signals elicited by individual chemokine receptors are attributed with a distinct identifying "fingerprint" or "signature."

4.7. Conclusions

PI3K is currently a major drug target for the pharmaceutical industry in inflammation and autoimmunity, where one aim is to block migration of leukocytes to the site of inflammatory lesion. The differing dependence of individual chemokine receptors on PI3K isoforms in different cell types and at different stages of activation makes it difficult to design a "one fits all" drug to inhibit inflammatory recruitment of cells. A complete understanding of the mechanisms guiding T-lymphocyte migration in the future will be essential in order to open new avenues for potential therapeutics in inflammatory and autoimmune disease settings.

References

1. Ward SG, Bacon K, Westwick J. Chemokines and T lymphocytes: more than an attraction. Immunity 1998;9(1):1–11.
2. Nibbs RJ, Kriehuber E, Ponath PD, et al. The beta-chemokine receptor D6 is expressed by lymphatic endothelium and a subset of vascular tumors. Am J Pathol 2001;158(3):867–77.
3. von Andrian UH, Mempel TR. Homing and cellular traffic in lymph nodes. Nat Rev Immunol 2003;3(11):867–878.
4. Miyasaka M, Tanaka T. Lymphocyte trafficking across high endothelial venules: dogmas and enigmas. Nat Rev Immunol 2004;4(5):360–370.
5. Ridley AJ, Schwartz MA, Burridge K, et al. Cell migration: integrating signals from front to back. Science 2003;302(5651):1704–1709.
6. Merlot S, Firtel RA. Leading the way: directional sensing through phosphatidylinositol 3-kinase and other signaling pathways. J Cell Sci 2003;116(Pt 17):3471–3478.
7. Hogg N, Laschinger M, Giles K, McDowall A. T-cell integrins: more than just sticking points. J Cell Sci 2003;116(Pt 23):4695–4705.
8. Vanhaesebroeck B, Leevers SJ, Ahmadi K, et al. Synthesis and function of 3-phosphorylated inositol lipids. Annu Rev Biochem 2001;70:535–602.
9. Foster FM, Traer CJ, Abraham SM, Fry MJ. The phosphoinositide (PI) 3-kinase family. J Cell Sci 2003;116(Pt 15):3037–3040.
10. Sotsios Y, Ward SG. Phosphoinositide 3-kinase: a key biochemical signal for cell migration in response to chemokines. Immunol Rev 2000;177:217–235.
11. Constantin G, Majeed M, Giagulli C, et al. Chemokines trigger immediate beta2 integrin affinity and mobility change: differential regulation and roles in lymphocyte arrest under flow. Immunity 2000;13:759–769.

12. Grabovsky V, Feigelson S, Chen C, et al. Subsecond induction of alpha4 integrin clustering by immobilized chemokines stimulates leukocyte tethering and rolling on endothelial vascular cell adhesion molecule 1 under flow conditions. J Exp Med 2000;192(4):495–506.
13. Han SB, Moratz C, Huang NN, et al. RGS1 and Galpha-i2 regulate the entrance of B lymphocytes into lymph nodes and B cell motility within lymph node follicles. Immunity 2005;22(3):343–354.
14. Ward SG. Do phosphoinositide 3-kinases direct lymphocyte navigation? Trends Immunol 2004;25(2):67–74.
15. Reif K, Okkenhaug K, Sasaki T, Penninger JM, Vanhaesebroeck B, Cyster JG. Cutting edge: differential roles for phosphoinositide 3-kinases, p110gamma and p110delta, in lymphocyte chemotaxis and homing. J Immunol 2004;173(4): 2236–2240.
16. Barber DF, Bartolome A, Hernandez C, et al. PI3Kgamma inhibition blocks glomerulonephritis and extends lifespan in a mouse model of systemic lupus. Nat Med 2005;11(9):933–935.
17. Nombela-Arrieta C, Lacalle RA, Montoya MC, et al. Differential requirements for DOCK2 and phosphoinositide-3-kinase gamma during T and B lymphocyte homing. Immunity 2004;21(3):429–441.
18. Curnock AP, Sotsios Y, Wright KL, Ward SG. Optimal chemotactic responses of leukemic T cells to stromal cell-derived factor-1 requires the activation of both class IA and IB phosphoinositide 3-kinases. J Immunol 2003;170(8):4021–4030.
19. Maffucci T, Cooke FT, Foster FM, Traer CJ, Fry MJ, Falasca M. Class II phosphoinositide 3-kinase defines a novel signaling pathway in cell migration. J Cell Biol 2005;169(5):789–799.
20. Domin J, Harper L, Aubyn D, et al. The class II phosphoinositide 3-kinase PI3K-C2beta regulates cell migration by a PtdIns(3)P dependent mechanism. J Cell Physiol 2005;205(3):452–462.
21. Terashima Y, Onai N, Murai M, et al. Pivotal function for cytoplasmic protein FROUNT in CCR2-mediated monocyte chemotaxis. Nat Immunol 2005;6(8): 827–835.
22. Cronshaw DG, Owen C, Brown Z, Ward SG. Activation of phosphoinositide 3-kinases by the CCR4 ligand macrophage-derived chemokine is a dispensable signal for T lymphocyte chemotaxis. J Immunol 2004;172(12):7761–7770.
23. Smit MJ, Verdijk P, van der Raaij-Helmer EM, et al. CXCR3-mediated chemotaxis of human T cells is regulated by a Gi- and phospholipase C-dependent pathway and not via activation of MEK/p44/p42 MAPK nor Akt/PI-3 kinase. Blood 2003;102(6):1959–1965.
24. Lacalle RA, Gomez-Mouton C, Barber DF, et al. PTEN regulates motility but not directionality during leukocyte chemotaxis. J Cell Sci 2004;117(Pt 25):6207–6215.
25. del Pozo M, Vicente-Manzanares M, Tejedor R, Serrador JM, Sanchez-Madrid F. Rho GTPases control migration and polarisation of adhesion molecules and cytoskeletal ERM components in T lymphocytes. Eur J Immunol 1999;29: 3609–3620.

26. Haddad E, Zugaza JL, Louache F, et al. The interaction between cdc42 and WASP is required for SDF-1-induced T lymphocyte chemotaxis. Blood 2001;97:33–38.
27. Raftopoulou M, Hall A. Cell migration: Rho GTPases lead the way. Dev Biol 2004;265(1):23–32.
28. Li Z, Dong X, Wang Z, et al. Regulation of PTEN by Rho small GTPases. Nat Cell Biol 2005;7(4):399–404.
29. Vielkind S, Gallagher-Gambarelli M, Gomez M, Hinton HJ, Cantrell DA. Integrin regulation by RhoA in thymocytes. J Immunol 2005;175(1):350–357.
30. Giagulli C, Scarpini E, Ottoboni L, et al. RhoA and PKCzeta control distinct modalities of LFA-1 activation by chemokines: critical role of LFA-1 affinity triggering in lymphocyte in vivo homing. Immunity 2004;20(1):25–35.
31. Shimonaka M, Katagiri K, Nakayama T, et al. Rap1 translates chemokine signals to integrin activation, cell polarization, and motility across vascular endothelium under flow. J Cell Biol 2003;161(2):417–427.
32. McLeod SJ, Li AH, Lee RL, Burgess AE, Gold MR. The Rap GTPases regulate B cell migration toward the chemokine stromal cell-derived factor-1 (CXCL12): potential role for Rap2 in promoting B cell migration. J Immunol 2002;169(3): 1365–1371.
33. Cullen PJ, Lockyer PJ. Integration of calcium and Ras signalling. Nat Rev Mol Cell Biol 2002;3(5):339–348.
34. Cronshaw DG, Kouroumalis A, Parry R, Webb A, Brown Z, Ward SG. Evidence that phospholipase C-dependent, calcium-independent mechanisms are required for directional migration of T lymphocytes in response to the CCR4 ligands CCL17 and CCL22. J Leukoc Biol 2006;79(6):1369–1380.
35. Fukui Y, Hashimoto O, Sanui T, et al. Hematopoietic cell-specific CDM family protein DOCK-2 is essential for lymphocyte migration. Nature 2001;12:826–830.
36. Sanui T, Inayoshi A, Noda M, et al. DOCK2 regulates Rac activation and cytoskeletal reorganization through interaction with ELMO1. Blood 2003;102(8): 2948–2950.
37. Nishihara H, Maeda M, Oda A, et al. DOCK2 associates with CrkL and regulates Rac1 in human leukemia cell lines. Blood 2002;100(12):3968–3974.
38. Lemmon MA, Ferguson KM. Signal-dependent membrane targeting by pleckstrin homology (PH) domains. Biochem J 2000;350(Pt 1):1–18.
39. Cote JF, Motoyama AB, Bush JA, Vuori K. A novel and evolutionarily conserved PtdIns(3,4,5)P_3-binding domain is necessary for DOCK180 signalling. Nat Cell Biol 2005;7(8):797–807.
40. Mellado M, Rodriguez-Frade JM, Manes S, Martinez AC. Chemokine signaling and functional responses: the role of receptor dimerization and TK pathway activation. Annu Rev Immunol 2001;19:397–421.
41. Stein JV, Soriano SF, M'Rini C, et al. CCR7-mediated physiological lymphocyte homing involves activation of a tyrosine kinase pathway. Blood 2003;101(1): 38–44.
42. Moriguchi M, Hissong BD, Gadina M, et al. CXCL12 signaling is independent of Jak2 and Jak3. J Biol Chem 2005;280(17):17408–1714.

43. Ottoson NC, Pribila JT, Chan AS, Shimizu Y. Cutting edge: T cell migration regulated by CXCR4 chemokine receptor signaling to ZAP-70 tyrosine kinase. J Immunol 2001;167(4):1857–1861.
44. Jordan MS, Singer AL, Koretzky GA. Adaptors as central mediators of signal transduction in immune cells. Nat Immunol 2003;4(2):110–116.
45. Berg LJ, Finkelstein LD, Lucas JA, Schwartzberg PL. Tec family kinases in T lymphocyte development and function. Annu Rev Immunol 2005;23:549–600.
46. Takesono A, Horai R, Mandai M, Dombroski D, Schwartzberg PL. Requirement for Tec kinases in chemokine-induced migration and activation of Cdc42 and Rac. Curr Biol 2004;14(10):917–922.
47. Fischer AM, Mercer JC, Iyer A, Ragin MJ, August A. Regulation of CXC chemokine receptor 4-mediated migration by the Tec family tyrosine kinase ITK. J Biol Chem 2004;279(28):29816–20.
48. Tan SL, Parker PJ. Emerging and diverse roles of protein kinase C in immune cell signalling. Biochem J 2003;376(Pt 3):545–552.
49. Li Z, Jiang H, Xie W, Zhang Z, Smrcka AV, Wu D. Roles of PLC-beta2 and -beta3 and PI3Kgamma in chemoattractant-mediated signal transduction. Science 2000;287(5455):1046–1049.
50. Volkov Y, Long A, McGrath S, Ni Eidhin D, Kelleher D. Crucial importance of PKC-beta(I) in LFA-1-mediated locomotion of activated T cells. Nat Immunol 2001;2(6):508–514.
51. Newton AC. Regulation of the ABC kinases by phosphorylation: protein kinase C as a paradigm. Biochem J 2003;370(Pt 2):361–371.
52. Iwabu A, Smith K, Allen FD, Lauffenburger DA, Wells A. Epidermal growth factor induces fibroblast contractility and motility via a protein kinase C delta-dependent pathway. J Biol Chem 2004;279(15):14551–14560.
53. Steinberg SF. Distinctive activation mechanisms and functions for protein kinase C-delta. Biochem J 2004;384(Pt 3):449–459.
54. Benes CH, Wu N, Elia AE, Dharia T, Cantley LC, Soltoff SP. The C2 domain of protein kinase C-delta is a phosphotyrosine binding domain. Cell 2005;121(2): 271–280.
55. Foxman EF, Kunkel EJ, Butcher EC. Integrating conflicting chemotactic signals. The role of memory in leukocyte navigation. J Cell Biol 1999;147(3):577–588.
56. Paoletti S, Petkovic V, Sebastiani S, Danelon MG, Uguccioni M, Gerber BO. A rich chemokine environment strongly enhances leukocyte migration and activities. Blood 2005;105(9):3405–3412.
57. Tian Y, New DC, Yung LY, et al. Differential chemokine activation of CC chemokine receptor 1-regulated pathways: ligand selective activation of Galpha 14-coupled pathways. Eur J Immunol 2004;34(3):785–795.
58. Tilton B, Ho L, Oberlin E, et al. Signal transduction by CXC chemokine receptor 4. Stromal cell-derived factor 1 stimulates prolonged protein kinase B and extracellular signal-regulated kinase 2 activation in T lymphocytes. J Exp Med 2000;192(3):313–324.

5

Chemokine Receptors and Neutrophil Trafficking

Teizo Yoshimura

Summary

Neutrophils are the most abundant leukocytes in circulation. They rapidly infiltrate sites of tissue injury and play a critical role in innate immune responses. In addition, they also contribute to the development of adaptive immune responses. Isolation of the human chemokine IL-8 and the cloning of its receptors CXCR1 and CXCR2, followed by the cloning of their orthologues or homologues in animals, have enabled researchers to elucidate the mechanisms of neutrophil trafficking during immune responses at a molecular level. Since then, there has been tremendous progress in understanding how the trafficking of neutrophils is regulated by the chemokine/chemokine receptor system under not only pathologic but also physiologic conditions. In this chapter, the roles of the chemokine receptors in regulating the trafficking of neutrophils are described.

Key Words: Chemokine; chemokine receptor; inflammation; neutrophils; trafficking.

5.1. Introduction

Neutrophils are the most abundant leukocytes in humans, comprising about two thirds of peripheral blood leukocytes. Upon tissue injury, they rapidly infiltrate injury sites and play an important role in innate immune responses. In addition, they also contribute to the development of adaptive immune responses by producing an array of cytokines and chemokines. Tissue infiltration of neutrophils is initiated by signals generated by the interaction between chemoattractants produced at sites of injury and their corresponding cell surface receptors. Classical chemoattractants, such as C5a, N-formyl-methionyl-leucyl-

From: *The Receptors: The Chemokine Receptors*
Edited by: J. K. Harrison and N. W. Lukacs © Humana Press Inc., Totowa, NJ

phenylalanine (fMLP), and leukotriene B4 (LTB4), attract neutrophils but they also attract monocytes; therefore, these molecules do not explain specific trafficking of neutrophils that is observed in an early stage of the inflammatory response. Identification of the chemokine/chemokine receptor system has allowed researchers to investigate the mechanisms regulating the selective trafficking of neutrophils during the inflammatory response in detail. The chemokine/chemokine receptor system also regulates the retention and release of neutrophils from the bone marrow. In this chapter, the roles of chemokine receptors in regulating the trafficking of neutrophils in physiologic and pathologic states are described.

5.2. Role for CXCR1 and CXCR2 in Tissue Infiltration of Neutrophils

As discussed elsewhere in this book, IL-8/CXCL8 is the first chemokine identified based on its activity to induce chemotactic movement of a particular leukocyte population, neutrophils *(1–3)*. The identification of IL-8 led to the cloning of the first two closely related human chemokine receptors, CXCR1 and CXCR2 *(4,5)*. CXCR1 is highly selective for IL-8/CXCL8, whereas CXCR2 is responsive to IL-8 and other CXC chemokines with a glutamic acid-leucine-arginine (ELR) motif, including growth-related oncogene (GRO)/CXCL1, neutrophil activating protein-2 (NAP-2)/CXCL7 and epithelial cell-derived neutrophil-activating factor-78 amino acids (ENA-78)/CXCL5. Granulocytc-chemoattractant protein-2 (GCP-2)/CXCL6 is an equipotent agonist for both CXCR1 and CXCR2 *(6)*. Circulating human neutrophils express both CXCR1 and CXCR2. Upon tissue injury, IL-8 and other CXC chemokines with the ELR motif are produced locally by various cell types, including macrophages, fibroblasts, and endothelial cells, and the interaction of these chemokines with CXCR1 and/or CXCR2 triggers the extravasation and subsequent directed migration of circulating neutrophils to injury sites.

There have been efforts to distinguish the roles of CXCR1 and CXCR2. In vitro, cells transfected to express either CXCR1 or CXCR2 respond to their corresponding ligands and exhibit a similar level of chemotactic migration; thus, both receptors are capable of regulating the infiltration of neutrophils. However, studies characterizing the downstream signaling pathways of these receptors indicated that the interaction of CXCR1 or CXCR2 with their ligands may involve different signaling molecules *(7,8)*. Thus, these receptors may play different roles in regulating neutrophil functions.

In vivo roles of CXCR1 and CXCR2 and their ligands were investigated using rabbits in which the orthologues of IL-8 and its two receptors have been identified. Intravenous injection of a neutralizing anti-IL-8 antibody prevented neutrophil accumulation and activation in a model of reperfusion tissue injury

(9) or glomerulonephritis *(10)*. Administration of anti-IL-8 or anti-GRO antibody inhibited about 50% of the peak leukocyte accumulation at 9 hours (neutrophils greater than 95%) in a lipopolysaccharide (LPS)-induced arthritis model. Coadministration of both antibodies increased the inhibition up to 70% at 9 hours and also inhibited the initial phase of leukocyte accumulation (neutrophil greater than 99%), which was not affected by administration of a single antibody. Interestingly, anti-GRO, but not anti-IL-8, reduced the levels of IL-1β and IL-1 receptor antagonist at 9 hours *(11)*. These observations suggested that CXCR1 and CXCR2 may indeed play a differential role in the regulation of neutrophil trafficking. However, the relative in vivo roles of these chemokine receptors need further clarification.

In mice or rats, the most commonly used small animal model, neither IL-8 nor CXCR1 is present. Therefore, the majority of in vivo studies focused on the role of CXCR2. To define the role of CXCR2, this gene was disrupted in mice *(12)*. Strikingly, the number of neutrophils from CXCR2$^{-/-}$ mice that migrated to the peritoneum in response to intraperitoneal injection of thioglycollate was one fifth that of CXCR2$^{+/+}$ mice despite that the total number of circulating neutrophils in CXCR2$^{-/-}$ mice was greatly increased. This study established a nonredundant role of CXCR2 in neutrophil trafficking during inflammatory responses, and it is now clear that the interaction of CXCR2 and its ligands, such as macrophage inflammatory protein-2 (MIP-2) or KC, plays a major role in the recruitment of neutrophils in many rodent disease models. Recent studies demonstrated that blocking CXCR2 with CXCR2 antagonists resulted in a marked reduction in neutrophil recruitment and subsequent tissue injury in a mouse model of reperfusion injury or systemic inflammatory response syndrome *(13,14)*. However, it should also be noted that a small but significant level of neutrophil infiltration still occurred in the absence of CXCR2 in response to intraperitoneal injection of thioglycollate *(12)*, indicating the presence of other mechanisms involved in neutrophil recruitment.

Although the role for CXCR2 in the trafficking of neutrophils in rodents, such as mice and rats, has become clearer, it remains unclear how much CXCR1 contributes to neutrophil trafficking in humans. The major reason that makes it difficult to compare the role of these two receptors in vivo is the apparent absence of functional CXCR1 and its ligand IL-8 in these small animals, as noted above. There are two rat genes closely related to the human CXCR1 and CXCR2 genes, one of which was identified as the rat orthologue of the human CXCR2 gene *(15)*. The other gene coded for a protein most similar to human CXCR1 with amino acid sequence identity of 71%, and the gene was therefore termed the rat CXCR1 gene. However, the transcript for the rat CXCR1 gene was mainly detected in macrophages rather than in neutrophils. Furthermore, human embryonic kidney (HEK) 293 cells expressing rat CXCR1-like protein (coded by the rat CXCR1 gene) did not show a detectable change in calcium influx in response to either rat MIP-2 or KC, and neither the ligand nor the function of this receptor has been identified. The mouse CXCR1 gene was recently identified *(16,17)*. This gene was mapped to mouse chromosome 1 along with the CXCR2 gene, and its genomic organization is very similar to that of the human CXCR1 gene. The transcript of the mouse CXCR1 gene was detected in many tissues, including the lung, spleen, and kidney, as well as in neutrophils and monocytes/macrophages by Northern blotting and RT-PCR *(16,17)*. To investigate whether mouse CXCR1-like protein (coded by the mouse CXCR1 gene) was a chemokine receptor, recombinant mouse CXCR1-like protein was coexpressed with recombinant heterodimeric G protein subunits $G\alpha_{i2}\beta_1\gamma_3$ in insect cells, and receptor-mediated G protein activation was assayed by measuring the effect of chemokines on the binding of [^{35}S] guanosine triphosphate (GTP)[S] to insect cell membranes. None of the CXC chemokines known to activate mouse CXCR2 or members of the CXCR1/2 family from other mammalian species activated the receptor *(17)*; thus, the ligand and the function of this receptor also remain unidentified. The information on the rat and mouse CXCR1 gene suggests that the CXCR1 gene may have evolved in these species, but the evolution of CXCR1 is incomplete. Generation and characterization of CXCR1 knockout mice will provide us with a definitive answer whether CXCR1 identified in these species are indeed the orthologues of human CXCR1 and play a role in the trafficking of neutrophils.

We previously demonstrated the presence of IL-8 in the guinea-pig *(18)*, leading us to hypothesize that its specific receptor CXCR1 is also present in this species. Recently, we obtained two guinea-pig genomic DNA clones coding for the potential guinea-pig orthologue of CXCR1 or CXCR2, respectively *(19)*. Transcripts for these genes were highly expressed in neutrophils, but not in macrophages, and became undetectable after stimulation of neutrophils with LPS. Functionally, both guinea-pig IL-8 and human IL-8 induced cell migration and extracellular signal-regulated kinase (ERK) phosphorylation in HEK 293 cells expressing either receptor, whereas human GRO activated only cells expressing one of the receptors, indicating that the protein coded by the clone was indeed guinea-pig CXCR1. Thus, functional CXCR1 is present in the guinea-pig. As we expected, ^{125}I-labeled human IL-8 bound to guinea-pig CXCR1 and addition of unlabeled human IL-8 completely abolished the binding. Interestingly, however, unlabeled guinea-pig IL-8 failed to compete against ^{125}I-labeled human IL-8. Thus, the avidity of human IL-8 for guinea-pig CXCR1 appears to be higher than that of guinea-pig IL-8, suggesting further tuning of the IL-8/CXCR1 axis in human. Identification and characterization of CXCR1 in the guinea-pig will allow us to use this small animal model to define the relative role of the IL-8/CXCR1 interaction in the trafficking of neutrophils and perhaps to examine the efficacy of antagonists in vivo.

5.3. Role for CC Chemokine Receptors in the Trafficking of Neutrophils

In tissue-infiltrating neutrophils, the expression of CXCR1 and CXCR2 is downregulated *(20)*. This downregulation can be induced by two mechanisms: internalization of receptors after their interaction with ligands and decreased expression of their transcripts. In vitro, LPS *(21,22)* and tumor necrosis factor (TNF)-α *(23,24)* can downregulate the expression of both receptors at the mRNA level. Conversely, the expression of several CC chemokine receptors can be upregulated in response to proinflammatory stimuli, and the interaction of these receptors with their ligands may also control the trafficking of neutrophils.

CCR1 is a promiscuous chemokine receptor bound by several chemokines, such as MIP-1α/CCL3, MIP-1β/CCL4, and RANTES/CCL5 *(25–28)*. Although these chemokines are generally known to attract monocytes but not neutrophils in vitro *(29)*, a previous report indicated that human MIP-1α but not MIP-1β, could induce a small but detectable level of calcium flux in human neutrophils *(30)*. When recombinant human MIP-1α was injected into human skin to examine the in vivo role of this chemokine, rapid infiltration of neutrophils was observed prior to the appearance of other cell types, such as monocytes and lymphocytes *(31)*.

Expression of CCR1 on human neutrophils has been evaluated in detail *(32,33)*. Human neutrophils constitutively expressed a low level of CCR1 mRNA, but they did not respond to CCR1 ligands, including MIP-1α, monocyte chemoattractant protein (MCP)-3, and RANTES. However, the expression of CCR1 mRNA could be markedly upregulated within 1 hour after treatment with cytokines such as interferon (IFN)-γ *(32)* or granulocyte-macrophage colony-stimulating factor (GM-CSF) *(33)*, and cells treated with these cytokines functionally responded to stimulation with multiple CCR1 ligands and exhibited migration and calcium mobilization; therefore, CCR1 expressed on human neutrophils was functional. In vivo, neutrophils infiltrating joints of patients with inflammatory joint diseases expressed low but significant levels of CCR1 *(20)*. These results support the hypothesis that CCR1, a receptor for MIP-1α, may play a role in the trafficking of neutrophils.

There is better evidence for a role for CCR1 in the trafficking of neutrophils in mice. In mice deficient in CCR1, neutrophil infiltration in response to intraperitoneal injection of thioglycollate was markedly reduced *(25)*. Reduction of neutrophil recruitment was also found when acute pancreatitis was induced in CCR1-deficient mice *(34)*. During sepsis after cecal ligation and puncture, blood neutrophils expressed functional CCR1 *(35)*. These studies strongly indicated that CCR1 plays a role in neutrophil trafficking in this species.

However, there is a significant difference between humans and rodents in the level of CCR1 expression on neutrophils; rodent, but not human, neutrophils express abundant CCR1, and the CCR1 ligand MIP-1α is a powerful granulocyte-activating chemokine. This led to the proposal that MIP-1α and CCR1 may subserve the human role of IL-8 and CXCR1 in rodents *(34)*. The assignment of biological activities to human chemokines and chemokine receptors based on studies in mice must be made with caution *(34)*.

CCR2 is the only established functional receptor for MCP-1/CCL2 expressed on hematopoietic cells, and the interaction of MCP-1 with CCR2 is critical for the recruitment of monocytes in many human diseases and animal disease models *(36)*. Expression of CCR2 was observed on murine neutrophils isolated from chronic inflammatory sites or blood neutrophils isolated during sepsis after cecal ligation and puncture in vivo *(35)*. These neutrophils exhibited a chemotactic response to the agonist MCP-1/CCL2. A role for CCR2 in neutrophil trafficking was also demonstrated using CCR2 knockout mice *(37)*. However, other studies using CCR2 knockout mice did not detect impaired neutrophil recruitment in response to intraperitoneal thioglycollate injection *(38,39)*. Tissue-infiltrating inflammatory monocytes/macrophages produce neutrophil-attracting chemokines, such as MIP-2 or KC. Therefore, reduced monocyte infiltration in CCR2 knockout mice could affect the recruitment of neutrophils to sites of inflammation. A recent study indicated constitutive CCR2 expression on 97% of circulating mouse neutrophils *(40)*. However, CCR2 is not expressed on circulating human neutrophils, and upregulation of CCR2 expression on human neutrophils has not been shown in vitro, suggesting that CCR2 expression by neutrophils may only occur in mice.

Cytokine-stimulated neutrophils were shown to express increased levels of CCR6 *(41)*. TNF-α stimulation induced high levels of CCR6 mRNA expression in neutrophils, whereas IFN-γ induced low levels, and the two cytokines together exhibited considerable synergy. Approximately 160 binding sites for CCR6 ligand (CCL20) were detected *(42)* with an equilibrium dissociation constant of 1.6 nM on neutrophils activated with TNF-α plus IFN-γ. Although the K_d of CCL20 binding to neutrophils was in the range previously reported for CCL20 binding to CCR6 transfected cells, the number of CCR6 binding sites on neutrophils was considerably lower compared with that on immature dendritic cells (DCs), which expressed approximately 42,000 binding sites per cell *(43)*. Nevertheless, cytokine-induced CCR6 on neutrophils was functional as shown by chemotactic responses of these cells to CCL20 *(41)*. CCR6 is expressed on immature DCs during DC maturation and is involved in the trafficking of immature DCs *(44)*, suggesting that neutrophils expressing this receptor might be in the process of acquiring features characteristic of DCs. In fact, cytokine-activated neutrophils have been demonstrated to express major histocompatibility complex (MHC) class II *(45)* and CD83 *(41)*, molecules expressed on mature DCs. However, CCR7, normally expressed on mature DCs could not be detected on neutrophils even after 4 days of incubation with TNF-α plus IFN-γ *(41)*. The ability of cytokine-activated neutrophils to present antigens to T cells remains controversial. This indicates that these cytokine-activated neutrophils do not parallel the maturation program of DCs *(41,46)*. It was previously shown that highly purified, lactoferrin positive, immediate precursors of end-stage neutrophils, but not fully mature neutrophils, could be reverted in their functional maturation program and driven to acquire the characteristics of DCs *(47)*. Thus, neutrophil precursors, under appropriate circumstances, can become DC-like cells and play a role in antigen presentation and T-cell proliferation. However, mature neutrophils most likely indirectly contribute to the induction of adaptive immunity through the production of cytokines and chemokines.

CC-chemokine receptor-like 2 (CCRL2) is a 344-amino-acid protein *(48)* with ~40% identity to other known chemokine receptors, and secondary structure of the protein predicts a seven-transmembrane receptor. The human *Ccrl2* gene resides on chromosome 3 in close proximity to other known chemokine receptors, such as CCR5, CCR2, and CCR3, leading to the prediction that CCRL2 is also a G protein–coupled chemokine receptor; however, this has not been demonstrated. A study using flow cytometry with a mouse monoclonal antibody indicated constitutive, high-level CCRL2 expression on human circulating, resting neutrophils *(49)*. In contrast, our in vitro study indicated that expression of CCRL2 is rapidly upregulated in human neutrophils in the presence of proinflammatory mediators, such as LPS, TNF-α, or TNF-α in combination with IFN-γ or GM-CSF *(50)*. CCRL2 was expressed on all neutrophils and a portion of macrophages infiltrating the synovium of patients with rheumatoid arthritis (RA). HEK 293 cells expressing CCRL2 migrated toward a fraction of the RA fluid, suggesting that an agonist(s) of CCRL2 is present in the synovial fluids of RA patients.

The murine orthologue of human CCRL2 is thought to be L-CCR *(51)* (49.3% amino acid identity). L-CCR has been shown to be expressed on microglial cells and to bind MCP-1 *(52)*. Because CCRL2 has considerable homology to CCR2, and high levels of MCP-1 are found in RA fluids, MCP-1 may be a ligand for CCRL2. However, we did not detect the migration of CCRL2-expressing HEK 293 cells in response to human recombinant MCP-1. Additionally, MCP-1 neutralizing antibodies did not block the chemotactic activity found in the RA fluid. Therefore, the active component in RA fluids that induced migration in CCRL2-expressing HEK cells is distinct from

MCP-1. It is possible that MCP-1 may bind the CCRL2 receptor without inducing migration or, alternatively, the murine and human receptors bind different ligands. Although the active component in the RA fractions was not identified, it was less than 30 kd and was inhibited by pertussis toxin *(50)*, which is consistent with the idea that a chemokine(s) activates the CCRL2 receptor. RA fluid contains a large number of proinflammatory mediators, including high concentrations of MIP-1α, RANTES *(30)*, and MIP-3α/CCL20 *(53)*. Thus, CCRL2 appears to be a functional receptor, and this ligand-receptor system is potentially involved in the trafficking or activation of neutrophils in RA *(50)*. Further studies are required to identify the ligand and the function for this receptor.

In addition to the receptors described above, CCR3 expression can be induced in neutrophils by IFN-γ *(32)* or GM-CSF *(33)*. CCR5 expression has been detected on circulating neutrophils *(40,54)*. Low levels of CX3CR1 expression could be detected on neutrophils *(55)*, and infiltrating inflammatory leukocytes, including neutrophils, displayed elevated levels of CX3CR1 in a rat glomerulonephritis model *(56)*. However, some of the results are still controversial, and the role for these chemokine receptors in neutrophil trafficking remains unclear.

5.4. Role for CXCR4 in Neutrophil Retention and Mobilization

The bone marrow (BM) contains a large reserve of mature neutrophils. Because neutrophils have a short half-life ($t_{1/2}$ = 6 hours), their rate of release from this tissue is a major determinant of the number of circulating neutrophils. In humans, approximately 10^{11} neutrophils are released from the BM per day *(57)*. Mature neutrophils reside within the hematopoietic compartment of the BM and are intimately associated with stromal cells and/or components of extracellular matrix. In response to inflammatory stimuli, the number of circulating neutrophils increases due to an acute release of neutrophils from the BM. A number of factors, including C5a, LTB4, TNF-α, IL-8, granulocyte colony-stimulating factor (G-CSF), and GM-CSF *(58–61)*, can promote rapid mobilization of neutrophils from the BM reserve. In addition to factors that directly stimulate neutrophil release, there appears to be another mechanism promoting the retention of neutrophils within the BM, and the balance between these opposing signals determines the rate of neutrophil release from the BM during the inflammatory responses.

The murine stromal-derived factor-1α (SDF-1)/CXCL12 was originally cloned from a cDNA library derived from the bone marrow stromal cell line ST2 by a method to clone cDNAs that carry specific amino terminal signal sequences, such as those encoding intercellular signaling-transducing molecules

(62). SDF-1α was later found to be identical to pre-B-cell growth-stimulating factor (PBSF) *(63)*. CXCR4 is the major receptor for SDF-1α *(64)*. When the gene for SDF-1α or CXCR4 was disrupted in mice, both SDF-1α–deficient and CXCR4-deficient mice died perinatally and displayed identical defects in neuron migration, organ vascularization, and hematopoiesis *(65–68)*. Interestingly, myelopoiesis, in addition to B lymphopoiesis, was decreased in fetal liver and virtually absent in the BM *(66)*. It was later found that the SDF-1α/CXCR4 interaction regulates B lymphopoiesis and myelopoiesis by confining precursors within the supportive fetal liver and BM microenvironment for further maturation *(69)*, establishing the role for the SDF-1α/CXCR4 interaction in the retention of hematopoietic precursors.

Although hematopoietic stem cells express CXCR4, studies characterizing the expression of CXCR4 by circulating neutrophils have yielded conflicting results. Several in vitro studies have shown that circulating neutrophils express CXCR4 and migrate or flux calcium in response to SDF-1α *(64,70)*, whereas another study did not observe these effects *(71)*. According to recent studies, it is likely that circulating resting neutrophils express a very low level of CXCR4, but CXCR4 expression becomes apparent during in vitro incubation *(57,72)*, suggesting that SDF-1α/CXCR4 interaction may play a role in the trafficking of mature neutrophils. Intravenous injection of CXCR4 antagonists, including anti-CXCR4 and AMD-3100, or a CXCR4 agonistic peptide into mice reduced marrow retention of mature neutrophils and mobilized them from the BM *(57,73,74)*. Interestingly, intravenous injection of the CXCR2 ligand KC also led to the mobilization of neutrophils independently of CXCR4, and co-injection of KC and CXCR4 antagonists augmented the KC effect. The functional responses of freshly isolated human and murine neutrophils to the CXCR2 agonist KC were significantly attenuated by SDF-1α. As a consequence, KC-induced mobilization of neutrophils from the BM was dramatically enhanced by blocking the effects of endogenous SDF-1α with a specific CXCR4 antagonist. Similarly, the response of mouse neutrophils to SDF-1α diminished when cells were pretreated with KC *(74)*. These studies have indicated a dynamic relationship between CXCR2 and CXCR4 in the release of neutrophils from the BM.

As described above, as neutrophils age, they upregulate the expression of CXCR4 and acquire the ability to migrate toward SDF-1α *(72)*. This spontaneous CXCR4 expression can be suppressed most potently by IFN-γ and also by IFN-α, G-CSF, and GM-CSF, likely due to their anti-apoptotic effect on neutrophils. When senescent mouse neutrophils expressing CXCR4 were injected intravenously into recipient mice, they preferentially homed to the BM in a CXCR4-dependent manner, suggesting SDF-1α/CXCR4 interaction also plays a role in the clearance of neutrophils from the circulation *(57)*.

5.5. Conclusions

As described above, the chemokine/chemokine receptor system plays a critical role in every aspect of neutrophil trafficking (Fig. 1). Neutrophil precursors are retained in the BM due to the interaction between CXCR4 and SDF-1α

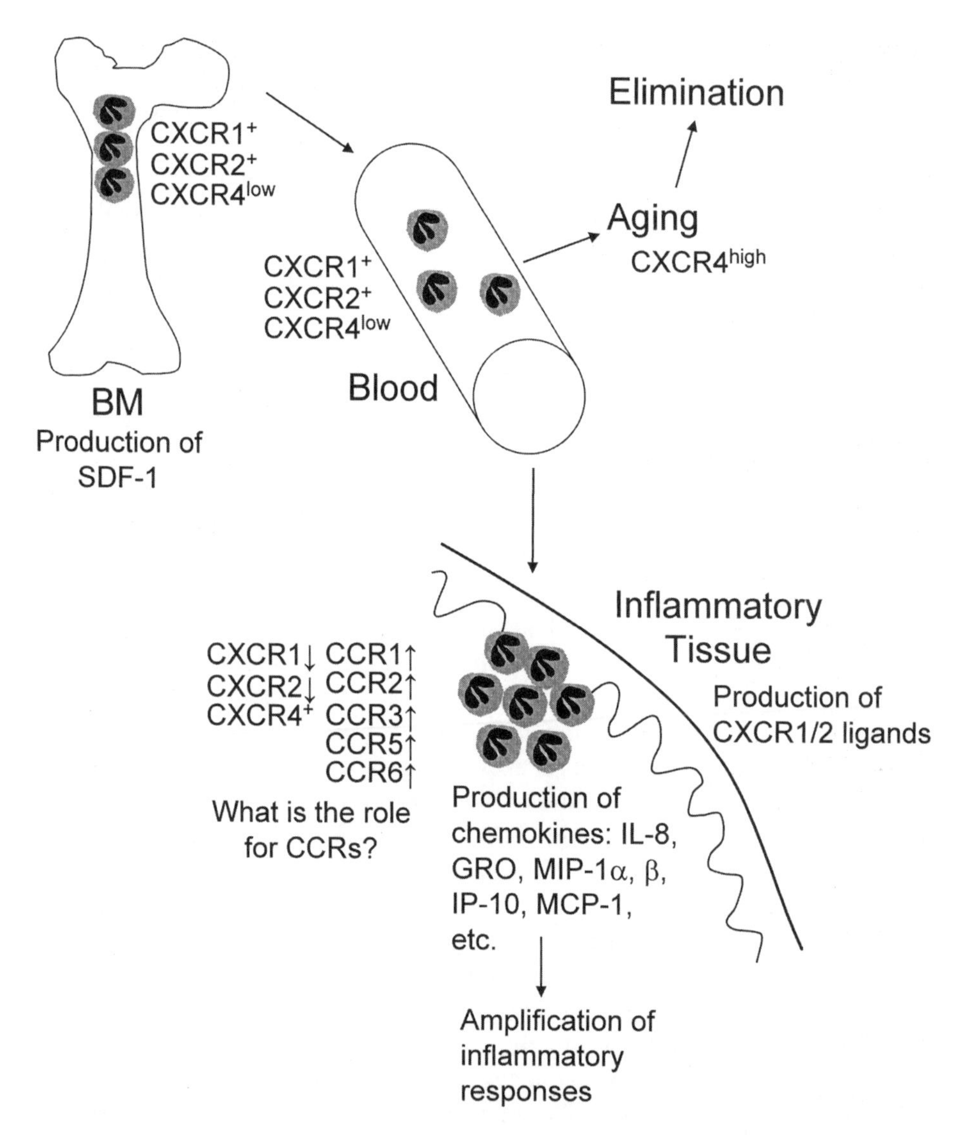

Fig. 1. Critical roles of the chemokine/chemokine receptor system in the trafficking of neutrophils under physiologic and pathologic conditions.

produced by stroma cells. As neutrophil precursors mature, they lose a significant portion of CXCR4 and are released into the circulation. CXCR1 (in humans) and CXCR2 ligands released during the inflammatory responses accelerate neutrophil release from the BM and also recruit neutrophils into injured tissues. These tissue-infiltrating, inflammatory neutrophils are capable of producing an array of CXC and CC chemokines, amplifying the trafficking of not only neutrophils but also monocytes and lymphocytes. Aged neutrophils express a high level of CXCR4, home to the BM, and are permanently cleared by macrophages. Despite considerable progress, however, important questions remain to be answered. (i) Is there any advantage of having CXCR1? In the case of angiogenesis, the activity of ELR^+ CXCR chemokines, including IL-8, is based on the sole expression of CXCR2 on endothelial cells *(75)*. However, human neutrophils express both CXCR1 and CXCR2 at a high level. There must be a reason for the IL-8/CXCR1 system to have evolved. (ii) The role of CXCR2 has been extensively studied using mice as a model, and there is no doubt that the interaction of CXCR2 with its ligands plays a major role in neutrophil trafficking. However, it remains unclear whether CXCR2 plays a dominant role in humans as in mice, and the results obtained using mice need to be validated using animal models in which both IL-8 and CXCR1 are present. (iii) What is the role for CC chemokine receptors in neutrophil trafficking? The results using mice have indicated significant contribution of CC chemokine receptors, especially CCR1, to neutrophil trafficking. However, there is no clear evidence that CCR1 plays a role in neutrophil trafficking in humans. The role for other CC chemokine receptors remains either controversial or unknown. It is the author's hope that studies of neutrophil trafficking focusing on the chemokine/chemokine receptor system will lead us to a better therapy of many human diseases in which this system plays a critical role.

Acknowledgments

The author is grateful to Dr. Joost J. Oppenheim for his review of this manuscript.

References

1. Yoshimura T, Matsushima K, Tanaka S, et al. Purification of a human monocyte-derived neutrophil chemotactic factor that has peptide sequence similarity to other host defense cytokines. Proc Natl Acad Sci U S A 1987;84:9233–9237.
2. Schroder JM, Mrowietz U, Morita E, Christophers E. Purification and partial biochemical characterization of a human monocyte-derived, neutrophil-activating peptide that lacks interleukin 1 activity. J Immunol 1987;139:3474–3483.

3. Walz A, Peveri P, Aschauer H, Baggiolini M. Purification and amino acid sequencing of NAF, a novel neutrophil-activating factor produced by monocytes. Biochem Biophys Res Commun 1987;149:755–761.
4. Holmes WE, Lee J, Kuang WJ, Rice GC, Wood WI. Structure and functional expression of a human interleukin-8 receptor. Science 1991;253:1278–1280.
5. Murphy PM, Tiffany HL. Cloning of complementary DNA encoding a functional human interleukin-8 receptor. Science 1991;253:1280–1283.
6. Murphy PM, Baggiolini M, Charo IF, et al. International union of pharmacology. XXII. Nomenclature for chemokine receptors. Pharmacol Rev 2000;52:145–176.
7. Cohen-Hillel E, Yron I, Meshel T, Soria G, Attal H, Ben-Baruch A. CXCL8-induced FAK phosphorylation via CXCR1 and CXCR2: Cytoskeleton- and integrin-related mechanisms converge with FAK regulatory pathways in a receptor-specific manner. Cytokine 2006;33:1–16.
8. Richardson RM, Pridgen BC, Haribabu B, Ali H, Snyderman R. Differential cross-regulation of the human chemokine receptors CXCR1 and CXCR2. Evidence for time-dependent signal generation. J Biol Chem 1998;273:23830–23836.
9. Sekido N, Mukaida N, Harada A, Nakanishi I, Watanabe I, Matsushima K. Prevention of lung perfusion injury in rabbits by a monoclonal antibody against interleukin-8. Nature 1993;365:654–657.
10. Wada T, Tomosugi N, Naito T, et al. Prevention of proteinuria by the administration of anti-interleukin 8 antibody in experimental acute immune complex-induced glomerulonephritis. J Exp Med 1994;180:1135–1140.
11. Matsukawa A, Yoshimura T, Fujiwara K, Maeda T, Ohkawara S, Yoshinaga M. Involvement of growth related protein (GRO) in lipopolysaccharide-induced rabbit arthritis: Cooperation between GRO and interleukin-8 (IL-8) and interrelated regulation among TNFα, IL-1, IL-8, and GRO. Lab Invest 1999;79:591–600.
12. Cacalano G, Lee J, Kikly K, et al. Neutrophil and B cell expansion in mice that lack the murine IL-8 receptor homolog. Science 1994;265:682–684.
13. Bertini R, Allegretti M, Bizzarri C, et al. Noncompetitive allosteric inhibitors of the inflammatory chemokine receptors CXCR1 and CXCR2: prevention of reperfusion injury. Proc Natl Acad Sci U S A 2004;101:11791–11796.
14. Kaneider NC, Agarwal A, Leger AJ, Kuliopulos A. Reversing systemic inflammatory response syndrome with chemokine receptor pepducins. Nat Med 2005;11: 661–665.
15. Dunstan C-AN, Salafranca MN, Adhikari S, Xia Y, Feng L, Harrison JK. Identification of two rat genes orthologous to the human interleukin-8 receptors. J Biol Chem 1996;271:32770–32776.
16. Fu W, Zhang Y, Zhang J, Chen WF. Cloning and characterization of mouse homolog of the CXC chemokine receptor CXCR1. Cytokine 2005;31:9–17.
17. Balabanian K, Lagane B, Infantino S, et al. The chemokine SDF-1/CXCL12 binds to and signals through the orphan receptor RDC1 in T lymphocytes. J Biol Chem 2005;280:35760–35766.
18. Yoshimura T, Johnson DG. cDNA cloning and expression of guinea pig neutrophil attractant protein-1 (NAP-1): NAP-1 is highly conserved in guinea pig. J Immunol 1993;151:6225–6236.

19. Takahashi M, Jeevan A, Sawant K, McMurray DN, Yoshimura T. Cloning and characterization of guinea pig CXCR1. Mol Immunol 2007;44:878–888.
20. Bruhl H, Wagner K, Kellner H, Schattenkirchner M., Schlondorff D, Mack M. Surface expression of CC- and CXC-chemokine receptors on leucocyte subsets in inflammatory joint diseases. Clin Exp Immunol 2001;126:551–559.
21. Khandaker MH, Xu L, Rahimpour R, et al. CXCR1 and CXCR2 are rapidly down-modulated by bacterial endotoxin through a unique agonist-independent, tyrosine kinase-dependent mechanism. J Immunol 1998;161:1930–1938.
22. Lloyd AR, Biragyn A, Johnston JA, et al. Granulocyte-colony stimulating factor and lipopolysaccharide regulate the expression of interleukin 8 receptors on polymorphonuclear leukocytes. J Biol Chem 1995;270:28188–28192.
23. Jawa RS, Quaid GA, Williams MA, et al. Tumor necrosis factor alpha regulates CXC chemokine receptor expression and function. Shock 1999;11:385–390.
24. Tikhonov I, Doroshenko T, Chaly Y, Smolnikova V, Pauza CD, Voitenok N. Down-regulation of CXCR1 and CXCR2 expression on human neutrophils upon activation of whole blood by S. aureus is mediated by TNF-α. Clin Exp Immunol 2001;125:414–422.
25. Gao J-L, Wynn TA, Chang Y, et al. Imparied host defense, hematopoiesis, granulomatous inflammation and type 1-type 2 cytokine balance in mice lacking CC chemokine receptor 1. J Exp Med 1997;185:1959–1968.
26. Neote K, Mak JY, Kolakowski LF Jr, Schall TJ. Functional and biochemical analysis of the cloned Duffy antigen: identity with the red blood cell chemokine receptor. Blood 1994;84:44–52.
27. Gao JL, Kuhns DB, Tiffany HL, et al. Structure and functional expression of the human macrophage inflammatory protein 1 alpha/RANTES receptor. J Exp Med 1993;177:1421–1427.
28. Neote K, DiGregorio D, Mak JY, Horuk R, Schall TJ. Molecular cloning, functional expression, and signaling characteristics of a C-C chemokine receptor. Cell 1993;72:415–425.
29. Zhang S, Youn BS, Gao JL, Murphy PM, Kwon BS. Differential effects of leukotactin-1 and macrophage inflammatory protein-1 alpha on neutrophils mediated by CCR1. J Immunol 1999;162:4938–4942.
30. McColl SR, Hachicha M, Levasseur S, Neote K, Schall TJ. Uncoupling of early signal transduction events from effector function in human peripheral blood neutrophils in response to recombinant macrophage inflammatory proteins-1 alpha and -1 beta. J Immunol 1993;150:4550–4560.
31. Lee SC, Brummet ME, Shahabuddin S, et al. Cutaneous injection of human subjects with macrophage inflammatory protein-1 alpha induces significant recruitment of neutrophils and monocytes. J Immunol 2000;164:3392–3401.
32. Bonecchi R, Polentarutti N, Luini W, et al. Up-regulation of CCR1 and CCR3 and induction of chemotaxis to CC chemokines by IFN-gamma in human neutrophils. J Immunol 1999;162:474–479.
33. Cheng SS, Lai JJ, Lukacs NW, Kunkel SL. Granulocyte-macrophage colony stimulating factor up-regulates CCR1 in human neutrophils. J Immunol 2001;166: 1178–1184.

34. Gerard C, Frossard JL, Bhatia M, et al. Targeted disruption of the beta-chemokine receptor CCR1 protects against pancreatitis-associated lung injury. J Clin Invest 1997;100:2022–2027.
35. Speyer CL, Gao H, Rancilio NJ, et al. Novel chemokine responsiveness and mobilization of neutrophils during sepsis. Am J Pathol 2004;165:2187–2196.
36. Charo IF, Taubman MB. Chemokines in the pathogenesis of vascular disease. Circ Res 2004;95:858–866.
37. Maus U, von Grote K, Kuziel WA, et al. The role of CC chemokine receptor 2 in alveolar monocyte and neutrophil immigration in intact mice. Am J Respir Crit Care Med 2002;166:268–273.
38. Boring L, Gosling J, Cleary M, Charo IF. Decreased lesion formation in $CCR2^{-/-}$ mice reveals a role for chemokines in the initiation of atherosclerosis. Nature 1998;394:894–897.
39. Kurihara T, Warr G, Loy J, Bravo R. Defects in macrophage recruitment and host defense in mice lacking the CCR2 chemokine receptor. J Exp Med 1997;186: 1757–1762.
40. Reichel CA, Khandoga A, Anders H-J, Schlondorff D, Luckow B, Krombach F. Chemokine receptors CCR1, CCR2, and CCR5 mediate neutrophil migration to postischemic tissue. J Leukoc Biol 2006;79:114–122.
41. Yamashiro S, Wang J-M, Gong W-H, Yan D, Kamohara H, Yoshimura T. Expression of CCR6 and CD83 by cytokine-activated human neutrophils. Blood 2000;96:3958–3963.
42. Greaves DR, Wang W, Dairaghi DJ, et al. CCR6, a CC chemokine receptor that interacts with macrophage inflammatory protein 3alpha and is highly expressed in human dendritic cells. J Exp Med 1997;186:837–844.
43. Yang D, Howard OMZ, Chen Q, Oppenheim JJ. Cutting edge: Immature dendritic cells generated from monocytes in the presence of TGF-beta1 express functional C-C chemokine receptor 6. J Immunol 1999;163:1737–1741.
44. Sozzani S, Allavena P, D'Amico G, et al. Differential regulation of chemokine receptors during dendritic cell maturation: a model for their trafficking properties. J Immunol 1998;161:1083–1086.
45. Gosselin EJ, Wardwell K, Rigby WF, Guyre PM. Induction of MHC class II on human polymorphonuclear neutrophils by granulocyte/macrophage colony-stimulating factor, IFN-gamma, and IL-3. J Immunol 1993;151:1482–1490.
46. Yamashiro S, Kamohara H, Yoshimura T. Alteration in the responsiveness to TNF-α is crucial for maximal expression of MCP-1 in human neutrophils. Immunology 2000;101:97–103.
47. Oehler L, Majdic O, Pickl WF, et al. Neutrophil granulocyte-commited cells can be driven to acquire dendritic cell characteristics. J Exp Med 1998;187:1019–1028.
48. Fan P, Kyaw H, Su K, et al. Cloning and characterization of a novel human chemokine receptor. Biochem Biophys Res Commun 1998;243:264–268.

49. Migeotte I, Franssen J-D, Goriely S, Willems F, Parmentier M. Distribution and regulation of expression of the putative human chemokine receptor HCR in leukocyte populations. Eur J Immunol 2002;32:494–501.
50. Galligan C, Matsuyama W, Matsukawa A, et al. Up-regulated expression and activation of the orphan chemokine receptor, CCRL2, in rheumatoid arthritis. Arthritis Rheum 2004;50:1806–1814.
51. Shimada T, Matsumoto M, Tatsumi Y, Kanamaru A, Akira S. A novel lipopolysaccharide inducible C-C chemokine receptor related gene in murine macrophages. FEBS Lett 1998;425:490–494.
52. Zuurman MW, Heeroma J, Brouwer N, Boddeke HW, Biber K. LPS-induced expression of a novel chemokine receptor (L-CCR) in mouse glial cells in vitro and in vivo. Glia 2003;41:327–336.
53. Ruth JH, Shahrara S, Park CC, et al. Role of macrophage inflammatory protein-3alpha and its ligand CCR6 in rheumatoid arthritis. Lab Invest 2003;83: 579–588.
54. Ottonello L, Montecucco F, Bertolotto M, et al. CCL3 (MIP-1α) induces in vitro migration of GM-CSF-primed human neutrophils via CCR5-dependent activation of ERK 1/2. Cell Signal 2005;17:355–363.
55. Imai T, Hieshima K, Haskell C, et al. Identification and molecular characterization of fractalkine receptor CX3CR1, which mediates both leukocyte migration and adhesion. Cell 1997;91:521–530.
56. Feng L, Chen S, Garcia GE, et al. Prevention of crescentic glomerulonephritis by immunoneutralization of the fractalkine receptor CX3CR1 rapid communication. Kidney Int 1999;56:612–620.
57. Martin C, Burdon PC, Bridger G, Gutierrez-Ramos JC, Williams TJ, Rankin SM. Chemokines acting via CXCR2 and CXCR4 control the release of neutrophils from the bone marrow and their return following senescence. Immunity 2003;19: 583–593.
58. Dale DC, Liles WC, Llewellyn C, Price TH. Effects of granulocyte-macrophage colony-stimulating factor (GM-CSF) on neutrophil kinetics and function in normal human volunteers. Am J Hematol 1998;57:7–15.
59. Jagels MA, Hugli TE. Neutrophil chemotactic factors promote leukocytosis. J Immunol 1992;148:1119–1128.
60. Sato N, Sawada K, Takahashi TA, et al. A time course study for optimal harvest of peripheral blood progenitor cells by granulocyte colony-stimulating factor in healthy volunteers. Exp Hematol 1994;22:973–978.
61. Terashima T, English D, Hogg JC, van Eeden SF. Release of polymorphonuclear leukocytes from the bone marrow by interleukin-8. Blood 1998;92:1062–1069.
62. Tashiro K, Tada H, Heilker R, Shirozu M, Nakano T, Honjo T. Signal sequence trap: a cloning strategy for secreted proteins and type I membrane proteins. Science 1993;261:600–603.
63. Nagasawa T, Kikutani H, Kishimoto T. Molecular cloning and structure of a pre-B-cell growth-stimulating factor. Proc Natl Acad Sci U S A 1994;91: 2305–2309.

64. Oberlin E, Amara A, Bachelerie F, et al. The CXC chemokine SDF-1 is the ligand for LESTR/fusin and prevents infection by T-cell-line-adapted HIV-1. Nature 1996;382:833–835.
65. Ma Q, Jones D, Borghesani PR, et al. Impaired B-lymphopoiesis, myelopoiesis, and derailed cerebellar neuron migration in CXCR4- and SDF-1-deficient mice. Proc Natl Acad Sci U S A 1998;95:9448–9453.
66. Nagasawa T, Hirota S, Tachibana K, et al. Defects of B-cell lymphopoiesis and bone-marrow myelopoiesis in mice lacking the CXC chemokine PBSF/SDF-1. Nature 1996;382:635–638.
67. Tachibana K, Hirota S, Iizasa H, et al. The chemokine receptor CXCR4 is essential for vascularization of the gastrointestinal tract. Nature 1998;393:591–594.
68. Zou YR, Kottmann AH, Kuroda M, Taniuchi I, Littman DR. Function of the chemokine receptor CXCR4 in haematopoiesis and in cerebellar development. Nature 1998;393:595–599.
69. Ma Q, Jones D, Springer TA. The chemokine receptor CXCR4 is required for the retention of B lineage and granulocytic precursors within the bone marrow microenvironment. Immunity 1999;10:463–471.
70. Ueda H, Siani MA, Gong W, Thompson DA, Brown GG, Wang JM. Chemically synthesized SDF-1alpha analogue, N33A, is a potent chemotactic agent for CXCR4/Fusin/LESTR-expressing human leukocytes. J Biol Chem 1997;272:24966–24970.
71. Bleul CC, Fuhlbrigge RC, Casasnovas JM, Aiuti A, Springer TA. A highly efficacious lymphocyte chemoattractant, stromal cell-derived factor 1 (SDF-1). J Exp Med 1996;184:1101–1109.
72. Nagase H, Miyamasu M, Yamaguchi M, et al. Cytokine-mediated regulation of CXCR4 expression in human neutrophils. J Leukoc Biol 2002;71:711–717.
73. Pelus LMBH, Fukuda S, Wong D, Merzouk A, Salari H. The CXCR4 agonist peptide, CTCE-0021, rapidly mobilizes polymorphonuclear neutrophils and hematopoietic progenitor cells into peripheral blood and synergizes with granulocyte colony-stimulating factor. Exp Hematol 2005;33:295–307.
74. Suratt BT, Petty JM, Young SK, et al. Role of the CXCR4/SDF-1 chemokine axis in circulating neutrophil homeostasis. Blood 2004;104:565–571.
75. Addison CL, Daniel TO, Burdick MD, et al. The CXC chemokine receptor 2, CXCR2, is the putative receptor for ELR+ CXC chemokine-induced angiogenic activity. J Immunol 2000;165:5269–5277.

6

Chemokine Receptors and Dendritic Cell Trafficking

Hiroyuki Yoneyama, Kenjiro Matsuno, and Kouji Matsushima

Summary

Dendritic cell (DC) networks dictate peripheral tolerance and immunity in lymph nodes (LNs). The type, timing, location, and interaction of LN-recruited DC subtypes are pivotal and regulated by chemokines. We propose a concept that any DC subtype including myeloid and plasmacytoid DCs (mDCs and pDCs) is fundamentally categorized by three stages depending on the function and anatomical position: naïve DC, primed DC, and effector DC. Naïve mDC precursors are recruited to inflamed tissues in response to CCR1 and CCR5 ligands to become primed mDCs, remobilized to draining LNs in response to CCR7 ligands, and activated to become effector mDCs to undergo antigen-presenting function. In contrast, pDC precursors directly migrate to LNs in a CXCR3-dependent manner. LN-recruited, primed pDCs are activated to become effector pDCs that produce large amounts of cytokines and chemokines. Concerted recruitment and adequate network formation of distinct effector DCs are pivotal to determine the type and efficacy of immune response.

Key Words: Chemokine; dendritic cell; migration; inflammation; lymph node.

6.1. The Concept of Migration-Dependent Dendritic Cell Activation

Dendritic cells (DCs) are bone marrow–derived professional antigen-presenting cells (APCs) *(1)*. The function of DCs depends on their maturation stages: progenitors in the bone marrow, precursors in the blood, immature DCs in peripheral tissues, antigen-transporting DCs in the afferent lymphatics, and

From: *The Receptors: The Chemokine Receptors*
Edited by: J. K. Harrison and N. W. Lukacs © Humana Press Inc., Totowa, NJ

mature APCs in lymph nodes (LNs) *(2)*. In a steady-state life cycle, a small number of DCs continually scans self-components at peripheral tissues and maintains peripheral tolerance to self after migrating to LNs *(3)*. When exposed to danger signals, accelerated traffic of DCs occurs, and these newly recruited DCs strongly promote cell-mediated immunity in LNs *(4–7)*. This trafficking pattern is quite distinct from that of T lymphocytes, which continually circulate from blood to LNs via high endothelial venules (HEVs). Naïve lymphocytes are exposed to antigens presented by DCs in the LNs, leading to their proliferation and differentiation into appropriate memory/primed T cells. Primed T cells exit the LNs and recirculate to home to peripheral tissues. It is well established that LNs are *induction sites* whereas peripheral tissues are *effector sites* for T cells. In this review, we propose novel concepts that LNs are effector sites whereas peripheral tissues are induction sites for DCs and that DCs are also classified by naïve, primed, and effector stages depending on the anatomical positioning (Fig.1). Because DCs encounter and ingest self or nonself antigens

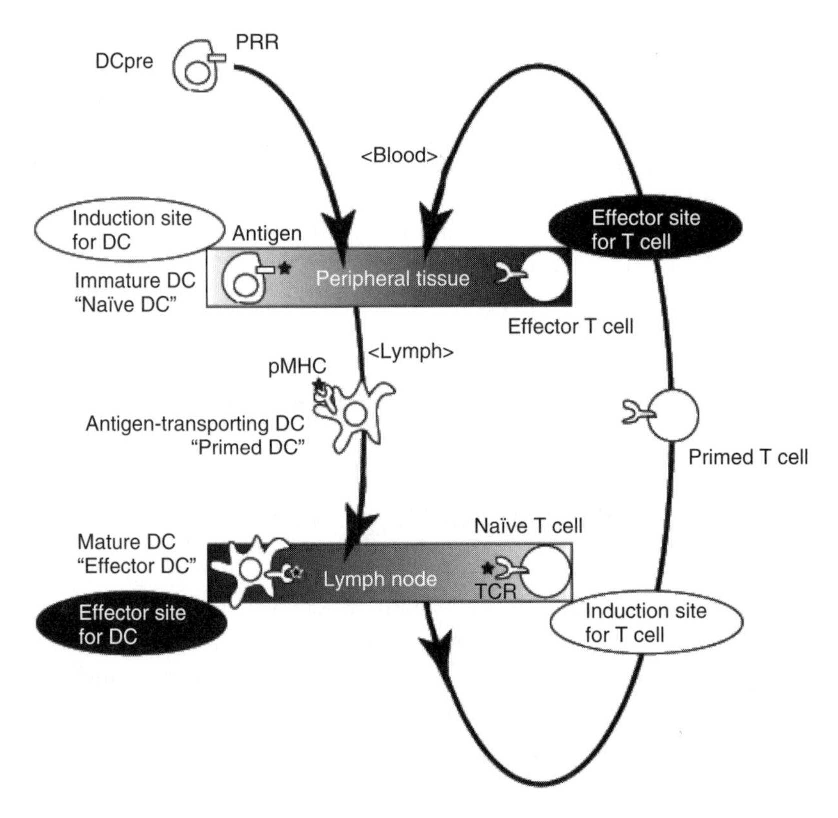

Fig. 1. A model for functional migration of DCs and T cells. DCs are classified into naïve, primed, and effector DCs, DCpre, and DC precursor. pMHC, peptide-major histocompatibility complex.

initially in peripheral tissues, the bone marrow, blood, and tissue DCs before encountering antigens are characterized as *naïve DCs*. Once recognizing and processing peripheral antigens, naïve DCs upregulate cell-surface peptide-MHC complex to become *primed DCs*, which locate within peripheral tissues, afferent lymphatics, and LNs. Finally, primed DCs acquire antigen-presenting ability to naïve T lymphocytes in T-cell zones of LNs. Because the most important effector function for DCs is antigen presentation, we call these fractions *effector DCs*. Effector DCs flexibly determine the type of immune responses in the LNs by summing up the information that they obtain from tissue environments and exogenous factors.

6.2. Two Subsets of Naïve Dendritic Cells with Distinct Migration and Functional Potentials

6.2.1. Dendritic Cell Subtypes

The migration pathway described above has been demonstrated only by DCs of myeloid origin. Recent investigations have highlighted the networks of several DC subtypes in the generation of immune responses *(8)*. In mice, there are at least three major functional subtypes of DCs in LNs, myeloid DCs (mDCs; CD11b$^+$B220$^-$CD11c$^+$), CD8α$^+$ DCs (CD8α$^+$B220$^-$CD11c$^+$), and plasmacytoid DCs (pDCs; B220$^+$CD11c$^+$), which induce distinct types of effector T lymphocytes. mDCs are derived from blood precursors, recruited to inflamed tissues, and induce antigen-specific CD4$^+$ helper T lymphocytes *(6,7,9)*. CD8α$^+$ DCs show poor migration potential, reside in LNs, and are responsible for cross-priming antiviral cytotoxic T lymphocytes (CTLs) *(10)*. pDCs function poorly as APCs for naïve T lymphocytes but produce large amounts of antiviral interferon-α (IFN-α) *(8,11)*. In this review, we will focus on the migration and function of mDCs and pDCs and show the new concepts also applicable to the life cycle of pDCs.

6.2.2. Chemokine Receptors in Naïve Circulating Dendritic Cell Precursors

We and others previously identified blood MHCII$^-$CD11c$^+$ cells as circulating DC precursors in mice *(6,12,13)*. Blood MHCII$^-$CD11c$^+$ cells are classified into only two subtypes: B220$^-$CD11c$^+$ mDC precursors and B220$^+$CD11c$^+$ pDC precursors. The numbers of both mDC and pDC precursors in naïve mouse blood were extremely low but increased greatly in response to danger signals such as bacterial and viral infection *(13,14)* (Fig. 2).

Chemokines and their receptors control the mobilization of naïve DC precursors into the circulation in inflammation. Both DC precursor types express a common set of chemokine receptors such as CCR1, CCR5, and CXCR4,

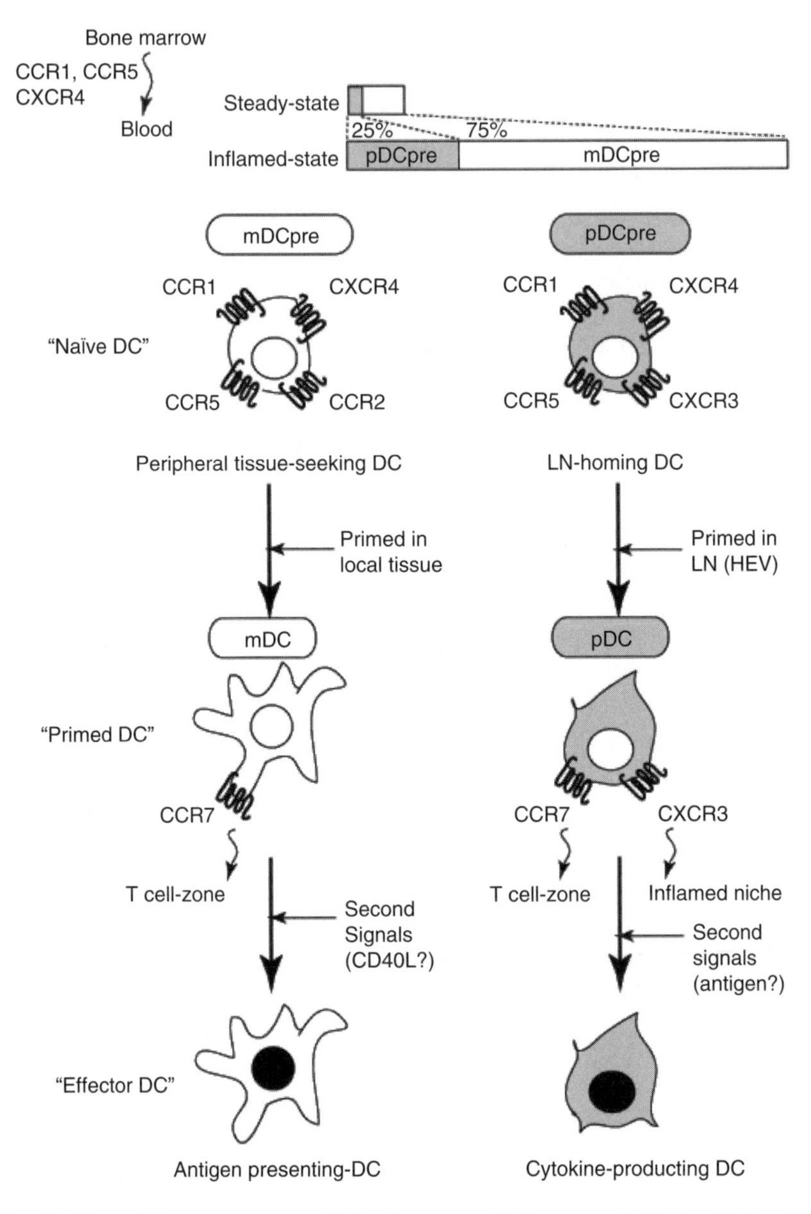

Fig. 2. Chemokine receptors on DC subtypes. mDCpre, myeloid dendritic cell precursor; pDCpre, plasmacytoid dendritic cell precursor.

whereas CXCR3 is selectively expressed by pDC precursors *(13)*. We have established that CCR1/CCR5 and their ligand macrophage inflammatory protein-1 (MIP-1)-α/CCL3 are pivotally involved in the mobilization of both mDC and pDC precursors into the circulation in *Propionibacterium acnes* infection *(14)*. *Propionibacterium acnes*–induced mobilization of both DC precursors is significantly inhibited by anti-CCL3 antibody and by gene-knockout for CCR1/CCR5. Moreover, a single intravenous administration of recombinant CCL3 to naïve mice induces a significant number of blood mDC and pDC precursors at a ratio of ~3 : 1. DC precursors rapidly increase in the circulation, peak at 16 hours after administration of CCL3, and decrease thereafter, suggesting that DC precursors are mobilized from the bone marrow pool in response to blood CCL3. It is interesting that both mDC and pDC precursors show parallel mobilization into the circulation at a ratio of ~3 : 1 (Fig. 1). This may be explained by the fact the expression levels of CCR1 and CCR5 are three times higher on mDC precursors than those on pDC precursors.

However, two naïve DC subtypes show distinct migration patterns after appearing in the blood. mDC precursors preferentially migrate to sites of inflammation in peripheral tissues in response to CCL3 *(13)*. When naïve mDC precursors encounter exogenous antigens via pattern recognition receptors (PRRs) such as Toll-like receptors (TLRs), these cells are activated to become *primed mDCs* and contribute to form local inflammatory niches. We propose naïve mDC precursors as *peripheral tissue–seeking DCs*, whose migration pathway licenses these cells to acquire the antigen-presenting function. In contrast, naïve pDC precursors predominately enter the LNs directly from the circulation in response to CXCL9 and E-selectin *(13)*. As will be discussed later, naïve pDC precursors are primed by activated HEV cells and become primed pDCs. We therefore propose naïve pDC precursors as *LN-homing DCs*, which follow a lymphocyte-like migration pathway (i.e., to enter the LNs directly without scanning exogenous antigens of peripheral inflamed tissues).

6.2.3. Chemokine Receptors in Primed Dendritic Cells

Interestingly, both naïve mDCs and pDCs downregulate CCR1, CCR5, and CXCR4 but upregulate CCR7 after priming *(13)*. For primed mDCs, CCR7 plays a pivotal role in the migration from peripheral tissues into the T-cell zones of draining LNs. Primed mDCs themselves produce CCL19/Epstein-Barr virus-inducible gene 1 ligand chemokine (ELC) and CCL21/secondary lymphoid organ chemokine (SLC), which attract naïve $CCR7^+CD4^+$ T cells *(7)*. Primed mDCs may require further undefined signals to become effector mDCs and present antigens to T cells.

For primed pDCs, CCR7 contributes to the retention of these cells within the T-cell zones of LNs. In inflamed LNs, many clusters between effector mDCs and T cells are formed in these zones *(7)*. CXCR3 plays a vital role to retain primed pDCs within these clusters, as cluster-forming cells produce high amounts of CXCL10/interferon inducible protein-10 (IP-10), a CXCR3 ligand. Within the clusters, primed pDCs become *effector DCs* probably by signals such as viral antigen–derived factors. Although primed pDCs function poorly as APCs in vivo, these cells produce large amounts of cytokines such as IFN-α and IL-12 and chemokines such as CCL21 and CXCL10 at a single cell level *(15,16)*. By this, effector pDCs may attract activated T cells and create local cytokine fields within the clusters. We therefore propose effector pDCs as *cytokine-producing DCs* in inflamed niches.

6.3. The Impact of Migration of Dendritic Cells on Their Effector Functions

6.3.1. Steady-State Migration

Recent findings of DC-mediated peripheral tolerance have changed our view of DC trafficking *(3)*. The significance of steady-state migration of mDCs is the establishment of peripheral tolerance to self-components before exposure to danger signals. In a physiologic condition, a small percentage of naïve mDC and pDC precursors exit in the circulation. It is speculated that these cells migrate into the peripheral and lymphoid organs to become tissue-resident immature DCs (naïve DCs). Because ~10% of tissue-resident DCs renew every day, a constant supply of naïve DCs from the circulation and a constant exit of self-antigen–primed DCs to afferent lymphatics is essential to maintain tissue DC networks *(2)*. The number of tissue-resident DCs is strictly preserved during steady-state migration, suggesting that certain molecules other than simple anatomical factors like lymph flow control the migration properties. Although chemokines like CCL21/CCR7, integrins, cell cycle regulators, and lipid mediators have been considered to regulate the steady-state migration, it still remains controversial *(17)*. It is increasingly important to identify some molecules regulating the traffic in view of peripheral tolerance. Concerning naïve pDC precursors, it is difficult to detect the steady-state migration due to the extremely low frequency of the cells. At the effector sites, (i.e., LNs), the pair of self-antigen–primed mDCs and resident pDCs (naïve or primed) induces peripheral tolerance (Fig. 3A).

6.3.2. Inflammation-Dependent Migration

In response to danger signals, a significant number of both mDC and pDC precursors appear de novo into the circulation *(5,13)*. However, their migration pathways are flexible and are largely dependent on tissue microenvironments. In a cutaneous herpes simplex virus (HSV)-1 infection model, newly appeared

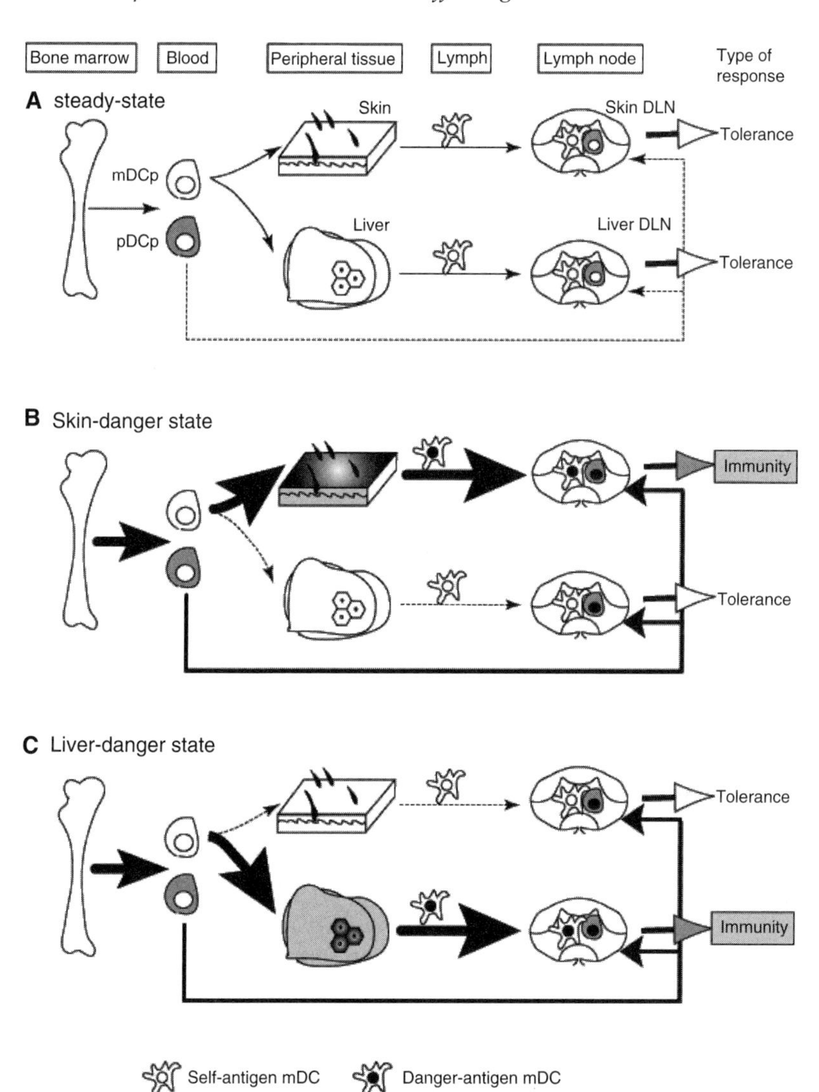

Fig. 3. DC network dictates peripheral tolerance and immunity. The type and activation of DCs recruited to effector lymph nodes determine the efficacy of immune responses. (a) Steady-state migration. (b) Skin infection model. (c) Liver inflammation model.

mDC precursors preferentially migrate to inflamed dermis through CCL3 and CCR1/CCR5 interactions (Fig. 3B). After receiving HSV-derived antigens, primed mDCs are remobilized to the draining LNs (DLNs) through CCL21 and CCR7 interactions. At the same time, newly appeared pDC precursors directly enter the skin DLNs as well as other LNs, for example, liver DLNs, both via

CXCL9 and CXCR3 interactions (Fig. 3B). In a *P. acnes*–induced granulomatous liver disease model, mDC precursors rapidly migrate to the hepatic sinusoids using the same chemokine/chemokine receptors described above and subsequently move to the liver DLNs (Fig. 3C). pDC precursors also migrate to liver DLNs as well as other LNs such as skin DLNs (Fig. 3C). At the effector sites, the pair of exogenous antigen–primed mDCs and HEV-primed pDCs induces immunity in both disease models (Fig. 3B and C). In contrast, the pair of self-antigen–primed mDCs and HEV-primed pDCs does not induce immunity (Fig. 3B and C). Thus, HEV-primed pDCs themselves may fail to induce strong immunity in the absence of exogenous antigen-presenting effector mDCs *(16)*.

6.4. Regulation by Chemokines of Dendritic Cell Migration–Dependent Immunity

To better understand the contribution of DC-related chemokines in the generation of immune responses, we investigated the effects of blocking antibodies against various chemokines on the traffic and effector functions of DCs. In the liver disease model, adoptive transfer experiments revealed that CCL21 blockade significantly reduced the number of liver DLN–migrated mDCs *(6)* (Fig. 4A). Functionally, effective T-cell responses in the DLNs did not occur by blocking CCL21. Instead, a large number of exogenous antigen-primed mDCs resided in the local inflamed niches leading to the chronic disease status (Fig. 4A). Therefore, adequate migration of primed mDCs to the DLNs is required not only for the establishment of immunity but also for the creation of protective inflammatory niches *(4)*. This also indicates that primed DCs are flexible in time and in distribution and thereby capable of affecting disease outcomes.

The role of pDC migration was investigated using the cutaneous HSV-1 infection model *(16)*. Adoptive transfer experiments revealed that blocking antibody against CXCL9, but not CCL21, dramatically reduced the number of skin DLN– as well as liver DLN–recruited pDCs (Fig. 4B). Surprisingly, at the effector sites for DCs (skin DLNs), T cells failed to mount anti-HSV immunity in spite of the presence of HSV antigen–primed mDCs. Ex vivo APC assays revealed a surprising finding that the primed mDCs failed to differentiate into effector mDCs possibly due to the virally infected microenvironment, and thus failed to undergo their effector antigen-presenting function. In addition, HSV was not fully eliminated, and chronic inflammation persisted at inflamed dermis. This new finding provides some insights into DC biology. First, antigen-primed mDCs further need to receive certain *ON signals* to become effector DCs, which effectively induce T-cell immunity. Second, although effector pDCs

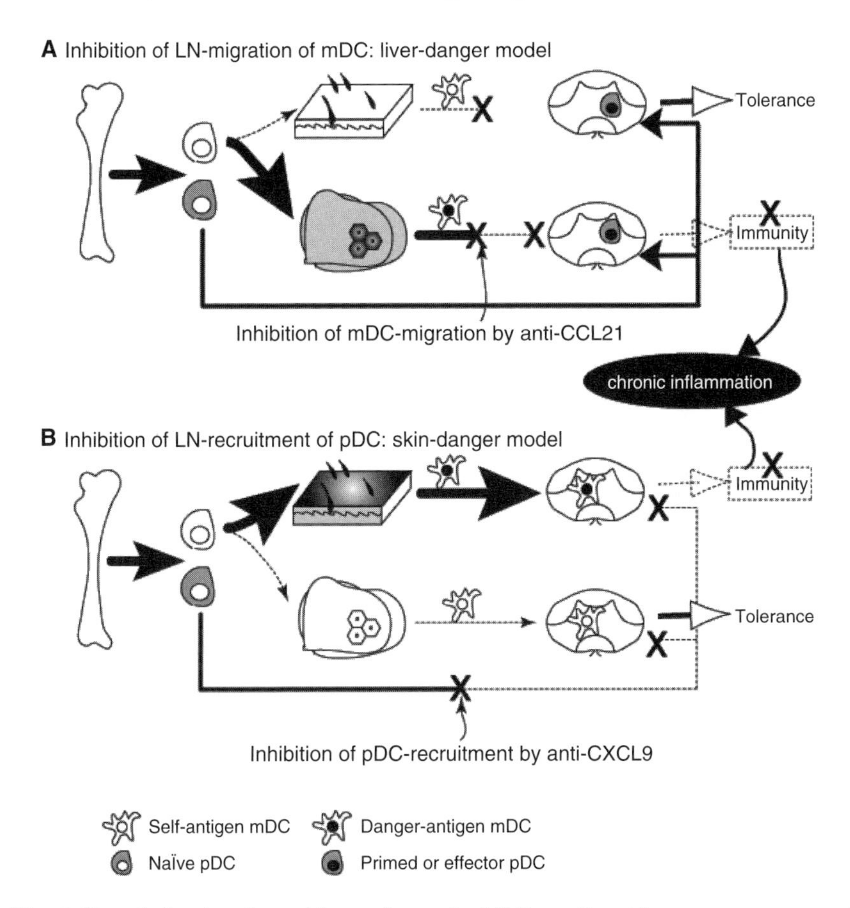

Fig. 4. Regulation by chemokines of recruited DC-mediated immune responses. (a) The effect of CCL21 on mDC migration in liver disease model. (b) The effect of CXCL9 on pDC migration in skin infection model.

themselves function poorly as APCs, these cells provide a critical cue to primed mDCs to differentiate into effector mDCs. Third, concerted recruitment of both mDCs and pDCs is needed to mount effective immunity and is strictly regulated by step-by-step navigation of appropriate chemokines/chemokine receptors.

6.5. Induction of Primed pDCs by Trans-HEV Migration

In contrast with mDCs, which are clearly primed by antigens at local inflamed tissues (Figs. 1 and 2), the site of pDC priming is not fully established. We consider that an activated HEV cell is a good candidate to create primed pDCs *(13)*. Because naïve pDC precursors preferentially bind to activated, but not

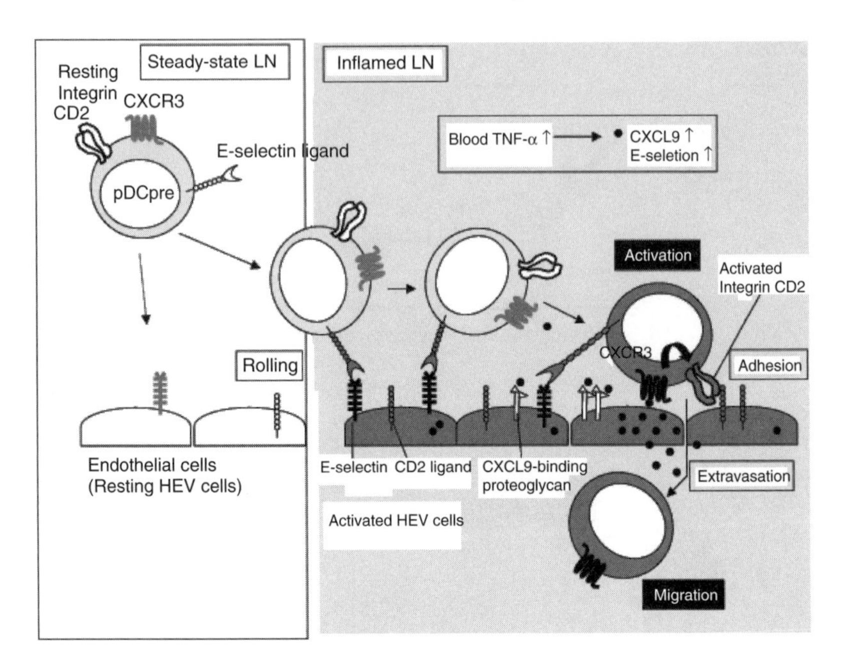

Fig. 5. A model for trans-HEV migration of blood pDC precursors. pDCpre, plasmacytoid dendritic cell precursor.

resting, HEVs in a CXCL9-dependent manner, naïve pDCs receive some signals through CXCR3 from activated HEVs (Fig. 5). Actually in mice, pDCs remodel their surface CD2 molecules during trans-HEV migration also in a CXCR3-dependent manner. In addition, coculture experiments between naïve pDCs and HEV cells revealed that naïve pDCs upregulate CCR7 after interaction with TNF-α–activated HEV cells (unpublished observation). When primed pDCs are exposed to HSV antigens, these cells rapidly produce large amounts of IFN-α and IL-12 in vitro. Therefore, danger signal–mediated trans-HEV migration primes pDCs and tightly links sequential functions of effector pDCs.

6.6. Concerted Recruitment of mDCs and pDCs: The Role of Chemokines

DC traffic in response to danger signals is dynamic, controlled systemically by cytokines and locally by chemokines *(5)*. The migration pathway is summarized as follows (Fig. 6; steps 1 to 8 in the figure correspond with i to viii in the following text). (i) Danger signals induce local production of TNF-α. (ii) TNF-α released into the circulation promotes systemic inflammation via activation of macrophages and endothelial cells. (iii) Serum TNF-α upregulates

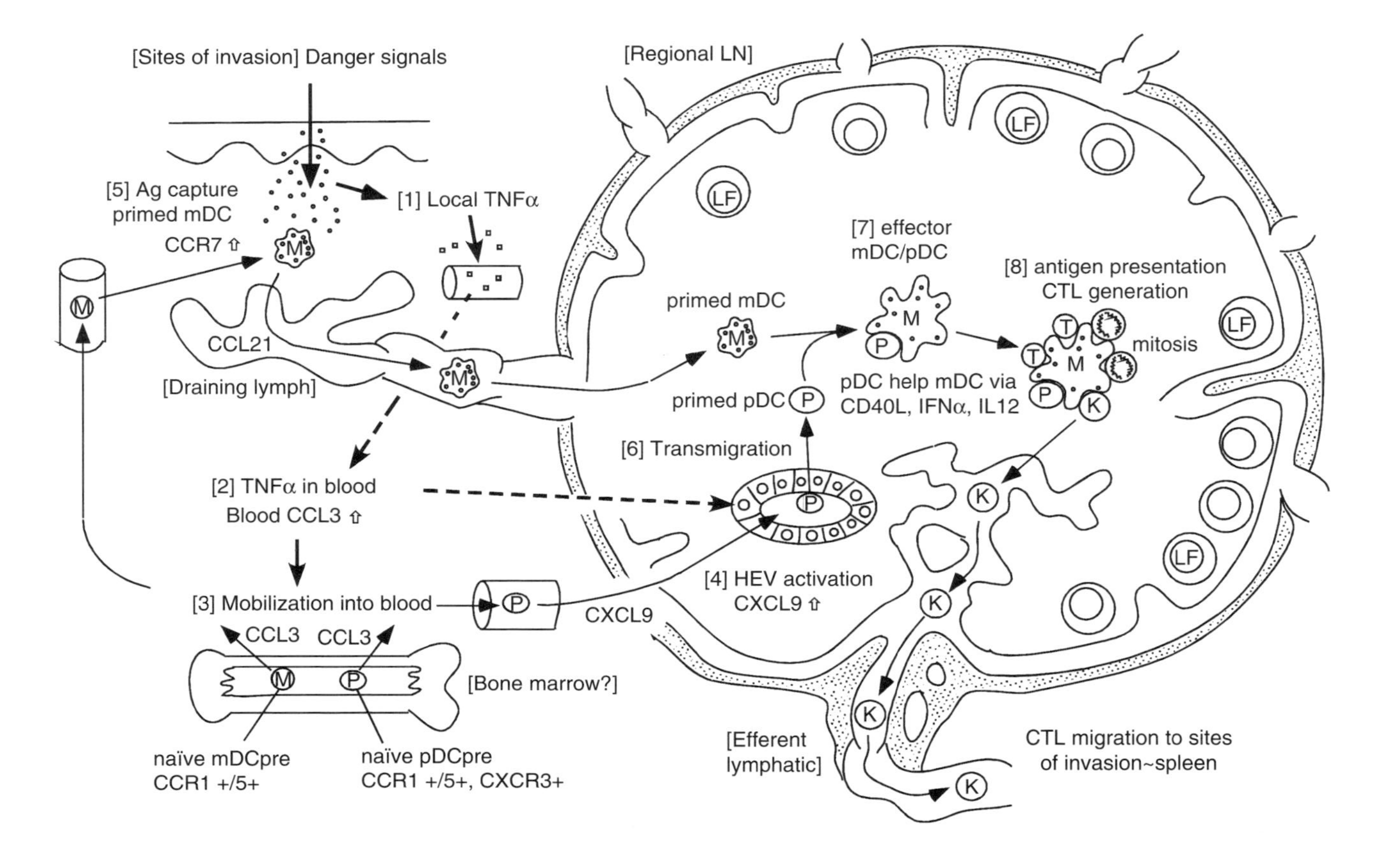

Fig. 6. A summary illustration of concerted recruitment of mDCs and pDCs. K, CTL; LF, lymph follicle; mDCpre, myeloid dendritic cell precursor; pDCpre, plasmacytoid dendritic cell precursor.

CCL3 and induces mobilization of both mDC and pDC precursors (both are positive for CCR1 and CCR5) into the circulation, possibly from bone marrow. (iv) Serum TNF-α also upregulates CXCL9 in systemic LN HEVs. (v) Naïve mDC precursors are recruited to sites of invasion, capture antigens to become primed mDCs, and consequently migrate into the draining LNs through afferent lymphatics. Local TNF-α accelerates influx of naïve mDC precursors via CCL3 as well as efflux of primed mDCs via CCL21. (vi) In contrast, naïve pDC precursors transmigrate across TNF-α stimulated HEVs in a CXCL9-dependent manner to become primed pDCs. Primed mDCs in the LNs attract and form clusters with primed pDCs. (vii) Primed pDCs further differentiate into effector pDCs possibly through interaction with antigen-derived factors. Therefore, naïve mDC precursors are *peripheral tissue–seeking DCs* and (viii) finally act as effector APCs to prime tissue-derived antigen-specific T cells. In contrast, naïve pDC precursors are *LN-homing DCs* and finally become effector pDCs that produce IFN-α, a direct antiviral cytokine that also plays an indirect role in cross-priming CTLs. Effector pDCs provide help signals like CD40 ligand to primed mDCs to generate CTLs *(16)* as well as interact with B cells to induce antibody-forming cells (unpublished observation).

Chemokine-mediated DC migration determines disease outcome. In the future, steady-state migration and more complicated migration within effector sites (LNs) will be clarified also by chemokine research. Such studies would provide novel therapeutic strategies against various immune diseases.

References

1. Banchereau J, Briere F, Caux C, et al. Immunobiology of dendritic cells. Annu Rev Immunol 2000;18:767–811.
2. Matsuno K, Ezaki T. Dendritic cell dynamics in the liver and hepatic lymph. Int Rev Cytol 2000;197:83–136.
3. Steinman RM, Nussenzweig MC. Avoiding horror autotoxicus: the importance of dendritic cells in peripheral T cell tolerance. Proc Natl Acad Sci U S A 2002;99(1):351–358.
4. Yoneyama H, Ichida T. Recruitment of dendritic cells to pathological niches in inflamed liver. Med Mol Morphol 2005;38(3):136–141.
5. Yoneyama H, Matsuno K, Matsushimaa K. Migration of dendritic cells. Int J Hematol 2005;81(3):204–207.
6. Yoneyama H, Matsuno K, Zhang Y, et al. Regulation by chemokines of circulating dendritic cell precursors, and the formation of portal tract-associated lymphoid tissue, in a granulomatous liver disease. J Exp Med 2001;193(1):35–49.
7. Yoneyama H, Narumi S, Zhang Y, et al. Pivotal role of dendritic cell-derived CXCL10 in the retention of T helper cell 1 lymphocytes in secondary lymph nodes. J Exp Med 2002;195(10):1257–1266.

8. Shortman K, Liu YJ. Mouse and human dendritic cell subtypes. Nat Rev Immunol 2002;2(3):151–161.
9. Zhao X, Deak E, Soderberg K, et al. Vaginal submucosal dendritic cells, but not Langerhans cells, induce protective Th1 responses to herpes simplex virus-2. J Exp Med 2003;197(2):153–162.
10. Allan RS, Smith CM, Belz GT, et al. Epidermal viral immunity induced by CD8alpha+ dendritic cells but not by Langerhans cells. Science 2003;301(5641):1925–1928.
11. Cella M, Jarrossay D, Facchetti F, et al. Plasmacytoid monocytes migrate to inflamed lymph nodes and produce large amounts of type I interferon. Nat Med 1999;5(8):919–923.
12. O'Keeffe M, Hochrein H, Vremec D, et al. Dendritic cell precursor populations of mouse blood: identification of the murine homologues of human blood plasmacytoid pre-DC2 and CD11c+ DC1 precursors. Blood 2003;101(4):1453–1459.
13. Yoneyama H, Matsuno K, Zhang Y, et al. Evidence for recruitment of plasmacytoid dendritic cell precursors to inflamed lymph nodes through high endothelial venules. Int Immunol 2004;16(7):915–928.
14. Zhang Y, Yoneyama H, Wang Y, et al. Mobilization of dendritic cell precursors into the circulation by administration of MIP-1alpha in mice. J Natl Cancer Inst 2004;96(3):201–209.
15. Dalod M, Hamilton T, Salomon R, et al. Dendritic cell responses to early murine cytomegalovirus infection: subset functional specialization and differential regulation by interferon alpha/beta. J Exp Med 2003;197(7):885–898.
16. Yoneyama H, Matsuno K, Toda E, et al. Plasmacytoid DCs help lymph node DCs to induce anti-HSV CTLs. J Exp Med 2005;202(3):425–435.
17. Randolph GJ. Dendritic cell migration to lymph nodes: cytokines, chemokines, and lipid mediators. Semin Immunol 2001;13(5):267–274.

7

Chemokine Receptors and Lymphocyte Trafficking

Michael N. Hedrick and Joshua M. Farber

Summary

The ordered movement of lymphocytes through and positioning within lymphoid organs and peripheral sites is controlled by adhesion molecules together with chemokines and their receptors. Chemokine-mediated lymphocyte migration is critical for establishing the architecture of lymphoid organs and many aspects of lymphocyte function, including lymphocyte development, activation, and effector activity. Of the 19 chemokine receptors described in humans, all have been reported to be expressed on lymphocytes, and the expression pattern of chemokine receptors can itself be used to define and characterize lymphocyte subsets. For example, by using chemokine receptors, memory T cells can be split into distinct populations, such as central and effector memory T cells; and T helper 1 (Th1) and T helper 2 (Th2) cells exhibit distinguishable patterns of chemokine receptor expression, which can be used to study Th1/Th2 differentiation. Beyond their physiologic roles, chemokine receptors on lymphocytes are exploited by pathogens, such as in the use of CCR5 and CXCR4 by HIV-1 as coreceptors for viral entry. In this chapter, we will focus on the roles of chemokine receptors in lymphocyte trafficking in lymphoid organs and peripheral tissues and how understanding the chemokine system has shed light on larger issues in lymphocyte biology. We will discuss the roles of chemokines and chemokine receptors during the life cycles of lymphocytes—from early development through the acquisition of memory and effector functions.

Key Words: Chemokines; lymphocyte development; T-cell differentiation; cell trafficking; memory; thymus; bone marrow.

From: *The Receptors: The Chemokine Receptors*
Edited by: J. K. Harrison and N. W. Lukacs © Humana Press Inc., Totowa, NJ

7.1. Chemokines and Primary Lymphoid Organs

7.1.1. *Thymus and Developing T Cells*

T cells are derived from the common lymphoid progenitors that arise in the bone marrow and migrate to the thymus. Thymocyte precursors are thought to enter at the corticomedullary junction and travel to the subcapsular region before migrating back to the medulla and exiting the thymus. The earliest T cells identified in the thymus are so-called double-negative (DN) thymocytes, as they express neither CD4 nor CD8, and the DN cells can be divided into the progressively more differentiated subsets DN 1, 2, 3, and 4. DN thymocytes undergo rearrangements of the alpha and beta chains of the T-cell receptor (TCR), and during this stage are strongly attracted to CXCL12 stromal cell-derived factor-1 (SDF-1) due to high-level expression of CXCR4, the specific receptor for CXCL12 *(1–4)*. CXCR4 may aid in the entry of the precursors into the thymus, because CXCR4 is expressed by progenitors in the bone marrow *(1)*, and may also serve to arrest the cells in the subcapsular region during rearrangement of the TCR. As the thymocytes develop into double-positive (DP; $CD4^{+}CD8^{+}$) and single-positive (SP) cells, they progressively lose responsiveness to CXCL12 *(5,6)*. Because mice deficient in CXCR4 seem to have normal thymocyte development, with only a marginal defect in thymocyte progenitor entry into the thymus, CXCR4 cannot be the sole determinant of progenitor cell entry or critical steps in thymocyte migration *(7–9)*.

There is reasonably good evidence that the receptor CCR9 regulates thymocyte trafficking *(10–13)*. CCR9, the specific receptor for CCL25 thymus-expressed chemokine (TECK), has been reported on a subset of bone marrow cells that display a phenotype similar to the progenitor cells that populate the thymus *(10,12)*. CCL25 is expressed by thymic endothelial cells in both the cortex and medullary regions of the thymus *(14)*. Although CCR9 is not found on thymocytes until the DN3 stage, competitive adoptive transfer of bone marrow has revealed a role for CCR9 in supporting T-cell development at the earliest identifiable (DN1) stage as well as independent roles at subsequent developmental stages *(10,15)*. Forced, early expression of CCR9 resulted in a significant but partial block at the DN stage *(16)*.

DP thymocytes express a functional TCR and migrate to the cortex of the thymus to undergo positive selection. This migration may be mediated by the cells' increasing responsiveness to CCL25 *(11)*. During the process of positive selection, the engagement of the TCR enhances the ability of the DP thymocytes to migrate to CCL25, but at the same time this engagement induces the downregulation of CCR9 expression *(11)*. These data suggest that CCR9 may be used for retention during the DP stage and downregulated quickly for passage into the next developmental stage, at least for CD4 SP cells *(11)*. For CD8 SP cells, some authors have reported persistent expression of CCR9, which is maintained on naïve CD8$^+$ T cells in the periphery *(15)*. In any case, downregulation of CCR9 is not required for T cells to exit the thymus *(16)*. Together, the data are consistent with a role for CCR9 in the homing of T-cell progenitors to the thymus as well as for subsequent developmental steps.

The final stage of T-lymphocyte development occurs in the medulla of the thymus, which is the progression into the SP stage. CCR7 has been implicated in the exit of T cells from the thymus, and in the final steps of maturation, SP T cells gain responsiveness to CCL19 secondary lymphoid-tissue chemokine (SLC), one of the two ligands for CCR7 [the other being CCL21 Epstein-Barr virus-induced molecule-1 ligand chemokine (ELC)] *(11,17–19)*. CCL19 is highly expressed on vessels/cells at the corticomedullary junction, the site of thymocyte egress into the periphery *(17)*. Neonatal CCR7$^{-/-}$ mice show a marked impairment of thymocyte egress *(17)*, consistent with a role for CCL19 and CCR7 in this process.

However, another ligand-receptor pair, sphingosine-1-phosphate and sphingosine-1-phosphate receptor 1 (S1P$_1$), has been suggested to play a vital role in exiting the thymus *(20,21)*. S1P$_1$, formerly known as Edg1, is a member of a five-receptor family, all of which are responsive to sphingosine-1-phosphate *(20,21)*. Like chemokine receptors, S1P$_1$ is a G protein–coupled receptor, and S1P$_1$ signaling is inhibited by pertussis toxin *(21,22)*. Both mature T cells and SP thymocytes respond to sphingosine-1-phosphate and express S1P$_1$ *(21)*, and S1P$_1$ is upregulated just before T cells exit the thymus *(20,21)*. S1P$_1^{-/-}$ mice lack mature T lymphocytes in the periphery, and the T cells remain sequestered in the thymus *(20)*. Deficiency of S1P$_1$ in progenitor cells does not inhibit the B cells from exiting the bone marrow, suggesting a specific effect on thymocyte egress *(21,23)*.

7.1.2. Bone Marrow and Developing B Cells

In the fetal liver and bone marrow, B cells develop through three main stages as pro–B cells, pre–B cells, and immature B cells, with final steps of maturation in the periphery. During the pro–B cell stage, the cells receive signals from the stromal cells of the bone marrow to rearrange the μ-chain of the B-cell receptor and begin to express B220 and MHC II. Once the cells express a functional μ-chain on the surface with a surrogate light chain, they have entered the pre–B cell stage. Finally, full recombination of both chains of the B-cell receptor signals that the cell is now an immature B cell and is ready for entry into the periphery. One chemokine-receptor pair with the best-described role in B-cell development is CXCL12-CXCR4. CXCL12 was originally described as a protein supporting B-cell development, produced by bone marrow stromal cells *(24–26)*. Although mice deficient in either CXCL12 or CXCR4 die early in gestation, they display a unique phenotype with respect to B-cell development *(8,27–29)*. Deficiency in CXCR4 severely compromises lymphopoiesis and myelopoiesis in the fetal liver and results in very low numbers of B-cell progenitors in the bone marrow *(8)*. Fetal liver cells from CXCR4-deficient mice are unable to reconstitute B-cell production in the bone marrow of irradiated mice *(7)*, suggesting that the CXCL12-CXCR4 pair regulates the maturation of B cells by controlling the localization of B-cell progenitors.

7.2. Chemokines and Secondary Lymphoid Organs

7.2.1. CCR7 and Secondary Lymphoid Organs

After development in the primary lymphoid organs and egress of mature lymphocytes into the periphery, lymphocytes recirculate through and survey secondary lymphoid organs for their cognate antigens *(30)*. Naïve lymphocytes express CCR7, as well as L-selectin (CD62L) and the integrin lymphocyte function-associated antigen-1 (LFA-1) (CD11a/CD18), and respond to the chemokines CCL19 and CCL21 *(31–34)*. The interaction of CCR7 and CCL21 induces integrin activation and adhesion of cells to lymph-node high endothelial venules (HEV) *(35)*. Mice deficient in CCR7 have impaired entry of both $CD4^+$ and $CD8^+$ T cells into lymph nodes and an overall increase in the circulating population of $CD4^+$ T cells *(36)*. Moreover, without CCR7, T and B cells lose their ordered distribution within secondary lymphoid organs *(36)*. In the spleen, in which lymphocyte entry is not through HEV, T cells from CCR7 knockout mice did not localize appropriately to the periarteriolar lymphatic sheath (PALS), and $CCR7^{-/-}$ B cells left the PALS for B-cell follicles prematurely compared with wild-type cells *(36)*. This dysregulation of lymphocyte compartmentalization presumably contributes to the delay in antibody responses found in the $CCR7^{-/-}$ mice *(36)*.

7.2.2. CXCR5 and Secondary Lymphoid Organs

Another key chemokine-receptor pair involved in the localization of B and T lymphocytes within the secondary lymphoid organs is CXCL13 (BLC) and its specific receptor, CXCR5 *(37–39)*. Pro–B cells show limited responsiveness to CXCL13, and this responsiveness is lost during the progression to the pre–B cell stage *(40)*. Strong responses to CXCL13 are regained once the cells have become immature B cells in the periphery *(40)*. Mice deficient in CXCR5 display malformed germinal centers in the spleen, which may account for these mice producing low-affinity IgM and showing compromised isotype switching *(41)*. CXCR5-deficient mice also fail to form lymph nodes and have severely impaired Peyer's patch formation *(41)*, and CXCL13 has been shown to participate in a positive feedback loop for the formation of lymphoid follicles *(42)*. CXCR5 is also expressed on a subset of memory $CD4^+$ T cells that have been suggested to provide B-cell help *(37,43,44)*. $CD4^+$ $CXCR5^+$ T cells are highly enriched in reactive lymphoid organs such as tonsil *(43)* compared with peripheral blood and home to the B-cell zones, colocalizing with CXCL13 expression *(37,39,43)*. T cells from CXCR5 knockout mice fail to enter the B-cell zone with the same efficiency as those from wild-type mice *(45)*, and $CD4^+$ $CXCR5^+$ T cells have been shown to be strong inducers of isotype switching *(44)*. Together, the data demonstrate that CXCR5 and CXCL13 are compartment-specific factors that are critical for the recruitment of B and T cells in the formation and function of B-cell follicles.

7.2.3. Sphingosine-1-phosphate Receptor 1 and Egress from Lymph Nodes

The receptor $S1P_1$, which is important for mature thymocyte egress, also plays a role in migration through lymph nodes. Exiting lymph nodes, like emigration from the thymus, requires $S1P_1$. Mice deficient in $S1P_1$ retain lymphocytes in the secondary lymphoid organs *(21)*. $S1P_1$ is also required for proper B-cell migration in the secondary lymphoid organs *(23)*. Treatment with FTY720, a pharmacologic inhibitor of sphingosine-1-phosphate receptor signaling, or absence of $S1P_1$ in mice leads to failure of B-cell migration to the marginal zone *(23)*.

7.3. Chemokines and Lymphocytes in the Periphery

7.3.1. Central/Effector Memory T Cells

For T cells, activation by antigen leads to the production of effector/memory cells, which show profound differences in patterns of migration compared with naïve cells, related to changes in expression of adhesion molecules and chemokine receptors. As described above, T cells require CCR7 to enter lymph nodes and position themselves appropriately within T-cell zones. As expected, therefore, CCR7 is found on all naïve T cells. However, expression of CCR7 is lost on a subpopulation of memory cells, and CCR7 expression has been used to propose a division of memory T cells into *central* (T_{CM}) and *effector* (T_{EM}) *(46–49)*. Human T_{CM} are $CD45RO^+$ cells that would be presumed to be able to enter non-activated lymph nodes on account of the cells' expression of CCR7 and CD62L (46,47). These cells were characterized as containing telomeres with lengths intermediate between those of naïve and T_{EM}, able to produce IL-2 but not effector cytokines, and able to give rise to T_{EM}-like cells after activation in vitro *(46,47)*. By contrast, T_{EM} were described as $CD45RO^+$ cells lacking CCR7, which would prevent them from entering lymph nodes and consign them to migrating into peripheral sites. T_{EM} were characterized as producing effector cytokines and, in the case of $CD8^+$ cells, high levels of perforin *(46,50)*.

The validity of this division within memory populations has been supported by descriptions of two populations of $CD4^+$ memory T cells in mice, one found in lymph nodes and producing high levels of IL-2 in response to antigen activation, and the other found in peripheral tissue and able to produce high levels of effector cytokines *(51)*. Additional work has shown decreasing numbers of T-cell receptor gene rearrangement excision circles and decreasing levels of T-cell receptor–induced calcium signals from naïve to T_{CM} to T_{EM} $CD4^+$ T cells, supporting the proposed pathway of differentiation *(52)*. The interesting, but thus far unproved proposition has been that the T_{CM} function as long-lived precursors for the generation of additional effector cells and T_{EM}, and that such a role is linked to their ability to traffic through lymph nodes. There is support for a model in which undifferentiated cells serve as repositories of long-term memory in a report that IFN-γ–negative but not IFN-γ–producing $CD4^+$ T cells isolated from cultures activated under T helper 1 (Th1) conditions in vitro were able to survive long-term after adoptive transfer into mice and give rise to Th1 cells upon repeat challenge *(53)*. Other work, however, has also described long-lived and presumably highly differentiated cells in peripheral tissues *(51)*. It is worth noting that although CCR7 and CD62L are necessary for efficient trafficking into noninflamed lymph nodes, the HEV of activated lymph nodes acquires characteristics of inflamed endothelium and supports recruitment of $CD62L^-$ memory cells *(54)*. Therefore, any proposal for a role for lymph node trafficking in the function of T_{CM} must account for the fact that the special access of T_{CM} to lymph nodes pertains only during homeostatic but not inflammatory conditions.

The work described above notwithstanding, there have been a number of reports questioning the observations underlying the T_{CM}/T_{EM} paradigm. CCR7 and effector cytokines can be coexpressed by mouse $CD4^+$ and $CD8^+$ T cells with no significant difference between cytokine-expressing and non-expressing cells in their migration to CCL21 *(49,50,55,56)*. $CCR7^+$ and $CCR7^-$ $CD8^+$ T cells were equally able to kill target cells, suggesting that effector memory cells were found within both subsets *(49)*. Despite the apparent discrepancies between the model as initially proposed and a number of subsequent reports, an inverse correlation between levels of CCR7 expression and production of effector cytokines has been generally observed *(49,50)*, along with enrichment for the $CCR7^-$ phenotype among antigen-specific cells infiltrating tissue sites (56,57). Discordant findings among groups may be due, at least in part, to the arbitrariness inherent in defining subsets as "positive" or "negative" based on the use of different antibodies and ligand-based reagents, in some cases across species.

7.3.2. Early Memory Cells That Display an Effector Phenotype

We have recently presented data demonstrating that $CD4^+$ T cells do not need to go through a nonpolarized $CD45RO^+$ stage before being able to produce effector cytokines *(52)*. We described cells from human adult peripheral blood that displayed the surface phenotype characteristic of naïve cells ($CD4RO^-CD45RA^+CD62L^+CD11a^{dim}CD27^+$) except for their expression of either CXCR3 or CCR4. These subsets had levels of T-cell receptor gene rearrangement excision circles and of T-cell receptor signaling intermediate between the $CXCR3^-$ /$CCR5^-$ and T_{CM} ($CD45RO^+CD62L^+$) cells, yet contained cells able to produce IFN-γ or IL-4, respectively, analogous to the standard memory populations expressing these receptors *(52)*. These are *effector memory* cells with a very *central* phenotype, which we believe represent the earliest identifiable memory cells. The existence of these cells indicates that the T_{CM}/T_{EM} paradigm, although elegant and informative, is an oversimplification and, importantly, that Th1/Th2 polarization in vivo does not require a sequence of preceding stages of T-cell differentiation, such as a nonpolarized, $CD45RO^+$ T_{CM} stage. Instead, Th1/Th2 polarization can occur very early after T-cell activation under appropriate conditions, and the association of the T_{EM} phenotype with high frequencies of effector $cytokine^+$ cells is likely the consequence of an increasing probability of encountering polarizing conditions as cells undergo greater numbers of divisions/activation events *(52)*.

7.3.3. Th1/Th2 Cells

A major challenge in understanding the roles of chemokine receptors on lymphocytes is to discover the organizing principles that determine patterns of chemokine receptor expression. The T_{CM}/T_{EM} paradigm provides a basis for understanding the pattern of CCR7 expression. Although other chemokine receptors show non-uniform expression when analyzed on T_{CM} versus T_{EM} subsets, this paradigm cannot provide a framework that is sufficiently informative for explaining the complexity of chemokine receptor expression on T cells. With regard to $CD4^+$ T cells, the Th1/Th2 paradigm is also clearly relevant. A variety of chemokine receptors have been reported to be expressed differentially on Th1 versus Th2 cells *(48,58–64)*. Although some reports suggest preferential expression of CCR7 on Th1 cells (65), most investigators have reported no specific association between CCR7 and Th phenotype *(31,34)*.

Several groups have reported the preferential expression of CXCR3 and CCR5, as well as CCR2, on Th1 cells, with the strongest and most consistent relationship shown for CXCR3 *(34,48,52,64,66–73)*. The association of CXCR3 with the Th1 phenotype is consistent with the strong induction of CXCR3 ligands, CXCL9–11, by IFN-γ. The expression of CXCR3 on Th1 T cells is dependent on the presence of IFN-γ. Activated T cells from mice deficient in IFN-γ are unable to express CXCR3 and fail to migrate to CXCL9–11 *(73)*. There are data indicating that CCR5 and CCR2 expression is not independently associated with the Th1 phenotype, but that the association is due to coexpression of CXCR3 on a high proportion of $CCR5^+$ and $CCR2^+$ cells *(71)*. However, CCR5 has been reported to be induced by IL-12 on activated T cells *(74,75)*, and to be a target gene of T-bet *(76)*, the driving cytokine and transcription factor, respectively, for Th1 differentiation. Whatever the strength of the association between Th1 polarization and CCR5, CCR5-null (CCR5δ32) humans are able to mount a productive Th1 phenotype immune response, with normal levels of IFN-γ produced from T cells *(77)*. Consistent with the observations above, both CXCR3 and CCR5 have been found to be highly expressed in inflammatory diseases and tissues associated with type 1 responses *(78)*.

CCR4 and CCR8 have been described as receptors preferentially expressed on Th2 cells *(58,69,72,79,80)*. Just as for the Th1-associated chemokine receptors, despite the strong preferential association of CCR4 and CCR8 with Th2 cells, there is no evidence that these Th2-associated receptors are necessary for Th2 cell polarization *(60,81)*.

7.3.4. Regulatory T Cells

There has been a great deal of recent interest in a subset of $CD4^+$ regulatory T cells (Tregs) that depend for their production on the transcription factor FoxP3 and that are characterized by high surface expression of CD25 (reviewed in Ref. 82). Tregs inhibit the activation of T cells and innate immune cells, apparently to suppress autoimmunity. To modulate responses in the periphery, Tregs must home to the site of inflammation, and Tregs have been shown to express a variety of chemokine receptors that would serve this function, including CCR4, CCR5, CCR6, and CCR8 *(83–86)*. Consistent with a role for these chemokine receptors in Tregs function, deficiency of CCR5 on these cells exacerbated experimental graft versus host disease *(85)* in mice. Chemokine receptors important in defining and characterizing pathways of T-cell differentiation are listed in Table 1.

7.3.5. Effector B Cells

Like memory/effector T cells, *effector* B cells (i.e., plasmablasts/plasma cells) depend on chemokine receptors for their migration and positioning. B-cell differentiation into plasma cells is associated with downregulation of CCR7 and CXCR5 and increased responsiveness to the CXCR4 ligand, CXCL12, which has been shown to be important for proper localization of plasma cells in spleen and bone marrow *(87)*. Subsets of memory B cells, plasmablasts, and plasma cells have been reported to depend on particular chemokine receptors analogous to effector/memory T-cell subsets. CXCR3 is induced on memory B cells and plasma cells in response to IFN-γ and is found preferentially on cells expressing IgG1 *(88)*. CCR9 and CCR10 have been found on cells producing IgA, consistent with the trafficking of these cells to mucosal sites *(89–91)* (see below). Both CCR9 and CCR10 have been shown to be important for trafficking of IgA-producing cells to the small intestine, whereas only CCR10 is important for these cells' trafficking to the colon *(92)*, and CCR10 has been shown to be important for the accumulation of IgA plasma cells in the lactating

Table 1
Chemokine Receptors on Memory T-Cell Subsets

Cell types	Receptors	Roles
Early memory cells	CXCR3, CCR4	Used to identify these cells, which have an otherwise naïve phenotype
Central memory (T_{CM})	CCR7	Used to define these cells and allows entry into secondary lymphoid organs
Effector memory (T_{EM})	CCR7 (absent)	Lack CCR7 and are thereby excluded from noninflamed secondary lymphoid organs
T helper 1 (Th1)	CXCR3, CCR5	Th1 cells are a subset of $CXCR3^+$ cells
T helper 2 (Th2)	CCR4, CCR8	Th2 cells are a subset of $CCR4^+$ cells
Regulatory T (Treg)	CCR4, CCR5, CCR6, CCR8	Tregs found in $CCR4^+CCR5^+$ subset

mammary gland *(90)*. Memory B cells produced during the response to rotavirus, an intestinal pathogen, have been reported to lose CCR9 and CCR10 and upregulate CCR6, which has been postulated to mediate a switch from gut-specific trafficking to widespread dissemination to mucosal sites *(91)*.

7.3.6. T Cells Homing to Skin

For memory/effector T cells, just as described above for naïve T cells and implied in the discussion of B-cell trafficking, some chemokine/receptor groups are dedicated to compartment-specific recruitment. The skin and the gut represent dichotomous compartments that have been intensively studied with regard to trafficking of memory lymphocytes, and cells specific for those compartments express cutaneous lymphocyte antigen (CLA) and the integrin $\alpha_4\beta_7$, respectively (reviewed in Ref. 93). The majority of skin-homing lymphocytes express CCR4 and to a lesser extent CCR10 *(94–96)*. CCL17 thymus and activation-regulated chemokine (TARC), a ligand for CCR4, is expressed by cutaneous endothelial cells and keratinocytes in the skin, and this expression is enhanced by IFN-γ and TNF-α *(94,97,98)*. CCL27, a ligand for CCR10, is expressed by keratinocytes and is upregulated by proinflammatory cytokines *(95,97,99)*. Intradermal injections of CCL27 can induce the migration of $CCR10^+$ CLA^+ T cells to the skin *(95)*, and anti-CCL27 has been reported to diminish contact hypersensitivity in mice *(95)*. Mice deficient in CCR4 also have decreased numbers of E-selectin ligand $(CLA)^+$ T cells, which suggests that CCR4 is required for the development and/or survival of skin-homing T cells *(100)*. Also in mouse studies, recruitment to inflamed skin could be supported by either CCR4 or CCR10 *(94)*, although data on expression of CCR4 and CCR10 on CD4$^+$ T cells in inflamed human skin suggest a dominant role for CCR4, as only a small percentage of cells express CCR10 *(96)*.

7.3.7. T Cells Homing to Gut/Mucosa

With regard to the gut, CCR9 is the best-described receptor involved in the migration of T cells specifically to this compartment, and within the gut, preferentially to the small bowel *(101)*. CCR9 has been shown to be important but not indispensable for development of small intestine intraepithelial lymphocytes *(10,102–104)*, as well as for recruitment to the small bowel of T cells activated in the mesenteric lymph nodes *(105)*. For the T cells activated in the mesenteric lymph nodes, it has been shown that dendritic cells from Peyer's patches provide the information to induce CCR9 and the $\alpha_4\beta_7$ integrin, thereby imprinting those T cells to traffic to the gut *(106)*. The second chemokine receptor with a specific role in the gut is CCR10, as described above for B cells. Regarding gut T cells, however, CCR10 is remarkable for its absence *(89)*. Thus, although CCR10 is important for trafficking of T cells to skin and of B cells to mucosal sites including the intestine, this receptor seems to have no role in T-cell trafficking to gut. Chemokine receptors important in tissue- and/or compartment-specific migration are listed in Table 2.

Table 2
Chemokine and Sphingosine-1-phosphate Receptors in Tissue-Specific Lymphocyte Migration

Organs	Receptors	Roles
Thymus	CXCR4	Progenitor entry and/or localization of DN thymocytes
	CCR9	Progenitor entry and/or DP thymocytes' migration to cortex
	CCR7	Egress by mature cells
	$S1P_1$	Egress by mature cells
Bone marrow	CXCR4	B-cell lymphopoiesis
Secondary lymphoid organs	CCR7	Entry and localization in T-cell zones
	CXCR5	Organization of B-cell follicles
	$S1P_1$	Egress
Skin	CCR4	Homing of effector/memory T-cells
	CCR10	Homing of effector/memory/T-cells
Gut	CCR9	Homing of memory T and B cells
	CCR10	Homing of IgA-secreting plasma cells

7.4. Conclusions

Chemokines and chemokine receptors play multiple roles in lymphocyte biology. Determining these roles will contribute to our understanding not only of lymphocyte migration per se but also of broader aspects of lymphocyte biology that all depend on the cells' proper movement and localization, including lymphocyte development, activation, differentiation, and the exercise of effector functions. Recent work has revealed roles for a number of individual receptors and their ligands in vivo. Beyond the characterizations of individual receptors, we need to understand the regulation and biological consequences of the complex patterns of chemokine receptor expression on lymphocytes. We have summarized above some of the informative principles underlying these patterns: the "time line" of lymphocyte development and differentiation from progenitors to naïve to effector/memory cells, the polarization into Th1 or Th2 cells in the case of $CD4^+$ T cells, and the imprinting of tissue- and/or compartment–specific homing. Future work should include the continued characterization of lymphocyte subsets defined by the expression of both individual and combinations of chemokine receptors. Such studies should help us understand not only the functions of the chemokine system in host defense and immune-mediated disease but also fundamental aspects of lymphocyte biology that can be used to design effective vaccines and therapies.

References

1. Hernandez-Lopez C, Varas A, Sacedon R, et al. Stromal cell-derived factor 1/CXCR4 signaling is critical for early human T-cell development. Blood 2002;99:546–554.
2. Berkowitz RD, Beckerman KP, Schall TJ, McCune JM. CXCR4 and CCR5 expression delineates targets for HIV-1 disruption of T cell differentiation. J Immunol 1998;161:3702–3710.
3. Zaitseva MB, Lee S, Rabin RL, et al. CXCR4 and CCR5 on human thymocytes: biological function and role in HIV-1 infection. J Immunol 1998;161:3103–3113.
4. Berkowitz RD, Alexander S, McCune JM. Causal relationships between HIV-1 coreceptor utilization, tropism, and pathogenesis in human thymus. AIDS Res Hum Retroviruses 2000;16:1039–1045.
5. Kim CH, Broxmeyer HE. Chemokines: signal lamps for trafficking of T and B cells for development and effector function. J Leukoc Biol 1999;65:6–15.
6. Suzuki G, Nakata Y, Dan Y, et al. Loss of SDF-1 receptor expression during positive selection in the thymus. Int Immunol 1998;10:1049–1056.
7. Ma Q, Jones D, Springer TA. The chemokine receptor CXCR4 is required for the retention of B lineage and granulocytic precursors within the bone marrow microenvironment. Immunity 1999;10:463–471.

8. Ma Q, Jones D, Borghesani PR, et al. Impaired B-lymphopoiesis, myelopoiesis, and derailed cerebellar neuron migration in CXCR4$^-$ and SDF-1-deficient mice. Proc Natl Acad Sci U S A 1998;95:9448–9453.
9. Tachibana K, Hirota S, Iizasa H, et al. The chemokine receptor CXCR4 is essential for vascularization of the gastrointestinal tract. Nature 1998;393:591–594.
10. Uehara S, Grinberg A, Farber JM, Love PE. A role for CCR9 in T lymphocyte development and migration. J Immunol 2002;168:2811–2819.
11. Uehara S, Song K, Farber JM, Love PE. Characterization of CCR9 expression and CCL25/thymus-expressed chemokine responsiveness during T cell development: CD3(high)CD69$^+$ thymocytes and gammadeltaTCR$^+$ thymocytes preferentially respond to CCL25. J Immunol 2002;168:134–142.
12. Youn BS, Kim CH, Smith FO, Broxmeyer HE. TECK, an efficacious chemoattractant for human thymocytes, uses GPR-9-6/CCR9 as a specific receptor. Blood 1999;94:2533–2536.
13. Norment AM, Bogatzki LY, Gantner BN, Bevan MJ. Murine CCR9, a chemokine receptor for thymus-expressed chemokine that is up-regulated following pre-TCR signaling. J Immunol 2000;164:639–648.
14. Wurbel MA, Philippe JM, Nguyen C, et al. The chemokine TECK is expressed by thymic and intestinal epithelial cells and attracts double- and single-positive thymocytes expressing the TECK receptor CCR9. Eur J Immunol 2000; 30:262–271.
15. Wurbel MA, Malissen B, Campbell JJ. Complex regulation of CCR9 at multiple discrete stages of T cell development. Eur J Immunol 2006;36:73–81.
16. Uehara S, Hayes SM, Li L, et al. Premature expression of chemokine receptor CCR9 impairs T cell development. J Immunol 2006;176:75–84.
17. Ueno T, Hara K, Willis MS, et al. Role for CCR7 ligands in the emigration of newly generated T lymphocytes from the neonatal thymus. Immunity 2002; 16:205–218.
18. Ueno T, Saito F, Gray DH, et al. CCR7 signals are essential for cortex-medulla migration of developing thymocytes. J Exp Med 2004;200:493–505.
19. Misslitz A, Pabst O, Hintzen G, et al. Thymic T cell development and progenitor localization depend on CCR7. J Exp Med 2004;200:481–491.
20. Allende ML, Dreier JL, Mandala S, Proia RL. Expression of the sphingosine 1-phosphate receptor, S1P1, on T-cells controls thymic emigration. J Biol Chem 2004;279:15396–401.
21. Matloubian M, Lo CG, Cinamon G, et al. Lymphocyte egress from thymus and peripheral lymphoid organs is dependent on S1P receptor 1. Nature 2004; 427:355–360.
22. Bassi R, Anelli V, Giussani P, Tettamanti G, Viani P, Riboni L. Sphingosine-1-phosphate is released by cerebellar astrocytes in response to bFGF and induces astrocyte proliferation through G(i)-protein-coupled receptors. Glia 2006;53: 621–630.
23. Cinamon G, Matloubian M, Lesneski MJ, et al. Sphingosine 1-phosphate receptor 1 promotes B cell localization in the splenic marginal zone. Nat Immunol 2004; 5:713–720.

24. Nagasawa T, Nakajima T, Tachibana K, et al. Molecular cloning and characterization of a murine pre-B-cell growth-stimulating factor/stromal cell-derived factor 1 receptor, a murine homolog of the human immunodeficiency virus 1 entry coreceptor fusin. Proc Natl Acad Sci U S A 1996;93:14726–14729.
25. D'Apuzzo M, Rolink A, Loetscher M, et al. The chemokine SDF-1, stromal cell-derived factor 1, attracts early stage B cell precursors via the chemokine receptor CXCR4. Eur J Immunol 1997;27:1788–1793.
26. Moser B, Loetscher M, Piali L, Loetscher P. Lymphocyte responses to chemokines. Int Rev Immunol 1998;16:323–344.
27. Nagasawa T, Hirota S, Tachibana K, et al. Defects of B-cell lymphopoiesis and bone-marrow myelopoiesis in mice lacking the CXC chemokine PBSF/SDF-1. Nature 1996;382:635–638.
28. Nagasawa T, Kikutani H, Kishimoto T. Molecular cloning and structure of a pre-B-cell growth-stimulating factor. Proc Natl Acad Sci U S A 1994;91: 2305–2309.
29. Egawa T, Kawabata K, Kawamoto H, et al. The earliest stages of B cell development require a chemokine stromal cell-derived factor/pre-B cell growth-stimulating factor. Immunity 2001;15:323–334.
30. Swain SL. Regulation of the development of helper T cell subsets. Immunol Res 1991;10:177–182.
31. Calabresi PA, Allie R, Mullen KM, Yun SH, Georgantas RW 3rd, Whartenby KA. Kinetics of CCR7 expression differ between primary activation and effector memory states of T(H)1 and T(H)2 cells. J Neuroimmunol 2003;139: 58–65.
32. Campbell JJ, Murphy KE, Kunkel EJ, et al. CCR7 expression and memory T cell diversity in humans. J Immunol 2001;166:877–884.
33. Sallusto F, Kremmer E, Palermo B, et al. Switch in chemokine receptor expression upon TCR stimulation reveals novel homing potential for recently activated T cells. Eur J Immunol 1999;29:2037–2045.
34. Langenkamp A, Nagata K, Murphy K, Wu L, Lanzavecchia A, Sallusto F. Kinetics and expression patterns of chemokine receptors in human $CD4^+$ T lymphocytes primed by myeloid or plasmacytoid dendritic cells. Eur J Immunol 2003;33: 474–482.
35. Campbell JJ, Bowman EP, Murphy K, et al. 6-C-kine (SLC), a lymphocyte adhesion-triggering chemokine expressed by high endothelium, is an agonist for the MIP-3beta receptor CCR7. J Cell Biol 1998;141:1053–1059.
36. Forster R, Schubel A, Breitfeld D, et al. CCR7 coordinates the primary immune response by establishing functional microenvironments in secondary lymphoid organs. Cell 1999;99:23–33.
37. Breitfeld D, Ohl L, Kremmer E, et al. Follicular B helper T cells express CXC chemokine receptor 5, localize to B cell follicles, and support immunoglobulin production. J Exp Med 2000;192:1545–1552.
38. Voigt I, Camacho SA, de Boer BA, Lipp M, Forster R, Berek C. CXCR5-deficient mice develop functional germinal centers in the splenic T cell zone. Eur J Immunol 2000;30:560–567.

39. Allen CD, Ansel KM, Low C, et al. Germinal center dark and light zone organization is mediated by CXCR4 and CXCR5. Nat Immunol 2004;5:943–952.
40. Bowman EP, Campbell JJ, Soler D, et al. Developmental switches in chemokine response profiles during B cell differentiation and maturation. J Exp Med 2000;191:1303–1318.
41. Forster R, Mattis AE, Kremmer E, Wolf E, Brem G, Lipp M. A putative chemokine receptor, BLR1, directs B cell migration to defined lymphoid organs and specific anatomic compartments of the spleen. Cell 1996;87:1037–1047.
42. Ansel KM, Ngo VN, Hyman PL, et al. A chemokine-driven positive feedback loop organizes lymphoid follicles. Nature 2000;406:309–314.
43. Kim CH, Rott LS, Clark-Lewis I, Campbell DJ, Wu L, Butcher EC. Subspecialization of $CXCR5^+$ T cells: B helper activity is focused in a germinal center-localized subset of $CXCR5^+$ T cells. J Exp Med 2001;193:1373–1381.
44. Schaerli P, Willimann K, Lang AB, Lipp M, Loetscher P, Moser B. CXC chemokine receptor 5 expression defines follicular homing T cells with B cell helper function. J Exp Med 2000;192:1553–1562.
45. Hardtke S, Ohl L, Forster R. Balanced expression of CXCR5 and CCR7 on follicular T helper cells determines their transient positioning to lymph node follicles and is essential for efficient B-cell help. Blood 2005;106:1924–1931.
46. Sallusto F, Langenkamp A, Geginat J, Lanzavecchia A. Functional subsets of memory T cells identified by CCR7 expression. Curr Top Microbiol Immunol 2000;251:167–171.
47. Sallusto F, Lenig D, Forster R, Lipp M, Lanzavecchia A. Two subsets of memory T lymphocytes with distinct homing potentials and effector functions. Nature 1999;401:708–712.
48. Rivino L, Messi M, Jarrossay D, Lanzavecchia A, Sallusto F, Geginat J. Chemokine receptor expression identifies Pre-T helper (Th)1, Pre-Th2, and nonpolarized cells among human $CD4^+$ central memory T cells. J Exp Med 2004;200:725–735.
49. Unsoeld H, Krautwald S, Voehringer D, Kunzendorf U, Pircher H. Cutting edge: $CCR7^+$ and $CCR7^-$ memory T cells do not differ in immediate effector cell function. J Immunol 2002;169:638–641.
50. Unsoeld H, Pircher H. Complex memory T-cell phenotypes revealed by coexpression of CD62L and CCR7. J Virol 2005;79:4510–4513.
51. Reinhardt RL, Khoruts A, Merica R, Zell T, Jenkins MK. Visualizing the generation of memory CD4 T cells in the whole body. Nature 2001;410:101–105.
52. Song K, Rabin RL, Hill BJ, et al. Characterization of subsets of $CD4^+$ memory T cells reveals early branched pathways of T cell differentiation in humans. Proc Natl Acad Sci U S A 2005;102:7916–791621.
53. Wu CY, Kirman JR, Rotte MJ, et al. Distinct lineages of T(H)1 cells have differential capacities for memory cell generation in vivo. Nat Immunol 2002;3:852–858.

54. Mackay CR, Marston W, Dudler L. Altered patterns of T cell migration through lymph nodes and skin following antigen challenge. Eur J Immunol 1992;22: 2205–2210.
55. Debes GF, Hopken UE, Hamann A. In vivo differentiated cytokine-producing CD4(+) T cells express functional CCR7. J Immunol 2002;168:5441–5447.
56. Debes GF, Bonhagen K, Wolff T, et al. CC chemokine receptor 7 expression by effector/memory CD4$^+$ T cells depends on antigen specificity and tissue localization during influenza A virus infection. J Virol 2004;78:7528–7535.
57. Roman E, Miller E, Harmsen A, et al. CD4 effector T cell subsets in the response to influenza: heterogeneity, migration, and function. J Exp Med 2002;196: 957–968.
58. D'Ambrosio D, Iellem A, Bonecchi R, et al. Selective up-regulation of chemokine receptors CCR4 and CCR8 upon activation of polarized human type 2 Th cells. J Immunol 1998;161:5111–5115.
59. Messi M, Giacchetto I, Nagata K, Lanzavecchia A, Natoli G, Sallusto F. Memory and flexibility of cytokine gene expression as separable properties of human T(H)1 and T(H)2 lymphocytes. Nat Immunol 2003;4:78–86.
60. Andrew DP, Ruffing N, Kim CH, et al. C-C chemokine receptor 4 expression defines a major subset of circulating nonintestinal memory T cells of both Th1 and Th2 potential. J Immunol 2001;166:103–111.
61. Colantonio L, Rossi B, Constantin G, D'Ambrosio D. Integration and independent acquisition of specialized skin- versus gut-homing and Th1 versus Th2 cytokine synthesis phenotypes in human CD4$^+$ T cells. Eur J Immunol 2004;34: 2419–2429.
62. Kim CH, Kunkel EJ, Boisvert J, et al. Bonzo/CXCR6 expression defines type 1-polarized T-cell subsets with extralymphoid tissue homing potential. J Clin Invest 2001;107:595–601.
63. Sallusto F, Mackay CR, Lanzavecchia A. Selective expression of the eotaxin receptor CCR3 by human T helper 2 cells. Science 1997;277:2005–2007.
64. Sallusto F, Lenig D, Mackay CR, Lanzavecchia A. Flexible programs of chemokine receptor expression on human polarized T helper 1 and 2 lymphocytes. J Exp Med 1998;187:875–883.
65. Randolph DA, Huang G, Carruthers CJ, Bromley LE, Chaplin DD. The role of CCR7 in TH1 and TH2 cell localization and delivery of B cell help in vivo. Science 1999;286:2159–2162.
66. Rabin RL, Alston MA, Sircus JC, et al. CXCR3 is induced early on the pathway of CD4$^+$ T cell differentiation and bridges central and peripheral functions. J Immunol 2003;171:2812–2824.
67. Agace WW, Roberts AI, Wu L, Greineder C, Ebert EC, Parker CM. Human intestinal lamina propria and intraepithelial lymphocytes express receptors specific for chemokines induced by inflammation. Eur J Immunol 2000;30: 819–826.
68. Balashov KE, Rottman JB, Weiner HL, Hancock WW. CCR5(+) and CXCR3(+) T cells are increased in multiple sclerosis and their ligands MIP-1alpha and IP-10

are expressed in demyelinating brain lesions. Proc Natl Acad Sci USA 1999; 96:6873–6878.

69. Yamamoto J, Adachi Y, Onoue Y, et al. Differential expression of the chemokine receptors by the Th1- and Th2-type effector populations within circulating CD4+ T cells. J Leukoc Biol 2000;68:568–574.
70. Romagnani P, Maggi L, Mazzinghi B, et al. CXCR3-mediated opposite effects of CXCL10 and CXCL4 on TH1 or TH2 cytokine production. J Allergy Clin Immunol 2005;116:1372–1379.
71. Kim CH, Rott L, Kunkel EJ, et al. Rules of chemokine receptor association with T cell polarization in vivo. J Clin Invest 2001;108:1331–1339.
72. Syrbe U, Siveke J, Hamann A. Th1/Th2 subsets: distinct differences in homing and chemokine receptor expression? Springer Semin Immunopathol 1999;21: 263–1285.
73. Nakajima C, Mukai T, Yamaguchi N, et al. Induction of the chemokine receptor CXCR3 on TCR-stimulated T cells: dependence on the release from persistent TCR-triggering and requirement for IFN-gamma stimulation. Eur J Immunol 2002;32:1792–1801.
74. Yang YF, Tomura M, Iwasaki M, et al. IL-12 as well as IL-2 upregulates CCR5 expression on T cell receptor-triggered human CD4+ and CD8+ T cells. J Clin Immunol 2001;21:116–125.
75. Iwasaki M, Mukai T, Gao P, et al. A critical role for IL-12 in CCR5 induction on T cell receptor-triggered mouse CD4(+) and CD8(+) T cells. Eur J Immunol 2001;31:2411–2420.
76. Matsuda JL, Zhang Q, Ndonye R, Richardson SK, Howell AR, Gapin L. T-bet concomitantly controls migration, survival and effector functions during the development of Vα 14i NKT cells. Blood 2006;107:2795–2805.
77. Odum N, Bregenholt S, Eriksen KW, et al. The CC-chemokine receptor 5 (CCR5) is a marker of, but not essential for the development of human Th1 cells. Tissue Antigens 1999;54:572–577.
78. Qin S, Rottman JB, Myers P, et al. The chemokine receptors CXCR3 and CCR5 mark subsets of T cells associated with certain inflammatory reactions. J Clin Invest 1998;101:746–754.
79. Bonecchi R, Bianchi G, Bordignon PP, et al. Differential expression of chemokine receptors and chemotactic responsiveness of type 1 T helper cells (Th1s) and Th2s. J Exp Med 1998;187:129–134.
80. Vestergaard C, Deleuran M, Gesser B, Gronhoj Larsen C. Expression of the T-helper 2-specific chemokine receptor CCR4 on CCR10-positive lymphocytes in atopic dermatitis skin but not in psoriasis skin. Br J Dermatol 2003;149: 457–463.
81. Chung CD, Kuo F, Kumer J, et al. CCR8 is not essential for the development of inflammation in a mouse model of allergic airway disease. J Immunol 2003;170: 581–587.
82. Sakaguchi S. Naturally arising Foxp3-expressing CD25+CD4+ regulatory T cells in immunological tolerance to self and non-self. Nat Immunol 2005;6:345–352.

83. Iellem A, Mariani M, Lang R, et al. Unique chemotactic response profile and specific expression of chemokine receptors CCR4 and CCR8 by CD4(+)CD25(+) regulatory T cells. J Exp Med 2001;194:847–853.
84. Bystry RS, Aluvihare V, Welch KA, Kallikourdis M, Betz AG. B cells and professional APCs recruit regulatory T cells via CCL4. Nat Immunol 2001;2:1126–1132.
85. Wysocki CA, Jiang Q, Panoskaltsis-Mortari A, et al. Critical role for CCR5 in the function of donor CD4+CD25+ regulatory T cells during acute graft-versus-host disease. Blood 2005;106:3300–3307.
86. Kleinewietfeld M, Puentes F, Borsellino G, Battistini L, Rotzschke O, Falk K. CCR6 expression defines regulatory effector/memory-like cells within the CD25(+)CD4+ T-cell subset. Blood 2005;105:2877–2886.
87. Hargreaves DC, Hyman PL, Lu TT, et al. A coordinated change in chemokine responsiveness guides plasma cell movements. J Exp Med 2001;194: 45–56.
88. Muehlinghaus G, Cigliano L, Huehn S, et al. Regulation of CXCR3 and CXCR4 expression during terminal differentiation of memory B cells into plasma cells. Blood 2005;105:3965–3971.
89. Kunkel EJ, Kim CH, Lazarus NH, et al. CCR10 expression is a common feature of circulating and mucosal epithelial tissue IgA Ab-secreting cells. J Clin Invest 2003;111:1001–1010.
90. Wilson E, Butcher EC. CCL28 controls immunoglobulin (Ig)A plasma cell accumulation in the lactating mammary gland and IgA antibody transfer to the neonate. J Exp Med 2004;200:805–809.
91. Jaimes MC, Rojas OL, Kunkel EJ, et al. Maturation and trafficking markers on rotavirus-specific B cells during acute infection and convalescence in children. J Virol 2004;78:10967–10976.
92. Hieshima K, Kawasaki Y, Hanamoto H, et al. CC chemokine ligands 25 and 28 play essential roles in intestinal extravasation of IgA antibody-secreting cells. J Immunol 2004;173:3668–3675.
93. Springer TA. Traffic signals for lymphocyte recirculation and leukocyte emigration: the multistep paradigm. Cell 1994;76:301–314.
94. Reiss Y, Proudfoot AE, Power CA, Campbell JJ, Butcher EC. CC chemokine receptor (CCR)4 and the CCR10 ligand cutaneous T cell-attracting chemokine (CTACK) in lymphocyte trafficking to inflamed skin. J Exp Med 2001;194: 1541–1547.
95. Homey B, Alenius H, Muller A, et al. CCL27-CCR10 interactions regulate T cell-mediated skin inflammation. Nat Med 2002;8:157–165.
96. Soler D, Humphreys TL, Spinola SM, Campbell JJ. CCR4 versus CCR10 in human cutaneous TH lymphocyte trafficking. Blood 2003;101:1677–1682.
97. Vestergaard C, Johansen C, Christensen U, Just H, Hohwy T, Deleuran M. TARC augments TNF-alpha-induced CTACK production in keratinocytes. Exp Dermatol 2004;13:551–557.

98. Vestergaard C, Deleuran M, Gesser B, Larsen CG. Thymus- and activation-regulated chemokine (TARC/CCL17) induces a Th2-dominated inflammatory reaction on intradermal injection in mice. Exp Dermatol 2004;13:265–271.
99. Humphreys TL, Baldridge LA, Billings SD, Campbell JJ, Spinola SM. Trafficking pathways and characterization of CD4 and CD8 cells recruited to the skin of humans experimentally infected with Haemophilus ducreyi. Infect Immun 2005;73:3896–902.
100. Baekkevold ES, Wurbel MA, Kivisakk P, et al. A role for CCR4 in development of mature circulating cutaneous T helper memory cell populations. J Exp Med 2005;201:1045–1051.
101. Kunkel EJ, Campbell JJ, Haraldsen G, et al. Lymphocyte CC chemokine receptor 9 and epithelial thymus-expressed chemokine (TECK) expression distinguish the small intestinal immune compartment: Epithelial expression of tissue-specific chemokines as an organizing principle in regional immunity. J Exp Med 2000;192:761–768.
102. Wurbel MA, Malissen M, Guy-Grand D, et al. Mice lacking the CCR9 CC-chemokine receptor show a mild impairment of early T- and B-cell development and a reduction in T-cell receptor gammadelta(+) gut intraepithelial lymphocytes. Blood 2001;98:2626–2632.
103. Onai N, Kitabatake M, Zhang YY, Ishikawa H, Ishikawa S, Matsushima K. Pivotal role of CCL25 (TECK)-CCR9 in the formation of gut cryptopatches and consequent appearance of intestinal intraepithelial T lymphocytes. Int Immunol 2002;14:687–694.
104. Marsal J, Svensson M, Ericsson A, et al. Involvement of CCL25 (TECK) in the generation of the murine small-intestinal CD8alpha alpha^{+}CD3^{+} intraepithelial lymphocyte compartment. Eur J Immunol 2002;32:3488–3497.
105. Svensson M, Marsal J, Ericsson A, et al. CCL25 mediates the localization of recently activated CD8alphabeta(+) lymphocytes to the small-intestinal mucosa. J Clin Invest 2002;110:1113–1121.
106. Mora JR, Bono MR, Manjunath N, et al. Selective imprinting of gut-homing T cells by Peyer's patch dendritic cells. Nature 2003;424:88–93.

8

Chemokines in Trafficking of Hematopoietic Stem and Progenitor Cells and Hematopoiesis

Chang H. Kim

Summary

Chemokines regulate the process of hematopoiesis by controlling trafficking, proliferation/survival, and differentiation of hematopoietic stem and progenitor cells. Unique expression of the chemokine receptor CXCR4 in combination with adhesion molecules [very late antigen (VLA)-4, lymphocyte function-associated antigen-1 (LFA-1), and VLA-5] on early hematopoietic stem and progenitor cells makes their homing to bone marrow and seeding the stem cell niche possible. In bone marrow, the CXCL12-CXCR4 chemokine axis promotes the survival and retention of hematopoietic stem and early progenitor cells. In certain conditions, hematopoietic stem and progenitor cells are released from bone marrow to the blood circulation, a process called mobilization. Many agents that mobilize marrow progenitor cells induce or activate certain proteases of neutrophils. This leads to degradation of CXCL12 and, therefore, weakens the chemotactic activity of bone marrow. Whereas CXCL12 plays a positive role in survival and proliferation of hematopoietic stem and progenitor cells, many other chemokines suppress these processes. During their differentiation into mature cells, hematopoietic progenitor cells upregulate cell lineage–specific chemokine receptors to migrate to appropriate tissue sites. This cell type–specific switch in chemokine receptors and adhesion molecules is important for tissue-specific migration of hematopoietic cells, a process important for their differentiation or effector function in the periphery.

Key Words: Stem cells; chemokines; hematopoiesis; chemotaxis; T cells; B cells.

From: *The Receptors: The Chemokine Receptors*
Edited by: J. K. Harrison and N. W. Lukacs © Humana Press Inc., Totowa, NJ

8.1. Introduction

Early hematopoietic activity is first detected in the yolk sac and aorta-gonad-mesonephros (AGM) and later in the liver during embryonic development *(1)*. After birth, bone marrow becomes the central organ for hematopoiesis. Spleen is another site where hematopoietic activity is detected in the red pulp area in some species such as mice. Hematopoietic stem and progenitor cells (HSPCs) in bone marrow produce massive numbers of new blood cells to replace old blood cells. Hematopoietic stem cells are cells that, by definition, have the self-renewal capacity and totipotency within the hematopoietic lineage. Based on the results of long-term in vivo repopulation studies that at least ~10^5 marrow cells are required to repopulate the hematopoietic system of an ablated mouse recipient *(2,3)*, it is reasonable to estimate that only a few in 10^5 cells in bone marrow would fit the definition of stem cells. As shown in Fig. 1, hematopoietic stem cells have the phenotype of c-Kit$^+$Thy-1^{low}Lin$^-$CD34$^+$CD33$^-$ in humans, and the mouse counterparts have the phenotype of c-Kit$^+$Thy-1^{low}Lin$^-$Sca-1$^+$ *(4)*. Survival, proliferation, and differentiation of HSPCs are tightly regulated by a number of factors including cytokines produced by a

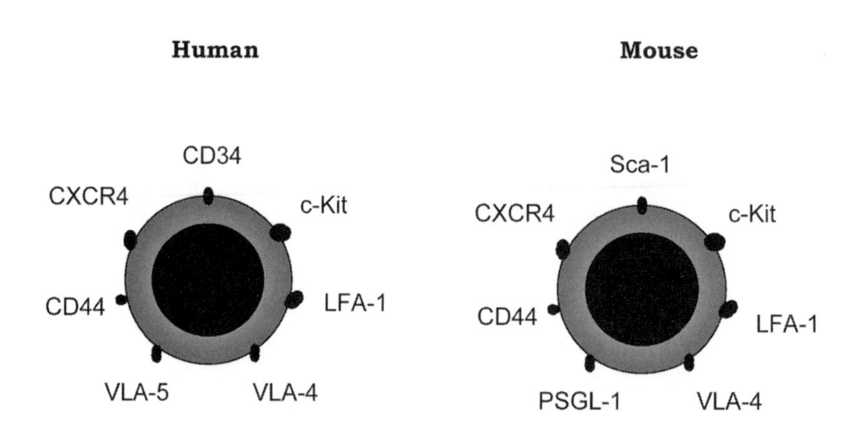

Fig. 1. Surface phenotype of HSPCs. Primitive HSPCs have the phenotype of c-Kit$^+$Thy-1^{low}Lin$^-$CD34$^+$CD33$^-$ in humans, and the mouse counterparts have the phenotype of c-Kit$^+$Thy-1^{low}Lin$^-$Sca-1$^+$. Primitive HSPCs express CXCR4 as the major chemokine receptor and various adhesion molecules such as VLA-4, VLA-5, LFA-1, P-selectin glycoprotein ligand-1 (PSGL-1), and CD44 for migration to the stem cell niche.

variety of cell types. HSPCs are localized in a specialized microenvironment called the stem cell niche constituted by spindle-shaped osteoblastic cells *(5)*. These spindle-shaped osteoblastic cells and other stromal cells in the stem cell niche of bone marrow produce large numbers of growth factors such as stem cell factor (SCF), granulocyte-colony stimulating factor (G-CSF), granulocyte macrophage colony stimulating factor (GM-CSF), macrophage colony stimulating factor (M-CSF), interleukin-1 (IL-1), IL-6, IL-7, transtorming growth factor-β (TGF-β) and macrophage inflammatory protein 1-α (MIP-1α). Successful hematopoiesis relies on HSPC migration into bone marrow, retention within the stem cell niche, survival, proliferation, and differentiation followed by release of mature blood cells to the periphery. All of these processes are regulated at some levels by a family of small secretory proteins called chemokines. Chemokines regulate the migration of HSPCs into bone marrow by inducing chemotaxis and by activating integrins. Chemokines also have other functions such as regulating cell proliferation/cycling, survival/apoptosis, and cell differentiation. In this chapter, we will discuss the roles of chemokines in regulation of the proliferation, survival, and trafficking of hematopoietic stem and progenitors.

8.2. Regulation of Survival and Proliferation of Hematopoietic Stem and Progenitor Cells by Chemokines

In 1989, it was reported that chemokines CCL3 (also called MIP-1α) and CCL4 (also called regulated upon activation, normal T-cell expressed and secreted (RANTES)) have an enhancing effect on colony formation of late myeloid progenitors induced by M-CSF or GM-CSF *(6)*. However, when stimulated by multiple cytokines such as GM-CSF, SCF, and erythropoietin (EPO), it was found that CCL3 inhibited proliferation of early progenitors such as colony-forming unit granulocyte erythrocyte monocytes megakaryocyte (CFU-GEMM), colony-forming unit granulocyte monocytes (CFU-GM), and burst-forming unit–erythrocyte (BFU-E) *(7)*. In this regard, CCL3 was the first chemokine found to suppress hematopoiesis *(8)*. Graham et al. were able to demonstrate that CCL3 can inhibit proliferation of primitive HSPCs (i.e., colony-forming unit–spleen at day 12, or CFU-S d12) *(8)*. The suppressive effect of CCL3 for primitive HSPCs appears to be dominant over the enhancing effect of CCL3 for the late progenitors, as in vivo administration of CCL3 resulted in suppression of hematopoiesis *(9)*. CCL3 has a protective effect on the HSPC compartment during chemotherapy as evidenced by significant improvement in the kinetics of neutrophil recovery *(10)*. CCL3, when injected into mice, rapidly decreases the cycling rates of HSPCs within 24 hours and the absolute numbers of myeloid progenitor cells in the marrow and spleen by 48 hours. CCL3 has multiple receptors such as CCR1, CCR4, and CCR5 *(11)*. The enhancing effect of CCL3 on myelopoiesis is through CCR1, but the inhibitory effect is not dependent on CCR1 expression *(12)*. Since the discovery of CCL3 as a suppressive agent for HSPCs, it has been determined that many of the ~45 members of the chemokine protein family also have the suppressive activity for HSPCs *(13)*. Such chemokines include CCL1, CCL2, CCL3, CCL6, CCL11, CCL13, CCL15, CCL16, CCL19, CCL20, CCL21, CCL23, CCL24, CCL25, CXCL2, CXCL4, CXCL5, CXCL6, CXCL8, CXCL9, CXCL10, and CL1. In contrast, the following chemokines do not suppress the proliferation and colony formation of HSPCs: CCL4, CCL5, CCL7, CCL8, CCL12, CCL14, CCL17, CCL18, CCL22, CXCL1, CXCL3, CXCL7, CXCL11, CXCL12, CXCL13, and CX3CL1. Together, the suppressive chemokines can act in synergy for greater suppression *(14)*. Several chemokine receptors that mediate the suppressive function of these chemokines have been identified. For example, the myelosuppressive function of CCL2 is mediated through its major receptor CCR2, and CXCL8 suppresses myelopoiesis through CXCR2 *(15,16)*. It is ironic that CCL11, which suppresses the proliferation of myeloid progenitors induced by multiple growth factors, acts as a colony-stimulating factor by itself for granulocytes and macrophages *(17)*. Therefore, the exact outcome of the suppression or activation of HSPCs by chemokines appears to be dependent on cytokine milieu and types of HSPCs.

CXCL12 was originally cloned from the cDNAs of a bone marrow stromal cell line and is highly expressed by bone marrow stromal cells *(18)*. As mentioned above, CXCL12 does not inhibit the myelopoiesis induced by multiple growth factors. On the contrary, CXCL12 plays a positive role in hematopoiesis by providing survival signals. In fact, CXCL12-deficient mice display multiple defects in hematopoiesis and die perinatally *(19)*. The abnormality in hematopoiesis includes reduced B-cell progenitors and myeloid progenitors. CXCR4 is the only receptor for CXCL12, and therefore it is not surprising that the mice deficient in CXCR4 are also defective in hematopoiesis and die perinatally *(20–22)*. Additionally, mice deficient in CXCR4 or its ligand CXCL12 are defective in vascular development. Whereas the numbers of myeloid progenitors were reduced only in bone marrow, the numbers of B-cell progenitors in $CXCL12^{-/-}$ embryos were severely reduced in both fetal liver and bone marrow. This is because CXCL12 and CXCR4 are required for generation of the earliest unipotent B-cell precursor population with the phenotype of c-$Kit^{+}Lin^{-}CD19^{-}$ IL-$7R\alpha^{+}AA4.1^{+}$ in fetal liver *(23)*. Consistently, CXCL12 was identified as a soluble mediator that supports the proliferation of a stromal cell–dependent pre–B cell clone *(24)*. Together with IL-7, CXCL12 synergistically augments the growth of bone marrow B-cell progenitors *(24)*. Also, CXCL12 synergizes with thrombopoietin to enhance the development of megakaryocytic progenitor cells *(25)*. Kawabata et al. found that $CXCR4^{+/+}$, but not $CXCR4^{-/-}$, fetal liver cells were able to repopulate the long-term lymphoid and myeloid compartments *(26)*. The positive effect of CXCL12 on the survival of HSPCs is further supported by the fact that CXCL12-expressing transgenic HSPCs or CXCL12-treated normal mouse or human HSPCs had a greater repopulating potential in mice after bone marrow transplantation *(27–29)*. Regulation of the proliferation and survival of HSPCs by chemokines is summarized in Fig. 2.

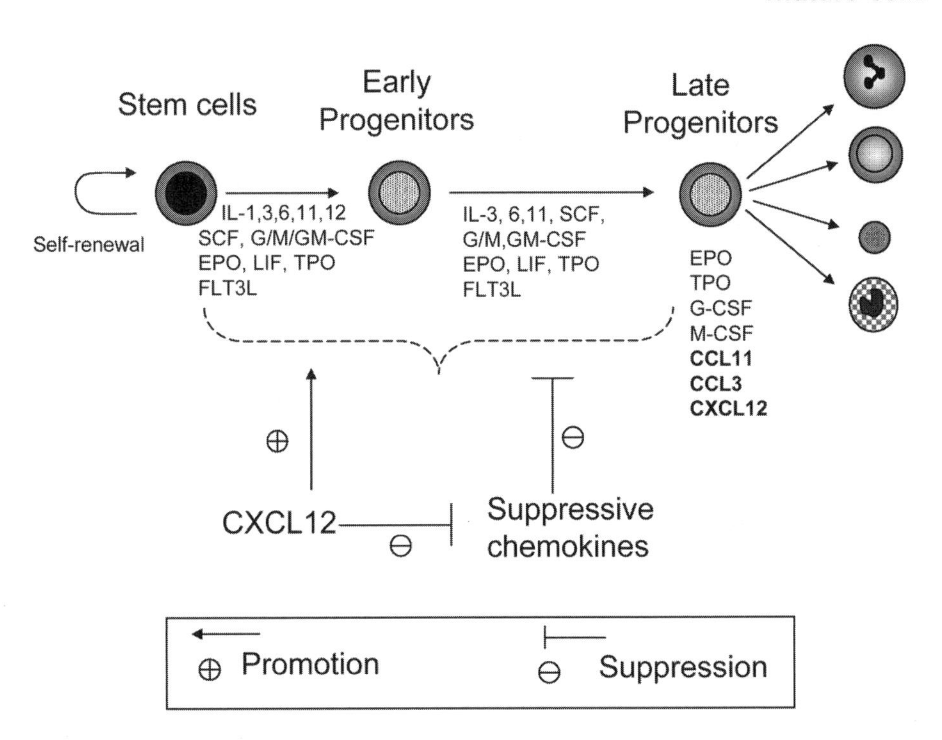

Fig. 2. Chemokines in regulation of the survival and proliferation of HSPCs. Proliferation and differentiation of HSPCs are finely controlled by many cytokines/growth factors in marrow. CXCL12 provides survival signals for HSPCs of both myeloid and lymphoid lineages. It is particularly important for the development of B-cell progenitors. In contrast, a number of chemokines, such as CCL1, CCL2, CCL3, CCL6, CCL11, CCL13, CCL15, CCL16, CCL19, CCL20, CCL21, CCL23, CCL24, CCL25, CXCL2, CXCL4, CXCL5, CXCL6, CXCL8, CXCL9, CXCL10, and CL1, suppress the proliferation of HSPCs induced by multiple hematopoietic growth and differentiation factors, which are indicated in the diagram. In addition to their positive effect on survival of HSPCs, CXCL12 antagonizes the suppressive activities of the myelosuppressive chemokines.

8.3. Regulation of the Homing and Mobilization of Hematopoietic Stem and Progenitor Cells

Homing of HSPCs refers to migration of HSPCs to bone marrow during development or after transplantation. In Tavassoli and Hardy's early review *(30)* published in 1990 and entitled "Molecular basis of homing of intravenously transplanted stem cells to the marrow," it was speculated that "Homing is likely to be a complex phenomenon involving multiple interactions at the molecular level. Membrane-associated molecules, as well as extracellular matrix, may be involved. Each of these molecules may act additively or synergistically with others." At the time, it was not clear exactly what regulates the homing of HSPCs to bone marrow. Since then, we have witnessed significant progress in our understanding of the process. It turned out that CXCL12, described earlier in this chapter for its positive role in survival and proliferation of HSPCs, is also a critical player in HSPC homing to bone marrow. Auiti et al. reported that CXCL12 is a chemoattractant for HSPCs *(31)*. CXCL12 induces in vitro chemotaxis of human bone marrow–derived $CD34^+$ cells with a broad activity for primitive ($CD34^+CD38^-$ or $CD34^+$HLA-DR^-) HSPC cells and the cells committed to erythroid, lymphoid, and myeloid lineages (BFU-E, CFU-GM, and CFU-GEMM progenitors). IL-3, a hematopoietic cytokine, enhances the chemotactic responsiveness of $CD34^+$ cells to CXCL12. They also found that the $CD34^+$ cells in peripheral blood have lower sensitivity to CXCL12 in chemotaxis than the $CD34^+$ progenitors in bone marrow, implying a possible role of CXCL12 in mobilization of $CD34^+$ cells out of bone marrow.

Stem cell factor (SCF) is another chemoattractant for HSPCs *(32)*. Along with CXCL12, SCF is produced by bone marrow stromal cells *(33)*. SCF has a largely chemokinetic activity for HSPCs, inducing random migration regardless of the polarity of chemoattractant gradient, whereas CXCL12 has a chemotactic activity for HSPCs, inducing directional migration *(34)*. The activities of CXCL12 and SCF in chemotaxis of HSPCs are at least additive. A negative concentration gradient of CXCL12 and SCF had a potent inhibitory effect on chemotaxis of HSPCs, and the two chemoattractants were synergistic in mobilizing HSPCs in an in vitro setting *(34)*. Consistently, human bone marrow plasma is highly chemotactic for HSPCs *(34)*. Plasma obtained from marrow aspirates, but not peripheral blood, had strong chemotactic activity for HSPCs. Most of the activity is likely to be that of CXCL12, because CXCL12 in a negative gradient can specifically antagonize the chemotactic activity of bone marrow plasma. It was demonstrated that CXCL12 is critical for the engraftment of human stem cells into the bone marrow of murine severe combined immunodeficient (SCID) mice *(35)*. Cytokines such as SCF and IL-6 induce CXCR4 expression on $CD34^+$ cells, thus potentiating the HSPC response to

CXCL12. In addition to chemoattractants, adhesion molecules play critical roles in HSPC homing to bone marrow. It should be appreciated that the CXCL12-dependent homing of HSPCs to marrow also involves adhesion molecules.

The first step of blood cell homing into most tissue sites including marrow is a weak on-and-off interaction called *rolling*, which is followed by chemokine receptor activation, leading to activation of integrin molecules on migrating cells *(36,37)*. It is now well documented that adhesion molecules such as selectins and vascular cell adhesion molecule (VCAM)-1 are involved in HSPC migration to bone marrow *(38,39)*. Rolling of HSPCs on the vascular endothelium of murine bone marrow is mediated by P-selectin, E-selectin, and VCAM-1 (also called CD106) *(40)*. VCAM-1 is a ligand of $\alpha_4\beta_1$ (an integrin molecule composed of CD49d and CD29; also called very late antigen 4). CXCL12, expressed on bone marrow endothelial cells, activates the CXCR4 on HSPCs so that HSPCs can firmly adhere to endothelial cells through interaction among the adhesion molecules of HSPCs [i.e., very late antigen (VLA)-4, lymphocyte function-associated antigen-1 (LFA-1), and VLA-5] and their counterreceptors on marrow endothelial cells *(41,42)*. Once within the marrow, interaction between adhesion molecules and their ligands such as VLA-4/VCAM-1, VLA-5/fibronectin, and β_2-integrins/intercellular cell adhesion molecule-1 is important for attachment of HSPCs to marrow stromal cells *(43)*.

The concentration of CXCL12 present in bone marrow is estimated to be greater than ~100 ng/mL *(34)*, whereas its concentration in peripheral blood is very low (~0.5 ng/mL) *(35)*. The minimum concentration of SDF-1 active for chemotaxis of HSPCs is ~50 ng/mL, and therefore the difference in CXCL12 concentrations between the bone marrow and peripheral blood would drive HSPCs into bone marrow. Conversely, disruption of the concentration difference would mobilize HSPCs to the periphery. Granulocyte colony-stimulating factor (G-CSF) is a strong mobilizer of marrow HSPCs into the blood circulation. A mechanism by which G-CSF mobilizes HSPCs is to induce the release of neutrophil elastase and cathepsin G to degrade CXCL12, CXCR4, and other molecules such as c-Kit and VCAM-1 *(44,45)*. Matrix metalloproteinase-9 (MMP-9) is another protease implicated in mobilization of marrow HSPCs *(46)*. The targets of these proteases also include vascular endothelial (VE)-cadherin and G-CSF itself *(47,48)*. The important role of neutrophils in G-CSF–induced mobilization is well supported by the fact that marrow HSPCs are not mobilized in the absence of neutrophils or their major chemokine receptor CXCR2 *(43,49,50)*. CD26/dipeptidylpeptidase IV (DPPIV), a membrane-bound extracellular peptidase, can cleave dipeptides from the N-terminus of chemokine polypeptide chains after a proline or an alanine residue *(51)*. For example, CD26/DPPIV can cut off the N-terminal dipeptides of CXCL12 and CCL22

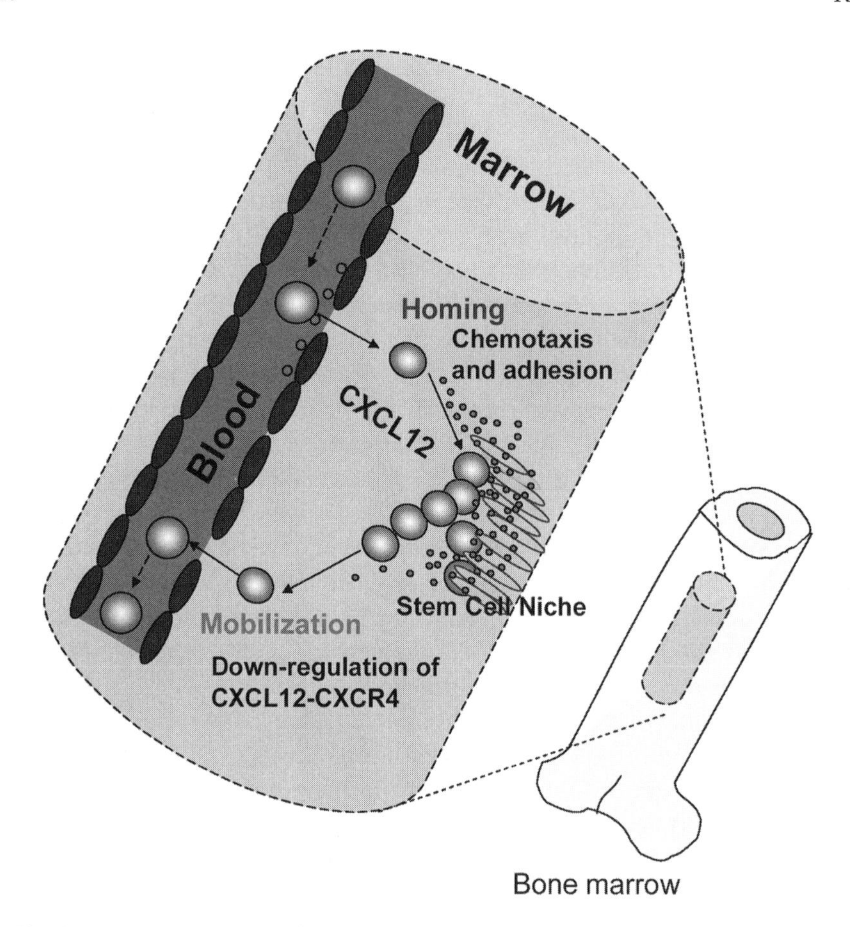

Fig. 3. Regulation of the homing and mobilization of HSPCs by chemokines. Bone marrow stromal cells and endothelial cells express CXCL12 to recruit HSPCs and position them in a specialized microenvironment within bone marrow called *stem cell niche*. Chemotaxis to CXCL12 is important not only for homing but also for regulation of the interaction between HSPCs and stromal cells that produce appropriate growth factors. Homing of HSPCs to marrow is mediated by the CXCL12 presented on marrow endothelial cells. CXCL12 activates integrins expressed by HSPCs, such as VLA-4 and VLA-5, and induces chemotaxis of HSPCs. Some cytokines and chemokines such as G-CSF and IL-8 induce mobilization of HSPCs. This mobilization is mediated by induction or activation of proteases (elastases, cathepsins, metalloproteinases, and CD26) of neutrophils or stem and progenitor cells themselves in response to the mobilizing agents. The major target of the proteases is the CXCL12-CXCR4 axis. Degradation of CXCL12 or CXCR4 decreases the chemotactic and adhesive activity of bone marrow, releasing HSPCs to the blood circulation.

(52). The N-terminal–truncated CXCL12 is unable to induce chemotaxis of CD34$^+$ HSPC cells, and it even acts as an antagonist for the intact CXCL12 *(51)*. The HSPC mobilization induced by G-CSF was significantly reduced in CD26$^{-/-}$ mice versus wild-type mice, and inhibition of CD26 by a small molecule antagonist in wild-type mice also resulted in reduced mobilization of HSPCs *(53,54)*.

When injected into animals or humans, chemokines, such as IL-8, Gro-β, and MIP-1α, can mobilize marrow HSPCs. Similar to G-CSF, these chemokines induce the degradation of CXCL12, CXCR4, extracellular matrix, and adhesion molecules *(55–59)*. Thus, the activities of many mobilizing agents and the proteases that they induce or activate are focused on degradation of the CXCL12 in bone marrow *(45,60–62)*. Consistently, direct injection of CXCL12 or 1.1′-[1,4-phenylenebis (methylene)bis-1,4,8,11-tetraazacyclotetradecane (AMD3100) (an antagonist of CXCR4) induces HSPC mobilization in mice and humans *(63–65)*. Moreover, pertussis toxin, which inhibits G proteins such as G_i and G_o, important for signal transduction of chemokine receptors including CXCR4, induces mobilization of HSPCs *(66)*. Oral fingolimod (FTY720) is an agonist for sphingosine-1-phosphate (S1P) receptor. FTY720 binds the S1P receptor and downregulates the receptor *(67,68)*. FTY720 treatment induces lymphocyte trapping within thymus and lymph nodes and makes lymphocytes unable to emigrate from primary and secondary lymphoid tissues *(69)*. Although the mechanism is unclear, it was shown that treatment with FTY720 increases the chemotactic response of HSPCs to CXCL12, enhancing their homing to bone marrow *(70)*. The role of chemokines in homing and mobilization of HSPCs is summarized in Fig. 3.

8.4. Development of Hematopoietic Cell Lineage–Specific Migration Program

As described above, hematopoietic stem and early progenitor cells uniquely express only CXCR4 and respond to CXCL12 among the ~40 chemokines identified so far in chemotaxis *(31,34,71)*. Whereas many mature hematopoietic cell types still retain their responsiveness to CXCL12 after differentiation, some lineages of white blood cells do not respond to CXCL12 but vigorously respond to different chemokines. For example, neutrophils do not respond to CXCL12 (because they do not express CXCR4) but respond to CXCL1, CXCL2, and CXCL8 among others. Even for the cell types that retain their responsiveness to CXCL12, it is significantly decreased when compared with that of stem and early progenitor cells. This implies that during their differentiation, HSPCs undergo changes in expression or function of their chemokine receptors (called *chemokine receptor switch*), which is important to dispatch mature or maturing blood cells to appropriate tissue sites in the periphery (Fig. 4).

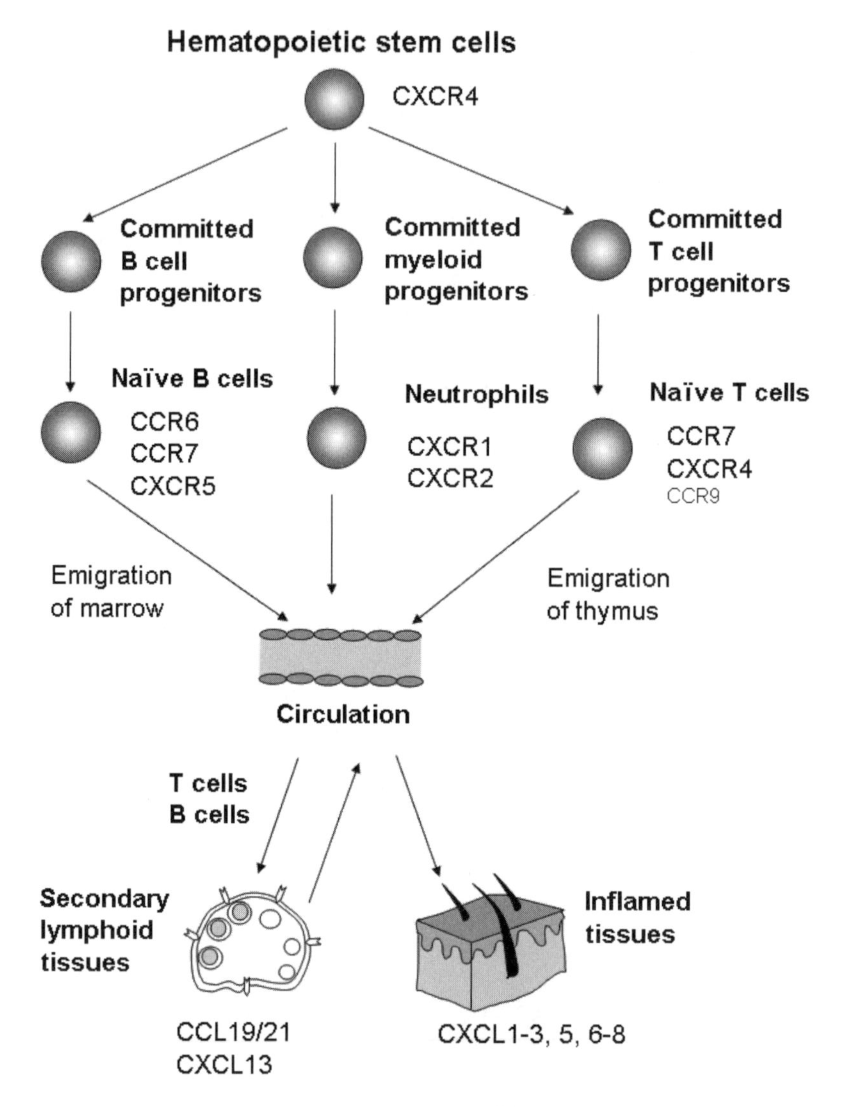

Fig. 4. Chemokine receptor switches occur during differentiation of HSPCs to diverse lineages of mature white blood cells. Three lineages of hematopoietic cells (T cells, B cells, and neutrophils) are shown as examples. Hematopoietic stem cells mainly express CXCR4, however, upon differentiation, upregulate lineage-specific or target tissue–specific chemokine receptors on mature white blood cells. For example, naïve B cells express CCR7, CXCR5, and CCR6 and migrate to B-cell follicles of secondary lymphoid tissues. Naïve T cells express CCR7 and CXCR4 (and CCR9 by some naïve CD8 T cells) to migrate to T-cell areas of secondary lymphoid tissues such as spleen, lymph nodes, and gut-associated lymphoid tissues. Neutrophils express CXCR1 and CXCR2 but do not express CCR7. This expression profile of chemokine receptors and chemotactic response keeps neutrophils in the blood circulation until they are recruited to sites of inflammation in response to the chemokine ligands of CXCR1 and CXCR2.

B cells are a good example of cells that undergo such a switch in chemokine receptors. B cells undergo maturation in bone marrow until they become naïve B cells. Chemokines regulate trafficking of B cells at various developmental stages. Committed B cell progenitors arise from hematopoietic stem cells that are highly responsive to CXCL12 and dependent on CXCL12 for their development *(23)*. Committed B cell progenitors undergo multistage differentiation in marrow: pre-pro B cells → pro B cells → pre B cells → immature B cells *(72)*. In humans, B cells are highly sensitive to CXCL12 at pro– and pre–B cell stages but become less sensitive at later stages *(73)*. This change is thought to weaken the interaction of B-cell progenitors with stromal cells expressing appropriate growth factors and to promote emigration of B cells out of bone marrow. While they lose their response to CXCL12, B-cell progenitors gain the responsiveness to a different group of chemokines such as CCL19 and CCL21 *(74,75)*. These chemokines are the ligands of CCR7 and are expressed by endothelial cells and stromal cells in the T zones of secondary lymphoid tissues *(76–78)*. In mice, the B-cell response to CCL19 and CCL21 is also increased as they become more mature in marrow. However, unlike in humans, the chemotactic response of mouse B-cell progenitors to CXCL12 is maintained at high levels at most stages of B-cell differentiation *(79)*. B-cell response to CXCL12 is slightly decreased at pro-B stage but is regained at pre-B stage *(80)*. The change in B-cell response to CXCL12 is not due to downregulation of CXCR4 but probably due to CXCR4 being unable to transmit signals as the result of cell differentiation *(80–82)*.

Whereas committed B-cell progenitors stay in marrow for their continuous development to naïve B cells, committed T-cell progenitors (called *prothymocytes*) leave bone marrow and migrate to thymus *(83)*. T-cell precursors expand and undergo maturation and selection processes to become naïve T cells in thymus. In a manner similar to B cells, thymocytes undergo discrete stages of development to become $CD4^+$ or $CD8^+$ naïve T cells. Based on expression of CD4 and CD8, prothymocytes undergo $CD4^-CD8^-$, $CD4^+CD8^+$, and $CD4^+CD8^-$ or $CD4^-CD8^+$ stages of development. T-cell precursors migrate into thymus through the postcapillary venules located deep in thymus near the corticomedullary junction, which is also the exit of thymus through which naïve T cells emigrate after development. Thymocyte subsets are differentially located in thymus: $CD4^-CD8^-$ cells are found in the subcapsular region, whereas $CD4^+CD8^+$ are located close to thymic medulla. $CD4^+CD8^-$ or $CD4^-CD8^+$ single-positive T cells are located in the medulla of thymus. Thymocytes undergo a developmental switch in chemotactic response to chemokines as described below *(84)*.

Immature $CD4^-CD8^-$ cells and $CD4^+CD8^+$ cells respond very well to CXCL12, whereas $CD4^+CD8^-$ or $CD4^-CD8^+$ cells respond poorly. Early thymocytes highly express the CCL12 receptor CXCR4 *(85–87),* and CXCL12 is important for their development to CD4$^+$CD8$^-$ cells *(88).* During this development, thymocytes decrease their response to CXCL12, whereas the response to CCL19 and CCL21 is increased. CD4$^+$CD8$^+$ T cells undergo a selection process called *positive selection* in thymus to select functional T cells that can recognize major histocompatibility complex (MHC) molecules. The chemotactic response to CCL19 and CCL21 is significantly increased after the positive-selection process. Upregulation of CCR7 expression is required for the thymocyte localization in the medulla and emigration of CD4$^+$CD8$^-$ or CD4$^-$CD8$^+$ single-positive T cells to the circulation. Artificial overexpression of CCR7 in CD4$^+$CD8$^+$ T cells misplaced them in the medulla of thymus, emphasizing the important role of CCL19 and its receptor CCR7 in thymocyte localization *(89).* Consistent with its role, CCL19 is specifically expressed in the medulla *(90),* and emigration of thymocytes is defective in CCR7-deficient mice *(91).* Other chemokines that attract thymocytes include CCL25/thymus-expressed chemokine (TECK), expressed by thymic dendritic cells *(92),* and CCL22/macrophage-derived chemokine (MDC) *(90,93).* A competitive transplantation study revealed that the bone marrow cells, isolated from CCR9$^{-/-}$ mice, were less efficient in repopulating the thymus of irradiated SCID mice compared with wild-type bone marrow cells *(94).* Most thymocytes express CCR9, which is more upregulated after pre-TCR signaling at the CD4$^+$CD8$^+$ T-cell stage *(95,96).* Expression of CCR9 is decreased in most CD4$^+$ or CD8$^+$ single-positive T cells but is retained by some naïve CD8$^+$ T cells and most γδ TCR$^+$ T cells *(94,97).* CCL22 is specifically expressed by thymic medullary epithelial cells, and it attracts single-positive T cells between the late cortical and early medullary stages *(90,93,98).* However, the functional importance of CCL22 in T-cell development in thymus is still unclear.

8.5. Conclusions

Among all the chemokines identified to date, it is probable that CXCL12 is most important for hematopoiesis. CXCL12 is required for recruitment, retention, and localization of HSPCs within bone marrow and thymus. Additionally, CXCL12 provides growth and survival signals for HSPCs. Whereas CXCL12 plays a positive role in survival and proliferation of HSPCs, many chemokines play negative roles in proliferation of myeloid progenitors. It is unclear at this time why these chemokines suppress the proliferation of myeloid progenitors. Based on the fact that expression of many of these suppressive chemokines is induced in inflammation or they are constitutively expressed in the periphery, it is speculated that they may limit unwanted proliferation of progenitor cells during inflammation or in the periphery. Hematopoietic cells switch their chemokine receptors or chemotactic responsiveness to chemokines during their development to emigrate from bone marrow and thymus. Chemokines that are differentially expressed in the periphery play important roles in recruiting and colocalizing immune cells, and this tissue-specific cell recruitment is critical for optimal activation and effector function of the recruited immune cells. It is now clear that many chemokines and receptors appear to play unique roles in regulating the trafficking and differentiation of hematopoietic cells in specific tissue sites and conditions.

Acknowledgments

This work was supported in part by grants from NIH-NIAID (AI063064), the Sidney Kimmel Foundation, and the American Heart Association to C.H.K.

References

1. Moore MA, Metcalf D. Ontogeny of the haemopoietic system: yolk sac origin of in vivo and in vitro colony forming cells in the developing mouse embryo. Br J Haematol 1970;18(3):279–296.
2. Molineux G, Pojda Z, Hampson IN, Lord BI, Dexter TM. Transplantation potential of peripheral blood stem cells induced by granulocyte colony-stimulating factor. Blood 1990;76(10):2153–2158.
3. Morrison SJ, Uchida N, Weissman IL. The biology of hematopoietic stem cells. Annu Rev Cell Dev Biol 1995;11:35–71.
4. Shizuru JA, Negrin RS, Weissman IL. Hematopoietic stem and progenitor cells: clinical and preclinical regeneration of the hematolymphoid system. Annu Rev Med 2005;56:509–538.
5. Suda T, Arai F, Hirao A. Hematopoietic stem cells and their niche. Trends Immunol 2005;26(8):426–433.
6. Broxmeyer HE, Sherry B, Lu L, et al. Myelopoietic enhancing effects of murine macrophage inflammatory proteins 1 and 2 on colony formation in vitro by murine and human bone marrow granulocyte/macrophage progenitor cells. J Exp Med 1989;170(5):1583–1594.
7. Broxmeyer HE, Sherry B, Lu L, et al. Enhancing and suppressing effects of recombinant murine macrophage inflammatory proteins on colony formation in vitro by bone marrow myeloid progenitor cells. Blood 1990;76(6):1110–1116.
8. Graham GJ, Wright EG, Hewick R, et al. Identification and characterization of an inhibitor of haemopoietic stem cell proliferation. Nature 1990;344(6265): 442–444.
9. Dunlop DJ, Wright EG, Lorimore S, et al. Demonstration of stem cell inhibition and myeloprotective effects of SCI/rhMIP1 alpha in vivo. Blood 1992;79(9): 2221–2225.

10. Maze R, Sherry B, Kwon BS, Cerami A, Broxmeyer HE. Myelosuppressive effects in vivo of purified recombinant murine macrophage inflammatory protein-1 alpha. J Immunol 1992;149(3):1004–1009.
11. Rossi D, Zlotnik A. The biology of chemokines and their receptors. Annu Rev Immunol 2000;18:217–242.
12. Broxmeyer HE, Cooper S, Hangoc G, Gao JL, Murphy PM. Dominant myelopoietic effector functions mediated by chemokine receptor CCR1. J Exp Med 1999; 189(12):1987–1992.
13. Broxmeyer HE, Kim CH. Regulation of hematopoiesis in a sea of chemokine family members with a plethora of redundant activities. Exp Hematol 1999; 27(7):1113–1123.
14. Broxmeyer HE, Sherry B, Cooper S, et al. Comparative analysis of the human macrophage inflammatory protein family of cytokines (chemokines) on proliferation of human myeloid progenitor cells. Interacting effects involving suppression, synergistic suppression, and blocking of suppression. J Immunol 1993;150(8 Pt 1):3448–3458.
15. Boring L, Gosling J, Chensue SW, et al. Impaired monocyte migration and reduced type 1 (Th1) cytokine responses in C-C chemokine receptor 2 knockout mice. J Clin Invest 1997;100(10):2552–2561.
16. Broxmeyer HE, Cooper S, Cacalano G, Hague NL, Bailish E, Moore MW. Involvement of Interleukin (IL) 8 receptor in negative regulation of myeloid progenitor cells in vivo: evidence from mice lacking the murine IL-8 receptor homologue. J Exp Med 1996;184(5):1825–1832.
17. Peled A, Gonzalo JA, Lloyd C, Gutierrez-Ramos JC. The chemotactic cytokine eotaxin acts as a granulocyte-macrophage colony-stimulating factor during lung inflammation. Blood 1998;91(6):1909–1916.
18. Tashiro K, Tada H, Heilker R, Shirozu M, Nakano T, Honjo T. Signal sequence trap: a cloning strategy for secreted proteins and type I membrane proteins. Science 1993;261(5121):600–603.
19. Nagasawa T, Hirota S, Tachibana K, et al. Defects of B-cell lymphopoiesis and bone-marrow myelopoiesis in mice lacking the CXC chemokine PBSF/SDF-1. Nature 1996;382(6592):635–638.
20. Zou YR, Kottmann AH, Kuroda M, Taniuchi I, Littman DR. Function of the chemokine receptor CXCR4 in haematopoiesis and in cerebellar development. Nature 1998;393(6685):595–599.
21. Ma Q, Jones D, Borghesani PR, et al. Impaired B-lymphopoiesis, myelopoiesis, and derailed cerebellar neuron migration in CXCR4- and SDF-1-deficient mice. Proc Natl Acad Sci U S A 1998;95(16):9448–9453.
22. Tachibana K, Hirota S, Iizasa H, et al. The chemokine receptor CXCR4 is essential for vascularization of the gastrointestinal tract. Nature 1998;393(6685):591–594.
23. Egawa T, Kawabata K, Kawamoto H, et al. The earliest stages of B cell development require a chemokine stromal cell-derived factor/pre-B cell growth-stimulating factor. Immunity 2001;15(2):323–334.

24. Nagasawa T, Kikutani H, Kishimoto T. Molecular cloning and structure of a pre-B-cell growth-stimulating factor. Proc Natl Acad Sci U S A 1994;91(6): 2305–2309.
25. Hodohara K, Fujii N, Yamamoto N, Kaushansky K. Stromal cell-derived factor-1 (SDF-1) acts together with thrombopoietin to enhance the development of megakaryocytic progenitor cells (CFU-MK). Blood 2000;95(3):769–775.
26. Kawabata K, Ujikawa M, Egawa T, et al. A cell-autonomous requirement for CXCR4 in long-term lymphoid and myeloid reconstitution. Proc Natl Acad Sci U S A 1999;96(10):5663–5667.
27. Broxmeyer HE, Cooper S, Kohli L, et al. Transgenic expression of stromal cell-derived factor-1/CXC chemokine ligand 12 enhances myeloid progenitor cell survival/antiapoptosis in vitro in response to growth factor withdrawal and enhances myelopoiesis in vivo. J Immunol 2003;170(1):421–429.
28. Broxmeyer HE, Kohli L, Kim CH, et al. Stromal cell-derived factor-1/CXCL12 directly enhances survival/antiapoptosis of myeloid progenitor cells through CXCR4 and G(alpha)i proteins and enhances engraftment of competitive, repopulating stem cells. J Leukoc Biol 2003;73(5):630–638.
29. Plett PA, Frankovitz SM, Wolber FM, Abonour R, Orschell-Traycoff CM. Treatment of circulating $CD34^+$ cells with SDF-1alpha or anti-CXCR4 antibody enhances migration and NOD/SCID repopulating potential. Exp Hematol 2002;30(9): 1061–1069.
30. Tavassoli M, Hardy CL. Molecular basis of homing of intravenously transplanted stem cells to the marrow. Blood 1990;76(6):1059–7100.
31. Aiuti A, Webb IJ, Bleul C, Springer T, Gutierrez-Ramos JC. The chemokine SDF-1 is a chemoattractant for human $CD34^+$ hematopoietic progenitor cells and provides a new mechanism to explain the mobilization of $CD34^+$ progenitors to peripheral blood. J Exp Med 1997;185(1):111–120.
32. Okumura N, Tsuji K, Ebihara Y, et al. Chemotactic and chemokinetic activities of stem cell factor on murine hematopoietic progenitor cells. Blood 1996;87(10): 4100–4108.
33. Cherry, Yasumizu R, Toki J, et al. Production of hematopoietic stem cell-chemotactic factor by bone marrow stromal cells. Blood 1994;83(4):964–971.
34. Kim CH, Broxmeyer HE. In vitro behavior of hematopoietic progenitor cells under the influence of chemoattractants: stromal cell-derived factor-1, steel factor, and the bone marrow environment. Blood 1998;91(1):100–110.
35. Gazitt Y, Liu Q. Plasma levels of SDF-1 and expression of SDF-1 receptor on $CD34^+$ cells in mobilized peripheral blood of non-Hodgkin's lymphoma patients. Stem Cells 2001;19(1):37–45.
36. Johnston B, Butcher EC. Chemokines in rapid leukocyte adhesion triggering and migration. Semin Immunol 2002;14(2):83–92.
37. Kim CH. The greater chemotactic network for lymphocyte trafficking: chemokines and beyond. Curr Opin Hematol 2005;12(4):298–304.
38. Papayannopoulou T, Craddock C, Nakamoto B, Priestley GV, Wolf NS. The VLA4/VCAM-1 adhesion pathway defines contrasting mechanisms of lodgement

of transplanted murine hemopoietic progenitors between bone marrow and spleen. Proc Natl Acad Sci U S A 1995;92(21):9647–9651.
39. Mazo IB, Gutierrez-Ramos JC, Frenette PS, Hynes RO, Wagner DD, von Andrian UH. Hematopoietic progenitor cell rolling in bone marrow microvessels: parallel contributions by endothelial selectins and vascular cell adhesion molecule 1. J Exp Med 1998;188(3):465–474.
40. Frenette PS, Subbarao S, Mazo IB, von Andrian UH, Wagner DD. Endothelial selectins and vascular cell adhesion molecule-1 promote hematopoietic progenitor homing to bone marrow. Proc Natl Acad Sci U S A 1998;95(24):14423–14428.
41. Peled A, Grabovsky V, Habler L, et al. The chemokine SDF-1 stimulates integrin-mediated arrest of CD34(+) cells on vascular endothelium under shear flow. J Clin Invest 1999;104(9):1199–1211.
42. Peled A, Kollet O, Ponomaryov T, et al. The chemokine SDF-1 activates the integrins LFA-1, VLA-4, and VLA-5 on immature human CD34(+) cells: role in transendothelial/stromal migration and engraftment of NOD/SCID mice. Blood 2000;95(11):3289–3296.
43. Simmons PJ, Masinovsky B, Longenecker BM, Berenson R, Torok-Storb B, Gallatin WM. Vascular cell adhesion molecule-1 expressed by bone marrow stromal cells mediates the binding of hematopoietic progenitor cells. Blood 1992;80(2):388–395.
44. Levesque JP, Hendy J, Takamatsu Y, Williams B, Winkler IG, Simmons PJ. Mobilization by either cyclophosphamide or granulocyte colony-stimulating factor transforms the bone marrow into a highly proteolytic environment. Exp Hematol 2002;30(5):440–449.
45. Levesque JP, Hendy J, Takamatsu Y, Simmons PJ, Bendall LJ. Disruption of the CXCR4/CXCL12 chemotactic interaction during hematopoietic stem cell mobilization induced by GCSF or cyclophosphamide. J Clin Invest 2003;111(2):187–196.
46. Heissig B, Hattori K, Dias S, et al. Recruitment of stem and progenitor cells from the bone marrow niche requires MMP-9 mediated release of kit-ligand. Cell 2002;109(5):625–637.
47. El Ouriaghli F, Fujiwara H, Melenhorst JJ, Sconocchia G, Hensel N, Barrett AJ. Neutrophil elastase enzymatically antagonizes the in vitro action of G-CSF: implications for the regulation of granulopoiesis. Blood 2003;101(5):1752–1758.
48. Hermant B, Bibert S, Concord E, et al. Identification of proteases involved in the proteolysis of vascular endothelium cadherin during neutrophil transmigration. J Biol Chem 2003;278(16):14002–14012.
49. Thomas J, Liu F, Link DC. Mechanisms of mobilization of hematopoietic progenitors with granulocyte colony-stimulating factor. Curr Opin Hematol 2002;9(3):183–189.
50. Pelus LM, Horowitz D, Cooper SC, King AG. Peripheral blood stem cell mobilization. A role for CXC chemokines. Crit Rev Oncol Hematol 2002;43(3):257–275.

51. Christopherson KW 2nd, Hangoc G, Broxmeyer HE. Cell surface peptidase CD26/dipeptidylpeptidase IV regulates CXCL12/stromal cell-derived factor-1 alpha-mediated chemotaxis of human cord blood CD34+ progenitor cells. J Immunol 2002;169(12):7000–7008.
52. Lambeir AM, Proost P, Durinx C, et al. Kinetic investigation of chemokine truncation by CD26/dipeptidyl peptidase IV reveals a striking selectivity within the chemokine family. J Biol Chem 2001;276(32):29839–29845.
53. Christopherson KW 2nd, Cooper S, Broxmeyer HE. Cell surface peptidase CD26/DPPIV mediates G-CSF mobilization of mouse progenitor cells. Blood 2003;101(12):4680–4686.
54. Christopherson KW, Cooper S, Hangoc G, Broxmeyer HE. CD26 is essential for normal G-CSF-induced progenitor cell mobilization as determined by $CD26^{-/-}$ mice. Exp Hematol 2003;31(11):1126–1134.
55. Pruijt JF, Fibbe WE, Laterveer L, et al. Prevention of interleukin-8-induced mobilization of hematopoietic progenitor cells in rhesus monkeys by inhibitory antibodies against the metalloproteinase gelatinase B (MMP-9). Proc Natl Acad Sci U S A 1999;96(19):10863–10868.
56. Fibbe WE, Pruijt JF, van Kooyk Y, Figdor CG, Opdenakker G, Willemze R. The role of metalloproteinases and adhesion molecules in interleukin-8-induced stem-cell mobilization. Semin Hematol 2000;37(1 Suppl 2):19–24.
57. Pruijt JF, Verzaal P, van Os R, et al. Neutrophils are indispensable for hematopoietic stem cell mobilization induced by interleukin-8 in mice. Proc Natl Acad Sci U S A 2002;99(9):6228–6233.
58. Lapidot T, Petit I. Current understanding of stem cell mobilization: the roles of chemokines, proteolytic enzymes, adhesion molecules, cytokines, and stromal cells. Exp Hematol 2002;30(9):973–981.
59. Gazitt Y. Homing and mobilization of hematopoietic stem cells and hematopoietic cancer cells are mirror image processes, utilizing similar signaling pathways and occurring concurrently: circulating cancer cells constitute an ideal target for concurrent treatment with chemotherapy and antilineage-specific antibodies. Leukemia 2004;18(1):1–10.
60. Ponomaryov T, Peled A, Petit I, et al. Induction of the chemokine stromal-derived factor-1 following DNA damage improves human stem cell function. J Clin Invest 2000;106(11):1331–1339.
61. Valenzuela-Fernandez A, Planchenault T, Baleux F, et al. Leukocyte elastase negatively regulates Stromal cell-derived factor-1 (SDF-1)/CXCR4 binding and functions by amino-terminal processing of SDF-1 and CXCR4. J Biol Chem 2002;277(18):15677–15689.
62. van Os R, van Schie ML, Willemze R, Fibbe WE. Proteolytic enzyme levels are increased during granulocyte colony-stimulating factor-induced hematopoietic stem cell mobilization in human donors but do not predict the number of mobilized stem cells. J Hematother Stem Cell Res 2002;11(3):513–521.
63. Hattori K, Heissig B, Tashiro K, et al. Plasma elevation of stromal cell-derived factor-1 induces mobilization of mature and immature hematopoietic progenitor and stem cells. Blood 2001;97(11):3354–3360.

64. Liles WC, Broxmeyer HE, Rodger E, et al. Mobilization of hematopoietic progenitor cells in healthy volunteers by AMD3100, a CXCR4 antagonist. Blood 2003;102(8):2728–2730.
65. Broxmeyer HE, Orschell CM, Clapp DW, et al. Rapid mobilization of murine and human hematopoietic stem and progenitor cells with AMD3100, a CXCR4 antagonist. J Exp Med 2005;201(8):1307–1318.
66. Papayannopoulou T, Priestley GV, Bonig H, Nakamoto B. The role of G-protein signaling in hematopoietic stem/progenitor cell mobilization. Blood 2003; 101(12):4739–4747.
67. Goetzl EJ, Rosen H. Regulation of immunity by lysosphingolipids and their G protein-coupled receptors. J Clin Invest 2004;114(11):1531–1537.
68. Graler MH, Goetzl EJ. The immunosuppressant FTY720 down-regulates sphingosine 1-phosphate G-protein-coupled receptors. FASEB J 2004;18(3): 551–553.
69. Chiba K, Hoshino Y, Suzuki C, et al. FTY720, a novel immunosuppressant possessing unique mechanisms. I. Prolongation of skin allograft survival and synergistic effect in combination with cyclosporine in rats. Transplant Proc 1996; 28(2):1056–1059.
70. Kimura T, Boehmler AM, Seitz G, et al. The sphingosine 1-phosphate receptor agonist FTY720 supports CXCR4-dependent migration and bone marrow homing of human $CD34^+$ progenitor cells. Blood 2004;103(12):4478–4486.
71. Wright DE, Bowman EP, Wagers AJ, Butcher EC, Weissman IL. Hematopoietic stem cells are uniquely selective in their migratory response to chemokines. J Exp Med 2002;195(9):1145–1154.
72. Hardy RR, Hayakawa K. B cell development pathways. Annu Rev Immunol 2001;19:595–621.
73. D'Apuzzo M, Rolink A, Loetscher M, et al. The chemokine SDF-1, stromal cell-derived factor 1, attracts early stage B cell precursors via the chemokine receptor CXCR4. Eur J Immunol 1997;27(7):1788–1793.
74. Kim CH, Pelus LM, Appelbaum E, Johanson K, Anzai N, Broxmeyer HE. CCR7 ligands, SLC/6Ckine/Exodus2/TCA4 and CKbeta-11/MIP-3beta/ELC, are chemoattractants for CD56(+)CD16(−) NK cells and late stage lymphoid progenitors. Cell Immunol 1999;193(2):226–235.
75. Kim CH, Broxmeyer HE. Chemokines: signal lamps for trafficking of T and B cells for development and effector function. J Leukoc Biol 1999;65(1):6–15.
76. Yoshie O, Imai T, Nomiyama H. Novel lymphocyte-specific CC chemokines and their receptors. J Leukoc Biol 1997;62(5):634–644.
77. Ngo VN, Tang HL, Cyster JG. Epstein-Barr virus-induced molecule 1 ligand chemokine is expressed by dendritic cells in lymphoid tissues and strongly attracts naive T cells and activated B cells. J Exp Med 1998;188(1):181–191.
78. Gunn MD, Tangemann K, Tam C, Cyster JG, Rosen SD, Williams LT. A chemokine expressed in lymphoid high endothelial venules promotes the adhesion and chemotaxis of naive T lymphocytes. Proc Natl Acad Sci U S A 1998; 95(1):258–263.

79. Bowman EP, Campbell JJ, Soler D, et al. Developmental switches in chemokine response profiles during B cell differentiation and maturation. J Exp Med 2000;191(8):1303–1318.
80. Tokoyoda K, Egawa T, Sugiyama T, Choi BI, Nagasawa T. Cellular niches controlling B lymphocyte behavior within bone marrow during development. Immunity 2004;20(6):707–718.
81. Fedyk ER, Ryyan DH, Ritterman I, Springer TA. Maturation decreases responsiveness of human bone marrow B lineage cells to stromal-derived factor 1 (SDF-1). J Leukoc Biol 1999;66(4):667–673.
82. Honczarenko M, Douglas RS, Mathias C, Lee B, Ratajczak MZ, Silberstein LE. SDF-1 responsiveness does not correlate with CXCR4 expression levels of developing human bone marrow B cells. Blood 1999;94(9):2990–2998.
83. Foss DL, Donskoy E, Goldschneider I. The importation of hematogenous precursors by the thymus is a gated phenomenon in normal adult mice. J Exp Med 2001;193(3):365–374.
84. Kim CH, Pelus LM, White JR, Broxmeyer HE. Differential chemotactic behavior of developing T cells in response to thymic chemokines. Blood 1998;91(12):4434–4443.
85. Zaitseva MB, Lee S, Rabin RL, et al. CXCR4 and CCR5 on human thymocytes: biological function and role in HIV-1 infection. J Immunol 1998;161(6):3103–3113.
86. Berkowitz RD, Beckerman KP, Schall TJ, McCune JM. CXCR4 and CCR5 expression delineates targets for HIV-1 disruption of T cell differentiation. J Immunol 1998;161(7):3702–3710.
87. Zamarchi R, Allavena P, Borsetti A, et al. Expression and functional activity of CXCR-4 and CCR-5 chemokine receptors in human thymocytes. Clin Exp Immunol 2002;127(2):321–330.
88. Hernandez-Lopez C, Varas A, Sacedon R, et al. Stromal cell-derived factor 1/CXCR4 signaling is critical for early human T-cell development. Blood 2002;99(2):546–554.
89. Kwan J, Killeen N. CCR7 directs the migration of thymocytes into the thymic medulla. J Immunol 2004;172(7):3999–4007.
90. Annunziato F, Romagnani P, Cosmi L, et al. Macrophage-derived chemokine and EBI1-ligand chemokine attract human thymocytes in different stages of development and are produced by distinct subsets of medullary epithelial cells: possible implications for negative selection. J Immunol 2000;165(1):238–246.
91. Ueno T, Hara K, Willis MS, et al. Role for CCR7 ligands in the emigration of newly generated T lymphocytes from the neonatal thymus. Immunity 2002;16(2):205–218.
92. Vicari AP, Figueroa DJ, Hedrick JA, et al. TECK: a novel CC chemokine specifically expressed by thymic dendritic cells and potentially involved in T cell development. Immunity 1997;7(2):291–301.
93. Campbell JJ, Pan J, Butcher EC. Cutting edge: developmental switches in chemokine responses during T cell maturation. J Immunol 1999;163(5):2353–2357.

94. Uehara S, Song K, Farber JM, Love PE. Characterization of CCR9 expression and CCL25/thymus-expressed chemokine responsiveness during T cell development: CD3(high)CD69+ thymocytes and gammadeltaTCR+ thymocytes preferentially respond to CCL25. J Immunol 2002;168(1):134–142.
95. Norment AM, Bogatzki LY, Gantner BN, Bevan MJ. Murine CCR9, a chemokine receptor for thymus-expressed chemokine that is up-regulated following pre-TCR signaling. J Immunol 2000;164(2):639–648.
96. Youn BS, Kim CH, Smith FO, Broxmeyer HE. TECK, an efficacious chemoattractant for human thymocytes, uses GPR-9–6/CCR9 as a specific receptor. Blood 1999;94(7):2533–2536.
97. Carramolino L, Zaballos A, Kremer L, et al. Expression of CCR9 beta-chemokine receptor is modulated in thymocyte differentiation and is selectively maintained in CD8(+) T cells from secondary lymphoid organs. Blood 2001;97(4):850–857.
98. Chantry D, Romagnani P, Raport CJ, et al. Macrophage-derived chemokine is localized to thymic medullary epithelial cells and is a chemoattractant for CD3(+), CD4(+), CD8(low) thymocytes. Blood 1999;94(6):1890–1898.

9

Chemokines in Transplantation Biology

Peter Jon Nelson, Stephan Segerer, and Detlef Schlondorff

Summary

Chemokines exert multiple actions in inflammation, wound healing, and differentiation. Because these processes underlie aspects of the pathophysiology of allograft rejection, chemokine biology has been a major focus of transplantation studies. Although diverse sets of chemokines can be detected in allografts after ischemic reperfusion injury and acute and chronic rejection, recent studies now suggest that relatively few chemokine receptors play central roles in the development of transplant rejection. The antagonism of select chemokine receptors causes a reduction in leukocyte infiltration, prolongs graft function, and can synergize with other immune-suppressive regimens. Chemokine-targeted therapy may represent an important approach in the treatment of allograft rejection.

Key Words: Allograft; transplantation; chemokine receptor; acute rejection; chronic rejection; CCR1; CCR5; CXCR3; CXCR1; CXCR2.

9.1. Introduction

Chemokines were initially characterized by their ability to attract select subpopulations of leukocytes *(1–3)*. As the infiltration of leukocytes from the peripheral circulation into the transplanted organ is a defining characteristic of allograft rejection, the biology of chemokines has been a major focus in transplantation research *(4–6)*. In various chapters of this volume, one will find detailed descriptions of the general biology of chemokines and their receptors. This chapter will focus on the diverse roles of chemokines in the pathophysiologic processes linked to solid-organ allograft rejection.

Transplant biology is important not only because of its direct clinical relevance, but also because it can serve as a model system for pathophysiologic

From: *The Receptors: The Chemokine Receptors*
Edited by: J. K. Harrison and N. W. Lukacs © Humana Press Inc., Totowa, NJ

processes associated with acute and chronic inflammatory disease. In experimental models, the aggressiveness of the rejection process can be adjusted either by administering subtherapeutic amounts of immune-suppressive agents or by selection of the genetic backgrounds used for host and allograft. Finally, because it is a surgical procedure, the time point for the initiation of tissue damage is known allowing a precise monitoring of subsequent events.

The literature on chemokines in transplantation has been extensively reviewed in recent years *(7–9)*. In this chapter, we focus on a limited number of chemokine receptors where evidence for a functional role has been verified. From the plethora of chemokine receptors, this has been demonstrated for CXCR1/2 in reperfusion injury and for CCR1, CCR5, and CXCR3 during acute and chronic allograft rejection.

Before examining the potential roles of chemokines in the pathophysiology of transplant rejection, it is important to review the biology that underlies the rejection of solid organ allografts.

9.2. The Immunobiology of Vascularized Allografts

The transplantation of solid organs such as lung, liver, kidney, and heart is only possible in the context of suppression of the immune response, specifically the adaptive T-cell response *(10)*. The three basic effector mechanisms involved in allograft rejection are production of alloantibodies (against both major histocompatibility complex (MHC) and non-MHC antigens), delayed-type hypersensitivity, and T-cell toxicity. The pathophysiologic processes can further be categorized based in part on when the damage to the graft occurs and which immune effector processes elicit the damage *(11)*. Hyperacute rejection is immediate, irreversible, and largely mediated by preformed circulating antibodies directed against surface molecules on the endothelium of the graft *(12)*. This form of rejection does not have an obvious chemokine component and will not be discussed in this review *(13)*.

9.2.1. *Acute Rejection*

Acute cellular allograft rejection is the most common form of early rejection. It occurs weeks to months after transplantation and involves the recruitment of leukocytes to arteries (*endothelialitis*), or to the tubulointerstitium (*tubulitis*) in renal transplantation. Acute rejection leads to interstitial edema with swelling of the organ and a rapid deterioration of allograft function *(4)*. The severity of early injury to the allograft helps to determine the long-term function of the grafted tissue *(14)*. The earliest damage is generated principally through ischemia-reperfusion injury *(15,16)*. In response to ischemia, various inflammatory mediators are produced, including proinflammatory cytokines and chemokines *(5,8,9,17)*. These agents activate adjacent endothelial cells. Circulating leukocytes roll across the activated endothelium and then firmly adhere to the surface of the endothelial layer through integrin-mediated mechanisms *(17,18)*. The subsequent steps in leukocyte recruitment include diapedesis and extravasation into the allograft.

Once in the allograft, leukocytes are further activated by exposure to injured tissue, which enhances the local production of inflammatory mediators and increases damage to the allograft *(5)*. Infiltrating macrophages and dendritic cells phagocytose allogenic tissue, process it, and present it on the surface of MHC class II molecules *(19)*. Allogeneic resident macrophages and dendritic cells are potent alloantigen presenting cells that strongly promote the activation of alloreactive T cells in secondary lymphoid organs. The alloreactive effector T cells are then subsequently recruited to the allograft to promote the rejection processes through CD178 (fas ligand) or cytotoxin (perforin, granzyme) mediated killing. Acute allograft rejection can generally be overcome through the use of conventional immune suppressive therapies *(10)*.

9.2.2. Chronic Allograft Damage

Although the early survival of allogeneic transplants has increased dramatically in recent years, the incidence of chronic rejection has not decreased appreciably *(20)*. Chronic rejection occurs in all types of solid-organ transplants. In heart transplants, it is characterized by progressive coronary artery disease *(21,22)*; in lung transplants, as bronchiolitis obliterans *(5,23)*. Liver allografts appear to be less affected by chronic rejection, but when it does occur, biliary epithelium is lost leading to hyperbilirubinemia and graft failure *(24)*. Chronic allograft nephropathy is the general term used to describe a slow deterioration of renal allograft function that is characterized histologically by interstitial fibrosis, tubular atrophy, widespread arterial intimal fibrosis, and global glomerulosclerosis *(25,26)*. In many respects, the tissue damage seen during chronic rejection mirrors the pathophysiologic processes seen in other non-transplant-associated disorders *(5,26)*.

The etiology of chronic allograft dysfunction is still unclear. Multiple factors appear to be associated with this process. It is thought to occur through the interaction of immunologic and nonimmunologic factors. There is some evidence that chronic rejection may represent a low-grade or smoldering rejection *(21,24)*. Cold ischemia-reperfusion injury at the time of transplantation is thought to be an important contributing factor to its occurrence. To date, there is no standard treatment for chronic rejection.

9.3. Chemokines and the Directed Recruitment of Leukocytes into Allografts

Until recently, it has been difficult to dissect the relative role of each chemokine in the inflammatory processes leading to allograft rejection, especially as many chemokines and chemokine receptors are seemingly redundant *(2,8)*. Indeed, during allograft rejection, the expression of many chemokines and chemokine receptors can be detected in the allograft—only a few of which represent viable targets for therapeutic intervention *(5,6)*.

Chemokine regulation of leukocyte migration occurs within a complex milieu of chemotactic signals where several receptors may be triggered simultaneously or successively *(18)*. Within this complex environment of receptor cross-talk and desensitization, the migrating leukocyte must interpret a hierarchy of signals within the tissue to reach the site of damage and express its effector function *(3)*. Within this already baroque complexity of interactions, biology has superimposed natural antagonists, receptor dimerization, and the interplay of posttranslationally modified chemokines *(27)*. To identify which chemokine and chemokine-receptor interactions underlie the biology of allograft dysfunction, various strategies have been employed including the application of blocking antibodies, ligand or receptor knockout mice, functional antagonists based on chemokine proteins, gene transfer, and small-molecule antagonists developed for selected receptors *(5,28)*.

Studies using mouse models to identify the importance of chemokine receptors can be problematic. For example, the microsurgery procedures required for mouse transplantation models can be technically complicated particularly in renal transplantation. An additional problem is that for unknown reasons, many common mouse transplant models allow engraftment without immune suppression even in the presence of major histocompatibility complex (MHC) class I and class II disparity (H-J. Gröne, personal communication). Finally, many early studies using chemokine receptor knockouts were performed using mice with mixed genetic backgrounds. This has made subsequent verification of these early results difficult or impossible to corroborate.

9.3.1. Ischemia-Reperfusion Injury in the Transplanted Organ

Early infiltration of polymorphonuclear leukocytes and monocyte/macrophages into the transplanted organ occurs soon after reperfusion of the transplant is initiated. In the context of transplantation, oxygen deprivation and general tissue stress of the graft leads to the activation of proinflammatory cytokines and the upregulation of the adhesion molecules required for leukocyte recruitment *(5,25,29)*. Strategies directed toward limiting the early tissue damage associated with reperfusion injury would likely have a significant impact on transplant survival. The receptors CXCR1 and CXCR2 are thought to control the effector polymorphonuclear leukocytes that mediate ischemia-reperfusion damage *(30,31)*. Making direct comparisons of these receptors between mouse, rat, and humans is complicated by the lack of a murine homologue for CXCR1 (or the ligand CXCL8). In the mouse, KC is considered to be the functional orthologue of human CXCL8. Blocking KC with an antisera reduced leukocyte recruitment and downstream expression of T cell–attracting chemokines (like CXCL10) *(32)*.

In murine models of cardiac allograft rejection, both polymorphonuclear leukocyte infiltration and intragraft proinflammatory cytokine expression have been shown to be reduced either through treatment of wild-type animals with a blocking CXCR2 antisera or when hearts were transplanted into $CXCR2^{-/-}$ recipients *(33)*. Short-term CXCR2 costimulatory blockades were also found to be effective in limiting T cell–mediated rejection of cardiac allografts *(33)*.

Early lung ischemia-reperfusion injury is a common cause of mortality after lung transplantation and is a risk factor for the development of the chronic bronchiolitis obliterans syndrome. Ischemia-reperfusion injury in lung is characterized by edema, a neutrophilic infiltrate, and elevated levels of ELR^{+} CXC chemokines. A rat model of orthotopic lung transplantation with cold ischemia-reperfusion injury showed an increased expression of CXCL1, CXCL2/3, their shared receptor CXCR2, and neutrophil infiltration, in association with lung damage. Inhibition of CXCR2 ligand interactions led to a marked reduction in lung neutrophil sequestration and graft injury *(34)*.

Ischemia-reperfusion injury is also a major obstacle to long-term renal allograft survival. In a rat ischemia-reperfusion injury model, the induction of CXCL1 was associated with granulocyte recruitment. A temporary ischemia model in the mouse also demonstrated induction of CXCL2, and a blockade of CXCL2 and CXCL3 using blocking antibodies decreased early neutrophil influx, prevented deterioration of renal function, and decreased mortality *(32)*. The blocking antisera were given just before reperfusion. In a model of rat reperfusion injury, treatment of the recipient animal with repertaxin, a small-molecule inhibitor of CXCR1/CXCR2, was recently shown to be effective in preventing granulocyte infiltration and renal function impairment, both in syngeneic and in allogeneic settings *(30,31)*. Unfortunately, human renal allograft biopsies are rarely performed within hours and the first day after reperfusion, therefore our knowledge about reperfusion-induced chemokine and chemokine-receptor expression in the human disease is still limited.

9.3.2. Chemokines, Dendritic Cells, and Maturation of the Allogeneic T-Cell Response

Transplanted organs contain immature dendritic cells (DCs) and resident macrophage cells. Immature DCs normally express a series of chemokine receptors important for their trafficking into these tissues including CCR1, CCR5, CCR6, and CXCR1 *(35)*. These immature DCs constantly sample antigens from the surrounding tissue and upon their activation by inflammatory stimuli, undergo maturation *(5,36)*. They stop sampling antigens, downregulate their proinflammatory chemokine receptors, and upregulate the receptors CXCR4, CCR4, and CCR7 *(35)*. CCR7 normally helps to facilitate the egress of DCs from the tissue into the draining lymphatic system. However, after surgery, the lymphatic system is disrupted. Thus, in the transplanted organ, the mature DCs most likely egress the tissue through the peripheral circulation and enter lymphatic organs such as the spleen where they respond to the CCR7 ligands CCL21 and CCL19 and traffic to the T cell–dependent zones of the secondary lymphoid tissue. Immature T cells also express CCR7 and thus accumulate in the same regions *(37)*. A subset of $CXCR5^+$ cells migrate to the B-cell zone in response to expression of CXCL13 and become central memory cells *(38,39)*. As the T cells become activated, they downregulate CCR7 and upregulate inflammatory chemokines receptors such as CCR5 and CXCR3, and the cells leave the secondary lymphatics via the peripheral circulation and home to the allograft in response to the proinflammatory chemokines expressed by the damaged organ *(35)*. Based on these observations, one would assume that these homing receptors, CCR7 and CXCR5, may play an important role in the initiation of alloreactive T cells. Interestingly, the targeted deletion of the receptors does not appear to have a significant impact on the development of acute rejection *(40)*. Only a modest prolongation of heart allograft survival has been described in $CCR7^{-/-}$ recipients accompanied by a delay in the cellular infiltration of allografts. It has been demonstrated that human renal allografts contain areas of nodular accumulations of inflammatory cells *(41)*. These areas are associated with lymphatic neoangiogenesis, expression of the chemokine CCL21, and accumulation of $CCR7^+$ cells. Proliferation of B and T cells in these areas supports the hypothesis of active ongoing alloresponses in these areas *(41)*. Although CCR7- and CXCR5-dependent processes support allograft rejection, they appear to be dispensable for the rejection response *(40)*.

9.3.3. CCR1, CCR5, and CXCR3 and the Recruitment of Monocyte/Macrophage Cells and Effector T Cells

Allograft rejection is thought to be primarily the result of a T helper 1 (Th1)-type immune response. Th1-like T cells often express CXCR3 and CCR5. T helper 2 (Th2) cells can express CCR3, CCR4, and CCR8 (reviewed in Ref. 42). Interestingly, the ligands for CXCR3, namely CXCL9, CXCL10, and CXCL11, are also functional antagonists for CCR3 *(43)*. Other examples of this phenomenon can be seen in CCL7, which recruits through CCR1, CCR2, and CCR3 but is an antagonist for CCR5 *(44)*. This suggests that chemokine expression in tissues can have an important effect on the direction of the T-cell response, a relevant issue for the development of tolerance and alloengraftment

The chemokine receptor CCR1 is expressed by subpopulations of $CD3^+$, $CD4^+$, $CD8^+$, and $CD16^+$ leukocytes in the peripheral blood, and in humans, it binds CCL3, CCL5, CCL8, CCL7, as well as other chemokine ligands *(2)*. CCR5 is expressed by the same general classes of leukocytes and shares many ligands with CCR1 (CCR1 ligands: CCL5, CCL4, CCL3, CCL8) *(2)*. The receptors CCR1 and CCR5 are, however, differentially expressed on leukocyte subsets (Fig. 1) *(45)*. Peripheral blood monocytes express high levels of CCR1 and low levels of CCR5 *(45)*. By contrast, $CD45RO^+$ "memory" T-cells express lower levels of CCR1 and high levels of CCR5 *(45)*. The apparent redundancy of action of these receptors can be explained by specialized roles in leukocyte recruitment where the induced firm adhesion of monocytes and $CD45RO^+$ T cells was found to be mediated predominately through CCR1, whereas CCR5 appeared to be important in leukocyte spreading *(45)*. This suggests specialized involvement of apparently redundant receptors in distinct steps of leukocyte trafficking.

In transplantation studies, mice lacking $CCR1^{-/-}$ showed a prolongation of cardiac allograft survival *(46)*. Class II mismatched heart allografts were permanently accepted by $CCR1^{-/-}$ recipients, and allografts transplanted across class I and class II were found to be rejected more slowly than controls *(46)*. A subtherapeutic dose of cyclosporin A enabled permanent allograft acceptance in $CCR1^{-/-}$ recipients *(46)*. A CCR1 receptor blockade using the small-molecule CCR1 antagonist BX471 has shown efficacy in rat heterotopic heart transplant models and rabbit renal transplantation models *(47,48)*. In both models, treatment with BX471 prolonged survival and preserved the morphologic integrity of the transplanted organ. In the rat heart transplant model, BX471 was found to synergize with subtherapeutic doses of cyclosporin A *(47)*.

Interstitial fibrosis is a key feature of chronic renal allograft dysfunction. Unfortunately, there is a lack of good murine models for chronic renal allograft nephropathy. Unilateral ureteral ligation produces a rapid interstitial fibrosis that is often used as a model of renal fibrosis. Treatment with the CCR1 antagonist BX471 during ureter obstruction reduced leukocyte infiltration as well as markers of renal fibrosis *(49)*. Interestingly, the late onset of treatment was also found to be effective suggesting that a CCR1 blockade may help limit the

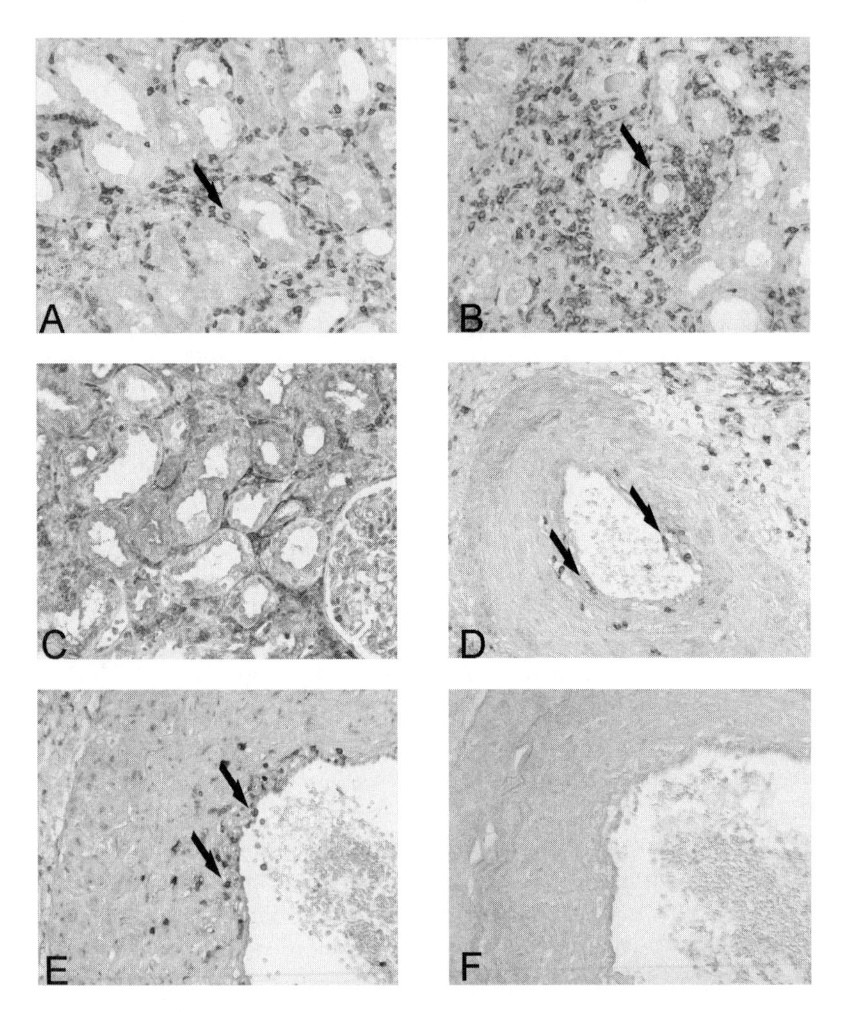

Fig. 1. This figure demonstrates the expression of a series of chemokine receptors linked to the development of allograft rejection in renal allografts undergoing rejection. (A) $CCR5^+$, (B) $CXCR3^+$, and (C) $CCR1^+$ infiltrating mononuclear cells can be seen in the tubulointerstitium and the tubular epithelium (arrows) during acute interstitial rejection. (D) $CXCR3^+$ and (E) $CCR5^+$ mononuclear cells, mostly T cells, can be seen infiltrating the subendothelial area (arrows) of an artery in vascular rejection. (F) Isotype control.

damage that occurs during the kidney damage associated with chronic rejection. The potential benefit of CCR1 blockade on renal fibrosis has also been confirmed in other models (e.g., models for lupus nephritis) *(50)*. To date, little is known about the expression of CCR1 during allograft rejection in humans. The only study on this issue demonstrated a low expression of CCR1. Additional studies are clearly needed *(51)*.

The amino-terminal regions of chemokines are important for receptor specificity. The addition of a single methionine residue to the amino-termini of CCL5 (met-RANTES) generates a functional antagonist for the CCL5 receptors CCR1 and CCR5 *(52)*. The use of met-RANTES in a rat model of mild acute renal allograft rejection showed a reduction in acute damage to endothelial and tubular compartments. The blockade of these receptors was found to synergize with subtherapeutic doses of cyclosporin A in an aggressive model of acute renal transplant rejection to significantly reduce damage to transplanted kidneys *(52)*. Consistent with these results, met-RANTES was found to prevent chronic allograft injury in a murine model of more chronic cardiac allograft vasculopathy *(53)*. Met-RANTES treatment significantly reduced intimal thickening and decreased the infiltration of CD4 and CD8 T lymphocytes and monocytes/macrophages into the transplanted hearts. In a recent study, ex vivo lentiviral gene transfer of an N-terminal deletion of CCL5 (RANTES 9-68) antagonist into rat hearts was found to significantly attenuate the inflammatory response and delay allograft rejection *(54)*. Thus, CCR1 and CCR5 appear to play significant roles in the development of chronic rejection.

As with any effective agent, chemokine antagonists may also have adverse effects related to specific chemokine function in other systems, to unspecific interactions, or to organ toxicity. For example, met-RANTES can aggravate glomerular damage and proteinuria in mice with immune complex glomerulonephritis *(55)*. These unexpected findings may be related to additional biological roles for the targeted receptors outside leukocyte recruitment.

An increase in expression of CCR5 is seen after the activation of monocytes and T cells. An increase of $CCR5^+$ infiltrating leukocytes is found during both acute and chronic phases of human renal allograft rejection (Fig. 1) *(56–58)*. The distribution of $CCR5^+$ leukocytes in areas of endothelialitis, tubulitis, and interstitial infiltrates mirrors the general expression of chemokines such as CCL5 and CCL4 *(59)*. CCR5 knockout animals disparate in MHC class II show reduced damage in transplanted hearts, and treatment with low-dose cyclosporin A augments the effects of a targeted deletion of CCR5 *(60,61)*.

One of the more interesting observations involving the biology of CCR5 in transplantation was found in a human study that suggested the presence of $CCR5^+$ leukocytes can influence the long-term survival of human renal allografts *(62)*. Approximately 1% of individuals of northern European heritage are homozygous for a null allele of CCR5 (CCR5Δ32). These individuals lack functional CCR5 because of a 32-base-pair deletion within the coding region of the gene. The prevalence of the CCR5Δ32 genotype was studied in a large cohort of human renal transplant recipients *(62)*. Patients identified as homozygous for CCR5Δ32 showed a dramatically prolonged allograft half-life compared with the heterozygous or wild-type allele. To date, these are the strongest supportive data for an important role for CCR5$^+$ leukocytes in human renal transplant nephropathy *(62)*. The development CCR5 antagonists as therapeutic agents for the suppression of R5 strains of HIV have provided the means to evaluate these agents in clinical transplant settings.

T cells upregulate expression of CXCR3 after activation. In addition, B cells and natural killer (NK) cells can also express CXCR3 *(2)*. A significant increase of CXCR3$^+$ cells is seen during acute allograft rejection paralleling the induction of the corresponding ligands CXCL10, CXCL9, and CXCL11 *(58,63,64)*. In human renal allografts, it has been demonstrated that CXCR3$^+$ cells (predominately CD4$^+$ and to a lesser extent CD8$^+$ T cells) infiltrate tubules during interstitial rejection (tubulitis) and arteries during vascular allograft rejection. The number of these cells increases significantly during rejection and decreases after successful rejection treatment. In a comparison between humoral and cellular rejection, it was demonstrated that only cellular rejection was associated with an accumulation of CXCR3$^+$ cells in renal allografts *(64)*. The measurement of CXCR3 ligands in the urine may help improve the surveillance of allograft recipients for rejection episodes *(65–67)*. Compelling experimental data on the potential role of CXCL10 and CXCR3 in allografts has been reported from knockout studies *(68,69)*. The data suggest a pivotal role of donor organ–derived CXCL10 and host CXCR3 expressing T cells in propagating the alloresponse. Mice deficient for CXCR3$^{-/-}$ were described as highly resistant to acute allograft rejection and, when treated with a transient, subtherapeutic dose of cyclosporin A, maintained their allografts *(68,69)*. These data provided a rationale for the use of therapeutic agents to block CXCR3. Unfortunately, these tantalizing results have not been supported by subsequent studies (H.G. Zerwes, personal communication), and the extent of the importance of this receptor in allograft damage still remains to be firmly established.

9.3.4. The Duffy Antigen/Receptor for Chemokines in Transplantation

The Duffy antigen/receptor for chemokines (DARC) is a chemokine-binding protein that lacks a G protein–coupled signaling domain. DARC binds members of both the CC and CXC chemokine families *(3)*. It has been proposed that the protein may function as a carrier for chemokines during their transcytosis from the basal side of endothelia to the luminal surface. Thus, DARC may contribute to leukocyte extravasation in high endothelial venules *(3)*. This protein is expressed on a low number of peritubular capillaries in the normal kidney, but the number of DARC$^+$ vessels increases during all forms of renal allograft rejection *(70,71)*. As DARC is a coreceptor for some malaria parasites, most people of Central African heritage do not express DARC on red blood cells (while they still demonstrate DARC expression on endothelial cells). The DARC genotype appears to have no significant impact on acute rejection and delayed graft function based on one study *(72),* whereas a second study found a lower allograft survival in Duffy (a-b-) patients in the context of delayed graft function *(73).* Both studies demonstrated no impact of erythrocyte DARC on acute rejection episodes, but the role of endothelial DARC in general and in transplants specifically remains to be defined.

9.4. Chemokine Polymorphisms as Predisposing or Prognostic Factors

Genetic variations including single nucleotide polymorphisms (SNPs) in chemokines or chemokine receptors have been linked to diverse inflammatory diseases *(28).* It is thought that these variations may change the tissue response to injury or the release of chemokines by infiltrating cells. A series of single nucleotide polymorphisms in chemokine and chemokine-receptor genes of the host have been linked to allograft dysfunction. The G allele of the CCL2/-2518 promoter polymorphism is associated with enhanced CCL2 expression in leukocytes and decreased renal allograft survival *(74).* A reduction in the risk of acute renal transplant rejection has been described in recipients with a CCR2-64I allele or with homozygosity for the CCR5 59029-A allele *(75).* Little is known about the role of donor polymorphisms on the outcome of allografts. It is hoped that with the information of polymorphisms, the immunosuppressive therapy might be tailored to low- or high-risk groups.

9.5. Conclusions

Recent developments have improved our understanding on the role of specific chemokine receptors in the pathophysiology of allograft dysfunction. The data are very encouraging regarding the potential for future therapies directed at CCR5, CCR1, CXCR1/2, and/or CXCR3 blockade. Clearly, blocking or modulating the expression of specific chemokines may be an important step in the control of the inflammatory processes leading to chronic transplant damage. Ultimately, chemokine blockades may be most advantageous when used in concert with conventional immune-suppressive agents.

References

1. Murphy PM. International Union of Pharmacology. Update on chemokine receptor nomenclature. Pharmacol Rev 2002;54:227–229.
2. Murphy PM, Baggiolini M, Charo IF, et al. International union of pharmacology. XXII. Nomenclature for chemokine receptors. Pharmacol Rev 2000;52:145–176.

3. Rot A, von Andrian UH. Chemokines in innate and adaptive host defense: basic chemokinese grammar for immune cells. Annu Rev Immunol 2004;22:891–928.
4. Racusen LC, Solez K, Colvin RB, et al. The Banff 97 working classification of renal allograft pathology. Kidney Int 1999;55:713–723.
5. Nelson PJ, Krensky AM. Chemokines, chemokine receptors, and allograft rejection. Immunity 2001;14:377–386.
6. Stein JV, Nombela-Arrieta C. Chemokine control of lymphocyte trafficking: a general overview. Immunology 2005;116:1–12.
7. Hancock WW, Gao W, Faia KL, Csizmadia V. Chemokines and their receptors in allograft rejection. Curr Opin Immunol 2000;12:511–516.
8. Segerer S, Nelson PJ, Schlondorff D. Chemokines, chemokine receptors, and renal disease: from basic science to pathophysiologic and therapeutic studies. J Am Soc Nephrol 2000;11:152–176.
9. Colvin BL, Thomson AW. Chemokines, their receptors, and transplant outcome. Transplantation 2002;74:149–155.
10. Halloran PF. Immunosuppressive drugs for kidney transplantation. N Engl J Med 2004;351:2715–2729.
11. Regele H, Bohmig GA. Tissue injury and repair in allografts: novel perspectives. Curr Opin Nephrol Hypertens 2003;12:259–266.
12. Bohmig GA, Exner M, Watschinger B, Regele H. Acute humoral renal allograft rejection. Curr Opin Urol 2002;12:95–99.
13. Bishop GA, Haxhinasto SA, Stunz LL, Hostager BS. Antigen-specific B-lymphocyte activation. Crit Rev Immunol 2003;23:149–197.
14. Ishikawa A, Flechner SM, Goldfarb DA, et al. Quantitative assessment of the first acute rejection as a predictor of renal transplant outcome. Transplantation 1999;68:1318–1324.
15. Land WG. The role of postischemic reperfusion injury and other nonantigen-dependent inflammatory pathways in transplantation. Transplantation 2005; 79:505–514.
16. Perico N, Cattaneo D, Sayegh MH, Remuzzi G. Delayed graft function in kidney transplantation. Lancet 2004;364:1814–1827.
17. Segerer S, Alpers CE. Chemokines and chemokine receptors in renal pathology. Curr Opin Nephrol Hypertens 2003;12:243–249.
18. Ley K. Arrest chemokines. Microcirculation 2003;10:289–95.
19. Raimondi G, Thomson AW. Dendritic cells, tolerance and therapy of organ allograft rejection. Contrib Nephrol 2005;146:105–120.
20. Afzali B, Taylor AL, Goldsmith DJ. What we CAN do about chronic allograft nephropathy: role of immunosuppressive modulations. Kidney Int 2005;68: 2429–2443.
21. Weis M, von Scheidt W. Cardiac allograft vasculopathy: a review. Circulation 1997;96:2069–2077.
22. Labarrere CA, Nelson DR, Park JW. Pathologic markers of allograft arteriopathy: insight into the pathophysiology of cardiac allograft chronic rejection. Curr Opin Cardiol 2001;16:110–117.

23. Neuringer IP, Chalermskulrat W, Aris R. Obliterative bronchiolitis or chronic lung allograft rejection: a basic science review. J Heart Lung Transplant 2005;24: 3–19.
24. Reding R, Davies HF. Revisiting liver transplant immunology: from the concept of immune engagement to the dualistic pathway paradigm. Liver Transpl 2004;10:1081–1086.
25. Joosten SA, Sijpkens YW, van Kooten C, Paul LC. Chronic renal allograft rejection: pathophysiologic considerations. Kidney Int 2005;68:1–13.
26. Meyers CM, Kirk AD. Workshop on late renal allograft dysfunction. Am J Transplant 2005;5:1600–1605.
27. Ogilvie P, Thelen S, Moepps B, et al. Unusual chemokine receptor antagonism involving a mitogen-activated protein kinase pathway. J Immunol 2004;172: 6715–6722.
28. Hancock WW. Chemokines and transplant immunobiology. J Am Soc Nephrol 2002;13:821–824.
29. Lopez-Neblina F, Toledo AH, Toledo-Pereyra LH. Molecular biology of apoptosis in ischemia and reperfusion. J Invest Surg 2005;18:335–350.
30. Cugini D, Azzollini N, Gagliardini E, et al. Inhibition of the chemokine receptor CXCR2 prevents kidney graft function deterioration due to ischemia/reperfusion. Kidney Int 2005;67:1753–1761.
31. Bertini R, Allegretti M, Bizzarri C, et al. Noncompetitive allosteric inhibitors of the inflammatory chemokine receptors CXCR1 and CXCR2: prevention of reperfusion injury. Proc Natl Acad Sci U S A 2004;101:11791–11796.
32. Morita K, Miura M, Paolone DR, et al. Early chemokine cascades in murine cardiac grafts regulate T cell recruitment and progression of acute allograft rejection. J Immunol 2001;167:2979–2984.
33. El-Sawy T, Belperio JA, Strieter RM, Remick DG, Fairchild RL. Inhibition of polymorphonuclear leukocyte-mediated graft damage synergizes with short-term costimulatory blockade to prevent cardiac allograft rejection. Circulation 2005; 112:320–331.
34. Belperio JA, Keane MP, Burdick MD, et al. CXCR2/CXCR2 ligand biology during lung transplant ischemia-reperfusion injury. J Immunol 2005;175:6931–6939.
35. Luster AD. The role of chemokines in linking innate and adaptive immunity. Curr Opin Immunol 2002;14:129–135.
36. Randolph GJ, Sanchez-Schmitz G, Angeli V. Factors and signals that govern the migration of dendritic cells via lymphatics: recent advances. Springer Semin Immunopathol 2005;26:273–287.
37. Hancock WW, Wang L, Ye Q, Han R, Lee I. Chemokines and their receptors as markers of allograft rejection and targets for immunosuppression. Curr Opin Immunol 2003;15:479–486.
38. Ohl L, Henning G, Krautwald S, et al. Cooperating mechanisms of CXCR5 and CCR7 in development and organization of secondary lymphoid organs. J Exp Med 2003;197:1199–1204.

39. Okada T, Ngo VN, Ekland EH, et al. Chemokine requirements for B cell entry to lymph nodes and Peyer's patches. J Exp Med 2002;196:65–75.
40. Beckmann JH, Yan S, Luhrs H, et al. Prolongation of allograft survival in ccr7-deficient mice. Transplantation 2004;77:1809–1814.
41. Kerjaschki D, Regele HM, Moosberger I, et al. Lymphatic neoangiogenesis in human kidney transplants is associated with immunologically active lymphocytic infiltrates. J Am Soc Nephrol 2004;15:603–612.
42. Mantovani A, Allavena P, Vecchi A, Sozzani S. Chemokines and chemokine receptors during activation and deactivation of monocytes and dendritic cells and in amplification of Th1 versus Th2 responses. Int J Clin Lab Res 1998;28: 77–82.
43. Fulkerson PC, Zhu H, Williams DA, Zimmermann N, Rothenberg ME. CXCL9 inhibits eosinophil responses by a CCR3- and Rac2-dependent mechanism. Blood 2005;106:436–443.
44. Blanpain C, Migeotte I, Lee B, et al. CCR5 binds multiple CC-chemokines: MCP-3 acts as a natural antagonist. Blood 1999;94:1899–1905.
45. Weber C, Weber KS, Klier C, et al. Specialized roles of the chemokine receptors CCR1 and CCR5 in the recruitment of monocytes and T(H)1-like/CD45RO(+) T cells. Blood 2001;97:1144–1146.
46. Gao W, Topham PS, King JA, et al. Targeting of the chemokine receptor CCR1 suppresses development of acute and chronic cardiac allograft rejection. J Clin Invest 2000;105:35–44.
47. Horuk R, Clayberger C, Krensky AM, et al. A non-peptide functional antagonist of the CCR1 chemokine receptor is effective in rat heart transplant rejection. J Biol Chem 2001;276:4199–4204.
48. Horuk R, Shurey S, Ng HP, et al. CCR1-specific non-peptide antagonist: efficacy in a rabbit allograft rejection model. Immunol Lett 2001;76:193–201.
49. Anders HJ, Vielhauer V, Frink M, et al. A chemokine receptor CCR-1 antagonist reduces renal fibrosis after unilateral ureter ligation. J Clin Invest 2002;109: 251–259.
50. Anders HJ, Banas B, Schlondorff D. Signaling danger: toll-like receptors and their potential roles in kidney disease. J Am Soc Nephrol 2004;15:854–867.
51. Ruster M, Sperschneider H, Funfstuck R, Stein G, Grone HJ. Differential expression of beta-chemokines MCP-1 and RANTES and their receptors CCR1, CCR2, CCR5 in acute rejection and chronic allograft nephropathy of human renal allografts. Clin Nephrol 2004;61:30–39.
52. Grone HJ, Weber C, Weber KS, et al. Met-RANTES reduces vascular and tubular damage during acute renal transplant rejection: blocking monocyte arrest and recruitment. FASEB J 1999;13:1371–1383.
53. Yun JJ, Whiting D, Fischbein MP, et al. Combined blockade of the chemokine receptors CCR1 and CCR5 attenuates chronic rejection. Circulation 2004;109: 932–937.
54. Vassalli G, Simeoni E, Li JP, Fleury S. Lentiviral gene transfer of the chemokine antagonist RANTES 9–68 prolongs heart graft survival. Transplantation 2006;81:240–246.

55. Anders HJ, Frink M, Linde Y, et al. CC chemokine ligand 5/RANTES chemokine antagonists aggravate glomerulonephritis despite reduction of glomerular leukocyte infiltration. J Immunol 2003;170:5658–5666.
56. Segerer S, Mack M, Regele H, Kerjaschki D, Schlondorff D. Expression of the C-C chemokine receptor 5 in human kidney diseases. Kidney Int 1999;56:52–64.
57. Segerer S, Cui Y, Eitner F, et al. Expression of chemokines and chemokine receptors during human renal transplant rejection. Am J Kidney Dis 2001;37: 518–531.
58. Panzer U, Reinking RR, Steinmetz OM, et al. CXCR3 and CCR5 positive T-cell recruitment in acute human renal allograft rejection. Transplantation 2004;78: 1341–1350.
59. Pattison J, Nelson PJ, Huie P, et al. RANTES chemokine expression in cell-mediated transplant rejection of the kidney. Lancet 1994;343:209–211.
60. Gao W, Faia KL, Csizmadia V, et al. Beneficial effects of targeting CCR5 in allograft recipients. Transplantation 2001;72:1199–1205.
61. Luckow B, Joergensen J, Chilla S, et al. Reduced intragraft mRNA expression of matrix metalloproteinases Mmp3, Mmp12, Mmp13 and Adam8, and diminished transplant arteriosclerosis in Ccr5-deficient mice. Eur J Immunol 2004;34: 2568–2578.
62. Fischereder M, Luckow B, Hocher B, et al. CC chemokine receptor 5 and renal-transplant survival. Lancet 2001;357:1758–1761.
63. Hoffmann U, Segerer S, Rummele P, et al. Expression of the chemokine receptor CXCR3 in human renal allografts—a prospective study. Nephrol Dial Transplant 2006;21:1373–1381.
64. Segerer S, Bohmig GA, Exner M, Kerjaschki D, Regele H, Schlondorff D. Role of CXCR3 in cellular but not humoral renal allograft rejection. Transpl Int 2005;18:676–680.
65. Hu H, Aizenstein BD, Puchalski A, Burmania JA, Hamawy MM, Knechtle SJ. Elevation of CXCR3-binding chemokines in urine indicates acute renal-allograft dysfunction. Am J Transplant 2004;4:432–437.
66. Kanmaz T, Feng P, Torrealba J, et al. Surveillance of acute rejection in baboon renal transplantation by elevation of interferon-gamma inducible protein-10 and monokine induced by interferon-gamma in urine. Transplantation 2004;78: 1002–1007.
67. Hauser IA, Spiegler S, Kiss E, et al. Prediction of acute renal allograft rejection by urinary monokine induced by IFN-gamma (MIG). J Am Soc Nephrol 2005; 16:1849–1858.
68. Hancock WW, Gao W, Csizmadia V, Faia KL, Shemmeri N, Luster AD. Donor-derived IP-10 initiates development of acute allograft rejection. J Exp Med 2001;193:975–980.
69. Hancock WW, Lu B, Gao W, et al. Requirement of the chemokine receptor CXCR3 for acute allograft rejection. J Exp Med 2000;192:1515–1520.
70. Segerer S, Regele H, Mack M, et al. The duffy antigen receptor for chemokines is up-regulated during acute renal transplant rejection and crescentic glomerulonephritis. Kidney Int 2000;58:1546–1556.

71. Segerer S, Bohmig GA, Exner M, et al. When renal allografts turn darc. Transplantation 2003;75:1030–1034.
72. Mange KC, Prak EL, Kamoun M, et al. Duffy antigen receptor and genetic susceptibility of African Americans to acute rejection and delayed function. Kidney Int 2004;66:1187–1192.
73. Akalin E, Neylan JF. The influence of Duffy blood group on renal allograft outcome in African Americans. Transplantation 2003;75:1496–1500.
74. Kruger B, Schroppel B, Ashkan R, et al. A Monocyte chemoattractant protein-1 (MCP-1) polymorphism and outcome after renal transplantation. J Am Soc Nephrol 2002;13:2585–2589.
75. Abdi R, Tran TB, Sahagun-Ruiz A, et al. Chemokine receptor polymorphism and risk of acute rejection in human renal transplantation. J Am Soc Nephrol 2002;13:754–758.

10

The Chemokine System and Arthritis

Marlon P. Quinones, Fabio Jimenez, Carlos A. Estrada, Hernan G. Martiniez, and Seema S. Ahuja

Summary

Rheumatoid arthritis (RA) affects 1% of the U.S. population, or 3 million people. RA is a chronic inflammatory disease characterized by synovial inflammation and hyperplasia, neovascularization, and progressive destruction of cartilage and bone. The cause(s) of RA remains poorly defined, but it is thought that both genetic and environmental factors, including molecular mimicry and host responses against cross-reactive antigens, lead to the development of autoimmune responses involving the joints. The goal of this chapter is to provide an overview of the contribution of members of the chemokine system to the immunopathogenesis of RA. This is of high clinical relevance because of the burgeoning interest in the use of chemokine and chemokine-receptor blockers in the treatment of a variety of inflammatory, autoimmune, and infectious disorders, including RA.

Key Words: Chemokines; chemokine receptors; rheumatoid arthritis; murine models.

10.1. Introduction

Rheumatoid arthritis (RA) affects 1% of the US population, or 3 million people *(1–3)*. Although women are 2.5 times more likely to get RA than men, some studies suggest the disease tends to be more severe in men. RA can affect people of all ages, but prevalence increases with age, approaching 5% in women over 55 years of age *(1–3)*. RA is a chronic inflammatory disease characterized by synovial inflammation and hyperplasia, neovascularization, and progressive destruction of cartilage and bone *(1,4,5)*. The cause(s) of RA remains poorly

From: *The Receptors: The Chemokine Receptors*
Edited by: J. K. Harrison and N. W. Lukacs © Humana Press Inc., Totowa, NJ

defined, but it is thought that both genetic and environmental factors, including molecular mimicry and host responses against cross-reactive antigens, lead to the development of autoimmune responses involving the joints *(1,6)*.

The goal of this chapter is to provide an overview of the contribution of members of the chemokine system to the immunopathogenesis of RA. This is of high clinical relevance because of the burgeoning interest in the use of chemokine and chemokine-receptor blockers in the treatment of a variety of inflammatory, autoimmune, and infectious disorders, including RA (Table 1). To begin to tackle the complexity of cellular and molecular processes taking place in a disease such as RA, it is conceptually helpful to deconstruct the phenotype in clusters of events happening over time *(7,8)*. In this light, it is possible that in RA there are at least four major pathogenic phases: an *induction phase*, in which the interaction between antigen-presenting cells (APC) [e.g., dendritic cells (DCs), B cells, or macrophages], and T cells leads to the generation of autoimmune responses; *an effector phase*, in which autoreactive T cells and antibodies (Abs) generated during the induction phase promote joint inflammation; *a resolution phase*, in which clearance of immune complexes by cells such as monocyte/macrophages and apoptosis of the autoreactive T cells leads to termination of the ongoing response; and finally, abnormalities in the resolution phase may lead to the characteristic chronicity seen in RA, ensuring the *maintenance* of the arthritic process. As we will show in this chapter, it is likely that chemokines and chemokine receptors play a key role in each of these pathogenic phases.

10.2. Role of Chemokines During the Different Phases of Rheumatoid Arthritis

A great deal of the data available thus far regarding the pathogenic mechanisms in arthritis has been derived from animal models. For instance, the identification of TNF-α as a therapeutic target, one of the effective immunotherapeutic agents in RA so far, was established in studies conducted in animal models *(5,9,10)*. Collagen type II (CII)-induced arthritis (CIA) in the arthritis-susceptible DBA/1j mouse strain is one of the most commonly used immunization-based models in which many of the pathologic features of human RA are recapitulated *(11–16)*. The role of chemokines during the four phases described above can be clearly appreciated in the CIA model.

10.2.1. Induction Phase

During this phase, chemokines and their receptors could play a very important role by promoting cellular encounters in lymphoid structures. Although a growing body of evidence supports the notion that DCs play a major role in the initiation and perpetuation of RA by presenting presumed arthritogenic antigens to autoreactive T cells *(17–20)*, the precise mechanisms by which DCs affect disease pathogenesis remain to be defined. DCs may capture the antigen (Ag) in the periphery (in the case of CIA, the Ag is injected intradermally) and migrate toward the draining lymph node (DLN). This migration is guided by chemokine gradients. In particular, upregulation of CCR7 in DCs, a phenomenon that occurs as part of their "maturation" process, confers responsiveness to high levels of the CCR7 ligand CCL21 that is upregulated in the lymphatic channels and in lymph nodes (LNs) *(21)*. Also, another CCR7 ligand, CCL19, is abundantly expressed in the T-cell zone of the LN, and this chemokine attracts Ag-transporting DCs to the T-cell zone *(21)*. In the T-cell zone, DCs encounter T cells responsive to CII epitopes.

One of the more important features of CIA is the production of anti-CII antibodies, which play an important role during the effector phase of the disease. After activation by DCs, CII-specific T cells encounter B cells that are present in the B-cell follicles of LNs. This interaction may depend on the upregulation of the chemokine receptor CXCR5, among others, in some activated CII-specific T cells, which leads to their migration toward the high concentration of the CXCR5 ligand and CXCL13 in the B-cell follicles *(21,22)*. Ag-specific B- and T-cell interaction may lead to the formation of germinal centers where B cells may become very efficient in producing Abs and begin their differentiation into plasma cells. Plasma cells commonly reside in the bone marrow (BM). Therefore, it is likely that some of the CII-specific B cells may upregulate CXCR4 and migrate toward the high levels of the CXCR4 ligand CXCL12 that are present in the BM microenvironment *(23)*.

10.2.2. From Effector Phase to Maintenance

At the conclusion of the induction phase in CIA, CII-specific T cells and plasma cells producing high levels of anti-CII antibodies would initiate the early effector phase of the disease. Recently, Wipke et al. *(24)* have proposed a four-step model to describe how certain autoantibodies promote the effector phase of arthritis (Fig. 1; see color plate): In the first stage, Abs in circulation may form immune complexes (ICs) with their cognate Ags. These ICs would bind to FcγRIII on neutrophils triggering the release of vasoactive mediators such as TNF-α, resulting in increase in vascular permeability at the level of the joints. Some IC-bound neutrophils would enter the perivascular space where they encounter other cells that may further amplify the disease process in the joints (e.g., mast cell). In the first stage, it is likely that neutrophil-attracting chemokines play a major role. In the second stage, additional Abs would bind to their target leading to activation of innate effector mechanisms and inflammation. For instance, Abs bound to the cartilage surface activate the alternative pathway of complement, producing the anaphylotoxin C5a; C5a then activates

Table 1
Chemokine or Chemokine-Receptor Antagonists in Clinical Trials

	Target	Name	Company	Status	Disease/comments	Ref.
Chemokines	CCL2	ABN912 (monoclonal antibody)	NOVARTIS	Phase I/II	Asthma. RA. In RA patients, there was increase dsynovial macrophages and CCL2 circulating levels.	95, 109
	CCL11 (Eotaxin)	Bertilimumab (CAT-213) Monoclonal IgG4	Cambridge AT	Phase II	Conjunctivitis, allergic rhinitis	110–113
	CXCL8 (IL-8)	Monoclonal Ab anti-IL8	XenoMouse Technology	Phase II	RA. No clinical benefits	114, 115
Chemokine receptors	CCR1	BX-471 (ZK-811752) (nonpeptide)	Berlex Biosciences/ Shering AG	Phase II Phase I	MS, psoriasis, eczema Alzheimer disease	111–113, 116, 117
		MLN-3897	Millenium Pharmaceuticals/ Aventis	Phase II	RA, MS, psoriasis	111, 112, 118
		MLN-3701	Millenium Pharmaceuticals	Phase I	RA	111, 112, 119
		Unknown	Pfizer	Phase I	RA. Reduction in joint inflammation, no. of macrophages and $CCR1^+$cells in synovium, decreases in overall cellularity, $CD4^+$ or $CD8^+$ T cells. However, no clinical benefits.	120

Table 1 *(Continued)*

CP-481,715	Pfizer		RA. Inhibits 90% of monocyte chemotactic activity present in 11 of 15 RA synovial fluid samples.		121
C-4462	Banyu (Merck)	Phase II	RA	122	
CCR2	MLN-1202 (antibody)	Millenium Pharmaceuticals	Phase II	RA	111, 112, 123
	INCB3284	Incyte Pharmaceuticals/ Pfizer	Phase IIa	RA, obese insulin-resistant. It also presents mild anti-CCR5 effect.	111, 112, 124
CCR3	GW-766994	GlaxoSmithKline	Phase II	Asthma, allergic rhinitis	111–113, 125
	DPC-168	Bristol-Myers Squibb	Phase I	Asthma	111, 112
CCR5	UK-427857 (MARAVIROC)	Pfizer	Phase II	HIV infection	111–113, 126–128
	ONO-4128	Ono Pharmaceutical/ GlaxoSmithKline	Phase II	HIV infection	111, 112
	Sch-351125 / Sch-417690	Schering-Plough	Phase I	HIV infection	111, 112
CXCR1/2	SB-332235	GlaxoSmithKline	Phase I	COPD, RA, psoriasis	111, 112
CXCR3	T0906487	Tularik-Amgen and Chemocentryx	Phase IIa	Psoriasis. Was suspended because no effect was found.	111, 129
CXCR4	AMD-3100	AnorMED	Phase II	Stem cell transplantation	111, 112
			Phase I	Repair cardiac tissue after myocardial infarction	111, 112
	AMD-070	AnorMED	Phase I	HIV infection	111, 112
	CTCE-0214	Chemokine Therapeutics	Phase II	Stem cell transplantation	111, 112

COPD, chronic obstructive pulmonary disease; MS, multiple sclerosis; RA, rheumatoid arthritis.

multiple cell types (neutrophils, mast cells, macrophages, and endothelial cells) to produce proinflammatory molecules (i.e., TNF-α, IL-1, and chemokines), thereby amplifying the inflammatory cascade (third stage). In the last stage, after prolonged inflammation, chronic changes in the joint occur including activation of osteoclasts, resulting in bone erosion, and formation of lymphoid tissue aggregates (see below). Locally, monocytes recruited from the circulation will differentiate into inflammatory macrophages. Also as part of the effector phase, CII-specific T cells may produce cytokines that promote the differentiation of bone-reabsorbing cells known as osteoclasts, for example receptor activator of nuclear factor kappa B ligand (RANKL) and TNF-α *(25,26)*. This chronic stage would in many cases imply the progression toward the maintenance phase of the disease (Fig. 1; see color plate).

Probably among all the phases of the disease, the effector and maintenance phases of RA are the ones for which we have the most information regarding the role of chemokines. Several studies have demonstrated that in the actively inflamed synovial membrane of patients with RA, their inflammatory synovial fluid, peripheral blood mononuclear cells (PBMCs), and serum/plasma samples, there is broad upregulation of the levels of several chemokines and their receptors (Table 2). In few studies, the specificity for RA in the upregulation of chemokines has been demonstrated by comparing the RA samples with those obtained from other processes affecting the joints such as osteoarthritis or infectious forms of arthritis. For instance, in synovial tissue derived from patients with rheumatoid arthritis, osteoarthritis, and reactive arthritis, abundant expression of CCR1, CXCR4, and CCR5 has been reported. Notably, tissue levels of CCL5 and CCL15 are higher in RA than osteoarthritis or reactive arthritis *(27)* (Table 3).

10.2.3. Formation of New Lymphoid Aggregates in the Inflamed Synovial Tissue

One of the more interesting processes that chemokines may drive during the effector and maintenance phases of RA is the formation of lymphoid aggregates in the synovium, known as lymphoid neogenesis. This involves the formation of organized B-cell follicles containing germinal centers and T-cell areas *(28)*. These newly formed lymphoid organs are embedded in the tissue of the joints and lack afferent lymph vessels. A key characteristic of the lymphoid structures in the joints is that in contrast with other structures such LNs, persistent antigenic stimulation is required to induce and maintain them *(28)*. Interestingly, formation of lymphoid aggregates has also been described in the joints of mice with CIA *(29)*.

The molecular mechanisms that underlie the formation and maintenance of lymphoid structures such as LNs and those in chronically inflamed synovium

Table 2
Chemokines and Chemokine Receptors Implicated in Rheumatoid Arthritis: Human Studies

Molecule	Location	Cell	Comments	Detection	Ref.
CCL2	Serum SF Synovium TB	EC SC FLS, CD68 OB, osteocytes, SC, EC, MNC	CCR2 ligand	IF, ELISA, IHC	38, 77, 130–136
CCL3	Synovium SF PBMCs TB FDL	EC Mo, PMN MDDC OB, MNC	CCR5 ligand	Real-time PCR (mRNA), IHC, ELISA	38, 77, 131, 134, 135, 137–141
CCL4	Synovium TB	 MNC	CCR5 ligand	IHC, ELISA	38, 77, 131
CCL5	Serum Synovium TB	FLS SC (nonendothelium) OB, MNC	Main CCR5 ligand	Real-time PCR (mRNA), ELISA, IHC, IF	77, 130, 131, 133, 142, 143
CCL17	SF PBMCs	 MDDC	CCR4 ligand. Th2 chemokine	Real-time PCR, ELISA	139, 144
CCL18	SF Synovium PBMCs	 MDDC	Expressed in DC. Can act as antagonist of CCL11 and CCL13 for binding to CCR3.	ELISA, real-time PCR (mRNA), IHC	139, 145
CCL19	Synovium PBMCs	EC, SC MDDC	CCR7 ligand	IF, real-time PCR (mRNA), IHC	32, 130, 139, 146

Table 2 *(Continued)*

Molecule	Location	Cell	Comments	Detection	Ref.
CCL20	Synovium, SF	EpC CD4 (CD45RO), FLS	Controls migration of immature DC through CCR6	Real-time PCR, RT-PCR, ISH, ELISA, IHC	32, 146–149
CCL21	Synovium, subcondral BM, serum, SF	CD3, CD20, EC	Involved in the organization of secondary LN (ectopic LN in synovium)	IHC, IF, RT-PCR, ISH (mRNA), ELISA	31, 32, 146, 150–153
CCL22	SF		Detection restricted to juvenile RA samples	ELISA	144
CCL28	Synovium	MLS	Ligand for CCR10 (present on B cells)	RT-PCR, FACS, IHC	154
CXCL1	Synovium TB	 OB, EC, SC, MNC		Real-time PCR (mRNA), IHC	131, 155
CXCL5	Synovium	CD14, EC		IHC, IF, ELISA	156
CXCL6	SF	FLS, CD68		ELISA, IHC	132
CXCL8	Serum, urine, SF, synovium FDL PBMCs TB	Mo MDDC OB, osteocytes, EC, SC, MNC	Neutrophil migration, acute inflammation, innate immunity	Real-time PCR (mRNA), ELISA, IHC, ISH	38, 131, 133, 134, 139, 140, 155–158
CXCL9	Synovium	FLS, STC	CXCR3 ligand	RT-PCR, FACS, IHC, ELISA, DNA microarray	38, 154, 155, 159, 160

Chemokines	CXCL10	Synovium	FLS, MLS, EC, EpC, STC	CXCR3 ligand	IHC, IF, real-time PCR (mRNA), ELISA, RT-PCR, DNA microarray	38, 142, 155, 159–163
		SF	pDC			
	CXCL11	Synovium	SC (nonendothelium), STC		Real-time PCR (mRNA), IF, ELISA, RT-PCR	130, 155, 160, 162
		SF	pDC			
	CXCL12	Synovium	EC, SC, FLS, MLS	Stromal derived factor. Involved in development of CNS and vasculature; B-cell lymphopoiesis.	Real-time PCR (mRNA), RT-PCR, Northern, ISH, IHC, IF, ELISA	108, 130, 162, 164–170
		SF	pDC			
		TB	Chondrocytes			
		Cartilage				
	CXCL13	Synovium, subcondral BM	CD3, CD20, FDC	Involved in the organization of secondary LN (ectopic LN in synovium)	RT-PCR, IHC, ISH (mRNA)	31, 151, 152, 171, 172
		Synovium	$CD11c^+$, CD68, CD14, EC			
	CXCL16	Synovium, SF	FLS, MLS, EC	Membrane bound form can act as adhesion molecule	IHC, IF, WB, real-time PCR, RT-PCR	173, 174
	XCL1	Synovium	CD4 ($CD28^-$), CD8 (CD28)	Secreted by T cells and involved in T-cell traffic. Th1 chemokine.	RT-PCR, ISH, ELISA	138, 175
		SF				
	CX_3CL1	Synovium	FLS, MLS, DC, EC		IHC, RT-PCR	176–178
		SF			FACS, ELISA	
Chemokine receptors	CCR1	PBMCs, SF	CD14		FACS, IHC	77
	CCR2	SF, synovium	$CD4^+$ ($CD45RA^-$), FLS, CD14		FACS, IF, RT-PCR, IHC	77, 130, 179–181
	CCRL2	SF	PMNs, Mø	Orphan receptor	IHC	182
	CCR3	PBMCs	CD3	Marker for Th2 cells	FACS, IHC	77, 180
		SF, synovium	CD14			

Table 2 *(Continued)*

	Molecule	Location	Cell	Comments	Detection	Ref.
Chemokine receptors	CCR4	SF, PBMCs	$CD4^+$ ($CD45RA^-$), (non- SF.MC) CD 3, CD4	Marker for Th2 cells	FACS, IHC	130, 144, 180, 183
	CCR5	PBMCs, SF, synovium	$CD4^+$ ($CD28^-$), $CD4^+$ (CD45RO), $CD4^+$ ($CD45RA^-$); FLS, CD3, CD4, CD8, CD14	T-cell chemotaxis (effector memory)	FACS, IF, RT-PCR, ISH	35, 47, 77, 130, 137, 138, 144, 150, 179–181, 184
	CCR6	Serum, Synovium.dral BM../ KJ,n? synovium, STC, SF	CD4 (CD45RO), DC		Real-time PCR, FACS, ISH, IHC	32, 146–148
	CCR7	PBMCs, SF, synovium	$CD4^+$ ($CD28^-$), $CD4^+$ (CD45RO), DC	Homing to LN and synovium (central-effector memory) Effector memory Central memory	FACS, IF, IHC	32, 35, 130, 146, 150
	CXCR1	Synovium	MLS		RT-PCR, WB, IHC	159
	CXCR2	Synovium			RT-PCR, WB	159
	CXCR3	Synovium, SF	CD3, MC, $CD4^+$ (CD45RO), plasma cells, FLS, pDC	Homing to inflamed tissue	DNA microarray, RT-PCR, FACS, IHC, IF, WB	130, 150, 154, 159, 161, 162, 179, 180

CXCR4	PBMCs, SF, synovium	$CD4^+$ ($CD28^-$), $CD4^+$ (CD45RO), $CD8^+$ (CD45RO), EC, pDC, MLS, FLS	Homing to LN and synovium (central-effector memory) Effector memory Central memory	FACS, IF, real-time PCR (mRNA), RT-PCR	35, 108, 130, 150, 162, 164–166, 179, 181, 184
CXCR6	Synovium, PBMCs, SF	CD4, CD8, $CD4^+$ ($CD45RA^-$)	Present on activated memory Th1 cells	FACS	130, 173
CX_3CR1	PBMCs, synovium, SF	CD3, CD4, CD4 ($CD28^-$), CD8, CD14, MLS, FLS, DC, EC	Interaction between ligand and receptor can mediate strong cell adhesion, and induce T-cell activation, cytotoxicity, and proliferation	IHC, RT-PCR., FACS	176–178, 180
XCR1	Synovium	FLS, MLS, T cells, MNC		RT-PCR, ISH	138, 175
DARC	Synovium	EC (venules only)	Possible role in transmigration of cells from blood to inflamed tissue	Northern, IHC	185, 186

DC, dendritic cell; EC, endothelial cell; EpC, epithelial cell; FDC, follicular dendritic cell; FDL, foot draining lymph; FLS, fibroblast-like synoviocyte; MC, mast cell; MDDC, monocyte-derived dendritic cell; MLS, macrophage-like synoviocyte; MNC, mononuclear cell; Mo, monocyte; Mø, macrophage; OB, osteoblast; pDC, plasmacytoid dendritic cell; PMNs, polymorphonuclear neutrophils; SC, stromal cell; STC, synovium tissue cell; TB, trabecular bone.

Table 3
Key Chemokines and Chemokine Receptors Relevant in Rheumatoid Arthritis Compared with Other Arthritides: Human Studies

Molecule	Sample	Detection	Comment	Ref.
CCL2	Serum	ELISA	Increased levels compared with OA samples.	133, 135, 187, 188
			Highest levels present in patients with follicular synovitis compared with those with diffuse synovitis.	
			Similar levels to inactive SLE but reduced when compared with active SLE.	
			Levels comparable with RP.	
		Multiplex IHC	Comparable levels found in early arthritis (<1 y), and late arthritis (>5 y).	
CCL3	Synovium, SF	ELISA	Higher levels compared with OA, gout, and traumatic joint injury.	38, 135, 141, 189
		IHC	Higher levels found in early arthritis (<1 y), in lining and sublining layer of synovium compared with late arthritis (>5 y).	
CCL4	Serum	Multiplex	Levels comparable with RP.	38, 187
	Synovium, SF	ELISA	Higher levels compared with OA and traumatic joint injury.	
CCL5	Serum	ELISA	Increased levels compared with OA samples.	133, 189, 190
			Highest levels present in patients with follicular synovitis compared with those with diffuse synovitis.	
	SF	ELISA	Higher levels compared with OA and gout.	
	Synovium	ISH (mRNA)	Increased expression in the cellular infiltrate compared with OA but not PA.	
CCL8	SF	ELISA	Higher levels compared with OA or other inflammatory arthritides.	132
CCL17	Synovium	Real-time PCR	Elevated levels compared with OA or normal controls.	139

CCL18	Serum, drain fluid, SF	ELISA	Higher levels present in SF of RA patients compared with serum of normal donors and drain fluids from lymph node resection of mammary carcinoma.	139, 145, 187, 191
			Higher levels compared with gout and OA.	
	Serum	Multiplex	Higher levels compared with normal controls but lower when compared with RP.	
CCL19	Synovium	Real-time PCR	Elevated levels compared with OA or normal controls.	139
CCL20	SE, SF	ELISA	Elevated levels compared with OA.	147–149
CXCL1	Synovium	ISH (mRNA)	Increased expression in the synovial lining layer compared with OA but not PA.	190
CXCL6	SF	ELISA	Higher levels compared with OA or other inflammatory arthritides.	132
CXCL8	Serum	ELISA	Increased levels compared with OA samples. Highest levels present in patients with follicular synovitis compared with those with diffuse synovitis.	38, 133, 139, 156, 189, 191–196
			Increased levels compared with AOSD and normal controls.	
		SPCIA	Decreased levels in ERA JA compared with RA.	
	SF	ELISA	Similar levels between polyarticular JA and RA.	
			Higher levels found in polyarticular JA compared with ERA JA.	
			Higher levels compared with OA and traumatic joint injury.	
			Comparable or higher levels with gout.	
			Elevated levels compared with OA or normal controls.	
	Synovium	Real-time PCR ELISA	Elevated levels compared with tissue obtained from amputation for diabetes or other etiologies.	
CXCL9	SF, synovium	DNA microarray, RT-PCR, ELISA, ISH (mRNA)	Higher levels compared with OA and traumatic joint injury.	38, 159, 160, 190
			Increased expression compared with OA but not PA.	

Table 3 *(Continued)*

	Molecule	Sample	Detection	Comment	Ref.
Chemokines	CXCL10	SF, synovium, serum	DNA microarray, RT-PCR, ELISA	Higher levels compared with OA and traumatic joint injury but not PA.	38, 159, 160, 162, 163
	CXCL11	SF, synovium	ELISA, RT-PCR	Higher levels compared with OA but not PA.	160, 162
	CXCL12	Serum	ELISA	Elevated levels in RA and OA, significant reduction after synovectomy.	166, 197
	XCL1	Synovium	RT-PCR (mRNA), ISH (mRNA), IHC	Higher levels compared with OA. Expression restricted to lymphocytic infiltrate in RA. Immunostaining shows strong staining in RA compared with OA and reactive arthritis but not PA.	175
	CX_3XL1	SF	ELISA	Higher levels compared with OA, JRA, PA, polyarthritis, and gout.	178
Chemokines	CCR4	PBMCs	FACS	Increased levels of CD4 CCR4 found in active RA, AS, and untreated SLE compared with controls.	183
	CCR5	SF	FACS	Increased ratio of CCR5/CCR3 of T cells ($CD8^-$) in RA compared with other arthritides.	198
	CXCR1	Synovium	DNA microarray, RT-PCR, WB	Higher levels compared with OA.	159
	CXCR2	Synovium	DNA microarray, RT-PCR	Higher levels compared with OA.	159
	CXCR3	Synovium	DNA microarray, RT-PCR, WB	Higher levels compared with OA.	159

AOSD, adult onset Still disease; AS, ankylosing spondylitis; CA, crystal-induced arthritis; ERA, enthesitis-related arthritis; JA, juvenile arthritis; PA, psoriatic arthritis; RA, rheumatoid arthritis; SE, synovium explants; SLE, systemic lupus erythematosus; SPCIA, solid phase 2 site chemiluminescent immunometric assay; RP, relapsing polychondritis.

are thought to be very similar *(28)* and mainly driven by the cytokine lymphotoxin and chemokines such as CCL19, CCL21, CXCL12, and CXCL13, which, as we reviewed above, regulate DC and lymphocyte homing and compartmentalization *(28)*. Chronic antigen stimulation leads to (i) persistent activation of innate and adaptive immune cells in the inflamed tissue; (ii) increased expression of lymphotoxin by activated B and T cells; and (iii) increased lymphoid chemokine expression by resident stromal cells, infiltrating macrophages, DCs, and other parenchymal cells. Synthesis of lymphoid chemokines might be induced by lymphotoxin itself. Recruitment of B cells, T cells, and DCs to the newly formed lymphoid structures is thought to be facilitated by acquisition of a high endothelial venule (HEV)-like phenotype by the activated endothelial cells in the joint. CCL19 and CCL21 produced by stromal cells (probably fibroblasts or fibroblast precursors) would favor the formation of T-cell areas *(28,30)*. Under the influence of lymphotoxin, stromal cells would acquire the phenotypic and functional properties of a follicular dendritic cell (FDC), including CXCL13 production, which, as mentioned earlier, promotes germinal-center organization *(28,31)*. Interestingly, a similar role for DC in disease initiation in the LN and perpetuation in the newly formed lymphoid tissue in the synovium has been postulated and would be directed by a similar array of chemokines and chemokine receptors (i.e., CCL21, CCL19, and CCR7) *(32)*.

10.2.4. Chemokines in the Migration of Arthritogenic T Cells

Naïve T cells migrate back and forth from the blood into the LN, and effector T cells migrate mainly from the blood into the inflamed tissues (i.e., the inflamed synovium) *(33,34)*. An increasing body of evidence suggests that after their activation by APCs in the LN, some T cells will become effector cells and migrate to the sites of inflammation, in the case of RA to the joints, whereas other T cells would become memory T cells. Some of these memory T cells have been termed *central memory T cells*, because they maintain the ability to migrate into the LN and expand rapidly after exposure to their cognate Ag *(33,34)*. T cells accumulating in RA-inflamed synovial tissue are likely to represent both short-lived effector memory cells and long-lived central memory T cells. The latter cell type finds a niche in LNs. By maintaining CCR7 expression, these end-differentiated T cells can still home to lymphoid organs, enhance their survival, support clonal expansion, and perpetuate autoreactivity *(35)*.

With regard to the specific accumulation of effector T cells in the inflamed RA joints, several papers have documented that T cells in the joints of individuals with RA preferentially express CCR5 *(36–38)*. Furthermore, RA *(39)* and CIA are thought to be a T helper 1 (Th1)-driven diseases *(12,40)*, and expression of CCR5 is considered to be the hallmark of Th1 differentiation, both in human and mice *(41,42)*. With this data in mind, it was surprising that absence of CCR5 in the DBA/1j background did not modify the course nor the incidence of CIA *(43)*, arguing against the current notion that CCR5 is a central player in RA pathogenesis. It is noteworthy, however, that in the context of a milder disease phenotype as can be seen in the B6x129 mice immunized with CII and complete Fruend's adjuvant (CFA), CCR5 gene inactivation has been shown to have a minor positive effect: whereas the incidence of arthritis was 24.5% in the wild type, the incidence in CCR5 knockout (KO) mice was 9.3% *(44)*. However, arthritis scores and days of onset and duration were not statistically different between the two groups *(44)*.

One explanation for the phenotype observed in the CCR5 null mice may be because CCR5 is not the only chemokine receptor whose expression is elevated in T cells present in the inflamed RA joints. For example, invariably both CCR5 and CXCR3 (and their ligands) are coelevated in RA *(36–38)*. Indeed, although pharmacologic antagonism of CCR5 leads to partial protection against the disease in the CIA model, it is important to note that the antagonist used in these studies blocks not only CCR5 but CXCR3 as well *(45)*, making it difficult to ascribe the protective effect solely to the antagonism of CCR5. Indeed, specific blockade of CCR5 with a monoclonal antibody (MC-68), either in early or late disease phase, did not significantly influence the arthritis score *(46)*.

Finally, Santiago et al. *(47)* demonstrated that lymphocyte chemoattraction induced by rheumatoid arthritis synovial fluid is independent of CCR5. They showed that T cells from individuals with CCR5 delta 32 mutation (a 32-base-pair frameshift deletion resulting in absent CCR5 expression) migrate normally to chemokines present in the synovial fluids of RA patients *(47)*.

10.3. Effects of Chemokine System Inactivation in Arthritis

The therapeutic goal in autoimmune diseases such as RA is to control disease, to establish remission, and eventually to cure. In theory, this goal can be achieved using either Ag-specific approaches, for example, elimination of self-reactive T cells (assuming that a finite number of key Ags can be identified as the target of the autoimmune process in RA), or the non-Ag-specific approaches, for example, blockade of cytokines as in the case of TNF-α neutralization. Currently, only the latter types of approaches have yielded clinical benefit, and it is in this category that approaches to block chemokines or receptors may be included. Despite their appeal in terms of effectiveness, non-Ag-specific approaches carry a higher risk of immunosuppression and opportunistic infections *(48)*.

Although there is a myriad of ongoing clinical trials testing the effects of chemokine/receptor blockers in RA (Table 1), to date one cannot yet predict how many of the current targets will prove to be clinically effective. So for now, one must rely on results obtained with testing the effects of the blockade of the chemokine system in animal models of arthritis. Indeed, in several experimental models of arthritis, blockade of chemokines and their receptors has been shown to be effective (Tables 4 and 5). This broad array of potential targets is somehow encouraging, but also, it is reminiscent of the state of affairs in other model systems of autoimmune diseases such as type I diabetes (TID). In experimental TID, there has been a surprisingly high rate of effective research treatments that when administered early in the natural history of the disease had profound effects. However, in the clinical arena, the great majority of these research treatments have failed to demonstrate clinical benefit *(49)*.

10.3.1. Study of Chemokines in Different Experimental Models of Arthritis

Different approaches have been used to probe the role of chemokines and their receptors in experimental models of arthritis; for instance, peptide antagonists or immunization of the host to promote the generation of "endogenous" neutralizing Abs via the use of chemokine plasmid DNA vaccination (Table 4 and Ref. 50), injection of neutralizing antibodies (Table 5), or mice that lack specific chemokine or chemokine receptors (Table 6).

In addition to CIA, there are several other models of arthritis that have also provided evidence for an important role of the chemokine system. Next, we will review some of the data available regarding the contribution of chemokines to disease pathogenesis in specific model systems beyond mere description of any given factor in the joints.

10.3.1.1 Adjuvant-Induced Arthritis

Rats develop arthritis that affects primarily diarthrodial, cartilaginous peripheral joints soon after injection of As emulsified in oil/adjuvants *(51)*. Adjuvant-induced arthritis commonly presents with a chronic relapsing disease course. It has been shown that early in the disease course, there is an increase in the expression of neutrophil-chemoattracting chemokines KC/Gro-α/MIP-2 (CXCL1/CXCL2), while later in the disease course monocyte-attracting chemokine CCL2 is upregulated *(52)*.

10.3.1.2. Collagen-Induced Arthritis

The incidence of arthritis induced in this model ranges from 40% to 80% depending on the method of induction, collagen preparation, and the adjuvants used for immunization *(11,12,14,15,46,53)*. Because, most commonly, disease is induced in the DBA/1j background, and the majority of the knockout mice available are on the C57BL/6 or other backgrounds, little is known regarding the effects of germ-line inactivation of chemokines in CIA. As mentioned

Table 4
Arthritis and Chemokine or Chemokine-Receptor Blockade Using Peptides/DNA

Target	Model of arthritis	Species	Peptide/DNA	Clinical effect	Potential mechanism/comments	Ref.
CCL2 (MCP-1)	CFA	MRL/lpr mice	Truncated CCL2, CCL2(9-76) (affect NH_2-terminal)	↓	Decrease cellular infiltration	56, 199
	AIA	Lewis rats	CCL2 naked DNA	↓ in acute and chronic phase	Production of self-specific antibody	50
CCL3 (MIP-1α)	AIA	Lewis rats	CCL3 naked DNA	↓ in acute and chronic phase	Production of self-specific antibody	50
CCL4 (MIP-1β)	AIA	Lewis rats	CCL4 naked DNA	↓ only in acute phase	Production of self-specific antibody	50
CCL5 (RANTES)	AIA	Lewis rats	CCL5 naked DNA	↓ in acute and chronic phase	Production of self-specific antibody	50
CXCL1 (GRO-α)	CFA	MRL/lpr mice	Truncated analogue of CXCL1, GRO-α(8-73).	↓	Less cell recruitment when treated with anti-CXCL1 + anti-CCL2 compared with anti-CCL2 alone	57
CXCL10 (IP-10)	AIA	Lewis rats	CXCL10 naked DNA	↓ developing and ongoing AIA	Production of self-specific antibody. Inhibits leukocyte migration, alters the Th1/Th2 pattern.	200
CX3CL1 (Fractalkine)	AIA	Lewis rats	CX3CL1 naked DNA	↔ in chronic phase	Similar arthritic score to wild-type mice	50

CCR1	CIA	Mice	A1B1, A4B7	↓	Reduced pannus formation and bone resorption	201
CCR2	AIA	Lewis rats	INCB3344	↓	Inhibition of inflammation and bone resorption	92
CCR5	CIA	DBA/1 mice	A nonpeptide CCR5 antagonist, TAK-779. Also inhibits CXCR3.	↓	Interfere with cell migration to joint lesions	45, 202
	CIA	DBA/1 mice	Met-RANTES. Also CCR1 antagonist.	↓	Reduce incidence of disease in a dose-dependent manner.	122, 203
	CIA	Rhesus monkeys	SCH-X	↓	Suppressed acute-phase reaction (reduction in C-reactive protein level and an altered antibody response toward type II collagen)	204
CXCR2	CXCL8 or LPS in knee joint	Rabbit	*N*-(3-(aminosulfonyl)-4-chloro-2-hydroxyphenyl)-*N′*-(2,3-dichlorophenyl) urea	↓	Decrease cell recruitment in synovial fluid	205

Table 4 *(Continued)*

Target	Model of arthritis	Species	Peptide/DNA	Clinical effect	Potential mechanism/comments	Ref.
CXCR4	CIA	DBA/1 mice	AMD3100 (JM3100)	↓	Reduce serum IL-6. No modified DTH. Inhibition of monocyte migration to CXCL12α.	206, 207
	CIA in IFNγR null mice	DBA/1 mice	AMD3100 (JM3100)	↓	Inhibition of monocyte migration to CXCL12α. Reduction of DTH response to collagen type II.	208
	CIA	DBA/1 mice	T140 analogue, 4F-benzoyl-TN14003.	↓	NS. Decrease levels of serum anti-bovine CII IgG2a antibody.	209

Note: Effect: ↓ decrease severity; ↑ increase severity; ↔ no effect.
AIA, adjuvant-induced arthritis; CFA, complete Freund's adjuvant; CIA, collagen-induced arthritis; DTH, delayed-type hypersensitivity; LPS, lipopolysaccharide.

Table 5
Arthritis and Results from the Use of Neutralizing Abs

Target	Model	Species	Antibody	Clinical effect	Potential mechanism/comments	Ref.
CCL2 (MCP-1)	AIA	Lewis rats	Monoclonal	↓	Decrease number of exudate macrophages	210
	CAIA	BALB/c mice		↓	Reduced arthritis score	211
CCL3 (MIP-1α)	CAIA	BALB/c mice		↓	Reduced arthritis score	211
	CIA	DBA/1 mice	Polyclonal	↓	Decrease cell infiltration	212
	AIA	Lewis rats	Polyclonal	↔		213
CCL5 (RANTES)	AIA	Lewis rats	Polyclonal	↓	Decrease cell infiltration	213
CXCL2 (MIP-2)	CIA	DBA/1 mice	Polyclonal	↓	Decrease cell infiltration	212
	CAIA	BALB/c mice		Mild ↓ or ↔		211
CXCL5 (ENA-78)	AIA	Lewis rats	Polyclonal	↓	Decrease macrophages and neutrophils recruitment. Effective only when treatment was done *before* development of disease.	115, 214

Table 5 *(Continued)*

	Target	Model	Species	Antibody	Clinical effect	Potential mechanism/comments	Ref.
	CXCL8 (KC, IL-8)	AIA	Lewis rats	Polyclonal	↔		213
	CX3CL1 (Fractalkine)	CIA	DBA/1 mice	Hamster anti-mouse FKN mAb (ICN Pharmaceuticals)	↓	Reduce infiltration of inflammatory cells in synovium and bone erosion. It did not affect production of serum anti-collagen type II (CII) IgG or IFN-γ by CII-stimulated splenic T cells.	215
Chemokine Receptors	CCR2	CIA	DBA/1 mice	Monoclonal	↓	$CCR2^{-}$ blockade at early time points markedly improved histologic and clinical signs of arthritis.	89, 91
					↑	Late CCR2 blockade worsens clinical and histologic outcome.	
	CCR5	CIA	DBA/1 mice	Monoclonal	↔		46

Note: Clinical effect: ↓ decrease severity; ↑ increase severity; ↔ no effect.
AIA, adjuvant-induced arthritis; CAIA, collagen Ab–induced arthritis; CIA, collagen-induced arthritis; DTH, delayed-type hypersensitivity.

Table 6
Arthritis and Chemokine-Deficient and Chemokine Receptor–Deficient Mice

	Gene deficient	Model of Arthritis	Strain	Clinical effect	Potential mechanism/comments	Ref.
Chemokines	CCL3 (MIP-1α)	CAIA	C57BL/6	↓	Lower SAP, TNF-α, and cell recruitment than wild type	216
Chemokine receptors	CCR2	CIA	DBA/1j	↑	Higher rheumatoid factor titers, enhanced T-cell activation, and monocyte/macrophage accumulation in the joints	43, 46
		CAIA	DBA/1j	↑	Prominent signs of chronic arthritis with pannus formation, and destructive bone and cartilage erosion, predominately of the distal joints	43, 46
		Lyme arthritis	C3H/HeJ,	↔	Similar to wild-type mice	217
		Lyme arthritis	C57BL/6J	↑ and ↔	Temporal worsening	217
		Mycobacterium avium infection	DBA/1j	↑	Hyperplasia, synovitis, synovial cyst formation, fibroblast proliferation, and osteoid deposition. Presence of neutrophils in the interarticular space. Higher levels of anti-CII Abs of the IgG1 subtype.	93
	CCR5	CIA	DBA/1	↔	Similar to wild-type mice	43, 44, 46
	CXCR2	Lyme arthritis	DBA	↔	Similar to wild-type mice	217
		Lyme arthritis	C3H/HeJ,	↓	Inhibition of neutrophils to migrate to joint	217
		Lyme arthritis	C57BL/6J	↓	Inhibition of neutrophils to migrate to joint	217

Note: Clinical effect: ↓ decrease severity; ↑ increase severity; ↔ no effect.
CAIA, collagen Ab–induced arthritis; CIA, collagen-induced arthritis; SAP, serum amyloid P-component.

earlier, DBA/1j CCR5 KO mice have a CIA phenotype that is indistinguishable from that of wild-type mice *(43)*. Interestingly, inactivation of CCR2 in the DBA/1j mice is not associated with protection but rather with worsening of the disease phenotype *(43)*. In the CIA model system, it has been shown that Ab blockade or chemical antagonism of several chemokines and chemokine receptors is highly effective in halting disease (Table 5).

10.3.1.3. MRL/lpr Mice

MRL/lpr mice exhibit a complex and fascinating disease phenotype. The autosomal recessive mutation *Ipr* (lymphoproliferation) that affects the apoptosis-inducing system FAS-FASL leads to the spontaneous development of antibodies to native DNA, severe immune complex disease, and massive lymphadenopathy within the first few months of life *(54)*. These mice spontaneously develop different autoimmune diseases such as systemic lupus erythematosus (SLE), Sjögren syndrome, and RA. These mice have circulating rheumatoid factor (RF) and develop histologic changes in their joints *(55)*. Development of arthritis is commonly accelerated via intradermal injection of complete Freund's adjuvant (CFA). An extensive amount of data have demonstrated that peptide-mediated blockade of CCL2 prior to *(56)* or after disease development *(57)* significantly ameliorates arthritis development or disease severity, respectively.

10.3.1.4. Other Models of Arthritis

There is a paucity of information regarding the role of chemokines in the more recently introduced models of arthritis: (a) The KRN mouse model, an experimental system in which arthritis develops in mice on an NOD background, transgenic for a T-cell receptor recognizing an epitope of bovine RNase (KRN mice) *(58,59)*. Notably, these mice produce Abs against glucose-6-phosphate isomerase (GPI), and Abs induced acute arthritis when injected in recipient mice. (b) TNF-α–based genetic murine models of RA: These genetic models of arthritis are based on the transgenic expression of human or murine TNF-α -Tg (Transgenic) *(60–63)*. These mice develop a chronic erosive polyarthritis with 100% phenotypic penetrance and timed disease onset. (c) Sakaguchi (SKG) mice *(64,65)*, in response to fungal cell wall products, develop T cell–mediated chronic autoimmune arthritis as a consequence of a mutation of the gene encoding a Src homology 2 (SH2) domain of ζ-associated protein of 70 kd (ZAP-70), a key signal transduction molecule in T cells. This mutation impairs positive and negative selection of T cells in the thymus, leading to thymic production of arthritogenic T cells. Clinically, joint swelling begins in small joints of the digits, progressing in a symmetrical fashion to larger joints including wrists and ankles. Histologically, the swollen joints show severe synovitis with formation of pannus invading and eroding adjacent cartilage and

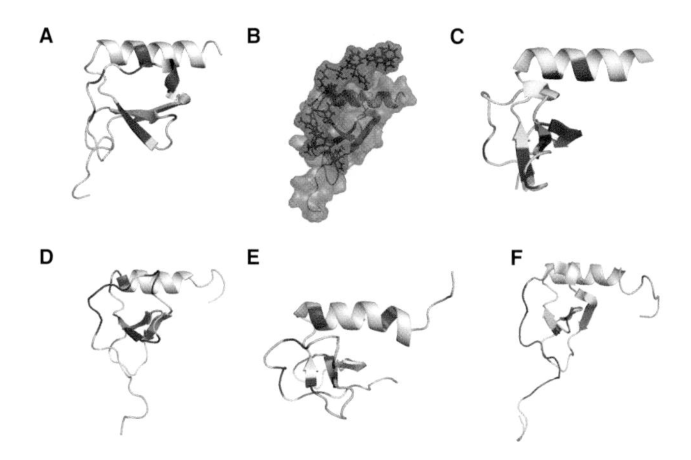

Fig. 2.4. Residues affected by titration of N-terminal receptor–based peptides for various chemokines (shown as monomers). (A, C to F) Chemokines are shown as cartoon representations with residues whose chemical shifts are affected by receptor-based peptides colored red. (A) CXCL8 titrated with CXCR1$^{1-40}$. (B) X-ray structure of CXCL8 complexed with CXCR1$^{9-29}$ modified peptide (IL-8) displayed as semitransparent surface over stick representation (red). (C) CXCL12 (SDF-1α) titrated with CXCR4$^{1-27}$ peptide. (D) CCL11 (eotaxin-1) and (E) CCL24 (eotaxin-2) titrated with CCR3$^{1-35}$ peptide. (F) CX3CL1 (fractalkine) titrated with CX3CR1$^{2-19}$ peptide.

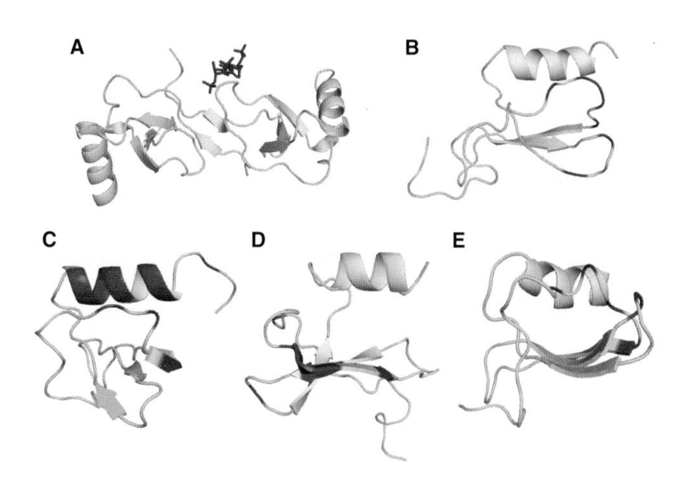

Fig. 2.5. Interactions of chemokines with heparin disaccharides. (A to C) Monomeric forms of chemokines are displayed as cartoons with residues found to interact with heparin colored red. (A) Monomer of crystal structure of CCL5 (RANTES) with heparin disaccharide I-S bound (red). (B) CXCL4 (PF4) with low-molecular-weight heparin ($MW < 9000\,d$). (C) CXCL8 with heparin disaccharide I-S. (D) CXCL12 (SDF-1α) with heparin disaccharide I-S. (E) CCL2, human IP-10, with conserved residues from murine IP-10 highlighted.

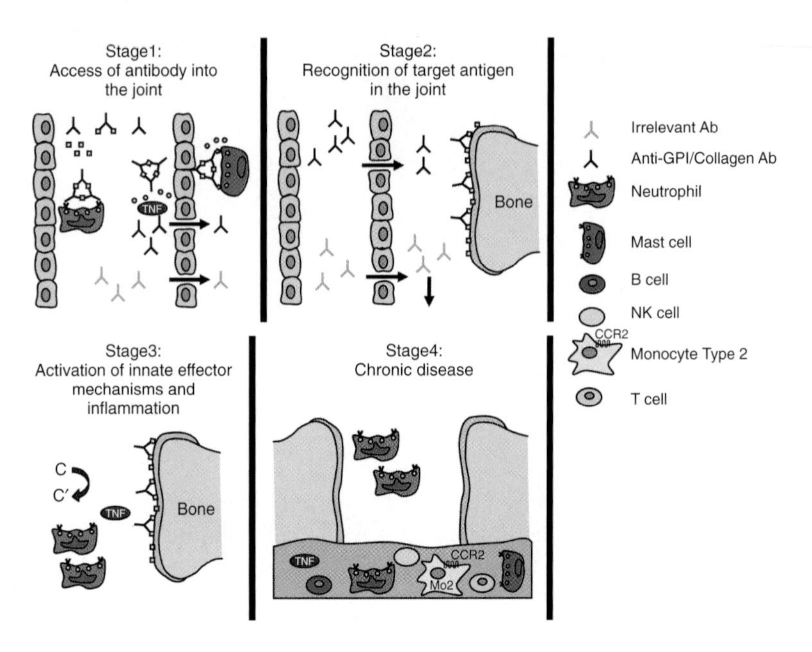

Fig. 10.1. Stages of effector phase. (Modified from Ref. 24.)

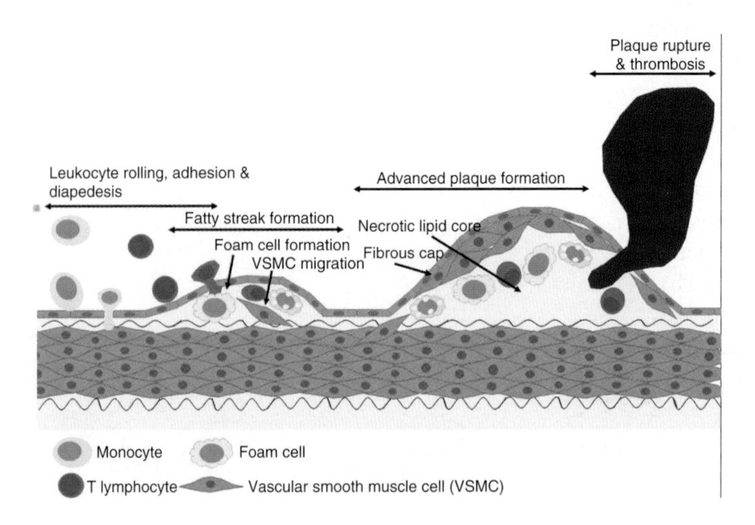

Fig. 11.1. Atherogenesis is a persistent inflammatory response that occurs in response to conditions that cause endothelial damage (e.g., hypercholesterolemia and oxLDL). After endothelial cells are activated, they elaborate cytokines, chemokines, and other mediators that recruit mononuclear cells (monocytes and T lymphocytes) to extravasate into the vessel wall where they are activated and release additional proinflammatory factors. Macrophages are able to take up oxLDL via scavenger receptors causing them to differentiate into foam cells and form a fatty streak that progresses to an atheroma with a necrotic lipid core and a fibrous cap. Chemokines can lead to weakening of the fibrous cap and eventual plaque rupture leading to thrombosis and occlusion of the involved vessel.

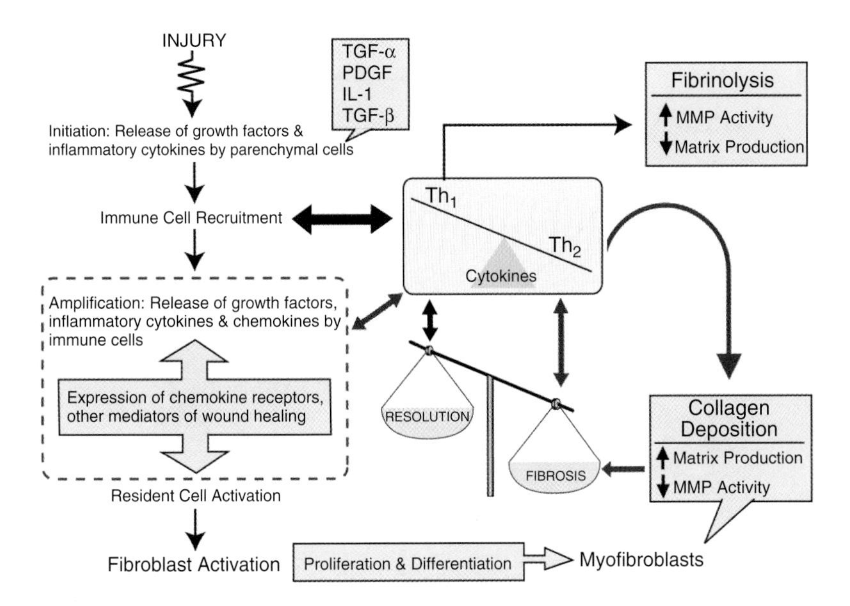

Fig. 14.1. The Th1/Th2 balance is central to the regulation of normal wound repair. Tissue injury results in the *initiation* of an inflammatory response, mediated by a variety of cells and their by-products. Immune cells are recruited and cross-regulate the Th1/ Th2 balance that occurs in response to the cytokine environment. This balance is in turn cross-regulated by the chemokine/chemokine-receptor expression profile, which functions to *amplify* the inflammatory process. Cells residing in the injured tissue release *profibrotic* mediators, which promote fibroblast activation, proliferation, and differentiation to the myofibroblast phenotype. Myofibroblasts produce collagen to repair damaged tissue, which is an event that is favored by the inhibition of MMP activity. The Th1/Th2 balance is central to whether a normal or aberrant wound-repair process is established: A Th1 environment promotes normal tissue resolution (fibrinolysis), whereas a Th2 environment maintains the progression of fibrotic disease (excessive collagen deposition).

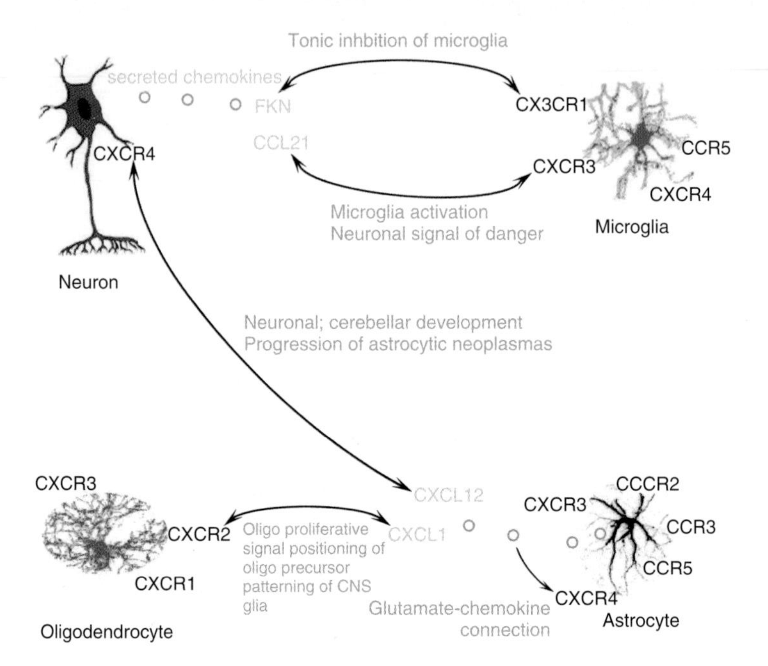

Fig. 17.1. Selected chemokine receptors in CNS cells and functional interactions. Neuron-astrocyte communication via CXCL12/CXCR4 is critically involved in neuronal positioning, and the proposed glutamate-chemokine pathway links astrocyte-microglia and astrocyte-neuron functions via the release of glutamate upon engagement of CXCR4 (with TNF-α secretion by activated microglia). CXCL1/CXCR2 roles in oligodendrocyte differentiation and migration create pathways for novel remyelination strategies. Constitutive shedding of neuronal fractalkine (FKN) and its interaction with microglial CX3CR1 inhibits microglial neurotoxicity. Neuronal release of CCL21 and the upregulation of CXCR3 in microglia represent an alternative inducible neuronal-microglia communication system.

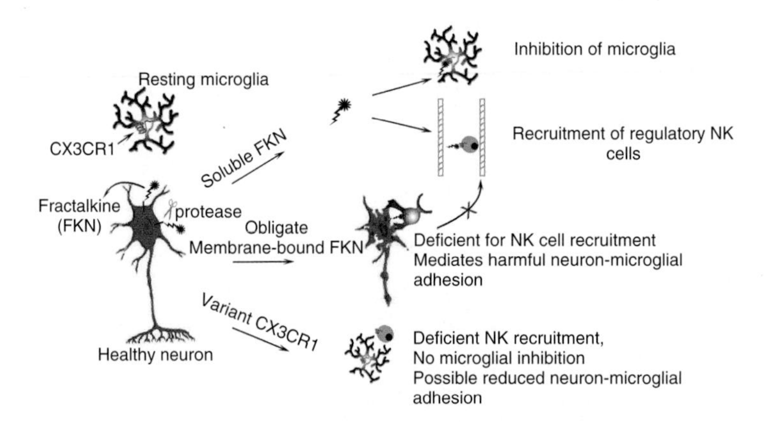

Fig. 17.2. Fractalkine-CX3CR1 actions in microglia and NK cells during neuroinflammation. Fractalkine and CX3CR1 provide key regulatory input for microglia and NK cells, the fractalkine-responsive cell types implicated in EAE. In particular, fractalkine and CX3CR1 mediate tonic inhibition of deleterious microglial activation, and recruit regulatory NK cells to the inflamed CNS during EAE.

subchondral bone. SKG mice develop extraarticular lesions, such as interstitial pneumonitis, vasculitides, and subcutaneous necrobiotic nodules not unlike rheumatoid nodules in RA. Serologically, they develop high titers of RF and autoantibodies specific for type II collagen. The role of chemokines in these models is being actively researched in the authors' laboratory.

10.4. Signaling Cascades Underlying the Effects of Chemokines in Arthritis

It is clear that chemokines mediate their effects via seven-transmembrane domain receptors, a subset of G protein–coupled receptors (GPCRs). Interestingly, recent approaches have allowed study of the potential downstream signaling cascades engaged by these chemokine receptors *(66)*. In rats, arthritis development is associated with increases in tyrosine phosphorylation of CCR1 and CCR2. More interestingly, in immunoprecipitation studies, Janus kinase (JAK-1), signal transducer and activator of transcription 1 (STAT-1), and STAT-3 were associated with CCR1, CCR2, and CCR5 expression at different time points during the disease course. Immunohistologic analysis revealed that CCR5, phosphorylated STAT-1, and phosphorylated STAT-3 are present in different cell types in the inflamed synovial tissue of the joints, including lining cells, macrophages, and endothelial cells *(66)*. Collectively, these studies suggest that effects of the chemokines in RA may be related to the ability of chemokine receptors to engage a wide variety of downstream signaling cascades.

10.5. Some Unsolved Issues Regarding Chemokines in Arthritis

10.5.1. CCR5, Genetics, and RA

As mentioned earlier, a 32-base pair (bp) deletion in the CCR5 gene (Δ32) results in a frameshift and premature termination in the region encoding the second extracellular loop of the CCR5 receptor, producing a nonfunctional receptor that remains intracellular *(67)*. About 1% of the Caucasian population is likely to be homozygous for Δ32 *(68)*. Notably, Gomez-Reino et al. *(69)* reported that in their cohort of 673 Spanish patients with RA, there were no patients homozygous for Δ32, but the frequency of this Δ32 mutation was normal in lupus patients and healthy controls. This striking observation was not replicated later on in four other studies performed in different populations in which RA patients homozygous for Δ32 were found, suggesting that CCR5 deficiency does not prevent rheumatoid arthritis *(70–73)*. A recent study of 516 RA patients and 985 healthy controls *(74)* also failed to find Δ32 homozygous RA patients, while 1.2% of the healthy controls were homozygous.

Although having a Δ32 allele may not totally prevent the development of RA, it may influence disease severity and progression. There are several contradictory reports that have addressed this question. In 163 Danish patients with RA, those carrying the Δ32 allele were more likely to be seronegative for RF (suggesting a more benign disease) and had less swollen joints *(71)*. This conclusion has been substantiated in another study *(72)*. In contrast with these results, a more detailed study using 5 years of follow-up data (including hand and foot radiographs, C-reactive protein (CRP) levels, and RF) from 438 patients found that the Δ32 allele does not predict disease progression *(73)*. This conclusion is also supported by two additional studies *(74,75)*.

Collectively, the data from genetic-association studies regarding the effect of Δ32 state on the risk of developing the disease, RA progression, severity, as well as the data from experimental models of RA, seem to suggest that CCR5 may play a minor role in RA pathogenesis and that according to the circumstances, its effects may or may not be noticeable (i.e., large enough cohort of specific ethnic origin or the use of B6X129 mice for CIA instead of the more susceptible DBA/1j mice).

10.5.2. Role of CCR2 in Experimental Arthritis

CCR2 is expressed by a large proportion of mononuclear cells present in inflamed joints and in synovial effusions of patients with different forms of arthritis *(76,77)* and in adjuvant-induced murine arthritis *(78)*. The cognate ligands of CCR2 (CCL2) and CCR5 (CCL3, CCL4, and CCL5) have been detected in the synovial fluid of patients with RA *(79–82)*, at sites of bone remodeling *(83)*, in articular cartilage *(84)*, and in inflamed joints in CIA *(85)*. Also, analogues of CCL2, neutralizing antibodies against CCL2, and DNA vaccines that encode these chemokines have all been shown to inhibit disease initiation or progression in rodent models of arthritis (Tables 4 and 5).

CCR2-null mice exhibit a profound defect in induced leukocyte migration, particularly monocyte/macrophage *(86,87)*, and DC migration associated with skewing toward T helper 2 (Th2) cytokine production *(88)*. Likewise, blockade of CCR2 with monoclonal antibody MC-21 markedly impairs monocytes/macrophage recruitment in vivo *(89)*. Because of these defective inflammatory responses and the fact that CIA can be ameliorated by administration of Th2 cytokines *(90)*, it was hypothesized that CCR2-null mice would be resistant to the development of CIA or its manifestations and that the blockade of CCR2 with monoclonal antibodies would result in a similar outcome. However, CCR2-null mice immunized with type II bovine collagen (CII) not only were fully susceptible to CIA but also showed accelerated and more severe disease than wild-type (WT) control mice *(43)*. CII-immunized CCR2-null mice had significantly higher serum levels of anti-collagen-specific IgG1 and IgG2a, increased levels of total IgE, and even positive titers of RF. Diseased joints in CCR2-null mice showed severe bone destruction and increased numbers of neutrophils and osteoclasts. The lymph nodes in these mice contained higher numbers of activated T and B cells, including an increase in the number of RANKL-bearing T cells. The severe disease phenotype of CCR2-null mice was transferred using bone marrow cells to lethally irradiated WT mice suggesting that absence of CCR2 in the hematopoietic compartment was responsible for the increase susceptibility of CCR2-null mice to CIA. Furthermore, in a complementary model of arthritis based on the intravenous administration of anti-CII antibodies, it was shown that WT and CCR2-null mice had comparable disease during the acute phases of the disease (first 2 weeks after anti-CII administration); the resolution phase of the disease that commonly begins 3 weeks after Ab injection was delayed in CCR2-null mice *(43)*.

Additional support for a negative regulatory role of CCR2 in arthritis was partially provided by blockade of CCR2 with specific monoclonal antibody *(91)*. Blockade during disease initiation (days 0 to 15) significantly reduced the severity of arthritis, whereas blockade of CCR2 in late phase of the disease (e.g., days 21 to 36) markedly aggravated clinical and histologic signs of arthritis. Blockade of CCR2 in early disease phases resulted in lower serum levels of anti-collagen-specific IgG, whereas the late blockade of CCR2 markedly increased the humoral immune response against collagen, induced a higher influx of monocytes and neutrophils into the joints, and increased bone and cartilage erosion.

Despite the bulk of evidence pointing to a negative regulatory role for CCR2 in experimental arthritis, results obtained using a novel rodent CCR2 antagonist, INCB3344 (Incyte Corporation Wilmington, Delaware), seem to challenge that notion *(92)*. INCB3344 was reported to be 100-fold more selective for CCR2 than for other closely related chemokine receptors. Administration of this small-molecule antagonist starting at day 9 after adjuvant treatment significantly reduced arthritis severity in rat adjuvant-induced arthritis. INCB3344 treatment was associated with 82% inhibition of joint inflammation and 64% reduction in bone reabsorption.

Interestingly, in a potentially new murine model, germ-line inactivation of CCR2 was also associated with increased susceptibility to arthritis development. Intradermal *Mycobacterium avium* challenge of WT and CCR2-null mice on DBA/1j and BALB/c background was associated with the development of clinically evident chronic persistent polyarthritis only in the CCR2-null DBA/1j mice but not in the WT mice nor the CCR2-null mice in the BALB/c background *(93)*. Increased susceptibility to infection was unlikely to be the only cause of arthritis as BALB/c CCR2-null mice had a high degree of susceptibility to the pathogen but did not develop arthritis. Histologic analysis revealed that the arthritic phenotype present in the DBA/1j CCR2-null mice was characterized by synovial hyperplasia, synovitis, synovial cyst formation, fibroblast proliferation, bone remodeling, osteoid deposition, and osteoclast activity. Interestingly, this disease phenotype was highly reminiscent of the autoimmune arthritis present in CCR2-null mice immunized with collagen *(43)*. Indeed, immunologic studies showed that compared with DBA/1j WT, BALB/c WT, or CCR2-null, DBA/1j CCR2-null mice had significantly higher titers of anti-collagen type II antibodies of IgG1 subtype.

How does one interpret the data available so far regarding the role of CCR2 in RA? Arguably, the more contrasting data relate to that between the severe CIA phenotype in CCR2-null mice and the antiarthritic effects of INCB3344 in the adjuvant-induced arthritis model. A working hypothesis is that the phenotype of the KO mouse is driven by immune abnormalities existing prior to immunization and that cannot be fully recapitulated after acute treatment with an antagonist that is initiated after the disease has begun. For instance, compared with the WT mice, CCR2-null mice have substantially lower numbers of specific monocyte subtypes in the blood *(94)* and DCs in the spleen *(88)*. These types of cellular changes are likely to be adaptations occurring as a consequence of the chronic absence of a key molecule in the system. If this hypothesis is correct, long-term administration of the CCR2 antagonist or neutralizing Ab may lead to a phenotype that may be concordant with that of CCR2-null mice.

Intriguingly, results from a recent clinical trial in RA patients suggest that the role of chemokines in RA may be proinflammatory. CCL2 is the major ligand for CCR2, and administration of a CCL2 blocking Ab was not associated with any clinical improvement in these subjects. On the contrary, these trials found an increase in the levels of C-reactive protein (a disease activity marker) and enhanced accumulation of macrophages in the joint *(95)*.

10.6. Effects of Chemokines in Rheumatoid Arthritis Beyond Cell Recruitment

It seems obvious to assume that by virtue of their powerful regulatory effects of leukocyte recruitment, many chemokine/chemokine receptors may play an important role in an inflammatory disease of the joints such as RA. However, chemokines may have other effects in disease pathogenesis that may not be related to leukocyte chemotaxis.

10.6.1. Chemokines and Osteoclastogenesis

Localized bone reabsorption is one of the more characteristic features of RA. This osteopenia is mediated in part by increased accumulation of osteoclasts. Interestingly, a handful of chemokines have been shown to have a profound effect on osteoclast differentiation, although it is unknown whether chemokines are mediators of osteoclastogenesis and bone loss in arthritis. The osteoclastogenic properties of the chemokine CCL3 have been extensively studied. CCL3 is present in inflamed joints in RA and experimental arthritis. Injection of recombinant CCL3 in vivo promotes a striking increase in osteoclast differentiation *(96)*. Moreover, in the context of multiple myeloma, a disease in which just like RA, there is significant degree of localized bone reabsorption, CCL3 has been shown to promote osteoclastogenesis *(97)*. CCL3 promotes osteoclast differentiation via CCR1 and CCR5 *(98)*, and its effects are not fully dependent on the actions of the master osteoclast differentiation factor RANKL *(99)*. Notably, chemokines such as CCL2 and CCL5 may also have a synergistic action on the ability of RANKL to promote osteoclast differentiation *(100,101)*.

10.6.2. Chemokines and the Cartilage

A pathologic feature of the RA joints is the evidence of significant destruction of cartilage *(102)*. Notably, chondrocytes themselves produce chemokines such as CXCL8, CXCL1, CCL2, CCL3, and CCL5 *(103)*, and their production of CXCL8, CXCL1, CCL3, and CCL4 is increased in RA *(104)*. Thus, it is likely that there is a role for chemokines and their receptors in cartilage degradation. Indeed, chemokines can induce the release of matrix-degrading enzymes such as matrix metalloproteinases 1, 3, and 13 and *N*-acetyl-beta-D-glucosaminidase. Furthermore, CXCL1 acting on CXCR2 can activate an apoptotic pathway in chondrocytes that leads to cell death *(105)*.

10.6.3. Chemokines and Angiogenesis

Endothelial cells are likely to be important in the pathogenesis of RA. Activation of endothelial cells during inflammatory conditions is associated with upregulation of adhesion molecules and chemokines that would promote leukocyte migration from the blood into the joint tissue *(106)*. Moreover, the RA synovium is rich in newly formed vessels as a consequence of increased angiogenesis. Different chemokines have angiogenic and angiostatic properties, and their levels in the joints may determine in part the magnitude of new blood vessel formation. For instance, CXCL8, CXCL5, CXCL1, and CTAP-III, chemokines that contain the ELR (glutamyl-leucyl-arginyl) amino acid sequence, are potent angiogenic factors. In contrast CXCL4, CXCL9, and CXCL10, chemokines that lack the ELR motif, inhibit neovascularization *(107)*. As an exception, the ELR-lacking CXCL12 is angiogenic potentially via its ability to influence migration and proliferation of endothelia progenitor cells *(107)*. Interestingly, it has been shown that synoviocyte-derived CXCL12 accumulates, and it is immobilized on heparan sulfate molecules of endothelial cells, where it can promote angiogenesis and inflammatory cell infiltration, supporting a multifaceted function for this chemokine in the pathogenesis of rheumatoid arthritis *(108)*.

10.7. Conclusions

Arbitrarily, it could be argued that in humans, definitive evidence that a molecule plays a unique role in a disease process can be provided by studies in which absence of disease, improvement in the course, or resolution of the disease is seen after specific inactivation of the molecule of interest (e.g., mutations or pharmacologic interventions). In this light, is there solid evidence for a nonredundant role of specific members of the chemokine system in RA? The answer is, not yet. Failures may be explained by different confounding factors such as target tested, the dosages of inhibitors used, presence of comorbidities, or the stage of the disease in which the intervention was performed. It is clear then that although chemokines can teach a lot about disease pathogenesis, it still remains to be demonstrated that they are relevant clinical targets.

Acknowledgments

This work was supported in part by the National Institutes of Health (NIH) grant (AI48644, AR052755) and VA Merit Award to S.S.A.; UTHSCA Department of Medicine/Pfizer Scholar Award, UTHSCSA Executive Research Committee (ERC) grant, and San Antonio Area Foundation awards to M.P.Q.

References

1. Jenkins JK, Hardy KJ, McMurray RW. The pathogenesis of rheumatoid arthritis: a guide to therapy. Am J Med Sci 2002;323(4):171–180.
2. Scott DL. Prognostic factors in early rheumatoid arthritis. Rheumatology (Oxford) 2000;39(Suppl 1):24–29.
3. Allaire SH, Prashker MJ, Meenan RF. The costs of rheumatoid arthritis. Pharmacoeconomics 1994;6(6):513–522.
4. Goldring SR. Pathogenesis of bone and cartilage destruction in rheumatoid arthritis. Rheumatology (Oxford) 2003;42(Suppl 2):ii11–6.
5. Feldmann M, Maini RN. Anti-TNF alpha therapy of rheumatoid arthritis: what have we learned? Annu Rev Immunol 2001;19:163–196.
6. Newkirk MM. Rheumatoid factors: host resistance or autoimmunity? Clin Immunol 2002;104(1):1–13.
7. Harris ED Jr. Rheumatoid arthritis. Pathophysiology and implications for therapy. N Engl J Med 1990;322(18):1277–1289.
8. Harris ED Jr. Pathogenesis of rheumatoid arthritis: its relevance to therapy in the ’90s. Trans Am Clin Climatol Assoc 1990;102:260–8; discussion 8–70.

9. Williams RO, Feldmann M, Maini RN. Anti-tumor necrosis factor ameliorates joint disease in murine collagen-induced arthritis. Proc Natl Acad Sci U S A 1992;89(20):9784–9788.
10. Van den Berg WB. Lessons from animal models of arthritis. Curr Rheumatol Rep 2002;4(3):232–239.
11. Anthony DD, Haqqi TM. Collagen-induced arthritis in mice: an animal model to study the pathogenesis of rheumatoid arthritis. Clin Exp Rheumatol 1999;17(2): 240–244.
12. Luross JA, Williams NA. The genetic and immunopathological processes underlying collagen-induced arthritis. Immunology 2001;103(4):407–416.
13. Rose CE Jr, Sung SS, Fu SM. Significant involvement of CCL2 (MCP-1) in inflammatory disorders of the lung. Microcirculation 2003;10(3–4):273–288.
14. Brahn E. Animal models of rheumatoid arthritis. Clues to etiology and treatment. Clin Orthop 1991(265):42–53.
15. Staines NA, Wooley PH. Collagen arthritis—what can it teach us? Br J Rheumatol 1994;33(9):798–807.
16. Brand DD, Kang AH, Rosloniec EF. Immunopathogenesis of collagen arthritis. Springer Semin Immunopathol 2003;25(1):3–18.
17. Pettit AR, Thomas R. Dendritic cells: the driving force behind autoimmunity in rheumatoid arthritis? Immunol Cell Biol 1999;77(5):420–427.
18. Pettit AR, Ahern MJ, Zehntner S, Smith MD, Thomas R. Comparison of differentiated dendritic cell infiltration of autoimmune and osteoarthritis synovial tissue. Arthritis Rheum 2001;44(1):105–110.
19. Thomas R, MacDonald KP, Pettit AR, Cavanagh LL, Padmanabha J, Zehntner S. Dendritic cells and the pathogenesis of rheumatoid arthritis. J Leukoc Biol 1999;66(2):286–292.
20. Leung BP, Conacher M, Hunter D, McInnes IB, Liew FY, Brewer JM. A novel dendritic cell-induced model of erosive inflammatory arthritis: distinct roles for dendritic cells in T cell activation and induction of local inflammation. J Immunol 2002;169(12):7071–7077.
21. Cyster JG. Chemokines and cell migration in secondary lymphoid organs. Science 1999;286(5447):2098–2102.
22. Vinuesa CG, Tangye SG, Moser B, Mackay CR. Follicular B helper T cells in antibody responses and autoimmunity. Nat Rev Immunol 2005;5(11):853–865.
23. Hargreaves DC, Hyman PL, Lu TT, et al. A coordinated change in chemokine responsiveness guides plasma cell movements. J Exp Med 2001;194(1):45–56.
24. Wipke BT, Wang Z, Nagengast W, Reichert DE, Allen PM. Staging the initiation of autoantibody-induced arthritis: a critical role for immune complexes. J Immunol 2004;172(12):7694–7702.
25. Gravallese EM. Bone destruction in arthritis. Ann Rheum Dis 2002;61(Suppl 2): ii84–86.
26. Kong YY, Feige U, Sarosi I, et al. Activated T cells regulate bone loss and joint destruction in adjuvant arthritis through osteoprotegerin ligand. Nature 1999; 402(6759):304–309.

27. Haringman JJ, Smeets TJ, Reinders-Blankert P, Tak PP. Chemokine and chemokine receptor expression in paired peripheral blood mononuclear cells and synovial tissue of patients with rheumatoid arthritis, osteoarthritis, and reactive arthritis. Ann Rheum Dis 2006;65(3):294–300.
28. Aloisi F, Pujol-Borrell R. Lymphoid neogenesis in chronic inflammatory diseases. Nat Rev Immunol 2006;6(3):205–217.
29. Zheng B, Ozen Z, Zhang X, et al. CXCL13 neutralization reduces the severity of collagen-induced arthritis. Arthritis Rheum 2005;52(2):620–626.
30. Buckley CD. Michael Mason prize essay 2003. Why do leucocytes accumulate within chronically inflamed joints? Rheumatology (Oxford) 2003;42(12): 1433–1444.
31. Takemura S, Braun A, Crowson C, et al. Lymphoid neogenesis in rheumatoid synovitis. J Immunol 2001;167(2):1072–1080.
32. Page G, Lebecque S, Miossec P. Anatomic localization of immature and mature dendritic cells in an ectopic lymphoid organ: correlation with selective chemokine expression in rheumatoid synovium. J Immunol 2002;168(10):5333–5341.
33. Sallusto F, Geginat J, Lanzavecchia A. Central memory and effector memory T cell subsets: function, generation, and maintenance. Annu Rev Immunol 2004;22: 745–763.
34. Ebert LM, Schaerli P, Moser B. Chemokine-mediated control of T cell traffic in lymphoid and peripheral tissues. Mol Immunol 2005;42(7):799–809.
35. Zhang X, Nakajima T, Goronzy JJ, Weyand CM. Tissue trafficking patterns of effector memory $CD4^+$ T cells in rheumatoid arthritis. Arthritis Rheum 2005; 52(12):3839–3849.
36. Qin S, Rottman JB, Myers P, et al. The chemokine receptors CXCR3 and CCR5 mark subsets of T cells associated with certain inflammatory reactions. J Clin Invest 1998;101(4):746–754.
37. Wedderburn LR, Robinson N, Patel A, Varsani H, Woo P. Selective recruitment of polarized T cells expressing CCR5 and CXCR3 to the inflamed joints of children with juvenile idiopathic arthritis. Arthritis Rheum 2000;43(4):765–774.
38. Patel DD, Zachariah JP, Whichard LP. CXCR3 and CCR5 ligands in rheumatoid arthritis synovium. Clin Immunol 2001;98(1):39–45.
39. Skapenko A, Leipe J, Lipsky PE, Schulze-Koops H. The role of the T cell in autoimmune inflammation. Arthritis Res Ther 2005;7(Suppl 2):S4–14.
40. Fournier C. Where do T cells stand in rheumatoid arthritis? Joint Bone Spine 2005;72(6):527–532.
41. Sallusto F, Lanzavecchia A. Understanding dendritic cell and T-lymphocyte traffic through the analysis of chemokine receptor expression. Immunol Rev 2000;177:134–140.
42. Luther SA, Cyster JG. Chemokines as regulators of T cell differentiation. Nat Immunol 2001;2(2):102–107.
43. Quinones MP, Ahuja SK, Jimenez F, et al. Experimental arthritis in CC chemokine receptor 2-null mice closely mimics severe human rheumatoid arthritis. J Clin Invest 2004;113(6):856–866.

44. Bao L, Zhu Y, Zhu J, Lindgren JU. Decreased IgG production but increased MIP-1beta expression in collagen-induced arthritis in C-C chemokine receptor 5-deficient mice. Cytokine 2005;31(1):64–71.
45. Gao P, Zhou XY, Yashiro-Ohtani Y, et al. The unique target specificity of a nonpeptide chemokine receptor antagonist: selective blockade of two Th1 chemokine receptors CCR5 and CXCR3. J Leukoc Biol 2003;73(2):273–280.
46. Quinones MP, Estrada CA, Kalkonde Y, et al. The complex role of the chemokine receptor CCR2 in collagen-induced arthritis: implications for therapeutic targeting of CCR2 in rheumatoid arthritis. J Mol Med 2005;83(9):672–681.
47. Santiago B, Galindo M, Rivero M, Brehmer MT, Mateo I, Pablos JL. The chemoattraction of lymphocytes by rheumatoid arthritis—synovial fluid is not dependent on the chemokine receptor CCR5. Rheumatol Int 2002;22(3):107–111.
48. Feldmann M, Steinman L. Design of effective immunotherapy for human autoimmunity. Nature 2005;435(7042):612–619.
49. Bach JF. Immunotherapy of type 1 diabetes: lessons for other autoimmune diseases. Arthritis Res 2002;4(Suppl 3):S3–15.
50. Youssef S, Maor G, Wildbaum G, Grabie N, Gour-Lavie A, Karin N. C-C chemokine-encoding DNA vaccines enhance breakdown of tolerance to their gene products and treat ongoing adjuvant arthritis. J Clin Invest 2000;106(3):361–371.
51. Holmdahl R, Lorentzen JC, Lu S, et al. Arthritis induced in rats with nonimmunogenic adjuvants as models for rheumatoid arthritis. Immunol Rev 2001;184:184–202.
52. Szekanecz Z, Halloran MM, Volin MV, et al. Temporal expression of inflammatory cytokines and chemokines in rat adjuvant-induced arthritis. Arthritis Rheum 2000;43(6):1266–1277.
53. Stuart JM, Watson WC, Kang AH. Collagen autoimmunity and arthritis. FASEB J 1988;2(14):2950–2956.
54. Steinberg AD, Roths JB, Murphy ED, Steinberg RT, Raveche ES. Effects of thymectomy or androgen administration upon the autoimmune disease of MRL/Mp-lpr/lpr mice. J Immunol 1980;125(2):871–873.
55. Bartlett RR, Popovic S, Raiss RX. Development of autoimmunity in MRL/lpr mice and the effects of drugs on this murine disease. Scand J Rheumatol Suppl 1988;75:290–299.
56. Gong JH, Ratkay LG, Waterfield JD, Clark-Lewis I. An antagonist of monocyte chemoattractant protein 1 (MCP-1) inhibits arthritis in the MRL-lpr mouse model. J Exp Med 1997;186(1):131–137.
57. Gong JH, Yan R, Waterfield JD, Clark-Lewis I. Post-onset inhibition of murine arthritis using combined chemokine antagonist therapy. Rheumatology (Oxford) 2004;43(1):39–42.
58. Kouskoff V, Korganow AS, Duchatelle V, Degott C, Benoist C, Mathis D. Organ-specific disease provoked by systemic autoimmunity. Cell 1996;87(5):811–822.
59. Kyburz D, Corr M. The KRN mouse model of inflammatory arthritis. Springer Semin Immunopathol 2003;25(1):79–90.
60. Redlich K, Hayer S, Ricci R, et al. Osteoclasts are essential for TNF-alpha-mediated joint destruction. J Clin Invest 2002;110(10):1419–1427.

61. Douni E, Akassoglou K, Alexopoulou L, et al. Transgenic and knockout analyses of the role of TNF in immune regulation and disease pathogenesis. J Inflamm 1995;47(1–2):27–38.
62. Keffer J, Probert L, Cazlaris H, et al. Transgenic mice expressing human tumour necrosis factor: a predictive genetic model of arthritis. EMBO J 1991;10(13):4025–4031.
63. Probert L, Keffer J, Corbella P, et al. Wasting, ischemia, and lymphoid abnormalities in mice expressing T cell-targeted human tumor necrosis factor transgenes. J Immunol 1993;151(4):1894–1906.
64. Yoshitomi H, Sakaguchi N, Kobayashi K, et al. A role for fungal β-glucans and their receptor Dectin-1 in the induction of autoimmune arthritis in genetically susceptible mice. J Exp Med 2005;201(6):949–960.
65. Sakaguchi N, Takahashi T, Hata H, et al. Altered thymic T-cell selection due to a mutation of the ZAP-70 gene causes autoimmune arthritis in mice. Nature 2003;426(6965):454–460.
66. Shahrara S, Amin MA, Woods JM, Haines GK, Koch AE. Chemokine receptor expression and in vivo signaling pathways in the joints of rats with adjuvant-induced arthritis. Arthritis Rheum 2003;48(12):3568–3583.
67. Liu R, Paxton WA, Choe S, et al. Homozygous defect in HIV-1 coreceptor accounts for resistance of some multiply-exposed individuals to HIV-1 infection. Cell 1996;86(3):367–377.
68. Dean M, Carrington M, Winkler C, et al. Genetic restriction of HIV-1 infection and progression to AIDS by a deletion allele of the CKR5 structural gene. Hemophilia Growth and Development Study, Multicenter AIDS Cohort Study, Multicenter Hemophilia Cohort Study, San Francisco City Cohort, ALIVE Study. Science 1996;273(5283):1856–1862.
69. Gomez-Reino JJ, Pablos JL, Carreira PE, et al. Association of rheumatoid arthritis with a functional chemokine receptor, CCR5. Arthritis Rheum 1999;42(5):989–992.
70. Cooke SP, Forrest G, Venables PJ, Hajeer A. The delta32 deletion of CCR5 receptor in rheumatoid arthritis. Arthritis Rheum 1998;41(6):1135–1136.
71. Garred P, Madsen HO, Petersen J, et al. CC chemokine receptor 5 polymorphism in rheumatoid arthritis. J Rheumatol 1998;25(8):1462–1465.
72. Zapico I, Coto E, Rodriguez A, Alvarez C, Torre JC, Alvarez V. CCR5 (chemokine receptor-5) DNA-polymorphism influences the severity of rheumatoid arthritis. Genes Immun 2000;1(4):288–289.
73. John S, Smith S, Morrison JF, et al. Genetic variation in CCR5 does not predict clinical outcome in inflammatory arthritis. Arthritis Rheum 2003;48(12):3615–3616.
74. Pokorny V, McQueen F, Yeoman S, et al. Evidence for negative association of the chemokine receptor CCR5 d32 polymorphism with rheumatoid arthritis. Ann Rheum Dis 2005;64(3):487–490.
75. Zuniga JA, Villarreal-Garza C, Flores E, et al. Biological relevance of the polymorphism in the CCR5 gene in refractory and non-refractory rheumatoid arthritis in Mexicans. Clin Exp Rheumatol 2003;21(3):351–354.
76. Bruhl H, Wagner K, Kellner H, Schattenkirchner M, Schlondorff D, Mack M. Surface expression of CC- and CXC-chemokine receptors on leucocyte subsets in inflammatory joint diseases. Clin Exp Immunol 2001;126(3):551–559.

77. Katschke KJ Jr, Rottman JB, Ruth JH, et al. Differential expression of chemokine receptors on peripheral blood, synovial fluid, and synovial tissue monocytes/macrophages in rheumatoid arthritis. Arthritis Rheum 2001;44(5):1022–1032.
78. Haas CS, Martinez RJ, Attia N, Haines GK 3rd, Campbell PL, Koch AE. Chemokine receptor expression in rat adjuvant-induced arthritis. Arthritis Rheum 2005;52(12):3718–3730.
79. Koch AE, Kunkel SL, Harlow LA, et al. Macrophage inflammatory protein-1 alpha. A novel chemotactic cytokine for macrophages in rheumatoid arthritis. J Clin Invest 1994;93(3):921–928.
80. Robinson E, Keystone EC, Schall TJ, Gillett N, Fish EN. Chemokine expression in rheumatoid arthritis (RA): evidence of RANTES and macrophage inflammatory protein (MIP)-1 beta production by synovial T cells. Clin Exp Immunol 1995;101(3):398–407.
81. Volin MV, Shah MR, Tokuhira M, Haines GK, Woods JM, Koch AE. RANTES expression and contribution to monocyte chemotaxis in arthritis. Clin Immunol Immunopathol 1998;89(1):44–53.
82. Hayashida K, Nanki T, Girschick H, Yavuz S, Ochi T, Lipsky PE. Synovial stromal cells from rheumatoid arthritis patients attract monocytes by producing MCP-1 and IL-8. Arthritis Res 2001;3(2):118–126.
83. Volejnikova S, Laskari M, Marks SC Jr, Graves DT. Monocyte recruitment and expression of monocyte chemoattractant protein-1 are developmentally regulated in remodeling bone in the mouse. Am J Pathol 1997;150(5):1711–1721.
84. Villiger PM, Terkeltaub R, Lotz M. Monocyte chemoattractant protein-1 (MCP-1) expression in human articular cartilage. Induction by peptide regulatory factors and differential effects of dexamethasone and retinoic acid. J Clin Invest 1992; 90(2):488–496.
85. Thornton S, Duwel LE, Boivin GP, Ma Y, Hirsch R. Association of the course of collagen-induced arthritis with distinct patterns of cytokine and chemokine messenger RNA expression. Arthritis Rheum 1999;42(6):1109–1118.
86. Kuziel WA, Morgan SJ, Dawson TC, et al. Severe reduction in leukocyte adhesion and monocyte extravasation in mice deficient in CC chemokine receptor 2. Proc Natl Acad Sci U S A 1997;94(22):12053–12058.
87. Boring L, Gosling J, Chensue SW, et al. Impaired monocyte migration and reduced type 1 (Th1) cytokine responses in C-C chemokine receptor 2 knockout mice. J Clin Invest 1997;100(10):2552–2561.
88. Sato N, Ahuja SK, Quinones M, et al. CC chemokine receptor (CCR)2 is required for langerhans cell migration and localization of T helper cell type 1 (Th1)-inducing dendritic cells. Absence of CCR2 shifts the Leishmania major-resistant phenotype to a susceptible state dominated by Th2 cytokines, b cell outgrowth, and sustained neutrophilic inflammation. J Exp Med 2000;192(2):205–218.
89. Mack M, Cihak J, Simonis C, et al. Expression and characterization of the chemokine receptors CCR2 and CCR5 in mice. J Immunol 2001;166(7):4697–4704.
90. Joosten LA, Lubberts E, Helsen MM, et al. Protection against cartilage and bone destruction by systemic interleukin-4 treatment in established murine type II collagen-induced arthritis. Arthritis Res 1999;1(1):81–91.

91. Bruhl H, Cihak J, Schneider MA, et al. Dual role of CCR2 during initiation and progression of collagen-induced arthritis: evidence for regulatory activity of CCR2+ T cells. J Immunol 2004;172(2):890–898.
92. Brodmerkel CM, Huber R, Covington M, et al. Discovery and pharmacological characterization of a novel rodent-active CCR2 antagonist, INCB3344. J Immunol 2005;175(8):5370–5378.
93. Quinones MP, Jimenez F, Martinez H, et al. CC chemokine receptor (CCR)-2 prevents arthritis development following infection by Mycobacterium avium. J Mol Med 2006;84(6):503–512.
94. Serbina NV, Pamer EG. Monocyte emigration from bone marrow during bacterial infection requires signals mediated by chemokine receptor CCR2. Nat Immunol 2006;7(3):311–317.
95. Haringman JJ, Gerlag DM, Smeets TJ, et al. A randomized placebo controlled trial with an anti-MCP-1 (CCL2) monoclonal antibody in patients with rheumatoid arthritis [Abstract 519]. Arthritis Rheum 2004;50(Suppl):S238.
96. Oyajobi BO, Franchin G, Williams PJ, et al. Dual effects of macrophage inflammatory protein-1alpha on osteolysis and tumor burden in the murine 5TGM1 model of myeloma bone disease. Blood 2003;102(1):311–319.
97. Roodman GD, Choi SJ. MIP-1 alpha and myeloma bone disease. Cancer Treat Res 2004;118:83–100.
98. Oba Y, Lee JW, Ehrlich LA, et al. MIP-1alpha utilizes both CCR1 and CCR5 to induce osteoclast formation and increase adhesion of myeloma cells to marrow stromal cells. Exp Hematol 2005;33(3):272–278.
99. Han JH, Choi SJ, Kurihara N, Koide M, Oba Y, Roodman GD. Macrophage inflammatory protein-1alpha is an osteoclastogenic factor in myeloma that is independent of receptor activator of nuclear factor kappaB ligand. Blood 2001; 97(11):3349–3353.
100. Kim MS, Day CJ, Morrison NA. MCP-1 is induced by receptor activator of nuclear factor-α B ligand, promotes human osteoclast fusion, and rescues granulocyte macrophage colony-stimulating factor suppression of osteoclast formation. J Biol Chem 2005;280(16):16163–16169.
101. Kim MS, Day CJ, Selinger CI, Magno CL, Stephens SR, Morrison NA. MCP-1-induced human osteoclast-like cells are tartrate-resistant acid phosphatase, NFATc1, and calcitonin receptor-positive but require receptor activator of NFkappaB ligand for bone resorption. J Biol Chem 2006;281(2):1274–1285.
102. Keen HI, Emery P. How should we manage early rheumatoid arthritis? From imaging to intervention. Curr Opin Rheumatol 2005;17(3):280–285.
103. Pulsatelli L, Dolzani P, Piacentini A, et al. Chemokine production by human chondrocytes. J Rheumatol 1999;26(9):1992–2001.
104. Borzi RM, Mazzetti I, Macor S, et al. Flow cytometric analysis of intracellular chemokines in chondrocytes in vivo: constitutive expression and enhancement in osteoarthritis and rheumatoid arthritis. FEBS Lett 1999;455(3):238–242.
105. Borzi RM, Mazzetti I, Marcu KB, Facchini A. Chemokines in cartilage degradation. Clin Orthop Relat Res 2004(427 Suppl):S53–61.

106. Middleton J, Americh L, Gayon R, et al. Endothelial cell phenotypes in the rheumatoid synovium: activated, angiogenic, apoptotic and leaky. Arthritis Res Ther 2004;6(2):60–72.
107. De Falco E, Porcelli D, Torella AR, et al. SDF-1 involvement in endothelial phenotype and ischemia-induced recruitment of bone marrow progenitor cells. Blood 2004;104(12):3472–3482.
108. Pablos JL, Santiago B, Galindo M, et al. Synoviocyte-derived CXCL12 is displayed on endothelium and induces angiogenesis in rheumatoid arthritis. J Immunol 2003;170(4):2147–2152.
109. Genovese MC. Biologic therapies in clinical development for the treatment of rheumatoid arthritis. J Clin Rheumatol 2005;11(3 Suppl):S45–54.
110. Ding C, Li J, Zhang X. Bertilimumab Cambridge Antibody Technology Group. Curr Opin Invest Drugs 2004;5(11):1213–1218.
111. Godessart N. Chemokine receptors: attractive targets for drug discovery. Ann N Y Acad Sci 2005;1051:647–657.
112. Terricabras E, Benjamim C, Godessart N. Drug discovery and chemokine receptor antagonists: eppur si muove! Autoimmun Rev 2004;3(7–8):550–556.
113. Bhalay G, Dunstan A. Chemokine Receptors and Drug Discovery—SMR meeting. 11 March 2004, Horsham, UK. IDrugs 2004;7(5):441–443.
114. Keystone EC. Abandoned therapies and unpublished trials in rheumatoid arthritis. Curr Opin Rheumatol 2003;15(3):253–258.
115. Haringman JJ, Ludikhuize J, Tak PP. Chemokines in joint disease: the key to inflammation? Ann Rheum Dis 2004;63(10):1186–1194.
116. Horuk R. Development and evaluation of pharmacological agents targeting chemokine receptors. Methods 2003;29(4):369–375.
117. Elices MJ. BX-471 Berlex. Curr Opin Invest Drugs 2002;3(6):865–869.
118. Millenium. R and D at Millenium, MLN3897. Available at http://www.mlnm.com/rd/inflammation/candidates/ccr1.asp.
119. Millenium. R and D at Millenium, MLN3701. Available at http://www.mlnm.com/rd/inflammation/candidates/mln3701.asp.
120. Haringman JJ, Kraan MC, Smeets TJ, Zwinderman KH, Tak PP. Chemokine blockade and chronic inflammatory disease: proof of concept in patients with rheumatoid arthritis. Ann Rheum Dis 2003;62(8):715–721.
121. Gladue RP, Zwillich SH, Clucas AT, Brown MF. CCR1 antagonists for the treatment of autoimmune diseases. Curr Opin Invest Drugs 2004;5(5):499–504.
122. Ribeiro S, Horuk R. The clinical potential of chemokine receptor antagonists. Pharmacol Ther 2005;107(1):44–58.
123. Millenium. R and D at Millenium, MLN1202. Available at http://www.mlnm.com/rd/inflammation/candidates/mln1202.asp.
124. Incyte. Incyte's Product Pipeline at Incyte Pharmaceuticals/Pfizer. Available at http://www.incyte.com/drugs_product_pipeline.html#chemokine.
125. Erin EM, Williams TJ, Barnes PJ, Hansel TT. Eotaxin receptor (CCR3) antagonism in asthma and allergic disease. Curr Drug Targets Inflamm Allergy 2002;1(2):201–214.

126. Dorr P, Westby M, Dobbs S, et al. Maraviroc (UK-427,857), a potent, orally bioavailable, and selective small-molecule inhibitor of chemokine receptor CCR5 with broad-spectrum anti-human immunodeficiency virus type 1 activity. Antimicrob Agents Chemother 2005;49(11):4721–4732.
127. Fatkenheuer G, Pozniak AL, Johnson MA, et al. Efficacy of short-term monotherapy with maraviroc, a new CCR5 antagonist, in patients infected with HIV-1. Nat Med 2005;11(11):1170–1172.
128. Rosario MC, Jacqmin P, Dorr P, van der Ryst E, Hitchcock C. A pharmacokinetic-pharmacodynamic disease model to predict in vivo antiviral activity of maraviroc. Clin Pharmacol Ther 2005;78(5):508–519.
129. Berry K. Evaluation of T0906487, a CXCR3 antagonist, in a phase 2a psoriasis trial. Inflammation Res 2004;Suppl 53:ps222.
130. Burman A, Haworth O, Hardie DL, et al. A chemokine-dependent stromal induction mechanism for aberrant lymphocyte accumulation and compromised lymphatic return in rheumatoid arthritis. J Immunol 2005;174(3):1693–1700.
131. Lisignoli G, Toneguzzi S, Grassi F, et al. Different chemokines are expressed in human arthritic bone biopsies: IFN-gamma and IL-6 differently modulate IL-8, MCP-1 and rantes production by arthritic osteoblasts. Cytokine 2002;20(5): 231–238.
132. Pierer M, Rethage J, Seibl R, et al. Chemokine secretion of rheumatoid arthritis synovial fibroblasts stimulated by Toll-like receptor 2 ligands. J Immunol 2004; 172(2):1256–1265.
133. Klimiuk PA, Sierakowski S, Latosiewicz R, et al. Histological patterns of synovitis and serum chemokines in patients with rheumatoid arthritis. J Rheumatol 2005;32(9):1666–1672.
134. Min DJ, Cho ML, Lee SH, et al. Augmented production of chemokines by the interaction of type II collagen-reactive T cells with rheumatoid synovial fibroblasts. Arthritis Rheum 2004;50(4):1146–1155.
135. Katrib A, Tak PP, Bertouch JV, et al. Expression of chemokines and matrix metalloproteinases in early rheumatoid arthritis. Rheumatology (Oxford) 2001; 40(9):988–994.
136. Ellingsen T, Buus A, Stengaard-Pedersen K. Plasma monocyte chemoattractant protein 1 is a marker for joint inflammation in rheumatoid arthritis. J Rheumatol 2001;28(1):41–46.
137. Wang CR, Liu MF. Regulation of CCR5 expression and MIP-1alpha production in CD4+ T cells from patients with rheumatoid arthritis. Clin Exp Immunol 2003;132(2):371–378.
138. Wang CR, Liu MF, Huang YH, Chen HC. Up-regulation of XCR1 expression in rheumatoid joints. Rheumatology (Oxford) 2004;43(5):569–573.
139. Radstake TR, van der Voort R, ten Brummelhuis M, et al. Increased expression of CCL18, CCL19, and CCL17 by dendritic cells from patients with rheumatoid arthritis, and regulation by Fc gamma receptors. Ann Rheum Dis 2005;64(3): 359–367.
140. Olszewski WL, Pazdur J, Kubasiewicz E, Zaleska M, Cooke CJ, Miller NE. Lymph draining from foot joints in rheumatoid arthritis provides insight into local

cytokine and chemokine production and transport to lymph nodes. Arthritis Rheum 2001;44(3):541–549.
141. Hanyuda M, Kasama T, Isozaki T, et al. Activated leucocytes express and secrete macrophage inflammatory protein-1alpha upon interaction with synovial fibroblasts of rheumatoid arthritis via a beta2-integrin/ICAM-1 mechanism. Rheumatology (Oxford) 2003;42(11):1390–1397.
142. Brentano F, Kyburz D, Schorr O, Gay R, Gay S. The role of Toll-like receptor signalling in the pathogenesis of arthritis. Cell Immunol 2005;233(2):90–96.
143. Ellingsen T, Buus A, Moller BK, Stengaard-Pedersen K. In vitro migration of mononuclear cells towards synovial fluid and plasma from rheumatoid arthritis patients correlates to RANTES synovial fluid levels and to clinical pain parameters. Scand J Rheumatol 2000;29(4):216–221.
144. Thompson SD, Luyrink LK, Graham TB, et al. Chemokine receptor CCR4 on $CD4^+$ T cells in juvenile rheumatoid arthritis synovial fluid defines a subset of cells with increased IL-4:IFN-gamma mRNA ratios. J Immunol 2001;166(11): 6899–6906.
145. van der Voort R, Kramer M, Lindhout E, et al. Novel monoclonal antibodies detect elevated levels of the chemokine CCL18/DC-CK1 in serum and body fluids in pathological conditions. J Leukoc Biol 2005;77(5):739–747.
146. Page G, Miossec P. Paired synovium and lymph nodes from rheumatoid arthritis patients differ in dendritic cell and chemokine expression. J Pathol 2004;204(1): 28–38.
147. Ruth JH, Shahrara S, Park CC, et al. Role of macrophage inflammatory protein-3alpha and its ligand CCR6 in rheumatoid arthritis. Lab Invest 2003;83(4): 579–588.
148. Matsui T, Akahoshi T, Namai R, et al. Selective recruitment of CCR6-expressing cells by increased production of MIP-3 alpha in rheumatoid arthritis. Clin Exp Immunol 2001;125(1):155–161.
149. Chabaud M, Page G, Miossec P. Enhancing effect of IL-1, IL-17, and TNF-alpha on macrophage inflammatory protein-3alpha production in rheumatoid arthritis: regulation by soluble receptors and Th2 cytokines. J Immunol 2001;167(10): 6015–6020.
150. Gattorno M, Prigione I, Morandi F, et al. Phenotypic and functional characterisation of $CCR7^+$ and $CCR7^-$ $CD4^+$ memory T cells homing to the joints in juvenile idiopathic arthritis. Arthritis Res Ther 2005;7(2):R256–267.
151. Bugatti S, Caporali R, Manzo A, Vitolo B, Pitzalis C, Montecucco C. Involvement of subchondral bone marrow in rheumatoid arthritis: lymphoid neogenesis and in situ relationship to subchondral bone marrow osteoclast recruitment. Arthritis Rheum 2005;52(11):3448–3459.
152. Manzo A, Paoletti S, Carulli M, et al. Systematic microanatomical analysis of CXCL13 and CCL21 in situ production and progressive lymphoid organization in rheumatoid synovitis. Eur J Immunol 2005;35(5):1347–1359.
153. Weninger W, Carlsen HS, Goodarzi M, et al. Naive T cell recruitment to nonlymphoid tissues: a role for endothelium-expressed CC chemokine ligand 21 in autoimmune disease and lymphoid neogenesis. J Immunol 2003;170(9):4638–4648.

154. Tsubaki T, Takegawa S, Hanamoto H, et al. Accumulation of plasma cells expressing CXCR3 in the synovial sublining regions of early rheumatoid arthritis in association with production of Mig/CXCL9 by synovial fibroblasts. Clin Exp Immunol 2005;141(2):363–371.
155. Fall N, Bove KE, Stringer K, et al. Association between lack of angiogenic response in muscle tissue and high expression of angiostatic ELR-negative CXC chemokines in patients with juvenile dermatomyositis: possible link to vasculopathy. Arthritis Rheum 2005;52(10):3175–3180.
156. Koch AE, Volin MV, Woods JM, et al. Regulation of angiogenesis by the C-X-C chemokines interleukin-8 and epithelial neutrophil activating peptide 78 in the rheumatoid joint. Arthritis Rheum 2001;44(1):31–40.
157. Taha AS, Grant V, Kelly RW. Urinalysis for interleukin-8 in the non-invasive diagnosis of acute and chronic inflammatory diseases. Postgrad Med J 2003; 79(929):159–163.
158. Kraan MC, Patel DD, Haringman JJ, et al. The development of clinical signs of rheumatoid synovial inflammation is associated with increased synthesis of the chemokine CXCL8 (interleukin-8). Arthritis Res 2001;3(1):65–71.
159. Ruschpler P, Lorenz P, Eichler W, et al. High CXCR3 expression in synovial mast cells associated with CXCL9 and CXCL10 expression in inflammatory synovial tissues of patients with rheumatoid arthritis. Arthritis Res Ther 2003;5(5): R241–252.
160. Ueno A, Yamamura M, Iwahashi M, et al. The production of CXCR3-agonistic chemokines by synovial fibroblasts from patients with rheumatoid arthritis. Rheumatol Int 2005;25(5):361–367.
161. Martini G, Zulian F, Calabrese F, et al. CXCR3/CXCL10 expression in the synovium of children with juvenile idiopathic arthritis. Arthritis Res Ther 2005; 7(2):R241–249.
162. Lande R, Giacomini E, Serafini B, et al. Characterization and recruitment of plasmacytoid dendritic cells in synovial fluid and tissue of patients with chronic inflammatory arthritis. J Immunol 2004;173(4):2815–2824.
163. Hanaoka R, Kasama T, Muramatsu M, et al. A novel mechanism for the regulation of IFN-gamma inducible protein-10 expression in rheumatoid arthritis. Arthritis Res Ther 2003;5(2):R74–81.
164. Bradfield PF, Amft N, Vernon-Wilson E, et al. Rheumatoid fibroblast-like synoviocytes overexpress the chemokine stromal cell-derived factor 1 (CXCL12), which supports distinct patterns and rates of $CD4^+$ and $CD8^+$ T cell migration within synovial tissue. Arthritis Rheum 2003;48(9):2472–2482.
165. Blades MC, Ingegnoli F, Wheller SK, et al. Stromal cell-derived factor 1 (CXCL12) induces monocyte migration into human synovium transplanted onto SCID Mice. Arthritis Rheum 2002;46(3):824–836.
166. Kanbe K, Takagishi K, Chen Q. Stimulation of matrix metalloprotease 3 release from human chondrocytes by the interaction of stromal cell-derived factor 1 and CXC chemokine receptor 4. Arthritis Rheum 2002;46(1):130–137.

167. van Oosterhout M, Levarht EW, Sont JK, Huizinga TW, Toes RE, van Laar JM. Clinical efficacy of infliximab plus methotrexate in DMARD naive and DMARD refractory rheumatoid arthritis is associated with decreased synovial expression of TNF alpha and IL18 but not CXCL12. Ann Rheum Dis 2005; 64(4):537–543.
168. Grassi F, Cristino S, Toneguzzi S, Piacentini A, Facchini A, Lisignoli G. CXCL12 chemokine up-regulates bone resorption and MMP-9 release by human osteoclasts: CXCL12 levels are increased in synovial and bone tissue of rheumatoid arthritis patients. J Cell Physiol 2004;199(2):244–251.
169. Watanabe N, Ando K, Yoshida S, et al. Gene expression profile analysis of rheumatoid synovial fibroblast cultures revealing the overexpression of genes responsible for tumor-like growth of rheumatoid synovium. Biochem Biophys Res Commun 2002;294(5):1121–1129.
170. Nanki T, Hayashida K, El-Gabalawy HS, et al. Stromal cell-derived factor-1-CXC chemokine receptor 4 interactions play a central role in $CD4^+$ T cell accumulation in rheumatoid arthritis synovium. J Immunol 2000;165(11):6590–6598.
171. Carlsen HS, Baekkevold ES, Morton HC, Haraldsen G, Brandtzaeg P. Monocyte-like and mature macrophages produce CXCL13 (B cell-attracting chemokine 1) in inflammatory lesions with lymphoid neogenesis. Blood 2004;104(10): 3021–3027.
172. Shi K, Hayashida K, Kaneko M, et al. Lymphoid chemokine B cell-attracting chemokine-1 (CXCL13) is expressed in germinal center of ectopic lymphoid follicles within the synovium of chronic arthritis patients. J Immunol 2001;166(1): 650–655.
173. van der Voort R, van Lieshout AW, Toonen LW, et al. Elevated CXCL16 expression by synovial macrophages recruits memory T cells into rheumatoid joints. Arthritis Rheum 2005;52(5):1381–1391.
174. Nanki T, Shimaoka T, Hayashida K, Taniguchi K, Yonehara S, Miyasaka N. Pathogenic role of the CXCL16-CXCR6 pathway in rheumatoid arthritis. Arthritis Rheum 2005;52(10):3004–3014.
175. Blaschke S, Middel P, Dorner BG, et al. Expression of activation-induced, T cell-derived, and chemokine-related cytokine/lymphotactin and its functional role in rheumatoid arthritis. Arthritis Rheum 2003;48(7):1858–1872.
176. Sawai H, Park YW, Roberson J, Imai T, Goronzy JJ, Weyand CM. T cell costimulation by fractalkine-expressing synoviocytes in rheumatoid arthritis. Arthritis Rheum 2005;52(5):1392–1401.
177. Nanki T, Imai T, Nagasaka K, et al. Migration of CX3CR1-positive T cells producing type 1 cytokines and cytotoxic molecules into the synovium of patients with rheumatoid arthritis. Arthritis Rheum 2002;46(11):2878–2883.
178. Ruth JH, Volin MV, Haines GK 3rd, et al. Fractalkine, a novel chemokine in rheumatoid arthritis and in rat adjuvant-induced arthritis. Arthritis Rheum 2001; 44(7):1568–1581.
179. Garcia-Vicuna R, Gomez-Gaviro MV, Dominguez-Luis MJ, et al. CC and CXC chemokine receptors mediate migration, proliferation, and matrix

metalloproteinase production by fibroblast-like synoviocytes from rheumatoid arthritis patients. Arthritis Rheum 2004;50(12):3866–3877.
180. Ruth JH, Rottman JB, Katschke KJ Jr, et al. Selective lymphocyte chemokine receptor expression in the rheumatoid joint. Arthritis Rheum 2001;44(12):2750–2760.
181. Nanki T, Nagasaka K, Hayashida K, Saita Y, Miyasaka N. Chemokines regulate IL-6 and IL-8 production by fibroblast-like synoviocytes from patients with rheumatoid arthritis. J Immunol 2001;167(9):5381–5385.
182. Galligan CL, Matsuyama W, Matsukawa A, et al. Up-regulated expression and activation of the orphan chemokine receptor, CCRL2, in rheumatoid arthritis. Arthritis Rheum 2004;50(6):1806–1814.
183. Yang PT, Kasai H, Zhao LJ, Xiao WG, Tanabe F, Ito M. Increased CCR4 expression on circulating CD4(+) T cells in ankylosing spondylitis, rheumatoid arthritis and systemic lupus erythematosus. Clin Exp Immunol 2004;138(2): 342–347.
184. Hisakawa N, Tanaka H, Hosono O, et al. Aberrant responsiveness to RANTES in synovial fluid T cells from patients with rheumatoid arthritis. J Rheumatol 2002;29(6):1124–1134.
185. Middleton J, Americh L, Gayon R, et al. A comparative study of endothelial cell markers expressed in chronically inflamed human tissues: MECA-79, Duffy antigen receptor for chemokines, von Willebrand factor, CD31, CD34, CD105 and CD146. J Pathol 2005;206(3):260–268.
186. Patterson AM, Siddall H, Chamberlain G, Gardner L, Middleton J. Expression of the duffy antigen/receptor for chemokines (DARC) by the inflamed synovial endothelium. J Pathol 2002;197(1):108–116.
187. Stabler T, Piette JC, Chevalier X, Marini-Portugal A, Kraus VB. Serum cytokine profiles in relapsing polychondritis suggest monocyte/macrophage activation. Arthritis Rheum 2004;50(11):3663–367.
188. Narumi S, Takeuchi T, Kobayashi Y, Konishi K. Serum levels of ifn-inducible PROTEIN-10 relating to the activity of systemic lupus erythematosus. Cytokine 2000;12(10):1561–1565.
189. McNearney T, Baethge BA, Cao S, Alam R, Lisse JR, Westlund KN. Excitatory amino acids, TNF-alpha, and chemokine levels in synovial fluids of patients with active arthropathies. Clin Exp Immunol 2004;137(3):621–627.
190. Konig A, Krenn V, Toksoy A, Gerhard N, Gillitzer R. Mig, GRO alpha and RANTES messenger RNA expression in lining layer, infiltrates and different leucocyte populations of synovial tissue from patients with rheumatoid arthritis, psoriatic arthritis and osteoarthritis. Virchows Arch 2000;436(5):449–458.
191. Schutyser E, Struyf S, Wuyts A, et al. Selective induction of CCL18/PARC by staphylococcal enterotoxins in mononuclear cells and enhanced levels in septic and rheumatoid arthritis. Eur J Immunol 2001;31(12):3755–3762.
192. Saxena N, Aggarwal A, Misra R. Elevated concentrations of monocyte derived cytokines in synovial fluid of children with enthesitis related arthritis and polyarticular types of juvenile idiopathic arthritis. J Rheumatol 2005;32(7):1349–1353.

193. Chen DY, Lan JL, Lin FJ, Hsieh TY. Proinflammatory cytokine profiles in sera and pathological tissues of patients with active untreated adult onset Still's disease. J Rheumatol 2004;31(11):2189–2198.
194. Rosengren S, Firestein GS, Boyle DL. Measurement of inflammatory biomarkers in synovial tissue extracts by enzyme-linked immunosorbent assay. Clin Diagn Lab Immunol 2003;10(6):1002–1010.
195. Ertenli I, Kiraz S, Calguneri M, et al. Synovial fluid cytokine levels in Behcet's disease. Clin Exp Rheumatol 2001;19(5 Suppl 24):S37–41.
196. Kaneko S, Satoh T, Chiba J, Ju C, Inoue K, Kagawa J. Interleukin-6 and interleukin-8 levels in serum and synovial fluid of patients with osteoarthritis. Cytokines Cell Mol Ther 2000;6(2):71–79.
197. Kanbe K, Takemura T, Takeuchi K, Chen Q, Takagishi K, Inoue K. Synovectomy reduces stromal-cell-derived factor-1 (SDF-1) which is involved in the destruction of cartilage in osteoarthritis and rheumatoid arthritis. J Bone Joint Surg Br 2004;86(2):296–300.
198. Nissinen R, Leirisalo-Repo M, Tiittanen M, et al. CCR3, CCR5, interleukin 4, and interferon-gamma expression on synovial and peripheral T cells and monocytes in patients with rheumatoid arthritis. J Rheumatol 2003;30(9):1928–1934.
199. Gong JH, Clark-Lewis I. Antagonists of monocyte chemoattractant protein 1 identified by modification of functionally critical NH2-terminal residues. J Exp Med 1995;181(2):631–640.
200. Salomon I, Netzer N, Wildbaum G, Schif-Zuck S, Maor G, Karin N. Targeting the function of IFN-gamma-inducible protein 10 suppresses ongoing adjuvant arthritis. J Immunol 2002;169(5):2685–2693.
201. Revesz L, Bollbuck B, Buhl T, et al. Novel CCR1 antagonists with oral activity in the mouse collagen induced arthritis. Bioorg Med Chem Lett 2005;15(23):5160–5164.
202. Yang YF, Mukai T, Gao P, et al. A non-peptide CCR5 antagonist inhibits collagen-induced arthritis by modulating T cell migration without affecting anti-collagen T cell responses. Eur J Immunol 2002;32(8):2124–2132.
203. Plater-Zyberk C, Hoogewerf AJ, Proudfoot AE, Power CA, Wells TN. Effect of a CC chemokine receptor antagonist on collagen induced arthritis in DBA/1 mice. Immunol Lett 1997;57(1–3):117–120.
204. Vierboom MP, Zavodny PJ, Chou CC, et al. Inhibition of the development of collagen-induced arthritis in rhesus monkeys by a small molecular weight antagonist of CCR5. Arthritis Rheum 2005;52(2):627–636.
205. Podolin PL, Bolognese BJ, Foley JJ, et al. A potent and selective nonpeptide antagonist of CXCR2 inhibits acute and chronic models of arthritis in the rabbit. J Immunol 2002;169(11):6435–6444.
206. Flomenberg N, DiPersio J, Calandra G. Role of CXCR4 chemokine receptor blockade using AMD3100 for mobilization of autologous hematopoietic progenitor cells. Acta Haematol 2005;114(4):198–205.
207. De Klerck B, Geboes L, Hatse S, et al. Pro-inflammatory properties of stromal cell-derived factor-1 (CXCL12) in collagen-induced arthritis. Arthritis Res Ther 2005;7(6):R1208–1220.

208. Matthys P, Hatse S, Vermeire K, et al. AMD3100, a potent and specific antagonist of the stromal cell-derived factor-1 chemokine receptor CXCR4, inhibits autoimmune joint inflammation in IFN-gamma receptor-deficient mice. J Immunol 2001;167(8):4686–4692.
209. Tamamura H, Fujisawa M, Hiramatsu K, et al. Identification of a CXCR4 antagonist, a T140 analog, as an anti-rheumatoid arthritis agent. FEBS Lett 2004; 569(1–3):99–104.
210. Ogata H, Takeya M, Yoshimura T, Takagi K, Takahashi K. The role of monocyte chemoattractant protein-1 (MCP-1) in the pathogenesis of collagen-induced arthritis in rats. J Pathol 1997;182(1):106–114.
211. Kagari T, Doi H, Shimozato T. The importance of IL-1 beta and TNF-alpha, and the noninvolvement of IL-6, in the development of monoclonal antibody-induced arthritis. J Immunol 2002;169(3):1459–1466.
212. Kasama T, Strieter RM, Lukacs NW, Lincoln PM, Burdick MD, Kunkel SL. Interleukin-10 expression and chemokine regulation during the evolution of murine type II collagen-induced arthritis. J Clin Invest 1995;95(6):2868–2876.
213. Barnes DA, Tse J, Kaufhold M, et al. Polyclonal antibody directed against human RANTES ameliorates disease in the Lewis rat adjuvant-induced arthritis model. J Clin Invest 1998;101(12):2910–2919.
214. Halloran MM, Woods JM, Strieter RM, et al. The role of an epithelial neutrophil-activating peptide-78-like protein in rat adjuvant-induced arthritis. J Immunol 1999;162(12):7492–7500.
215. Nanki T, Urasaki Y, Imai T, et al. Inhibition of fractalkine ameliorates murine collagen-induced arthritis. J Immunol 2004;173(11):7010–7016.
216. Chintalacharuvu SR, Wang JX, Giaconia JM, Venkataraman C. An essential role for CCL3 in the development of collagen antibody-induced arthritis. Immunol Lett 2005;100(2):202–204.
217. Brown CR, Blaho VA, Loiacono CM. Susceptibility to experimental Lyme arthritis correlates with KC and monocyte chemoattractant protein-1 production in joints and requires neutrophil recruitment via CXCR2. J Immunol 2003;171(2): 893–901.

11

Chemokine Receptors in Atherosclerosis

Maya R. Jerath, Mildred Kwan, Peng Liu, and Dhavalkumar D. Patel

Summary

Atherosclerosis is an inflammatory process that is strongly affected by chemokines that regulate the trafficking of inflammatory cells. A large amount of data from in vitro studies and murine models suggest an important role for chemokines in atherosclerosis. In man, genetic studies have revealed that specific polymorphisms in chemokine and chemokine receptor genes are associated with atherosclerotic diseases including coronary artery disease and carotid artery occlusive disease. Specifically, there are sound data supporting roles for the following receptors and their ligands: CCR2 and MCP-1 (CCL2); CX3CR1 and fractalkine (CX3CL1); CCR1, CCR5, and RANTES (CCL5); CXCR2 and IL-8 (CXCL8); CXCR6 and CXCL16; and CXCR3 and its ligands MIG (CXCL9), IP-10 (CXCL10) and I-TAC (CXCL11). These chemokines and their receptors participate in vascular inflammation via T-cell and monocyte chemoattraction, adhesion of monocytes to the vessel wall, and vascular smooth muscle cell migration and proliferation. Chemokines and chemokine receptors are being studied as potential therapeutic targets for the prevention or retardation of atherosclerotic disease. Although many of these therapies are promising, there are also limitations to specific chemokine-targeted therapy.

Key Words: Atherosclerosis; chemokine; chemokine receptor; inflammation; animal model; CCR2; MCP-1; CX3CR1; fractalkine.

11.1. Introduction

Atherosclerosis (AS) is a progressive disease characterized by the accumulation of lipids and the development of fibrosis in arterial walls. It is the pathophysiologic process behind cardiovascular disease whose clinical

From: *The Receptors: The Chemokine Receptors*
Edited by: J. K. Harrison and N. W. Lukacs © Humana Press Inc., Totowa, NJ

manifestations, including myocardial infarction, stroke, and peripheral vascular disease, are major causes of morbidity and mortality worldwide.

It is currently understood that atherosclerosis is inherently an inflammatory process and is strongly affected by chemokines, which help control the trafficking of inflammatory cells in the body *(1,2)*. The pathogenesis of atherosclerosis (Fig. 1; see color plate) is believed to occur as follows. Circulating low-density lipoproteins (LDLs) become oxidized by oxygen free radicals generated by a variety of established cardiovascular risk factors, including hypertension, smoking, and diabetes mellitus. Oxidized LDLs then damage the arterial endothelium and cause it to become activated. Upon activation, endothelium expresses cell surface adhesion molecules and chemokines that lead to the localization of circulating leukocytes to the site of injury. Interactions with adhesion molecules lead to arrest of mononuclear cells and subsequent diapedeses into the vessel wall. Monocytes mature into macrophages within the intima, the layer of the vessel wall between the endothelium and the internal elastic lamina. Macrophages phagocytose oxidized LDLs via scavenger receptors to form foam cells—the early lesion of atherosclerosis. Macrophages are also activated to secrete more cytokines, which results in lesion progression to a fatty streak. Further cytokine elaboration recruits vascular smooth muscle cells (VSMCs) to the lesion where they proliferate in response to secreted growth factors. In advanced plaques, VSMCs form the fibrous plaque. Thus, the lesion is complex and evolves over time. Physical disruption of the plaque may trigger spurts of accelerated growth. Atherosclerotic plaques may also become unstable through apoptosis of endothelial cells or activation of matrix metalloproteinases causing degradation of the subendothelial basement membrane. These latter two processes are triggered by inflammatory mediators (e.g., cytokines, chemokines, etc.) *(1,3)*. If a plaque ruptures, acute thrombosis can occur and lead to complete occlusion of the vessel and give rise to the clinical manifestations of the resultant ischemia (myocardial infarction if this occurs in the coronary circulation, stroke if in the cerebrovascular circulation, and arterial thrombosis if in the peripheral circulation).

Thus, inflammatory factors play a role at all stages of atherosclerosis: in its initiation, progression, and final clinical manifestations. This new understanding of the pathophysiology of atherosclerosis has partly resulted from the study of chemokines and their receptors *(4)*.

The basic biology of chemokines and their receptors is well covered in Chapters 2 and 3 of this book, and we will focus hereafter upon the roles of individual chemokines and receptors in atherosclerosis. The largest amount of data on the roles of chemokines in cardiovascular disease (CVD) has been obtained from in vitro studies and murine models, which will be discussed in detail. In man, genetic polymorphisms in chemokine and chemokine-receptor genes have pointed to an important role for specific chemokines in various atherosclerotic diseases including coronary artery disease and carotid artery occlusive disease. For properties see Table 1.

Table 1
Properties of Selected Molecules in Atherosclerosis That Affect Atherogenesis and Plaque Formation/Rupture

Molecule	Cell expression	Presumed role in atherosclerosis	Animal data in atherosclerosis	Human data in atherosclerosis
Chemokines				
CCL2 (MCP-1)	Endothelial cells, VSMCs, and fibroblasts	1. Chemoattractant for T cells and monocytes.	1. Hyperlipidemia in primates associated with elevated levels of CCL2. 2. Overexpressing CCL2 leads to increased macrophages and lipids in mouse vessel walls. 3. CCL2 deficiency associated with decreased atherosclerosis in susceptible mice. 4. Decreased intimal hyperplasia in femoral artery injury model seen with CCL2 deficiency.	1. Increased levels in human atherosclerotic plaques. 2. CCL2 levels higher in patients with acute coronary syndrome. 3. CCL2 levels in blood predict restenosis after angioplasty. 4. Polymorphism in promoter for CCL2 associated with increased CAD risk.

Table 1 *(Continued)*

Molecule	Cell expression	Presumed role in atherosclerosis	Animal data in atherosclerosis	Human data in atherosclerosis
CX3CL1 (FKN)	Endothelial cells, macrophages, VSMCs, and dendritic cells	1. Chemoattractant for monocytes, T cells and NK cells. 2. Integrin independent firm adhesion of monocytes, $CD8^+$ T cells.	1. CX3CL1 mRNA is upregulated in mice susceptible to atherosclerosis. 2. CX3CL1 deficiency results in decreased atherosclerosis in innominate arteries of susceptible mice.	1. CX3CL1 is found in macrophages in atherosclerotic lesions.
CCL5 (RANTES)	T cells, macrophages, and platelets	1. Chemoattractant for T cells and monocytes/ macrophages. 2. Firm adhesion of monocytes.	1. Blockade decreases plaque formation, intimal hyperplasia, and monocyte adhesion/ infiltration. 2. Increased secretion with MHC mismatched allografts.	1. Colocalization with T cells and macrophages in plaques. 2. Overexpression polymorphism of RANTES suggested to lead to increased CAD.
CXCL8 (IL-8)	Monocytes, macrophages/ foam cells, endothelial cells, and VSMCs	1. Chemoattractant for monocytes. 2. Adhesion of monocytes. 3. VSMC migration and proliferation. 4. Angiogenesis. 5. Dysregulation of MMPs leading to thinning of fibrous cap.	1. Blockade decreases lesion sizes (see CXCR2 in table, below). 2. Blockade decreases post ischemia-reperfusion myocardial necrosis.	1. Localization to atherosclerotic plaques. 2. Human explants induced angiogenesis in rat cornea. Blocked by IL-8 antibodies. 3. Levels correlate with serum homocysteine levels.

CXCL16	Macrophages, B cells, and dendritic cells	1. Chemoattractant for T cells. 2. Increase lipid uptake by macrophages.	1. Increased in atherosclerotic lesions in high-fat–fed Apo E knockout mice. 2. IFN-γ–dependent CXCL6 upregulation leads to increased oxLDL uptake in macrophages.	Increased levels in human atherosclerotic lesions.
CXCR3 ligands	Endothelial cells, macrophages, and VSMCs (IP-10)	Chemoattractant for activated T cells.		1. Increased levels in human atherosclerotic lesions. 2. Expressed in atheroma-associated cells.
Chemokine receptors				
CCR2	Macrophages and T lymphocytes	Receptor for CCL2 (see CCL2 in table, above).	1. CCR2 deficiency results in decreased atherosclerosis in vessels of susceptible mice. 2. Decreased intimal hyperplasia in femoral artery injury model of atherosclerosis seen with CCR2 deficiency.	CCR2 expression is increased in the monocytes of hypercholesterolemic patients.

Table 1 *(Continued)*

Molecule	Cell expression	Presumed role in atherosclerosis	Animal data in atherosclerosis	Human data in atherosclerosis
CX3CR1	Monocytes, NK cells, $CD8^+$ T cells, and VSMCs	Receptor for CX3CL1 (see CX3CL1 in table, above).	1. CX3CR1 deficiency leads to decreased atherosclerosis in susceptible mice. 2. CX3CR1 deficiency leads to decreased intimal hyperplasia in an injury model of atherosclerosis.	Individuals with a polymorphism in the CX3CR1 gene have decreased acute coronary events and decreased risk of coronary and internal carotid artery disease.
CCR5	T cells, macrophages, endothelial cells, and VSMCs	Receptor for RANTES (see RANTES in table, above).	Blockade decreases receptor expression.	Loss of function polymorphism suggests a decrease in early incidence of CAD.
CXCR2	Monocytes, macrophages, T cells, neutrophils, mast cells, and NK cells	Receptor for IL-8 (see IL-8 in table, above).	CXCR2 deficiency in susceptible mice decreased plaque formation.	

CXCR6	T cells	Recruitment of T cells to atherosclerotic lesions.	Increased levels in atherosclerotic lesions in high-fat–fed Apo E knockout mice.	Increased levels in human atherosclerotic lesions.
CXCR3	T cells, NK cells, endothelial cells, VSMCs, and macrophages	Recruitment of T cells to atherosclerotic lesions.	Apo E/CXCR3 double knockout mice on a high-fat diet displayed reduced atherosclerotic lesions during early stages of atherogenesis.	Expressed on T cells within human atherosclerotic lesions.

Apo E, apolipoprotein E; CAD, coronary artery disease; FKN, fractalkine; MHC, major histocompatibility complex; MCP-1, monocyte chemoattractant protein 1; MMPs, matrix metalloproteinases; NK, natural killer; oxLDL, oxidized LDL; RANTES, regulated on activation, normal T cell expressed and secreted; VSMCs, vascular smooth muscle cells.

11.2. Animal Models

Understanding of the importance of chemokines and their receptors in the atherosclerotic disease process has come in large part from the use of genetically deficient mice.

11.2.1. Hyperlipidemia Models

Two mouse models have been widely used to study atherosclerosis *(5)*. Apolipoprotein E (Apo E) knockout mice lack the apolipoprotein E ligand, resulting in hypercholesterolemia and increased atherosclerosis susceptibility. Low-density lipoprotein receptor (LDLR) knockout mice lack the LDL receptors, thus prolonging the plasma half-lives of lipoproteins VLDL and LDL *(6)*. A third less extensively used animal model, human apo B transgenic mice, which overproduce apo B–containing lipoproteins, have elevated plasma cholesterol and increased susceptibility to AS when fed a high-fat diet. These models are similar, and which one is superior is a matter of debate. We will focus on the Apo E model that has been most extensively used for studies of chemokines and their receptors. Apo E knockout mice exhibit many of the features of coronary artery disease when fed a high-fat, Western diet *(7)*. These include the initial monocyte invasion to susceptible regions of the aorta, foam cell lesion development, and the formation of a fibrous cap consisting of smooth muscle cells. The high-fat diet increases the rate at which atheromas develop. However, there is a concern in that the very high hypercholesterolemia in Apo E–deficient mice fed a high-fat diet causes a change in the immune response from T helper 1 (Th1)- to T helper 2 (Th2)-dependent help and thus may not represent human atherosclerosis that is associated with Th1 responses *(8,9)*. At late stages of vessel occlusion, vessels in animals fed a high-fat diet are occluded with xanthoma rather than the typical atherosclerotic plaque *(10,11)*. Realizing this potential confounder, many investigators are now studying Apo $E^{-/-}$ animals given a normal rodent chow diet. Although it takes longer for the plaques to develop, some investigators have suggested that evaluating the arteries in normal-chow-fed animals is one of the better ways to study atherosclerosis *(12–14)*.

In Apo E–deficient animals fed a normal chow diet, fatty streaks are first observed in the proximal aorta at 10 to 12 weeks *(15)*. The xanthoma that forms in the intima contains foam cells and is often called the early atherosclerotic lesion and is critically dependent on monocytes. Smooth muscle cells (SMCs) arrive in the intima at approximately 15 weeks and form a fibrous cap around 20 weeks *(16)*. By 36 weeks, lumen narrowing occurs in the external branches of the carotid artery (incidence ~75%), but the lumen size is maintained in the aorta. Lumen narrowing, or stenosis, does not correlate with plaque size but with adventitial inflammation and medial atrophy. Seo et al. *(17)* suggest that the stenotic process in advanced atherosclerotic vessels depends on medial smooth muscle cell apoptosis, possibly in response to inflammatory changes in the plaque or adventitia. By 42 weeks, plaques begin to rupture due to a loss of continuity of the fibrous cap with intraplaque hemorrhage *(18)*. This is possibly due to apoptosis of intimal SMC or macrophages. By 60 weeks, there is a virtual disappearance of the fibrous cap, being replaced by an extensive acellular collagenous mass. Calcification of advanced lesions occurs reproducibly between 45 and 75 weeks of age, with 100% incidence by 75 weeks *(19)*. The rate of progression in these late, and even early, stages of atherogenesis in Apo $E^{-/-}$ mice is most consistently seen in the innominate artery *(20,21)*, although evaluation of several arterial beds provides valuable knowledge.

11.2.2. Arterial Injury Models

Drawbacks to using mice genetically deficient in chemokines and receptors in the Apo $E^{-/-}$ model include (i) generating mice that are doubly deficient in the chemokine/receptor and Apo E and (ii) the long-term studies that are necessary. Many investigators have used endoluminal artery injury models in which endothelial denudation is induced by transluminal passage of a wire in the carotid or femoral artery, which results in platelet adhesion, recruitment of inflammatory cells, and neointimal formation after the injury in a predictable and timely manner. In this model, mouse femoral arteries display endothelial denudation with minimal or no damage to the internal elastic lamina (IEL). Within the first hours after injury, platelets and neutrophils readily adhere to the denuded surface with platelets predominating at day 1. Within 2 days, the endothelium has begun to regenerate. Five days after injury, there is an extensive accumulation of monocytes, platelets (and/or platelet antigens), and fibrin(ogen) along the luminal surface of the IEL. Monocytes also collect in the adventitia. At 14 days, VSMC accumulation in the intima is readily apparent. At 28 days, α-actin–expressing VSMCs predominate within the neointima, and substantial amounts of extracellular matrix are also present. Apoptosis of medial SMCs increases in the first 24 hours after injury and declines thereafter. Although the endoluminal injury model is more representative of restenosis after catheterization than atherosclerosis, it has provided rapid insights into the mechanisms of chemokine effects in the vasculature.

11.3. Chemokines and Receptors

The chemokine receptor/ligand pairs that have data to support their role in the pathogenesis of atherosclerosis and that will be discussed in this chapter include CCR2/MCP-1 (CCL2), CX3CR1/fractalkine (CX3CL1), CCR1/CCR5/

RANTES (CCL5), CXCR2/IL-8 (CXCL8), CXCR6/CXCL16, and CXCR3 and its ligands MIG (CXCL9), IP-10 (CXCL10), and I-TAC (CXCL11).

11.3.1. CCL2 (MCP-1) and CCR2

Recruitment of mononuclear leukocytes to the atherosclerotic lesion is a critical step in both the initial development and further progression of the plaque. Monocyte chemoattractant protein (MCP)-1, CCL2, is a member of the CC chemokine family and is a potent monocyte and lymphocyte chemoattractant *(22)*. It is produced by various cell types in the arterial wall including endothelial cells *(23,24)*, smooth muscle cells *(25)*, and fibroblasts *(23)*. CCL2 initiates signal transduction through binding to CC chemokine receptor 2 (CCR2), its only known receptor, which is highly expressed on macrophages and T lymphocytes (Fig. 2). There is a significant amount of literature supporting a major role for this chemokine-receptor pair in the pathogenesis of atherosclerosis (AS).

Increased CCL2 has been detected in macrophage-rich human atherosclerotic lesions *(26)* and in the blood of patients with acute coronary syndrome (implying an unstable plaque) *(27,28)*. It is found in the arteries of primates on a high-cholesterol diet *(29)* and is upregulated in vascular endothelial cells and

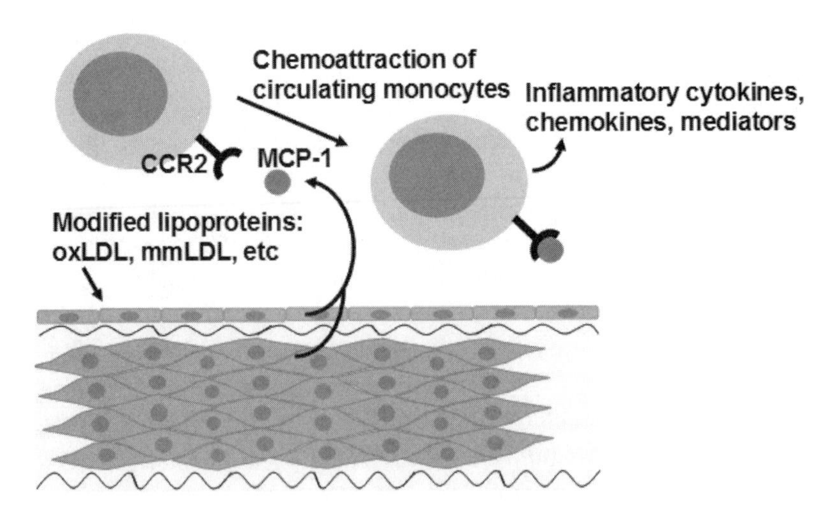

Fig. 2. The role of MCP-1 (CCL2)/CCR2 in atherosclerosis is thought to occur through the response of endothelial cells and vascular smooth muscle cells to oxidized lipoproteins. After injury by oxidized lipoproteins, MCP-1 is released and attracts CCR2-expressing monocytes to the site of injury and activates them to secrete inflammatory mediators.

smooth muscle cells exposed to minimally modified lipids *(30)*. Localization to the lesion, expression by cell types within the plaque, and upregulation in the setting of traditional atherosclerotic risk factors suggests that CCL2 expression plays a role in the development of AS. CCR2 expression has been found to be increased by 2.4-fold in monocytes isolated from hypercholesterolemic patients *(31)*. These monocytes exhibited increased chemotactic responses to CCL2. These studies suggest a possible mechanistic link between hyperlipidemia, foam cell formation, and atherosclerosis progression.

The roles of CCL2 in AS have been extensively evaluated using AS-susceptible mice with either over- or underexpression of CCL2. Aiello et al. *(32)* demonstrated by bone marrow transplantation studies that CCL2 overexpression in bone marrow–derived cells results in increased lipid accumulation, increased oxidized lipids, and increased macrophage markers relative to control mice at 20 weeks. CCL2 deficiency in LDLR$^{-/-}$ mice on a high-cholesterol diet for 12 to 14 weeks resulted in a 79% decrease in atherosclerotic lesion area at the aortic root and a similar decrease in the lesions in the thoracic and abdominal aorta *(33)*. CCL2 deficiency in apo B-transgenic mice fed a high-fat diet for 15 to 18 weeks resulted in a 60% to 70% decrease in aortic root lesion size *(34)*.

The findings with CCR2-deficient mice have been largely similar to that for CCL2. CCR2$^{-/-}$, Apo E$^{-/-}$ mice fed a high-fat diet had a significant decrease in macrophage infiltration of the aortic sinus at 5 weeks, decreased lesion area at the aortic root at 5, 9, and 13 weeks, and decreased thoracic/abdominal aortic lesion area at 13 weeks compared with Apo E$^{-/-}$ controls *(35)*. These reductions occurred despite similar cholesterol profiles in both groups of animals. When fed a normal chow diet, CCR2$^{-/-}$, Apo E$^{-/-}$ mice had one-third smaller aortic sinus lesions at 4 months compared with CCR2$^{+/-}$, Apo E$^{-/-}$ littermates *(36)*. CCR2 deficiency did not alter the peripheral blood cholesterol profile *(35,36)*. In the femoral arterial injury model, CCR2 deficiency resulted in a 61% reduction in intimal area and a 62% reduction in the intima:media ratio *(37)*. Five days after injury, the medial proliferation index, as determined by bromodeoxyuridine incorporation, was decreased by 60% in CCR2$^{-/-}$ mice *(37)*, a finding similar to that seen with CCL2-deficient mice *(38)*. These results demonstrate that CCL2/CCR2 plays an important role in mediating intimal hyperplasia and smooth muscle cell proliferation. In particular, CCL2/CCR2 may be an important target for inhibiting the response to acute arterial injury such as occurs after angioplasty, thus ameliorating the complication of restenosis. This would be particularly important in light of the results of a study of patients with restenosis after coronary angioplasty who were shown to have elevated levels of plasma CCL2 *(39)*.

Genetic data also support a role for CCR2 in atherosclerosis. The human CCR2 gene has a single nucleotide polymorphism (SNP) that results in a Val to Ile substitution at position 64 (V64I) *(40)* that is associated with decreased CCR2 function *(41–43)*. The CCR2 I64 isoform is associated with a decrease in the extent of coronary artery calcification (CAC) in a manner that is independent of traditional cardiovascular disease risk factors *(44)*. These data provide the final link suggesting that CCL2/CCR2 plays a role in AS in animal models and that it may play an important role in human disease.

11.3.2. CX3CL1/CX3CR1

CX3CR1 is the only known endogenous receptor for the chemokine fractalkine (FKN; CX3CL1) *(45)*. In man, CX3CR1 is found primarily on effector cells including monocytes and perforin/granzyme-expressing lymphocytes including natural killer cells and $CD8^+$ T cells *(45,46)*. CX3CR1 is also expressed in platelets and vascular smooth muscle cells *(47,48)*. Fractalkine, a cell surface molecule that can also be cleaved to form a soluble molecule, can be produced by several cell types, including activated endothelial cells, smooth cells, macrophages, and dendritic cells. Proinflammatory stimuli, such as lipopolysaccharide (LPS), TNF-α, and IL-1β for endothelial cells and IFN-γ and TNF-α for VSMCs, substantially upregulate fractalkine expression *(49–52)*.

Like other chemokines and their receptors, FKN and CX3CR1 function in chemotaxis *(45,53)*. The shed form of FKN is chemotactic for natural killer (NK) cells, T lymphocytes, monocytes, and microglia *(45,54,55)*. Unlike other chemokines and G protein–coupled receptors (GPCRs), FKN and CX3CR1 also act as cell adhesion molecules under both static and dynamic conditions. FKN can mediate the rapid capture and firm adhesion of monocytes, macrophages, NK cells, and $CD8^+$ T cells under physiologic flow conditions in an integrin and intracellular signaling independent manner *(56)*. These characteristics have led investigators to hypothesize that CX3CR1/CX3CL1 plays a central role in the pathophysiologic process of atherosclerosis (Fig. 3).

Evidence implicating CX3CR1 and fractalkine in atherosclerosis is growing. Immunohistochemical analysis of human atherosclerotic lesions demonstrate that macrophages in the intima as well as smooth muscle cells, mononuclear cells, and foam cells in the deep intima and media express CX3CL1, whereas normal arteries do not *(57,58)*.

Evaluation of CX3CL1 and CX3CR1 deficiency in animal models has provided further evidence for a role of these molecules in AS. Fractalkine mRNA is upregulated in the aortas of Apo $E^{-/-}$ mice fed a high-fat diet *(59)*. $CX3CL1^{-/-}$ mice on either an Apo $E^{-/-}$ or $LDLR^{-/-}$ background have decreased atherosclerotic lesions in the innominate artery *(60)*. A deficiency of the receptor CX3CR1 in Apo $E^{-/-}$ mice fed a high-fat diet results in decreased atherosclerosis compared with Apo $E^{-/-}$ controls *(61,62)*. Moreover, the fatty streaks of CX3CR1-deficient animals had decreased monocytes, suggesting that cellular trafficking of monocytes is affected by CX3CR1 *(62)*. In the femoral artery

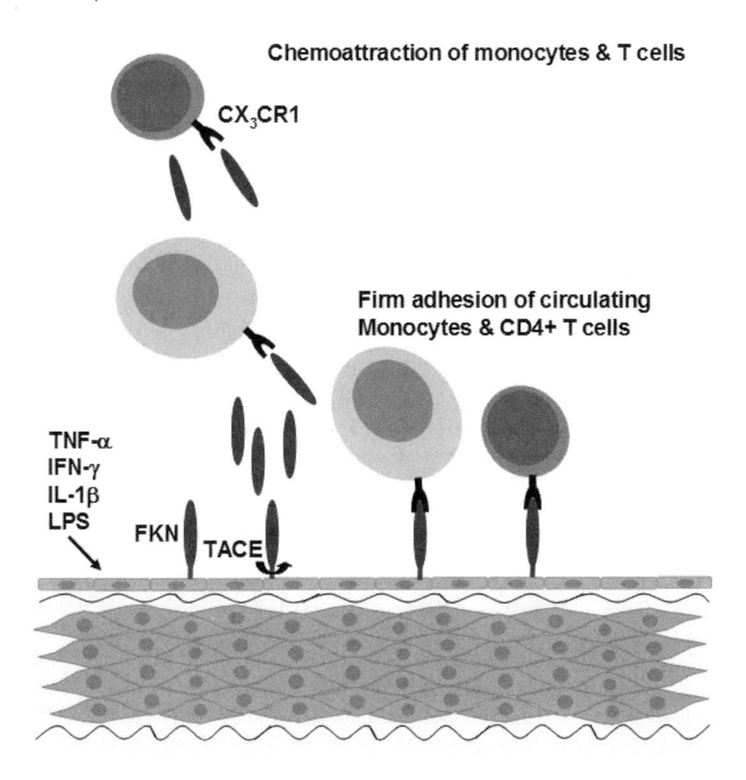

Fig. 3. The role of FKN/CX3R1 in inducing atherosclerosis may involve two mechanisms. The first is the release of soluble FKN through cleavage of membrane-expressed FKN by TNF-alpha converting enzyme (TACE). Soluble FKN is then able to attract circulating monocytes and effector T cells expressing CX3CR1. The second mechanism is that of firm adhesion of circulating cells to membrane-bound FKN.

injury model, CX3CL1 expression was found to be upregulated on endothelial and smooth muscle cells after injury, and mice lacking CX3CR1 had lower incidence and degree of intimal hyperplasia after injury *(63)*. The role of CX3CR1 in regulating vascular inflammation appeared to involve monocyte accumulation in the vessel wall and possibly VSMC proliferation and migration. Based on a recent report of the effect of IL-15 on CX3CR1/CX3CL1 expression and the response to arterial injury *(64)*, IL-15 may decrease neointimal formation in response to arterial injury via suppression of CX3CR1 signaling in VSMC. These studies in an injury model suggest that CX3CR1 could play a role not only in atherosclerosis but also in restenosis.

Perhaps the most compelling data supporting a role for CX3CR1 in AS comes from genetic evidence where two SNPs in human CX_3CR1, V249I and T280M, have been found. These SNPs are in linkage disequilibrium, and whereas the

I249 SNP can be found alone, the M280 SNP usually occurs in the context of I249. CX_3CR1-I249 has been extensively linked with protection from acute coronary events and carotid artery disease with improved endothelial-dependent coronary artery vasodilation in response to acetylcholine *(65,66)*, and CX_3CR1-M280 has been found to confer protection from coronary artery disease (CAD) as well as internal carotid artery occlusive disease *(67)*. Because of the linkage with I249 and the fact that M280 is a functional polymorphism resulting in a CX3CR1 receptor that is defective in adhesion *(68)*, it is likely that M280 is the SNP that results in protection from vascular diseases. Thus, in vitro, in vivo, and genetic data provide a compelling role for CX3CL1/CX3CR1 in the pathogenesis of atherosclerosis.

11.3.3. RANTES and CCR1/CCR5

RANTES (regulated on activation, normal T cell expressed and secreted), CCL5, is a CC chemokine first described as a T cell–derived factor. RANTES, produced by platelets *(12–16)*, T lymphocytes, and macrophages *(69)*, binds CCR1, CCR3, and CCR5 (70). CCR5, considered to be the primary RANTES receptor implicated in cardiovascular disease, is expressed by T lymphocytes *(71,72)*, macrophages *(71)*, coronary endothelium *(71,73)*, and vascular smooth muscle cells *(71,74,75)*. RANTES likely mediates monocyte and T-cell recruitment to atherosclerotic lesions (Fig. 4). This hypothesis has been supported by colocalization of RANTES with macrophages and T cells in plaques from human subjects *(3,69,76–78)* as well as in the Apo $E^{-/-}$ and $LDLR^{-/-}$ mouse models of atherosclerosis *(77,79,80)*. In addition, patients after acute myocardial infarction (MI) have increased levels of RANTES and numbers of infiltrating mononuclear cells when compared with controls *(81)*. These results suggest a role for RANTES in recruiting mononuclear cells leading to subsequent post MI myocardial damage. Interestingly, patients with hyperhomocysteinemia may have increased levels of RANTES suggesting a positive association between homocysteine and RANTES *(82)*.

Epidemiology also points to a possible role for RANTES in CAD *(83–86)*. Two polymorphisms have been identified in the RANTES gene promoter, -28C to G (-28G) and -403G to A (-403A), which reportedly cause increases in RANTES mRNA expression *(83,87,88)*. Liu et al. *(87)* showed that $CD4^+$ cells from subjects with the -28G/-403A haplotype secreted higher levels of RANTES when stimulated as compared with $CD4^+$ cells from either the -28G-only or -403A-only haplotype. The -403A polymorphism forms a new GATA transcription factor consensus binding site in the RANTES promoter leading to increased activity in in vitro assays *(88)*. Based on this molecular evidence, the effects of the -403A variant has been examined in the human population and has produced mixed results *(85,86)*. Szalai et al. *(86)* examined a small popula-

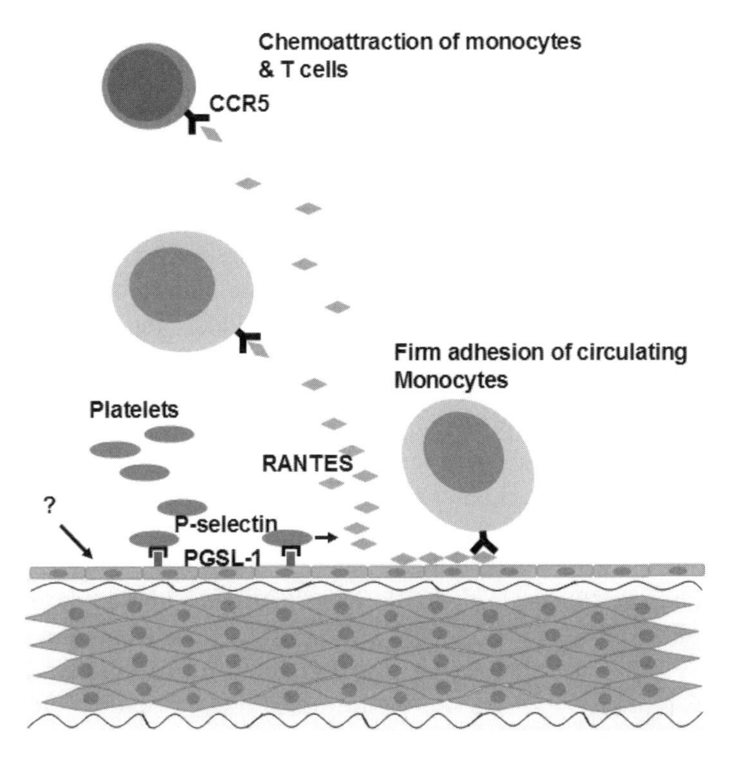

Fig. 4. The role of RANTES in recruiting monocytes and T lymphocytes to injured endothelium involves the binding of platelets to activated endothelium in a P-selectin–dependent manner with subsequent secretion of RANTES by bound platelets. RANTES can then attract CCR5-expressing monocytes and T cells to the damaged endothelium as well as causing monocyte firm adhesion to occur via binding to platelet-derived RANTES deposited on the endothelial cell layer.

tion of patients undergoing coronary artery bypass graft (CABG) surgery and were unable to show a significant difference in allelic frequency of -403A between the two groups. However, a more recent study by Simeoni et al. *(85)*, with a larger cohort, showed a significant increase in the allelic frequency of -403A in patients with mild and moderate/severe CAD when compared with controls. There was also a positive correlation between -403A and serum C-reactive protein (CRP) levels. The discrepancies between the two studies are not trivial and should lead to further examination of the -403A and -28G polymorphisms to discern whether these mutations truly have an effect on vascular inflammation.

The role of RANTES has been further elucidated through the use of mouse models via three approaches: (i) induction of hyperlipidemia with a high-fat

Western diet *(80)*, (ii) carotid artery endothelial injury *(89)*, and (iii) major histocompatibility complex (MHC) mismatched cardiac allograft *(90)*. Administration of Met-RANTES (an N-terminally modified RANTES that blocks CCR1 and CCR5) resulted in decreased plaque formation in the aorta and aortic roots, leukocyte infiltration, and expression of CCR2 and CCR5 *(80)*. In the carotid artery injury model in Apo $E^{-/-}$ mice *(77–79,91)*, RANTES was detected in the intima and media of damaged carotid arteries. Met-RANTES abrogated formation of neointimal tissue and monocyte infiltration of damaged/denuded endothelium *(78,79,91)*. In the cardiac allograft model, implantation of MHC I *(92)* or MHC II *(90)* mismatched allografts resulted in secretion of RANTES in donor hearts, which correlated with leukocyte infiltration. In addition, there was colocalization of RANTES with mononuclear cells within vessels that exhibited intimal thickening that was not observed in mice that received isografts *(90,92)*.

Although RANTES binds with high affinity to CCR1, CCR3, and CCR5 *(18)*, most of the data concerning the role for these receptors in conjunction with RANTES has focused on CCR5 *(83–86;93)*. A 32-base-pair deletion in the CCR5 gene (CCR5Δ32) leads to synthesis of a truncated protein that is not expressed as a cell surface molecule *(84)* and to protection from HIV-1. A study wherein CAD patients >55 years have a decreased allelic frequency of CCR5Δ32 when compared with CAD patients <60 years suggests that an absence of CCR5 protects from early CAD induction *(84)*. Another study found no CCR5Δ32 homozygous patients undergoing coronary artery bypass grafting. However, this report is controversial because there was no difference in frequency of CAD patients that were heterozygotes for CCR5Δ32 *(86)*. Finally, another study found no correlation between the CCR5Δ32 mutation and premature MI *(85)*. Thus the role of CCR5 in AS is unclear.

Animal models are equally confusing *(78,93–95)*. Kuziel et al. *(95)* showed that whereas the absence of CCR5 on monocytes in $CCR5^{-/-}$, Apo $E^{-/-}$ mice led to decreased binding of RANTES, CCR5 deficiency led to no changes in aorta and aortic root plaque size with no changes in the numbers of infiltrating macrophages and dendritic cells in the atherosclerotic lesions. This suggests that multiple chemokine receptors may be required for RANTES-mediated atherogenesis. Indeed, CCR1 and CCR5 have overlapping as well as distinct functions in RANTES-induced recruitment of monocytes and Th1/memory lymphocytes *(96)*. However, the hypothesis that CCR1 leads to atherosclerosis in concert with CCR5 is contradicted or at least complicated by a recent study where myeloablated $LDLR^{-/-}$ mice that received $CCR1^{-/-}$ bone marrow had similar incidence and size of fatty streaks (at 8 weeks on a high-fat diet) but had increased lipid staining and plaque area in the thoracic aorta and aortic sinus, respectively, with an increase in macrophage infiltration in atherosclerotic lesions at 12 weeks *(97)*. In addition, there was decreased VSMC accumulation and collagen suggesting the formation of unstable plaques. Deficiency of CCR1 in leukocytes also led to skewing toward a Th1 phenotype. Indeed, this study suggests that CCR1 has a protective effect against the formation of atherosclerotic plaques *(97)*. Thus, the roles of RANTES and its receptors in AS are not clear at this time.

11.3.4. IL-8/GRO and CXCR2

IL-8/CXCL8 is a member of the CXC subfamily that was initially characterized as being chemotactic for neutrophils *(94,98,99)*. It is now recognized that IL-8 is elaborated by many cell types including many that are involved in atherogenesis: monocytes and macrophages (including foam cells) *(3,100–104)*, endothelial cells *(3,100,101,104)*, and vascular smooth muscle cells *(100,105)*. IL-8 binds with high affinity to CXCR1 and CXCR2 *(106–108)*, both of which were first described on neutrophils *(101,108)*, but are also expressed by monocytes *(99,108–110)*, macrophages *(111)*, T lymphocytes *(108,112)*, mast cells, and NK cells *(108)*.

Involvement of IL-8/GRO-α in atherosclerosis was initially overlooked because neutrophils are rare in atherosclerotic lesions. It was also believed that IL-8 did not cause monocyte chemotaxis *(51,99)*. However, more recent data points to a probable role for IL-8 in AS (Fig. 5). New studies demonstrate that IL-8 is able to chemoattract monocytes and, in addition, is able to mediate adhesion of monocytes to endothelial cells *(113)*. IL-8 is also induced by many factors that may play a role in CAD including physical arterial injury, intralesional cytokines (TNF-α, IL-1β), complement, LDL, hypoxia, and bacterial components (LPS and possibly *Chlamydia pneumoniae*) *(100)*. IL-8 localizes to intimal lesions in macrophage regions *(3,114)*, more specifically to macrophage-derived foam cells *(102–104,115)*. Functional studies further developed the possible role for IL-8 in atherogenic inflammation. Not only does IL-8 recruit monocytes, but it also mediates firm adhesion of monocytes to endothelium *(116)* and may be able to modulate formation of advanced plaques by stimulating angiogenesis and smooth muscle cell migration and proliferation. IL-8 stimulation of human umbilical vein endothelial cells (HUVECs) causes chemotaxis of these cells, and treatment of avascular rat cornea with IL-8 led to the new formation of blood vessels that was abrogated by anti-IL-8 antibodies *(117)*. Importantly, implantation of directional coronary atherectomy tissue in rat cornea caused angiogenesis that was inhibited by anti-IL8 antibodies *(118)*. IL-8 may also lead to destabilization of advanced lesions through dysregulation of matrix metalloproteinases (MMPs) that weaken plaques through the degradation of the fibrous cap. Tissue inhibitor of metalloproteinases (TIMP-) 1 inhibits the action of MMPs, and IL-8 decreases TIMP-1 activity *(119)*. This

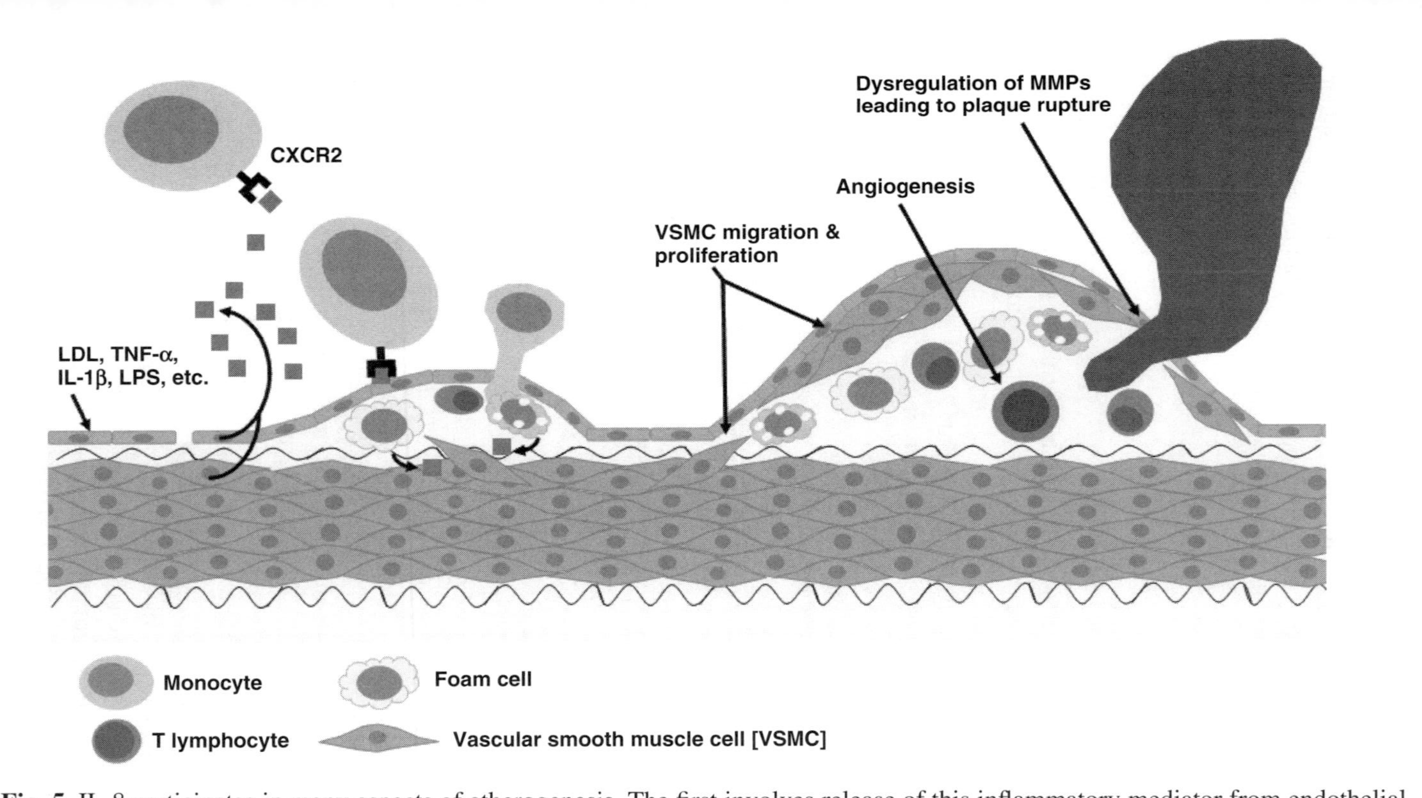

Fig. 5. IL-8 participates in many aspects of atherogenesis. The first involves release of this inflammatory mediator from endothelial cells and vascular smooth muscle cells that have been activated by LDL, proinflammatory cytokines (e.g., TNF-α, IL-1β), and bacterial products (e.g., LPS) leading to recruitment of CXCR2-expressing monocytes. Subsequently, macrophages and foam cells within the atheroma secrete additional IL-8 leading to vascular smooth muscle cell migration and proliferation and formation of the fibrous cap. IL-8 also induces angiogenesis and possibly neovascularization of the atheroma. Disruption of this microvasculature can also lead to thrombosis and vasculature occlusion. Finally, IL-8 has been shown to dysregulate the activity of matrix metalloproteinases (MMPs) leading to increased activity of these enzymes and thinning of the fibrous cap allowing for plaque rupture.

implies that IL-8 is secreted by damaged endothelial cells and/or vascular smooth muscle cells leading to monocyte recruitment and adhesion. Further, IL-8 is produced by these interactions potentially prolonging an inflammatory reaction. Thus, IL-8 may lead to destabilization and/or rupture of advanced atheromas through formation of plaque microvessels that can hemorrhage causing thrombotic arterial occlusion. Alternatively, dysregulation of MMPs could lead to thinning of the fibrous cap resulting in plaque rupture.

Research using animal models has further bolstered the idea that IL-8 may play an important role in CAD. $LDLR^{-/-}$ mice transplanted with $CXCR2^{-/-}$ bone marrow had significant decreases in lesion size (aorta and aortic valve sinus), size of necrotic lipid cores, and smooth muscle proliferation beyond $LDLR^{-/-}$ mice that received $CXCR2^{+/+}$ bone marrow *(111)*. In a canine model of coronary artery ischemia and reperfusion, IL-8 induction led to neutrophil recruitment to the region of injury likely contributing to further inflammation and damage after the initial insult *(120)*. This finding was confirmed in a rabbit model of ischemia-reperfusion where use of anti-IL-8 antibodies was able to decrease the degree of myocardial necrosis *(121)*.

Growing clinical data also points to the importance of IL-8 in atherogenesis. IL-8 has been found in atheromatous lesions from patients with atherosclerotic disease including carotid artery stenosis *(103)*, CAD *(118)*, abdominal aortic aneurysms (AAA) *(103,104,114)*, and peripheral vascular disease (PVD) *(104)*. Furthermore, studies using plaque explant samples have yielded more direct evidence for IL-8 involvement. Media from cultured AAA tissue induced IL-8–dependent human aortic endothelial cell (HAEC) chemotaxis *(122)*. Homocysteine, implicated as a possible biomarker for CAD, is also capable of inducing IL-8 *(123–125)* by direct stimulation of endothelial cells *(123,124)* and monocytes *(125)*. When patients with hyperhomocysteinemia were treated with low-dose folic acid, decreases in homocysteine levels correlated with decreases in IL-8 levels *(126)*. Statins significantly decrease serum levels of IL-6, IL-8, and MCP-1, as well as expression of IL-6, IL-8, and MCP-1 mRNA by peripheral blood monocytes and HUVECs *(127)*. Thus, IL-8 may be an underappreciated factor in the pathogenesis of atherosclerosis.

11.3.5. CXCL16 and CXCR6

CXCL16, a recently discovered transmembrane chemokine, has been detected in murine and human atherosclerotic lesions *(128,129)*. Although the precise role of CXCL16 in atherosclerosis remains unclear, studies suggest that it is a candidate molecule in atherosclerosis with multiple functions. Similar to fractalkine, CXCL16 acts as a chemoattractant and adhesion molecule for CXCR6-expressing T lymphocytes and NK cells to sites of inflammation and injury *(130–133)* independent of integrin activation *(134,135)*. In atherosclerotic lesions, both CXCL16 and CXCR6 are expressed in regions rich in macrophages and T lymphocytes *(128,129)*. Upon treatment with proinflammatory cytokine IFN-γ, CXCL16 expression was upregulated in atherosclerotic plaques in Apo E knockout mice fed normal chow *(129)*. These findings indicate that CXCL16/CXCR6 may be important in the development of atherosclerosis by promoting T-lymphocyte recruitment to inflamed lesions. CXCL16/CXCR6 also functions as a scavenger chemokine/receptor for phosphatidylserine and oxidized lipoprotein *(136)*. By increasing macrophage uptake of oxidized LDL *(129)*, CXCL16/CXCR6 may also have a role in foam-cell formation. It is too early to know whether CXCL16/CXCR6 plays a major role in AS, but ongoing studies, especially those using genetically deficient mice, will likely yield important information.

11.3.6. CXCR3 and Its Ligands

Although their role in AS is debatable, T lymphocytes are found in the atherosclerotic plaque *(1,2)*, and CXCR3 has been detected on the T cells found in atherosclerotic lesions *(137)*. CXCL9 monokine induced by gamma-interferon (MIG), CXCL10 interferon-gamma-induced protein (IP-10), and CXCL11 interferon-inducible T-cell alpha chemoattractant (I-TAC) are IFN-γ–inducible chemokines and potent chemoattractants for CXCR3-expressing T lymphocytes *(138,139)*, and their expression has been detected in atheroma-associated cells, including endothelial cells (ECs), SMCs, and macrophages *(137)*. The presence of this chemokine-receptor pair in atherosclerotic plaques suggests a possible role in atherogenesis, and there is additional support from murine models. A potential role for lymphocytes in AS has been suggested by the fact that lymphocyte-deficient RAG1-null mice on an LDLR$^{-/-}$ background have reduced atherosclerotic lesion development after 8 weeks of a Western-type diet but not at later time points *(140)*. Consistently, CXCR3/Apo E double knockout mice displayed reduced early atherosclerotic lesions compared with Apo E$^{-/-}$ mice *(141)*. This implicates CXCR3 and its ligands in the recruitment and retention of T lymphocytes during the early development of atherosclerosis.

11.4. Therapeutic Implications

From the preceding sections, it is clear that chemokines are important players in atherosclerotic disease and, as such, are being considered as possible targets in the treatment of this prevalent inflammatory condition. Under consideration at this time are both traditional nonspecific therapies [e.g., 3-hydroxy-3-methylglutaryl coenzyme A (HMG-CoA) reductase inhibitors, fibrates, etc.], as well as chemokine specific approaches *(142)*.

Based on the conventional understanding that CAD results from hyperlipidemia, atherosclerosis has been treated with good success by reduction of serum cholesterol as well as triglycerides through the use of HMG-CoA reductase inhibitors, better known as statins. However, it has been noted that statins likely have vasoprotective properties outside of control of serum lipids and have immunomodulating effects that also lead to regression/prevention of atherosclerotic disease. Statins inhibit HMG-CoA reductase preventing the synthesis of cholesterol from mevalonic acid, and this has long been believed to be the main mechanism of cardioprotection. There are now data suggesting that statins also have anti-inflammatory effects that include inhibition of leukocyte recruitment and adhesion *(143–147)*, reduction in chemokine production, upregulation of peroxisome proliferator-activated receptor (PPAR)-α and -γ *(144–149)*, downregulation of MMPs with concomitant upregulation of TIMP-1 *(145,146)*, subversion of Th response toward a Th2 response through inhibition of proinflammatory cytokine production (e.g., TNF-α and IL-1β) *(144–147)*, and several other proposed mechanisms. In particular, statins have been shown to directly affect chemokine secretion. In several studies, in vitro as well as in vivo, statins have been shown to decrease the expression of MCP-1 *(127,144–149)*, RANTES *(145,147)*, and IL-8 *(127,144,145,148)*. Statins have also been implicated indirectly in the downregulation of MCP-1 by decreasing oxidation of LDL and CRP *(146)*.

Among the other traditional treatments for hyperlipidemia, hypertension, and diabetes mellitus are other pharmacologic agents that have been shown to decrease chemokine levels. Fibrates may indirectly downregulate MCP-1 by inducing PPARs *(148)* and via reduction of CRP *(143)*, although it has also been shown in one study that fenofibrate increased MCP-1 and IL-8 secretion *(148)*. Medications that inhibit the renin-angiotensin axis have a role in blocking chemokine production. For example, the angiotensin-converting enzyme (ACE) inhibitor enalapril was able to decrease serum MCP-1 levels in patients with acute myocardial infarction *(150)*. Other more recent studies have shown that angiotensin receptor blockers (ARBs) also decreased plasma MCP-1 concentrations, and combination therapy with either an ACE inhibitor or ARB plus a statin led to further decreases when compared with ACE inhibitor/ARB monotherapy (reviewed in Ref. 143). Finally, the antidiabetic glitazones have been proven to reduce MCP-1 *(148)* and CRP *(143)* secretion and prevent release of some CXC chemokines from endothelial cells *(148)*.

Chemokine-targeted therapeutic approaches are under development. Much interest has been focused on the CCL2-CCR2 pathway because this is the best characterized chemokine-receptor pair in atherosclerosis. Anti-MCP-1 gene therapy by a novel approach involving transfection into skeletal muscle has been shown to inhibit atherogenesis and plaque destabilization in Apo E knockout mice *(151)*. Interestingly, it did so without lowering the serum lipid concentration, allowing for the possibility of even further control of atherosclerosis with the addition of a lipid-lowering agent. This therapy has also been shown to be effective in preventing restenosis in animal models of angioplasty *(152)*. Decreased neointimal formation was seen in primates after stent placement, and permission to try this in humans is being sought in Japan *(153)*. Additionally, a monoclonal anti-CCR2 antibody has shown decreased neointimal hyperplasia in monkeys *(154)*, perhaps paving the way for a new generation of medicated stents.

Several pharmaceutical companies are working on orally bioavailable small-molecule antagonists of CCR2. One company has published the details of their compound, and although it has not been reported to be tested in atherosclerosis models, the authors note that the compound finds application in a variety of diseases where macrophages are implicated *(155)*. Many methods to downregulate the CCR2-CCL2 axis exist, and they show promise as future therapies for atherosclerosis. One can particularly conceive of them being given around the time of angioplasty or chronically to someone with established atherosclerotic disease as secondary prevention.

The other pathway that has shown promise as a therapeutic target is RANTES-CCR1/CCR5. Global blockade of CCR1, CCR3, and CCR5 using a peptide antagonist, Met-RANTES, has been shown to reduce atherosclerotic lesion formation and macrophage infiltration in LDLR knockout mice *(80)* and neointima formation after carotid artery injury in Apo E knockout mice *(79)*. Similar results were seen in LDLR knockout mice using a small-molecule CCR5 antagonist originally developed for the treatment of HIV infection *(156)*. CCR1 antagonists are also under development, although they have not yet been reported in atherosclerosis models *(157)*.

Whereas many of these therapies are promising, there are limitations to specific chemokine-targeted therapy. First, as chemokines have many functions, targeting any one systemically may have undesired effects [e.g., increased susceptibility to infections or other immune-mediated diseases *(158–160)*]. Second, many chemokines clearly play a role in atherosclerosis, and it is not evident which would be the best one to target, especially because different ones play roles at varying stages of atherosclerosis. Finally, all the models used to study the roles of chemokines and to assess the efficacy of chemokine blockade are ones in which atherosclerosis is forming de novo. The pathophysiology of established disease such as in human atherosclerotic disease at the time of diagnosis may be very different and may not respond to these interventions. For example, whether chemokine-directed therapy can cause plaque regression is not known.

Restenosis after angioplasty is a situation where these therapies may find an early application. The current animal models evaluate the formation of atherosclerosis or a neointima in the same accelerated and artificially induced manner as occurs after balloon angioplasty or stent placement. Also, chronic dosing of these agents either through drug-eluting stents (local delivery of medication) or systemically in the peri-angioplasty period is far more feasible than lifelong administration as would be required to prevent de novo atherosclerosis that develops over decades in humans.

11.5. Conclusions

Our understanding of atherosclerosis as an inflammatory process has made great strides in the past decade, and it is clear from animal studies as well as clinical data that chemokines play an important role in atherosclerotic vascular disease.

Based on the data presented in this chapter, it is possible to construct a theoretical model of how the aforementioned chemokines act in concert with other inflammatory mediators to promote atheroma formation. In response to factors that cause vessel wall damage (e.g., modified lipoproteins, proinflammatory cytokines, bacterial products, etc.), endothelial cells elaborate inflammatory mediators including MCP-1, FKN, IL-8, (CXCL16, MIG, IP-10, and I-TAC), and VSMCs secrete MCP-1, IL-8, MIG, IP-10, and I-TAC. The expression of chemokines in turn attracts circulating monocytes and T lymphocytes to the site of vascular injury. Platelets are also recruited to activated endothelium induced to express PGSL-1 and bind via platelet P-selectin. This leads to release of RANTES from platelets and further trafficking of monocytes and T cells to damaged vasculature. Firm adhesion of monocytes may then be mediated by FKN, RANTES, and IL-8 while FKN, CXCL16, MIG, IP-10, and I-TAC lead to adhesion of T cells. These cells diapedese into the vessel wall where monocytes differentiate into tissue macrophages and interact with T cells leading to further activation and production of cytokines and chemokines and persistent inflammation. In addition, macrophages are able to take up oxidized LDL (oxLDL) via scavenger receptors leading to foam-cell formation and development of the fatty streak. Further foam-cell formation and eventual cell death leads to the progression of the lesion with inception of a necrotic lipid core. Macrophages and foam cells within the evolving lesion continue to produce IL-8, which in turn causes the migration and proliferation of VSMCs that form the fibrous cap of an advanced atherosclerotic lesion. IL-8 also induces angiogenesis and thus vascularization of atheromas with microvessels that can rupture and, conceivably, allow arterial thrombosis and occlusion. IL-8 may also play a role in plaque rupture by downregulating TIMP-1, allowing thinning of the fibrous cap.

The interplay of these chemokines in the atherosclerotic process is unclear, although there are some inferences that can be made about when certain chemokines are thought to be important in the process of plaque formation. CXCR3 and its ligands MIG, IP-10, and I-TAC are believed to play a role in early plaque formation based on data from CXCR3$^{-/-}$, Apo E$^{-/-}$ mice that showed a decrease in early lesion formation. This is in contrast with CCR2/Apo E double knockout mice that showed reduced lesions in early atherosclerotic plaques but exhibited greater reductions in atheroma formation at later points of disease. In addition to MCP-1/CCR2 and CXCR3 and its ligands, the involvement of IL-8 in the progression of the advanced atherosclerotic plaque can also be implied by its ability to induce VSMC migration/proliferation, angiogenesis, and dysregulation of MMPs. These data perhaps suggest that CXCR3 and its ligands MIG, IP-10, and I-TAC are crucial in recruiting T lymphocytes early in fatty streak formation while MCP-1 and IL-8 are necessary for persistent recruitment of monocytes and formation of the fibrous cap, respectively, during the late stages of disease with development of the advanced plaque. Prior to plaque rupture, IL-8 may also be responsible for destabilization of the atheroma leaving the affected artery more susceptible to thrombotic occlusion.

References

1. Libby P. Inflammation in atherosclerosis. Nature 2002;420(6917):868–874.
2. Ross R. Atherosclerosis—an inflammatory disease. N Engl J Med 1999;340(2): 115–126.
3. Reape TJ, Groot PH. Chemokines and atherosclerosis. Atherosclerosis 1999; 147(2):213–225.
4. Charo IF, Ransohoff RM. The many roles of chemokines and chemokine receptors in inflammation. N Engl J Med 2006;354(6):610–621.
5. Smith JD, Breslow JL. The emergence of mouse models of atherosclerosis and their relevance to clinical research [Review]. J Int Med 1997;242(2):99–109.
6. Ishibashi S, Goldstein JL, Brown MS, Herz J, Burns DK. Massive xanthomatosis and atherosclerosis in cholesterol-fed low density lipoprotein receptor-negative mice. J Clin Invest 1994;93(5):1885–1893.
7. Nakashima Y, Plump AS, Raines EW, Breslow JL, Ross R. Apo E-deficient mice develop lesions of all phases of atherosclerosis throughout the arterial tree. Arterioscler Thromb 1994;14(1):133–140.
8. Zhou XH, Paulsson G, Stemme S, Hansson GK. Hypercholesterolemia is associated with a T helper (Th) 1 Th2 switch of the autoimmune response in atherosclerotic apo E knockout mice. J Clin Invest 1998;101(8):1717–1725.
9. Geng YJ, Holm J, Nygren S, Bruzelius M, Stemme S, Hansson GK. Expression of the macrophage scavenger receptor in atheroma—relationship to immune activation and the T-cell cytokine interferon-gamma. Arterioscler Thromb Vasc Biol 1995;15(11):1995–2002.

10. Rosenfeld ME, Polinsky P, Virmani R, Kauser K, Rubanyi G, Schwartz SM. Advanced atherosclerotic lesions in the innominate artery of the Apo E knockout mouse. Arterioscler Thromb Vasc Biol 2000;20(12):2587–2592.
11. Caligiuri G, Levy B, Pernow J, Thoren P, Hansson GK. Myocardial infarction mediated by endothelin receptor signaling in hypercholesterolemic mice. Proc Natl Acad Sci U S A 1999;96(12):6920–6924.
12. Bea F, Blessing E, Bennett BJ, et al. Chronic inhibition of cyclooxygenase-2 does not alter plaque composition in a mouse model of advanced unstable atherosclerosis. Cardiovasc Res 2003;60(1):198–204.
13. Bea F, Blessing E, Shelley MI, Shultz JM, Rosenfeld ME. Simvastatin inhibits expression of tissue factor in advanced atherosclerotic lesions of apolipoprotein E deficient mice independently of lipid lowering: potential role of simvastatin-mediated inhibition of Egr-1 expression and activation. Atherosclerosis 2003; 167(2):187–194.
14. Rosenfeld ME, Polinsky P, Virmani R, Kauser K, Rubanyi G, Schwartz SM. Advanced atherosclerotic lesions in the innominate artery of the Apo E knockout mouse. Arterioscler Thromb Vasc Biol 2000;20(12):2587–2592.
15. Piedrahita JA, Zhang SH, Hagaman JR, Oliver PM, Maeda N. Generation of mice carrying a mutant apolipoprotein-E gene inactivated by gene targeting in embryonic stem-cells. Proc Natl Acad Sci U S A 1992;89(10):4471–4475.
16. Meir KS, Leitersdorf E. Atherosclerosis in the apolipoprotein E-deficient mouse—a decade of progress. Arterioscler Thromb Vasc Biol 2004;24(6):1006–1014.
17. Seo HS, Lombardi DM, Polinsky P, et al. Peripheral vascular stenosis in apolipoprotein E-deficient mice—potential roles of lipid deposition, medial atrophy, and adventitial inflammation. Arterioscler Thromb Vasc Biol 1997;17(12):3593–3601.
18. Rosenfeld ME, Polinsky P, Virmani R, Kauser K, Rubanyi G, Schwartz SM. Advanced atherosclerotic lesions in the innominate artery of the Apo E knockout mouse. Arterioscler Thromb Vasc Biol 2000;20(12):2587–2592.
19. Rattazzi M, Bennett BJ, Bea F, et al. Calcification of advanced atherosclerotic lesions in the innominate arteries of ApoE-deficient mice: potential role of chondrocyte-like cells. Arterioscler Thromb Vasc Biol 2005;25(7):1420–1425.
20. Rosenfeld ME, Polinsky P, Virmani R, Kauser K, Rubanyi G, Schwartz SM. Advanced atherosclerotic lesions in the innominate artery of the Apo E knockout mouse. Arterioscler Thromb Vasc Biol 2000;20(12):2587–2592.
21. Seo HS, Lombardi DM, Polinsky P, et al. Peripheral vascular stenosis in apolipoprotein E-deficient mice—potential roles of lipid deposition, medial atrophy, and adventitial inflammation. Arterioscler Thromb Vasc Biol 1997;17(12):3593–3601.
22. Boring L, Gosling J, Chensue SW, et al. Impaired monocyte migration and reduced type 1 (Th1) cytokine responses in C-C chemokine receptor 2 knockout mice. J Clin Invest 1997;100(10):2552–2561.
23. Rollins BJ, Yoshimura T, Leonard EJ, Pober JS. Cytokine-activated human endothelial cells synthesize and secrete a monocyte chemoattractant, MCP-1/JE. Am J Pathol 1990;136(6):1229–1233.

24. Li YS, Shyy YJ, Wright JG, Valente AJ, Cornhill JF, Kolattukudy PE. The expression of monocyte chemotactic protein (MCP-1) in human vascular endothelium in vitro and in vivo. Mol Cell Biochem 1993;126(1):61–68.
25. Valente AJ, Graves DT, Vialle-Valentin CE, Delgado R, Schwartz CJ. Purification of a monocyte chemotactic factor secreted by nonhuman primate vascular cells in culture. Biochemistry 1988;27(11):4162–4168.
26. Nelken NA, Coughlin SR, Gordon D, Wilcox JN. Monocyte chemoattractant protein-1 in human atheromatous plaques. J Clin Invest 1991;88(4): 1121–1127.
27. de Lemos JA, Morrow DA, Sabatine MS, et al. Association between plasma levels of monocyte chemoattractant protein-1 and long-term clinical outcomes in patients with acute coronary syndromes. Circulation 2003;107(5):690–65.
28. Matsumori A, Furukawa Y, Hashimoto T, et al. Plasma levels of the monocyte chemotactic and activating factor/monocyte chemoattractant protein-1 are elevated in patients with acute myocardial infarction. J Mol Cell Cardiol 1997;29(1):419–423.
29. Yu X, Dluz S, Graves DT, et al. Elevated expression of monocyte chemoattractant protein 1 by vascular smooth muscle cells in hypercholesterolemic primates. Proc Natl Acad Sci U S A 1992;89(15):6953–6957.
30. Cushing SD, Berliner JA, Valente AJ, et al. Minimally modified low density lipoprotein induces monocyte chemotactic protein 1 in human endothelial cells and smooth muscle cells. Proc Natl Acad Sci U S A 1990;87(13):5134–5138.
31. Han KH, Han KO, Green SR, Quehenberger O. Expression of the monocyte chemoattractant protein-1 receptor CCR2 is increased in hypercholesterolemia. Differential effects of plasma lipoproteins on monocyte function. J Lipid Res 1999;40(6):1053–1063.
32. Aiello RJ, Bourassa PA, Lindsey S, et al. Monocyte chemoattractant protein-1 accelerates atherosclerosis in apolipoprotein E-deficient mice. Arterioscler Thromb Vasc Biol 1999;19(6):1518–1525.
33. Gu L, Okada Y, Clinton SK, et al. Absence of monocyte chemoattractant protein-1 reduces atherosclerosis in low density lipoprotein receptor-deficient mice. Mol Cell 1998;2(2):275–281.
34. Gosling J, Slaymaker S, Gu L, et al. MCP-1 deficiency reduces susceptibility to atherosclerosis in mice that overexpress human apolipoprotein B. J Clin Invest 1999;103(6):773–778.
35. Boring L, Gosling J, Cleary M, Charo IF. Decreased lesion formation in $CCR2^{-/-}$ mice reveals a role for chemokines in the initiation of atherosclerosis. Nature 1998;394(6696):894–897.
36. Dawson TC, Kuziel WA, Osahar TA, Maeda N. Absence of CC chemokine receptor-2 reduces atherosclerosis in apolipoprotein E-deficient mice. Atherosclerosis 1999;143(1):205–211.
37. Roque M, Kim WJ, Gazdoin M, et al. CCR2 deficiency decreases intimal hyperplasia after arterial injury. Arterioscler Thromb Vasc Biol 2002;22(4): 554–559.

38. Kim WJ, Chereshnev I, Gazdoiu M, Fallon JT, Rollins BJ, Taubman MB. MCP-1 deficiency is associated with reduced intimal hyperplasia after arterial injury. Biochem Biophys Res Commun 2003;310(3):936–942.
39. Cipollone F, Marini M, Fazia M, et al. Elevated circulating levels of monocyte chemoattractant protein-1 in patients with restenosis after coronary angioplasty. Arterioscler Thromb Vasc Biol 2001;21(3):327–334.
40. Struyf F, Thoelen I, Charlier N, et al. Prevalence of CCR5 and CCR2 HIV-coreceptor gene polymorphisms in Belgium. Hum Hered 2000;50(5):304–307.
41. Smith MW, Dean M, Carrington M, et al. Contrasting genetic influence of CCR2 and CCR5 variants on HIV-1 infection and disease progression. Hemophilia Growth and Development Study (HGDS), Multicenter AIDS Cohort Study (MACS), Multicenter Hemophilia Cohort Study (MHCS), San Francisco City Cohort (SFCC), ALIVE Study. Science 1997;277(5328):959–965.
42. Hizawa N, Yamaguchi E, Furuya K, Jinushi E, Ito A, Kawakami Y. The role of the C-C chemokine receptor 2 gene polymorphism V64I (CCR2–64I) in sarcoidosis in a Japanese population. Am J Respir Crit Care Med 1999;159(6):2021–2023.
43. Abdi R, Tran TB, Sahagun-Ruiz A, et al. Chemokine receptor polymorphism and risk of acute rejection in human renal transplantation. J Am Soc Nephrol 2002;13(3):754–758.
44. Valdes AM, Wolfe ML, O'Brien EJ, et al. Val64Ile polymorphism in the C-C chemokine receptor 2 is associated with reduced coronary artery calcification. Arterioscler Thromb Vasc Biol 2002;22(11):1924–1928.
45. Imai T, Hieshima K, Haskell C, et al. Identification and molecular characterization of fractalkine receptor CX3CR1, which mediates both leukocyte migration and adhesion. Cell 1997;91(4):521–530.
46. Combadiere C, Gao J, Tiffany HL, Murphy PM. Gene cloning, RNA distribution, and functional expression of mCX3CR1, a mouse chemotactic receptor for the CX3C chemokine fractalkine. Biochem Biophys Res Commun 1998;253(3):728–732.
47. Lucas AD, Bursill C, Guzik TJ, Sadowski J, Channon KM, Greaves DR. Smooth muscle cells in human atherosclerotic plaques express the fractalkine receptor CX3CR1 and undergo chemotaxis to the CX3C chemokine fractalkine (CX3CL1). Circulation 2003;108(20):2498–2504.
48. Schafer A, Schulz C, Eigenthaler M, et al. Novel role of the membrane-bound chemokine fractalkine in platelet activation and adhesion. Blood 2004;103(2):407–412.
49. Harrison JK, Jiang Y, Wees EA, et al. Inflammatory agents regulate in vivo expression of fractalkine in endothelial cells of the rat heart. J Leukoc Biol 1999;66(6):937–944.
50. Garcia GE, Xia Y, Chen S, et al. NF-kappaB-dependent fractalkine induction in rat aortic endothelial cells stimulated by IL-1beta, TNF-alpha, and LPS. J Leukoc Biol 2000;67(4):577–584.

51. Ahn SY, Cho CH, Park KG, et al. Tumor necrosis factor-alpha induces fractalkine expression preferentially in arterial endothelial cells and mithramycin A suppresses TNF-alpha-induced fractalkine expression. Am J Pathol 2004;164(5): 1663–1672.
52. Ludwig A, Berkhout T, Moores K, Groot P, Chapman G. Fractalkine is expressed by smooth muscle cells in response to IFN-gamma and TNF-alpha and is modulated by metalloproteinase activity. J Immunol 2002;168(2):604–612.
53. Bazan JF, Bacon KB, Hardiman G, et al. A new class of membrane-bound chemokine with a CX3C motif. Nature 1997;385(6617):640–644.
54. Harrison JK, Jiang Y, Chen S, et al. Role for neuronally derived fractalkine in mediating interactions between neurons and CX3CR1-expressing microglia. Proc Natl Acad Sci U S A 1998;95(18):10896–10901.
55. Maciejewski-Lenoir D, Chen S, Feng L, Maki R, Bacon KB. Characterization of fractalkine in rat brain cells: migratory and activation signals for CX3CR$^{-/-}$ expressing microglia. J Immunol 1999;163(3):1628–1635.
56. Fong AM, Robinson LA, Steeber DA, et al. Fractalkine and CX3CR1 mediate a novel mechanism of leukocyte capture, firm adhesion, and activation under physiologic flow. J Exp Med 1998;188(8):1413–1419.
57. Greaves DR, Hakkinen T, Lucas AD, et al. Linked chromosome 16q13 chemokines, macrophage-derived chemokine, fractalkine, and thymus- and activation-regulated chemokine, are expressed in human atherosclerotic lesions. Arterioscler Thromb Vasc Biol 2001;21(6):923–929.
58. Wong BW, Wong D, McManus BM. Characterization of fractalkine (CX3CL1) and CX3CR1 in human coronary arteries with native atherosclerosis, diabetes mellitus, and transplant vascular disease. Cardiovasc Pathol 2002;11(6):332–338.
59. Lutgens E, Faber B, Schapira K, et al. Gene profiling in atherosclerosis reveals a key role for small inducible cytokines: validation using a novel monocyte chemoattractant protein monoclonal antibody. Circulation 2005;111(25):3443–3452.
60. Teupser D, Pavlides S, Tan M, Gutierrez-Ramos JC, Kolbeck R, Breslow JL. Major reduction of atherosclerosis in fractalkine (CX3CL1)-deficient mice is at the brachiocephalic artery, not the aortic root. Proc Natl Acad Sci U S A 2004; 101(51):17795–1800.
61. Combadiere C, Potteaux S, Gao JL, et al. Decreased atherosclerotic lesion formation in CX3CR1/apolipoprotein E double knockout mice. Circulation 2003;107(7): 1009–1016.
62. Lesnik P, Haskell CA, Charo IF. Decreased atherosclerosis in CX3CR1$^{-/-}$ mice reveals a role for fractalkine in atherogenesis. J Clin Invest 2003;111(3):333–340.
63. Liu P, Patil S, Rojas M, Fong AM, Smyth SS, Patel DD. CX3CR1 deficiency confers protection from intimal hyperplasia after arterial injury. Arterioscler Thromb Vasc Biol 2006;26:2056–2062.
64. Garton KJ, Gough PJ, Blobel CP, et al. Tumor necrosis factor-alpha-converting enzyme (ADAM17) mediates the cleavage and shedding of fractalkine (CX3CL1). J Biol Chem 2001;276(41):37993–38001.

65. Moatti D, Faure S, Fumeron F, et al. Polymorphism in the fractalkine receptor CX3CR1 as a genetic risk factor for coronary artery disease. Blood 2001; 97(7):1925–1928.
66. McDermott DH, Halcox JP, Schenke WH, et al. Association between polymorphism in the chemokine receptor CX3CR1 and coronary vascular endothelial dysfunction and atherosclerosis. Circ Res 2001;89(5):401–407.
67. Ghilardi G, Biondi ML, Turri O, Guagnellini E, Scorza R. Internal carotid artery occlusive disease and polymorphisms of fractalkine receptor CX3CR1: a genetic risk factor. Stroke 2004;35(6):1276–1279.
68. McDermott DH, Fong AM, Yang Q, et al. Chemokine receptor mutant CX3CR1-M280 has impaired adhesive function and correlates with protection from cardiovascular disease in humans. J Clin Invest 2003;111(8):1241–1250.
69. Wilcox JN, Nelken NA, Coughlin SR, Gordon D, Schall TJ. Local expression of inflammatory cytokines in human atherosclerotic plaques. J Atheroscler Thromb 1994;1(Suppl 1):S10–S13.
70. Nelson PJ, Krensky AM. Chemokines, lymphocytes and viruses: what goes around, comes around. Curr Opin Immunol 1998;10(3):265–270.
71. Rottman JB, Ganley KP, Williams K, Wu L, Mackay CR, Ringler DJ. Cellular localization of the chemokine receptor CCR5. Correlation to cellular targets of HIV-1 infection. Am J Pathol 1997;151(5):1341–1351.
72. Ward SG, Westwick J. Chemokines: understanding their role in T-lymphocyte biology. Biochem J 1998;333(Pt 3):457–470.
73. Berger O, Gan X, Gujuluva C, et al. CXC and CC chemokine receptors on coronary and brain endothelia. Mol Med 1999;5(12):795–805.
74. Hayes IM, Jordan NJ, Towers S, et al. Human vascular smooth muscle cells express receptors for CC chemokines. Arterioscler Thromb Vasc Biol 1998;18(3): 397–403.
75. Schecter AD, Calderon TM, Berman AB, et al. Human vascular smooth muscle cells possess functional CCR5. J Biol Chem 2000;275(8):5466–5471.
76. Gerard C, Rollins BJ. Chemokines and disease. Nat Immunol 2001;2(2):108–115.
77. von HP, Weber KS, Huo Y, et al. RANTES deposition by platelets triggers monocyte arrest on inflamed and atherosclerotic endothelium. Circulation 2001;103(13):1772–1777.
78. Weber C, Schober A, Zernecke A. Chemokines: key regulators of mononuclear cell recruitment in atherosclerotic vascular disease. Arterioscler Thromb Vasc Biol 2004;24(11):1997–2008.
79. Schober A, Manka D, von HP, et al. Deposition of platelet RANTES triggering monocyte recruitment requires P-selectin and is involved in neointima formation after arterial injury. Circulation 2002;106(12):1523–1529.
80. Veillard NR, Kwak B, Pelli G, et al. Antagonism of RANTES receptors reduces atherosclerotic plaque formation in mice. Circ Res 2004;94(2):253–261.

81. Parissis JT, Adamopoulos S, Venetsanou KF, Mentzikof DG, Karas SM, Kremastinos DT. Serum profiles of C-C chemokines in acute myocardial infarction: possible implication in postinfarction left ventricular remodeling. J Interferon Cytokine Res 2002;22(2):223–229.
82. Sun W, Wang G, Zhang ZM, Zeng XK, Wang X. Chemokine RANTES is upregulated in monocytes from patients with hyperhomocysteinemia. Acta Pharmacol Sin 2005;26(11):1317–1321.
83. Bursill CA, Channon KM, Greaves DR. The role of chemokines in atherosclerosis: recent evidence from experimental models and population genetics. Curr Opin Lipidol 2004;15(2):145–149.
84. Gonzalez P, Alvarez R, Batalla A, et al. Genetic variation at the chemokine receptors CCR5/CCR2 in myocardial infarction. Genes Immun 2001;2(4): 191–195.
85. Simeoni E, Winkelmann BR, Hoffmann MM, et al. Association of RANTES G-403A gene polymorphism with increased risk of coronary arteriosclerosis. Eur Heart J 2004;25(16):1438–1446.
86. Szalai C, Duba J, Prohaszka Z, et al. Involvement of polymorphisms in the chemokine system in the susceptibility for coronary artery disease (CAD). Coincidence of elevated Lp(a) and MCP-1 -2518 G/G genotype in CAD patients. Atherosclerosis 2001;158(1):233–239.
87. Liu H, Chao D, Nakayama EE, et al. Polymorphism in RANTES chemokine promoter affects HIV-1 disease progression. Proc Natl Acad Sci U S A 1999; 96(8):4581–4585.
88. Nickel RG, Casolaro V, Wahn U, et al. Atopic dermatitis is associated with a functional mutation in the promoter of the C-C chemokine RANTES. J Immunol 2000;164(3):1612–1616.
89. Huo Y, Schober A, Forlow SB, et al. Circulating activated platelets exacerbate atherosclerosis in mice deficient in apolipoprotein E. Nat Med 2003;9(1): 61–67.
90. Yun JJ, Fischbein MP, Laks H, et al. Rantes production during development of cardiac allograft vasculopathy. Transplantation 2001;71(11):1649–1656.
91. Weber C. Platelets and chemokines in atherosclerosis: partners in crime. Circ Res 2005;96(6):612–616.
92. Yun JJ, Fischbein MP, Laks H, et al. Early and late chemokine production correlates with cellular recruitment in cardiac allograft vasculopathy. Transplantation 2000;69(12):2515–2524.
93. Eriksson EE. Mechanisms of leukocyte recruitment to atherosclerotic lesions: future prospects. Curr Opin Lipidol 2004;15(5):553–558.
94. Charo IF, Taubman MB. Chemokines in the pathogenesis of vascular disease. Circ Res 2004;95(9):858–866.
95. Kuziel WA, Dawson TC, Quinones M, et al. CCR5 deficiency is not protective in the early stages of atherogenesis in apo E knockout mice. Atherosclerosis 2003;167(1):25–32.

96. Weber C, Weber KS, Klier C, et al. Specialized roles of the chemokine receptors CCR1 and CCR5 in the recruitment of monocytes and T(H)1-like/CD45RO(+) T cells. Blood 2001;97(4):1144–1146.
97. Potteaux S, Combadiere C, Esposito B, et al. Chemokine receptor CCR1 disruption in bone marrow cells enhances atherosclerotic lesion development and inflammation in mice. Mol Med 2005;11:16–20.
98. Mackay CR. Chemokines: immunology's high impact factors. Nat Immunol 2001;2(2):95–101.
99. Terkeltaub R, Boisvert WA, Curtiss LK. Chemokines and atherosclerosis. Curr Opin Lipidol 1998;9(5):397–405.
100. Schwartz D, Andalibi A, Chaverri-Almada L, et al. Role of the GRO family of chemokines in monocyte adhesion to MM-LDL-stimulated endothelium. J Clin Invest 1994;94(5):1968–1973.
101. Shin WS, Szuba A, Rockson SG. The role of chemokines in human cardiovascular pathology: enhanced biological insights. Atherosclerosis 2002;160(1): 91–102.
102. Wang N, Tabas I, Winchester R, Ravalli S, Rabbani LE, Tall A. Interleukin 8 is induced by cholesterol loading of macrophages and expressed by macrophage foam cells in human atheroma. J Biol Chem 1996;271(15):8837–8842.
103. Apostolopoulos J, Davenport P, Tipping PG. Interleukin-8 production by macrophages from atheromatous plaques. Arterioscler Thromb Vasc Biol 1996; 16(8):1007–1012.
104. Rus HG, Vlaicu R, Niculescu F. Interleukin-6 and interleukin-8 protein and gene expression in human arterial atherosclerotic wall. Atherosclerosis 1996;127(2): 263–271.
105. Wang JM, Sica A, Peri G, et al. Expression of monocyte chemotactic protein and interleukin-8 by cytokine-activated human vascular smooth muscle cells. Arterioscler Thromb 1991;11(5):1166–1174.
106. Lee J, Horuk R, Rice GC, Bennett GL, Camerato T, Wood WI. Characterization of two high affinity human interleukin-8 receptors. J Biol Chem 1992;267(23): 16283–16287.
107. Moser B, Schumacher C, von Tscharner V, Clark-Lewis I, Baggiolini M. Neutrophil-activating peptide 2 and gro/melanoma growth-stimulatory activity interact with neutrophil-activating peptide 1/interleukin 8 receptors on human neutrophils. J Biol Chem 1991;266(16):10666–10671.
108. Murdoch C, Finn A. Chemokine receptors and their role in inflammation and infectious diseases. Blood 2000;95(10):3032–3043.
109. Chuntharapai A, Lee J, Hebert CA, Kim KJ. Monoclonal antibodies detect different distribution patterns of IL-8 receptor A and IL-8 receptor B on human peripheral blood leukocytes. J Immunol 1994;153(12):5682–5688.
110. Morohashi H, Miyawaki T, Nomura H, et al. Expression of both types of human interleukin-8 receptors on mature neutrophils, monocytes, and natural killer cells. J Leukoc Biol 1995;57(1):180–187.

111. Boisvert WA, Santiago R, Curtiss LK, Terkeltaub RA. A leukocyte homologue of the IL-8 receptor CXCR-2 mediates the accumulation of macrophages in atherosclerotic lesions of LDL receptor-deficient mice. J Clin Invest 1998;101(2): 353–363.
112. Xu L, Kelvin DJ, Ye GQ, et al. Modulation of IL-8 receptor expression on purified human T lymphocytes is associated with changed chemotactic responses to IL-8. J Leukoc Biol 1995;57(2):335–342.
113. Gerszten RE, Garcia-Zepeda EA, Lim YC, et al. MCP-1 and IL-8 trigger firm adhesion of monocytes to vascular endothelium under flow conditions. Nature 1999;398(6729):718–723.
114. Koch AE, Kunkel SL, Pearce WH, et al. Enhanced production of the chemotactic cytokines interleukin-8 and monocyte chemoattractant protein-1 in human abdominal aortic aneurysms. Am J Pathol 1993;142(5):1423–1431.
115. Lucas AD, Greaves DR. Atherosclerosis: role of chemokines and macrophages. Expert Rev Mol Med 2001;2001:1–18.
116. Lukacs NW, Strieter RM, Elner V, Evanoff HL, Burdick MD, Kunkel SL. Production of chemokines, interleukin-8 and monocyte chemoattractant protein-1, during monocyte: endothelial cell interactions. Blood 1995;86(7):2767–2773.
117. Koch AE, Polverini PJ, Kunkel SL, et al. Interleukin-8 as a macrophage-derived mediator of angiogenesis. Science 1992;258(5089):1798–1801.
118. Simonini A, Moscucci M, Muller DW, et al. IL-8 is an angiogenic factor in human coronary atherectomy tissue. Circulation 2000;101(13):1519–1526.
119. Moreau M, Brocheriou I, Petit L, Ninio E, Chapman MJ, Rouis M. Interleukin-8 mediates downregulation of tissue inhibitor of metalloproteinase-1 expression in cholesterol-loaded human macrophages: relevance to stability of atherosclerotic plaque. Circulation 1999;99(3):420–426.
120. Kukielka GL, Smith CW, LaRosa GJ, et al. Interleukin-8 gene induction in the myocardium after ischemia and reperfusion in vivo. J Clin Invest 1995;95(1): 89–103.
121. Boyle EM, Jr., Kovacich JC, Hebert CA, et al. Inhibition of interleukin-8 blocks myocardial ischemia-reperfusion injury. J Thorac Cardiovasc Surg 1998;116(1): 114–121.
122. Szekanecz Z, Shah MR, Harlow LA, Pearce WH, Koch AE. Interleukin-8 and tumor necrosis factor-alpha are involved in human aortic endothelial cell migration. The possible role of these cytokines in human aortic aneurysmal blood vessel growth. Pathobiology 1994;62(3):134–139.
123. Poddar R, Sivasubramanian N, DiBello PM, Robinson K, Jacobsen DW. Homocysteine induces expression and secretion of monocyte chemoattractant protein-1 and interleukin-8 in human aortic endothelial cells: implications for vascular disease. Circulation 2001;103(22):2717–2723.
124. Geisel J, Hennen B, Hubner U, Knapp JP, Herrmann W. The impact of hyperhomocysteinemia as a cardiovascular risk factor in the prediction of coronary heart disease. Clin Chem Lab Med 2003;41(11):1513–1517.

125. Zeng X, Dai J, Remick DG, Wang X. Homocysteine mediated expression and secretion of monocyte chemoattractant protein-1 and interleukin-8 in human monocytes. Circ Res 2003;93(4):311–320.
126. Huo Y, Weber C, Forlow SB, et al. The chemokine KC, but not monocyte chemoattractant protein-1, triggers monocyte arrest on early atherosclerotic endothelium. J Clin Invest 2001;108(9):1307–1314.
127. Rezaie-Majd A, Maca T, Bucek RA, et al. Simvastatin reduces expression of cytokines interleukin-6, interleukin-8, and monocyte chemoattractant protein-1 in circulating monocytes from hypercholesterolemic patients. Arterioscler Thromb Vasc Biol 2002;22(7):1194–1199.
128. Minami M, Kume N, Shimaoka T, et al. Expression of scavenger receptor for phosphatidylserine and oxidized lipoprotein (SR-PSOX) in human atheroma. Ann N Y Acad Sci 2001;947:373–376.
129. Wuttge DM, Zhou X, Sheikine Y, et al. CXCL16/SR-PSOX is an interferon-gamma-regulated chemokine and scavenger receptor expressed in atherosclerotic lesions. Arterioscler Thromb Vasc Biol 2004;24(4):750–755.
130. Heydtmann M, Adams DH. Understanding selective trafficking of lymphocyte subsets. Gut 2002;50(2):150–152.
131. Kim CH, Kunkel EJ, Boisvert J, et al. Bonzo/CXCR6 expression defines type 1-polarized T-cell subsets with extralymphoid tissue homing potential. J Clin Invest 2001;107(5):595–601.
132. Shashkin P, Simpson D, Mishin V, Chesnutt B, Ley K. Expression of CXCL16 in human T cells. Arterioscler Thromb Vasc Biol 2003;23(1):148–149.
133. Unutmaz D, Xiang W, Sunshine MJ, Campbell J, Butcher E, Littman DR. The primate lentiviral receptor Bonzo/STRL33 is coordinately regulated with CCR5 and its expression pattern is conserved between human and mouse. J Immunol 2000;165(6):3284–3292.
134. Shimaoka T, Nakayama T, Fukumoto N, et al. Cell surface-anchored SR-PSOX/CXC chemokine ligand 16 mediates firm adhesion of CXC chemokine receptor 6-expressing cells. J Leukoc Biol 2004;75(2):267–274.
135. Chandrasekar B, Bysani S, Mummidi S. CXCL16 signals via Gi, phosphatidylinositol 3-kinase, Akt, I kappa B kinase, and nuclear factor-kappa B and induces cell-cell adhesion and aortic smooth muscle cell proliferation. J Biol Chem 2004; 279(5):3188–3196.
136. Shimaoka T, Kume N, Minami M, et al. Molecular cloning of a novel scavenger receptor for oxidized low density lipoprotein, SR-PSOX, on macrophages. J Biol Chem 2000;275(52):40663–40666.
137. Mach F, Sauty A, Iarossi AS, et al. Differential expression of three T lymphocyte-activating CXC chemokines by human atheroma-associated cells. J Clin Invest 1999;104(8):1041–1050.
138. Loetscher M, Gerber B, Loetscher P, et al. Chemokine receptor specific for IP10 and mig: structure, function, and expression in activated T-lymphocytes. J Exp Med 1996;184(3):963–969.

139. Piali L, Weber C, LaRosa G, et al. The chemokine receptor CXCR3 mediates rapid and shear-resistant adhesion-induction of effector T lymphocytes by the chemokines IP10 and Mig. Eur J Immunol 1998;28(3):961–972.
140. Song L, Leung C, Schindler C. Lymphocytes are important in early atherosclerosis. J Clin Invest 2001;108(2):251–259.
141. Veillard NR, Steffens S, Pelli G, et al. Differential influence of chemokine receptors CCR2 and CXCR3 in development of atherosclerosis in vivo. Circulation 2005;112(6):870–878.
142. Sheikine YA, Hansson GK. Chemokines as potential therapeutic targets in atherosclerosis. Curr Drug Targets 2006;7(1):13–27.
143. Koh KK, Quon MJ, Han SH, et al. Additive beneficial effects of fenofibrate combined with atorvastatin in the treatment of combined hyperlipidemia. J Am Coll Cardiol 2005;45(10):1649–1653.
144. McCarey DW, Sattar N, McInnes IB. Do the pleiotropic effects of statins in the vasculature predict a role in inflammatory diseases? Arthritis Res Ther 2005; 7(2):55–61.
145. Schonbeck U, Libby P. Inflammation, immunity, and HMG-CoA reductase inhibitors: statins as antiinflammatory agents? Circulation 2004;109(21 Suppl 1): II18-II26.
146. Shovman O, Levy Y, Gilburd B, Shoenfeld Y. Antiinflammatory and immunomodulatory properties of statins. Immunol Res 2002;25(3):271–285.
147. Weitz-Schmidt G. Statins as anti-inflammatory agents. Trends Pharmacol Sci 2002;23(10):482–486.
148. Burke-Gaffney A, Brooks AV, Bogle RG. Regulation of chemokine expression in atherosclerosis. Vascul Pharmacol 2002;38(5):283–292.
149. Koh KK, Han SH, Quon MJ. Inflammatory markers and the metabolic syndrome: insights from therapeutic interventions. J Am Coll Cardiol 2005;46(11):1978–1985.
150. Soejima H, Ogawa H, Yasue H, et al. Angiotensin-converting enzyme inhibition reduces monocyte chemoattractant protein-1 and tissue factor levels in patients with myocardial infarction. J Am Coll Cardiol 1999;34(4):983–988.
151. Ni W, Egashira K, Kitamoto S, et al. New anti-monocyte chemoattractant protein-1 gene therapy attenuates atherosclerosis in apolipoprotein E-knockout mice. Circulation 2001;103(16):2096–2101.
152. Kitamoto S, Egashira K. Anti-monocyte chemoattractant protein-1 gene therapy for cardiovascular diseases. Expert Rev Cardiovasc Ther 2003;1(3):393–400.
153. Egashira K. Molecular mechanisms mediating inflammation in vascular disease: special reference to monocyte chemoattractant protein-1. Hypertension 2003;41(3 Pt 2):834–841.
154. Horvath C, Welt FG, Nedelman M, Rao P, Rogers C. Targeting CCR2 or CD18 inhibits experimental in-stent restenosis in primates: inhibitory potential depends on type of injury and leukocytes targeted. Circ Res 2002;90(4):488–494.
155. Brodmerkel CM, Huber R, Covington M, et al. Discovery and pharmacological characterization of a novel rodent-active CCR2 antagonist, INCB3344. J Immunol 2005;175(8):5370–5378.

156. van Wanrooij EJ, Happe H, Hauer AD, et al. HIV entry inhibitor TAK-779 attenuates atherogenesis in low-density lipoprotein receptor-deficient mice. Arterioscler Thromb Vasc Biol 2005;25(12):2642–2647.
157. Saeki T, Naya A. CCR1 chemokine receptor antagonist. Curr Pharm Des 2003;9(15):1201–1208.
158. Niess JH, Brand S, Gu X, et al. CX3CR1-mediated dendritic cell access to the intestinal lumen and bacterial clearance. Science 2005;307(5707):254–258.
159. Quinones MP, Estrada CA, Kalkonde Y, et al. The complex role of the chemokine receptor CCR2 in collagen-induced arthritis: implications for therapeutic targeting of CCR2 in rheumatoid arthritis. J Mol Med 2005;83(9):672–681.
160. Quinones MP, Ahuja SK, Jimenez F, et al. Experimental arthritis in CC chemokine receptor 2-null mice closely mimics severe human rheumatoid arthritis. J Clin Invest 2004;113(6):856–866.

12

Chemokine Receptors in Allergic Lung Disease

Dennis M. Lindell and Nicholas W. Lukacs

Summary

This chapter is an attempt to integrate recent studies concerning the role of chemokine receptors in the initiation, development, and maintenance of allergic lung diseases collectively referred to as asthma. The pathogenesis of asthma involves the coordinated trafficking of inflammatory cells to the lungs and draining lymph nodes, as well as the activation of these inflammatory cells. Chemokine receptors and their ligands play a prominent role in directing the inflammation associated with allergic lung disease. T lymphocyte–mediated immune responses can be broadly categorized as being type 1 or type 2, based on the cell types present and the associated cytokines produced. Allergic lung disease is a predominately type 2–mediated disease. The chemokine receptors CCR4, CCR6, and CCR8 serve to promote the recruitment of type 2 T (T helper 2; Th2) cells, whereas CXCR3 antagonizes type 2 and promotes type 1 T (T helper 1; Th1) cells. The pathophysiologic manifestations of asthma, including excessive mucus production, eosinophilia, and airway hyperreactivity, are dependent upon the trafficking and activation of eosinophils, mast cells, and goblet cells. Roles for chemokine receptors, including CCR4, CCR2, and CXCR4, in the trafficking and activation of these cell types during allergic lung disease are discussed. Finally, the incidence of allergic lung disease is increasing, and the costs associated with it are substantial. Chemokine receptor expression and use by inflammatory cells during allergic lung disease makes chemokine receptors an attractive therapeutic target. Implications for drug development are discussed in the context of experimental results.

Key Words: Allergy; asthma; lung; pulmonary; T cell; mast cell; eosinophil; inflammation; dendritic cell; IgE; B cell.

From: *The Receptors: The Chemokine Receptors*
Edited by: J. K. Harrison and N. W. Lukacs © Humana Press Inc., Totowa, NJ

12.1. Introduction

Allergic lung disease encompasses the development, establishment, and maintenance of inflammation leading to asthma. Allergic asthma is variable but is characterized by the following: altered lung function, airway hyperreactivity (AHR), chronic inflammation, excessive mucus production, goblet cell hyperplasia, and increased IgE production. In chronic disease, significant remodeling of the airways occurs including smooth muscle hypertrophy and peribronchial fibrosis. These responses have been examined extensively in animal models.

Initial steps in the development of allergic lung disease involve the exposure of genetically susceptible individuals to allergen via inhalation. Initial exposure results in the priming and expansion of allergen-specific T cells, as well as expansion of B cells and allergen-specific antibody production. Repeated exposure in susceptible individuals results in the expansion and trafficking of T helper 2 (Th2) cells, mast cells, and eosinophils to the lungs (Fig. 1). Cytokine production, including IL-4, IL-5, and IL-13, promotes the production of allergen-specific IgE by B cells, airway hyperresponsiveness, mucus production, and inflammation. Binding of specific antigen by IgE bound to mast cells

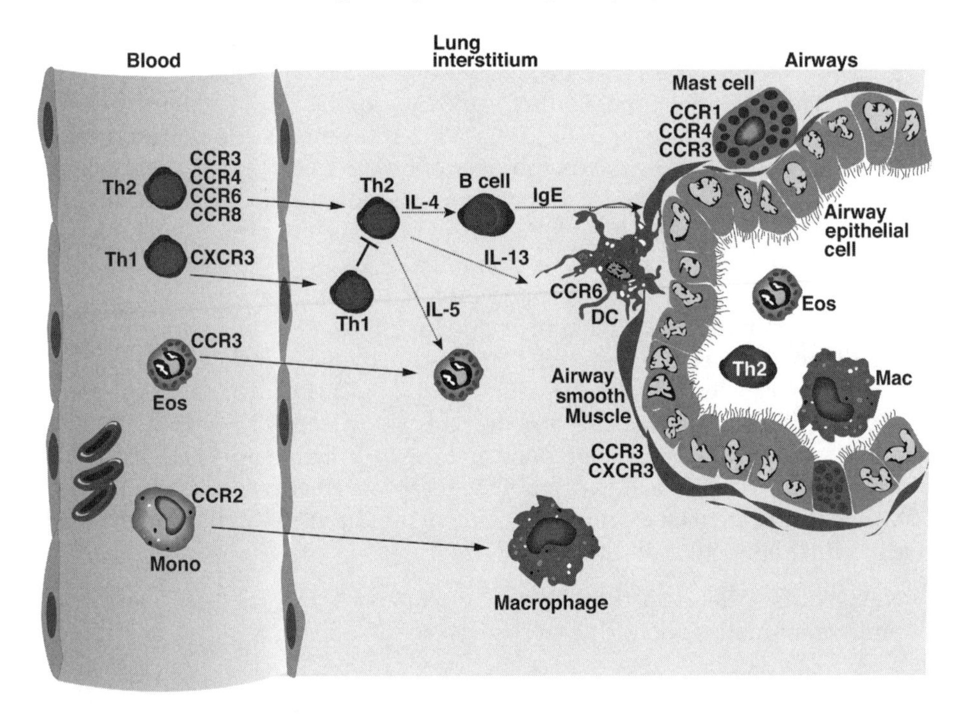

Fig.1. Expression of CC and CXC chemokine receptors by inflammatory and structural cells in allergic lung disease. Eos, eosinophil; Mac, macrophage; Mono, monocyte.

triggers acute allergic events. Each of these events is dependent upon the coordinated production and expression of chemokine receptor/ligand systems. Chemokine receptors and ligands play essential roles in the trafficking, activation, and effector functions of inflammatory and structural cells (Table 1).

12.2. Th1 versus Th2 Chemokine Receptor Profiles

Although there are exceptions to the rule *(1)*, differential chemokine receptor expression tends to be present under Th1 and Th2 inflammatory conditions. Under Th1 conditions, CXCR3 and, to a lesser extent, CCR5 predominate. Conversely, under Th2 conditions, CCR3, CCR4, and CCR8 are preferentially expressed. Thus, under Th1-promoting conditions, CXCR3 ligands including 10kDa interferon-gamma-induced protein (IP-10)/CXCL9, gamma interferon-induced monokine (MIG)/CXCL10, and interferon-inducible T-cell alpha chemoattractant (I-TAC)/CXCL11 preferentially attract Th1 T cells. CCR3, CCR4, and CCR8 ligands, including eotaxins/CCL 11, 24, 26, monocyte-derived chemokine (MDC)/CCL22, thymus associated regulatory chemokine (TARC)/CCL17, and T-cell activation protein 3 (TCA-3)/CCL1, preferentially attract Th2 cells, eosinophils, mast cells, and basophils.

Clinical evidence tends to support the Th1/Th2 differential chemokine receptor paradigm, represented primarily as CXCR3 versus CCR4. Pulmonary $CCR4^{+}CD4^{+}$ cells and levels of CCL17 and CCL22 were significantly increased in asthmatic children versus children with nonasthmatic chronic cough or without airway disease. In asthmatic children, $CCR4^{+}CD4^{+}$ cells correlated positively with levels of CCL17, CCL22, serum IgE levels, and negatively with FEV 1 (forced expiratory volume in the first second of exhalation, a measure of lung function). Conversely, $CXCR3^{+}CD8^{+}$ cells and levels of CXCL11 were significantly increased in children with nonatopic chronic cough compared with the other groups *(2)*. These results have been recapitulated in segmental allergen challenge studies. In asthmatics, a majority of T cells in post–allergen challenge biopsies expressed CCR4 *(3,4)*. Expression of the CCR4 ligands CCL22 and CCL17 was also upregulated on airway epithelial cells upon allergen challenge *(3)*. CCR4 expression was not found on T cells from patients with chronic obstructive pulmonary disease, which instead expressed CXCR3 *(4)*. Cumulatively, these results demonstrate an increase in CCR4-expressing T cells in allergic airways and increased production of CCL17 and CCL22. Furthermore, these results demonstrate that the increase in CCR4 expression by T cells in allergic lungs may serve to differentiate allergic from nonallergic inflammation.

One potential mechanism for the continued polarization of T-cell chemokine receptor expression during inflammatory lung disease is that chemokines can also serve as receptor antagonists. CXCR3 ligands act as antagonists of CCR3, and CCL11 can serve as an antagonist for CXCR3 *(5,6)*. Thus, Th1-type

Table 1
Expression of Chemokine Receptors by Leukocytes and Structural Cells in Allergic Lung Disease

	CCR1	CCR2	CCR3	CCR4	CCR6	CCR8	CXCR1	CXCR2	CXCR3	CXCR4
T cells, B cells	++	++	++	++	++	++			++	++
Eosinophils/basophils	++	+	++	++					++	++
Neutrophils	+	+	+				+	+		+
Mast cells	++	++	++	++	+		+	+		+
Epithelial cells			++		+				+	+
Airway smooth muscle cells		+	++						++	
Monocyte/macrophages	+	++				++				+
Fibroblasts		+	+					+		
Dendritic cells	+	+		+	++	+	+	+	+	+

Note: +, expressed by cell type; ++, evidence for role in allergic lung disease.

chemokines may simultaneously promote Th1 and antagonize Th2, similar to Th1 and Th2 cytokines.

12.2.1. CCR4

There is overwhelming evidence from humans that CCR4 and its principal ligands CCL17 and CCL22 are associated with allergic lung disease. The airway epithelium is thought to be the major producer of CCR4 ligands, although some studies have suggested that smooth muscle cells, mast cells, and even naïve $CD4^+$ T cells may contribute *(7)*. Increased expression of CCL17 and CCL22 mRNA has been demonstrated in asthmatic bronchial epithelium and submucosa *(8)*. CCL17 and/or CCL22 were elevated in the serum and bronchoalveolar lavage (BAL) of asthmatics relative to controls *(9,10)*. Additionally, bronchial epithelial cells from asthmatic patients produced more CCL17 *(11,12)*. Thus, the production of CCR4 ligands is increased in asthmatic patients. In BAL of allergic asthmatic patients given serial antigen, CCL22 and CCL17 were highly increased compared with saline-challenged areas *(13,14)*. The increased levels of CCL17 and CCL22 correlated with lymphocyte numbers and the levels of IL-5 and IL-13 (14). Thus, within individual lungs, production of CCL17 and CCL22 correlate with areas of allergic inflammation.

Evidence from animal studies targeting the CCR4 axis in allergic lung disease has been less clear. Using a guinea-pig ovalbumin model, a CCR4 blocking monoclonal antibody did not significantly affect the development of allergic inflammation or the production of CCL11 and CCL22 *(15)*. Only a modest decrease in $CCR4^+$ Th cells was observed *(15)*. On the other hand, neutralization of CCL22 attenuated interstitial eosinophilia in the mouse model of ovalbumin-induced allergic lung disease. This was accompanied by a decrease in AHR in CCL22-neutralized mice *(16)*. Monoclonal antibody–mediated neutralization of CCL17 resulted in attenuated allergic disease in the ovalbumin model including decreases in AHR, eosinophilia, and Th2-type cytokines *(17)*. In an adoptive transfer model using effector $CD4^+$ T cells, both CCR3 and CCR4 were involved in trafficking of Th2 cells to the lungs, with CCR3 mediating early recruitment of Th2 cells and CCR4 mediating the later phase *(18)*. Cumulatively, these results demonstrate that CCL17 and CCL22, in part via the recruitment of $CCR4^+$ Th2 cells, play a role in driving eosinophilia and airway hyperreactivity in the ovalbumin model of allergic lung disease. Similar to the results obtained in the ovalbumin models, in a live *Aspergillus* model of allergic lung disease, $CCR4^{-/-}$ mice had attenuated airway hyperresponsiveness and attenuated eosinophilia at later time points after fungal challenge *(19)*. Interestingly, $CCR4^{-/-}$ mice had increased airway neutrophils and macrophages early day3 (d3) after fungal challenge, which was accompanied by an increase in IL-4 and IL-5 *(19)*. Follow-up studies have demonstrated that both CCL17 and CCL22 impair the innate antifungal immune response, thereby promoting the maintenance of acquired Th2-mediated asthmatic disease *(20,21)*. These results may be explained in part by the expression of CCR4 by other cell types including dendritic cells, platelets, T regulatory cells, and mast cells *(22–25)*. Thus, even though expression data in human disease suggest that CCR4 and its ligands may be good targets for therapy, data from murine models of disease indicate that a diverse number of mechanisms may be controlled by CCR4.

12.2.2. CCR6

Another chemokine receptor associated with Th2 cells is CCR6. CCR6 is expressed by a subset of dendritic cells, primarily by immature dendritic cells, memory T cells, and B lymphocytes *(26–28)*. The only known chemokine ligand for CCR6, MIP-3α/CCL20, is expressed constitutively in some secondary lymphoid tissues, including murine spleen and human tonsils *(29,30)*. The predominant cell source of CCR6 ligands in the lungs is thought to be epithelial cells *(31)*. Primary human bronchoepithelial cells stimulated with type 2 cytokines (IL-4 and IL-13) or particulate matter produced macrophage inflammatory protein 3-α (MIP-3α)/CCL20, and this may be one mechanism for dendritic cell trafficking to the lung *(32)*. The role of CCR6/CCL20 has been investigated by our laboratory using a cockroach antigen–induced allergic airway disease. CCL20 is induced within hours of allergen challenge *(33)*. $CCR6^{-/-}$ mice were equivalently immunized, as assessed by T-cell cytokine production, but had attenuated airway hyperreactivity, fewer eosinophils, lower IL-5, and lower IgE levels than wild-type mice *(33)*. CCL20 was not directly chemotactic for eosinophils at doses comparable with those used for CCL11-mediated chemotaxis *(33)*, thus the effect on eosinophils was likely related to the lower IL-5. Type 2 cytokine and eosinophilia were reconstituted in $CCR6^{-/-}$ mice by adoptive transfer of T cells from sensitized wild-type mice, however airway hyperreactivity was unaffected. Thus, CCR6 plays a role in promoting allergic lung disease, in part due to its expression on T cells, but also likely in part due to T cell–independent effects *(34)*. Although the effects in $CCR6^{-/-}$ mice demonstrated striking reductions in allergen sensitized and challenged mice, additional research is needed to identify the exact mechanism of the altered phenotype, which may include recruitment of dendritic cell (DC) subsets to the lung.

12.2.3. CCR8

A primary chemokine receptor associated with Th2-type responses is CCR8, which is also upregulated on Th2 cells derived in vitro. CCR8 is expressed by thymocytes, monocytes, as well as Th2 cells, and serves as the receptor for I-309/TCA-3/CCL1. In humans, expression of CCR8 was found on T cells in endobronchial biopsies of asthmatic patients taken 24 hours after allergen challenge *(35)*. Additionally, increased expression of the CCR8 ligand CCL1 has been reported during allergic lung disease *(35)*. One study found impaired Th2 responses in CCR8$^{-/-}$ mice in ovalbumin (OVA)- and cockroach antigen–induced models of allergic lung disease, as well as *Schistosoma mansoni* soluble egg antigen (SEA)-induced Th2 granuloma model. In each of these models, decreases in both Th2-type cytokines and eosinophils were present *(36)*. Also using the OVA-induced model of allergic lung disease, two groups reported that CCR8$^{-/-}$ mice developed similar eosinophilia, lung inflammation, and Th2 cytokine levels compared with wild-type mice *(35,37)*. The reason for discordance in these findings is unclear, but recent studies point to a role for CCR8 on other cell types including CD4^{+}CD25^{+} T cells and dendritic cells *(38,39)*. Thus, the role of CCR8 in development of allergic responses may be due in part to regulation of Th1-type responses by the production of IL-10. Further analyses will likely examine the potential mechanism of T regulatory (Treg) cell development and CCR8 expression.

12.2.4. CXCR3

CXCR3 is expressed by a number of cell types but preferentially on Th1 and T-cell type 1 (Tc1) cells. CXCR3 ligands CXCL9-11 were all originally identified as IFN-γ–inducible chemokines, so it comes as no surprise that CXCR3 ligands predominate under Th1 conditions. CXCR3-expressing cells predominate under type 1 conditions such as pulmonary sarcoidosis and chronic obstructive pulmonary disease (COPD) *(3)*. The Th1/Th2 dogma would predict that the CXCR3 axis may serve to antagonize Th2 development, but in established allergic lung disease, the expression of CXCR3 and its ligands would be minimal. For the most part, published data support this view. For example, neutralization of CXCL9 in the airways at time of allergen challenge increased AHR, eosinophilia, and IL-4. Conversely, exogenous addition of CXCL9 reduced AHR and eosinophilia and increased IL-12 *(40)*. Similarly, neutralization of CXCL10 resulted in exaggerated AHR and eosinophilia associated with decreased numbers of CXCR3^{+}CD4^{+} T cells *(41)*. Interestingly, exogenous administration of CXCL10 to the airways attenuated early AHR but increased eosinophilia and later AHR *(41)*. Thus, modulation of CXCR3 ligands has both expected and unexpected effects on allergic lung disease. This is likely due to the ability to recruit both Th1-type cells and eosinophils to the airways during allergen challenge.

CXCR3 is expressed highly in the lungs. At least two studies have found high expression levels of CXCR3 on lung T cells in both normal and asthmatic patients compared with T cells in other sites *(42,43)*. After segmental allergen challenge in asthmatic patients, the CXCR3 ligand CXCL10 was upregulated (along with CCR4 ligands) at the site of allergen challenge, relative to saline-challenged sites

(13). In another segmental allergen challenge study, CXCR3 was upregulated on airway eosinophils compared with peripheral blood *(44)*. Thus evidence from clinical studies suggests that CXCR3 expression may be increased in allergic airways but may be increased in nonallergic airways as well.

The expression of CXCR3 is not limited to Th1/Tc1 type cells and eosinophils during lung disease. Airway epithelial cells express functional CXCR3, binding of which results in mitogen-activated protein kinase (MAPK) activation, DNA synthesis, and chemotaxis *(45)*. Additionally, CXCR3 was the most highly expressed chemokine receptor on lung mast cells *(46)*, and specifically mast cells in airway smooth muscle *(47)*. Smooth muscle cell–driven chemotaxis of mast cells was predominately CXCR3-mediated, and CXCL10 production by smooth muscle was increased in asthmatics relative to controls *(47)*. Together, these results suggest a role for CXCR3 and its ligands in promoting allergic airway disease independent of CXCR3 expression by T cells. It is noteworthy that genetic studies have identified a polymorphism in CXCR3, which is associated with the risk of asthma development *(48)*. The polymorphism is a G→A substitution, but the functional consequences of this polymorphism are not yet clear *(48)*. This finding brings up the possibility that targeting of particular chemokine receptors during allergic airway disease may depend upon the genotype of the patient as well as the cell type expressing the receptor.

12.3. CCR3

Eosinophils are a hallmark of allergic airway disease. Eosinophils are recruited to airways and surrounding interstitial space during allergic disease. Mechanisms of action of chemokines on eosinophils include chemotaxis, as well as direct activation and prosurvival signals. Eosinophils are thought to be significant producers of eicosanoids, including prostaglandins and leukotrienes, as well as proteolytic enzymes that result in damage and remodeling of airways *(49)*. The chemokine receptor that is predominately expressed on eosinophils is CCR3, which appears to mediate the most potent activating and recruitment responses. However, CCR3 is not only expressed by eosinophils but also by basophils, mast cells, and Th2 cells. Ligands for CCR3 include CCL 11, 24, 26 (eotaxin-1, -2, -3), as well as CCL 5, 7, 8, and 13. CCL11 induces chemotaxis and degranulation of eosinophils through the activation of extracellular signal-regulated kinase 2 (ERK2) and p38 mitogen–activated protein kinases *(50,51)*. Interestingly, CXCR3 ligands CXCL 9-11 can behave as natural antagonists for CCR3 *(5)*, providing evidence for competitive binding between chemokines that may regulate migration. The role of CCR3/CCL11 in the chemotaxis and activation of eosinophils during allergic lung disease has figured prominently in the literature and has been the focus of many drug discovery programs.

Clinically, the expression of CCL11 and CCR3 are elevated in atopic asthmatics. CCL11 mRNA is produced by epithelial and endothelial cells, with CCR3 being predominately localized to eosinophils *(52)*. Eotaxin levels were elevated in patients with acute asthma *(53)*. Additionally, $CCR3^+$ eosinophils were elevated in the bone marrow and peripheral blood of individuals with allergic asthma but not in atopic individuals in general *(54,55)*. Increased CCR3 expression was found on $CD4^+$ and double-negative T cells in peripheral blood of dust mite allergic individuals *(56)*. More recently, increased expression of CCR3 ligands CCL24 and CCL26 was found to be associated with persistent eosinophilic inflammation in the lungs of asthmatics *(57)*. Thus, it is likely that multiple CCR3 ligands play a contributory role to the recruitment of eosinophils in the lungs and airways.

Animal models of allergic lung disease have demonstrated that CCR3 plays a prominent role in the recruitment of eosinophils to the lungs. However, the earliest studies examining $CCL11^{-/-}$ mice indicated that there was little role for this ligand in allergic responses. In contrast, studies using $CCR3^{-/-}$ mice have demonstrated that CCR3 plays a critical role in the recruitment of eosinophils to the lungs during allergen challenge *(58)*. Surprisingly, more intraepithelial mast cells were present in $CCR3^{-/-}$ mice after allergen challenge, and $CCR3^{-/-}$ mice had exacerbated airway hyperreactivity *(58)*. Studies using a combination of eotaxin-$1^{-/-}$ mice, eotaxin-$2^{-/-}$, eotaxin-1,2 double knockouts, and $CCR3^{-/-}$ mice have demonstrated redundancy between eotaxin-1 and -2 for recruitment of eosinophils to the lungs *(59)*. Although recent studies using anti-IL-5 to deplete eosinophils in human asthmatics have failed to attenuate disease *(60–62)*, the short-term nature of the studies have allowed justification for the continuation of CCR3 antagonist programs.

CCR3 blockade has received considerable attention as a target for therapeutics *(63)*. A number of CCR3 antagonists including humanized antibodies and small-molecule inhibitors have been or are currently under development *(64–68)*. To date, 34 U.S. patents have been issued pertaining to CCR3 [U.S. PTO using ABST/(CCR3)]. Future clinical data will soon highlight the efficacy of this treatment strategy.

The expression of CCR3 is not limited to basophils, eosinophils, and Th2 cells. More recently, functional CCR3 has been described on airway epithelial cells *(69)*, as well as on primary bronchial epithelial cells from patients with inflammatory lung diseases including asthma *(69)*. Additionally, mast cells express CCR3 intracellularly, which can be mobilized to the surface upon activation via FcεR. The combination of FcεR plus eotaxin (and possibly other CCR3 ligands) led to enhanced IL-13 production *(70)*, thereby providing an additional source of proallergic and profibrotic mediators. Another recent study demonstrated the expression of CCR3 on asthmatic airway smooth muscle *(71)*;

CCR3 was functional on these cells as evidenced by calcium flux, and eotaxin was chemotactic for airway smooth muscle (ASM) cells in vitro *(71)*. Therefore, although CCR3 blockade may be useful for antagonizing the accumulation of eosinophils in the lungs during allergic lung disease, CCR3 antagonists are likely to have numerous effects on other cell types as well. These may offer additional beneficial effects for the treatment of asthma not provided by the previous studies using anti-IL-5, which targeted only eosinophils.

12.4. CCR1

CCR1 ligands play a role in the chemotaxis of eosinophils and other leukocytes during allergic lung disease *(72,73)*. CCR1 is expressed on several cell types including mast cells, neutrophils, monocytes, lymphocytes, and eosinophils and binds CCL3, CCL5, and CCL7, as well as a number of other chemokines. In a model of chronic fungal allergic lung disease, $CCR1^{-/-}$ mice exhibited an enhanced type 1 response relative to wild-type mice and were protected from airway remodeling but had similar airway hyperresponsiveness *(74)*. In the same model, a CCR1 antagonist attenuated airway inflammation, hyperresponsiveness, and remodeling *(75)*. Recent data from our laboratory have pointed to a critical role for $CD8^{+}$ T cells expressing CCR1 in a model of virus-induced exacerbation of allergic lung disease *(76–79)*. Thus, CCR1 appears to be most relevant during pathogen-associated disease that promotes exacerbated airway responses. Additionally, there is evidence that CCR1 expression by eosinophils and mast cells may play a role in allergic lung disease. Mast cells migrate to CCL5 in vitro *(80)*, and $CCR1^{+}$ mast cells were found in the lungs of asthmatic patients, with comparatively few found in normal controls *(25)*. Met-RANTES, a modified chemokine protein, inhibited eosinophil effector function, which was mediated at least in part via its effects on CCR1 blockade *(81,82)*. A number of other antibody-based and small-molecule antagonists have been developed for CCR1 as well *(64,68,83)*. Although CCR1 antagonists have not performed well in clinical trials of autoimmune diseases, they have yet to be applied to asthmatic disease. Because many asthmatic exacerbations are caused by infectious agents, CCR1 antagonists may provide significant therapeutic effects.

12.5. CCR2

One of the earliest identified receptors associated with disease progression was CCR2. CCR2 is expressed by many cell types, including monocytes, dendritic cells, activated T cells, natural killer (NK) cells, and basophils. CCR2 ligands include CCL2, CCL7, CCL12, and CCL13. In particular, CCR2 appears to play a critical role in a number of lung inflammatory diseases including asthma, acute respiratory distress syndrome (ARDS), COPD, and pulmonary fibrosis *(84–88)*. However, the roles of CCR2 and CCR2 ligands in allergic lung disease have been controversial when applied to multiple animal models.

In a model of allergic fungal lung disease, $CCR2^{-/-}$ mice exhibited decreased neutrophil recruitment, but exaggerated eosinophilia, and an enhanced type 2 response *(89)*. Likewise, neutralization of a primary CCR2 ligand, CCL2, in the first 14 days after fungal conidia challenge also impaired fungal clearance and increased AHR. Conversely, overexpression of MCP-1 exaggerated fungal clearance and decreased AHR *(90)*. However, neutralization of MCP-1 beginning at day 14 after conidia challenge resulted in attenuated AHR *(90)*. Together, these results demonstrate that CCR2 and CCL2/MCP-1 play a role in the protective response to remove fungus but also a pathogenic role in perpetuating inflammation. Thus, examination of appropriate models of disease as well as phases of disease may dictate different mechanistic roles of a specific receptor.

Examination of models of allergen-induced disease have demonstrated divergent results in CCR2/CCL2-related mechanisms. Using a fungal allergen–induced model of allergic lung disease, mice deficient in either MCP-1 or CCR2 developed eosinophilic airway inflammation, goblet cell hyperplasia, airway hyperreactivity, elevations in serum IgE, and airway fibrosis similar to those seen in wild-type mice *(91)*. In the ovalbumin model of allergic lung disease, one group reported exaggerated allergic responses in $CCR2^{-/-}$ mice *(92)*, whereas another reported that $CCR2^{-/-}$ mice had indistinguishable responses from wild-type mice *(93)*. In models of allergen-induced disease, animals treated with anti-CCL2 had significantly reduced airway parameters related to mast cell activation and lymphocyte recruitment. Interestingly, CCL2 appears to be able to direct the immune response toward a Th2 phenotype through downregulation of IL-12 and direct skewing of T cells for increased IL-4 production *(94)*. Thus, the role of CCR2 in allergic lung disease remains controversial and may depend upon a number of factors, including the experimental model (OVA; live *Aspergillus*; *Aspergillus* antigen) used to induce airway disease, as well as susceptibility of background strains of mice. CCR2 likely plays a role in modulating the innate immune response to viable organisms, as well as recruitment of mononuclear cells to the lungs. In the case of static antigens, such as *Aspergillus* antigen or ovalbumin, CCR2 seems to play a less prominent role. Comparatively few studies in humans have been published that either confirm or deny a role for CCR2 or its ligands in allergic lung disease, but those that do exist have not implicated CCR2 prominently *(4)*.

12.6. CXCR4

Classically regarded as primarily homeostatic based on its ubiquitous expression pattern and its role in stem cell trafficking, the CXCR4/CXCL12 stromal-derived factor 1(SDF-1) receptor/ligand pair also has a demonstrated role in allergic lung disease. One study reported that CXCR4 is increased on BAL eosinophils from patients with eosinophilic lung disease relative to eosinophils in peripheral blood *(95)*. However, another study suggested increases in CCR4, CCR9, and CXCR3, without increased CXCR4 on BAL eosinophils *(44)*. In animal models, antibody-mediated neutralization of CXCR4 or CXCL12 resulted in attenuated eosinophilia and reduced airway hyperresponsiveness in the OVA-induced allergic disease model *(96)*. Conversely, overexpression of CXCL12 exacerbated inflammation. More recently, a number of studies have demonstrated, using a specific CXCR4 antagonist *(97)*, attenuated allergic responses in allergic lung disease, as well as other inflammatory disease models *(98–100)*. In the cockroach antigen model of allergic lung disease, this was accompanied by decreases in eosinophilia and T cells in the lungs and a shift in balance of lung cytokines from type 2 to type 1 *(98)*. Therefore, in addition to its role in hematopoietic homeostasis, CXCR4 can play an important role in regulating allergic lung disease. It is likely that as CXCR4 can be found on multiple leukocyte populations, both innate and acquired immune cells, the targeting of this receptor may have an overall anti-inflammatory effect.

12.7. CXCR1/CXCR2

CXCR1 and CXCR2 are the predominant receptors used by neutrophils for chemotaxis. The potential role of IL-8/CXCL8 in human lung inflammation has received significant attention, with small-molecule antagonists developed recently *(101)*. Neutrophilic inflammation and increased CXCL8 production were found in a number of chronic lung diseases including ARDS, COPD, as well as some allergic asthma *(102)*. Additionally, genetic polymorphisms in CXCR1 have been identified that were present at increased frequency in children with asthma *(103)*. Although CXCR1 and CXCR2 ligands are thought to be produced primarily by epithelial cells, eosinophils can make CXCL5/epithelial-derived neutrophil-activating protein 78 (ENA-78) and CXCL1/growth-regulated protein alpha precursor (GRO-α), implicating a potential role for these chemokine systems in chronic allergic disease *(104,105)*. These findings suggest that eosinophils can perpetuate inflammation, contributing to the inflamed state of the lungs.

In the live *Aspergillus* model of allergic lung disease, CXCR2$^{-/-}$ mice had fewer eosinophils and T cells (but interestingly similar neutrophil numbers). These were accompanied by decreased lung levels of IL-4, IL-5, and CCL11 and an increase in CXCL9, CXCL10, and IFN-γ, suggesting a shift from type 2 to type 1 response *(106)*. In the mouse model of respiratory syncytial virus (RSV) infection, CXCR2 neutralization resulted in attenuated airway hyperreactivity and ablation of mucus production in response to viral infection in the lungs *(107)*. Viral clearance and neutrophil influx were similar between the two groups *(107)*, demonstrating that CXCR2 promotes the development of lung disease in response to viral or fungal infection, but this role does not appear to occur via neutrophil recruitment. Neutralization of CXCL1 and CXCL8 in a cockroach antigen model of allergic lung disease resulted in attenuated neutrophils and lymphocytes in the airways and a decrease in serum IgE *(108)*. Thus, CXCR2 may contribute to allergic lung disease via promoting both neutrophil dependent and/or neutrophil-independent mechanisms. Whether blocking CXCR2 has an effect on long-term disease or on specific aspects, such as mucus overproduction, may depend upon the nature of the exacerbating stimulus.

12.8. CX3CR1/CX3CL1

Fractalkine/CX3CL1 and its receptor CX3CR1 are the lone members of the CX3C family of chemokines. CX3CL1 can function both as an adhesion molecule in its membrane form and as a soluble chemokine, as it is attached to the cell membrane by an adhesion molecule-like stalk *(109,110)*. CX3CR1/CX3CL1 has been studied extensively in atherosclerosis and kidney disease, but it may play a role in allergic disease as well. Serum CX3CL1 and expression of CX3CR1 by T cells were increased in asthmatic patients *(111)*. Additionally, soluble CX3CL1 was upregulated in BAL after allergen challenge, as was membrane-bound CX3CL1 on endothelial and epithelial cells *(111)*. Thus, CX3CL1 may provide a significant bound molecule allowing a "solid phase" chemokine gradient to be readily established. It has also been reported that fractalkine is produced by airway smooth muscle cells and may contribute to mast cell recruitment in asthma *(112)*. Thus, there is some evidence that fractalkine/CX3CL1 and its receptor CX3CR1 is present during allergic asthmatic disease and may contribute to the pathogenesis of allergic lung disease.

12.9. Conclusions

There are several new frontiers that confront chemokine biologists. It is clear from the recent literature that chemokines have more functions than just chemotaxis. One area of challenge that has taken the lead is whether specific chemokine receptor systems have tissue-specific function. One ligand/receptor system that may prove to have lung-specific function is CCL18 and its receptor (which has yet to be identified). CCL18 was preferentially expressed in the lungs, associated with allergic disease, and was chemoattractant for Th2 cells and basophils *(113)*.

Studies of CCL18, however, will be hampered by the fact that human CCL18 has no mouse orthologue. A recent study that sought to determine chemotaxis of T cells generated under Th1 versus Th2 disease conditions in vivo found that the chemotactic responses of T cells were influenced strongly by the compartment from which they were obtained (lung vs. spleen). As would be predicted, T cells from the lungs of mice infected with *Nippostrongylus brasiliensis* (a Th2-mediated response) migrated better to the CCR4 ligand CCL17 than T cells from the lungs of influenza-infected mice *(114)*. However, T cells from the spleens of *N. brasiliensis*–infected mice migrated better to CCL17 than lung T cells *(114)*. These findings are an example of the dynamic nature of chemotaxis, in that a combination of receptors and ligands are likely responsible for trafficking to, and even within, tissues. It is unlikely that any particular tissue compartment uses a single receptor system; rather, it may depend upon the nature of the immune/inflammatory response generated in that tissue.

Another area that will certainly receive more attention in the future is the characterization of the proximal signals that mediate chemokine and chemokine-receptor expression. Recent studies have demonstrated that the production of CCL 11, 17, and 22 are signal transducer and activator of transcription (STAT) 6 dependent, whereas CXCL9 and CXCL10 production are dependent upon STAT1 and IFN-γ *(115)*. Although pattern recognition receptors (including Toll-like receptors, mannose receptor, etc.) have received considerable attention in host defense, their roles in allergic lung disease will likely be addressed more in the near future. Viral infection presents a likely situation for cross talk between innate immunity and allergic lung disease. Individuals with severe acute viral infection can present with signs and symptoms similar to allergic lung disease including inflammation, mucus production, and airway hyperresponsiveness. Studies from our laboratory and others have shown that viral infections can play a role in the exacerbation of allergic lung disease *(76–78)*. Respiratory viral infections, including adenovirus, rhinovirus, and respiratory syncytial virus, contribute to the exacerbation of existing disease. CCR1, together with its ligands, was one chemokine receptor system that has been implicated in this phenomenon.

Another area that has gained increased attention is the role of matrix metalloproteinases (MMPs) and other modifiers of chemokine activity in allergic lung disease. MMPs have been shown to play a role in allergic lung disease that was related to chemokines *(116,117)*. The focus, however, has been primarily on the role of MMPs in airway remodeling. Coupled with the idea that MMPs can modify chemokines, such as converting inactive to active forms, or vice versa, suggests that MMPs are likely to play a significant role in the direct regulation of chemokines during allergic lung disease, especially given the role for Th2 cytokines in driving MMP production *(118–120)*.

In conclusion, the role of chemokines and their receptors continue to be investigated in asthmatic diseases. The role of specific chemokine receptor/ligand systems will depend upon the nature of the inciting stimulus (i.e., non-infectious vs. infectious), as well as the intracellular (viral) versus extracellular (fungal) nature of microbes. Given the broad nature of the classification of asthmatic disease, a number of chemokine receptors as potential targets continue to be relevant.

Acknowledgments

The authors would like to acknowledge and thank Robin Kunkel for preparation of Fig. 1.

References

1. Morgan AJ, Symon FA, Berry MA, Pavord ID, Corrigan CJ, Wardlaw AJ. IL-4-expressing bronchoalveolar T cells from asthmatic and healthy subjects preferentially express CCR 3 and CCR 4. J Allergy Clin Immunol 2005;116(3): 594–600.
2. Hartl D, Griese M, Nicolai T, et al. Pulmonary chemokines and their receptors differentiate children with asthma and chronic cough. J Allergy Clin Immunol 2005;115(4):728–736.
3. Panina-Bordignon P, Papi A, Mariani M, et al. The C-C chemokine receptors CCR4 and CCR8 identify airway T cells of allergen-challenged atopic asthmatics. J Clin Invest 2001;107(11):1357–1364.
4. Kallinich T, Schmidt S, Hamelmann E, et al. Chemokine-receptor expression on T cells in lung compartments of challenged asthmatic patients. Clin Exp Allergy 2005;35(1):26–33.
5. Loetscher P, Pellegrino A, Gong JH, et al. The ligands of CXC chemokine receptor 3, I-TAC, Mig, and IP10, are natural antagonists for CCR3. J Biol Chem 2001;276(5):2986–2991.
6. Weng Y, Siciliano SJ, Waldburger KE, et al. Binding and functional properties of recombinant and endogenous CXCR3 chemokine receptors. J Biol Chem 1998;273(29):18288–18291.
7. Hirata H, Arima M, Cheng G, et al. Production of TARC and MDC by naive T cells in asthmatic patients. J Clin Immunol 2003;23(1):34–45.
8. Ying S, O'Connor B, Ratoff J, et al. Thymic stromal lymphopoietin expression is increased in asthmatic airways and correlates with expression of Th2-attracting chemokines and disease severity. J Immunol 2005;174(12):8183–8190.
9. Sekiya T, Yamada H, Yamaguchi M, et al. Increased levels of a TH2-type CC chemokine thymus and activation-regulated chemokine (TARC) in serum and induced sputum of asthmatics. Allergy 2002;57(2):173–177.

10. Lezcano-Meza D, Negrete-Garcia MC, Dante-Escobedo M, Teran LM. The monocyte-derived chemokine is released in the bronchoalveolar lavage fluid of steady-state asthmatics. Allergy 2003;58(11):1125–1130.
11. Sekiya T, Miyamasu M, Imanishi M, et al. Inducible expression of a Th2-type CC chemokine thymus- and activation-regulated chemokine by human bronchial epithelial cells. J Immunol 2000;165(4):2205–2213.
12. Berin MC, Eckmann L, Broide DH, Kagnoff MF. Regulated production of the T helper 2-type T-cell chemoattractant TARC by human bronchial epithelial cells in vitro and in human lung xenografts. Am J Respir Cell Mol Biol 2001;24(4): 382–389.
13. Bochner BS, Hudson SA, Xiao HQ, Liu MC. Release of both CCR4-active and CXCR3-active chemokines during human allergic pulmonary late-phase reactions. J Allergy Clin Immunol 2003;112(5):930–934.
14. Pilette C, Francis JN, Till SJ, Durham SR. CCR4 ligands are up-regulated in the airways of atopic asthmatics after segmental allergen challenge. Eur Respir J 2004;23(6):876–884.
15. Conroy DM, Jopling LA, Lloyd CM, et al. CCR4 blockade does not inhibit allergic airways inflammation. J Leukoc Biol 2003;74(4):558–563.
16. Gonzalo JA, Pan Y, Lloyd CM, et al. Mouse monocyte-derived chemokine is involved in airway hyperreactivity and lung inflammation. J Immunol 1999;163(1): 403–411.
17. Kawasaki S, Takizawa H, Yoneyama H, et al. Intervention of thymus and activation-regulated chemokine attenuates the development of allergic airway inflammation and hyperresponsiveness in mice. J Immunol 2001;166(3):2055–2062.
18. Lloyd CM, Delaney T, Nguyen T, et al. CC chemokine receptor (CCR)3/eotaxin is followed by CCR4/monocyte-derived chemokine in mediating pulmonary T helper lymphocyte type 2 recruitment after serial antigen challenge in vivo. J Exp Med 2000;191(2):265–274.
19. Schuh JM, Power CA, Proudfoot AE, Kunkel SL, Lukacs NW, Hogaboam CM. Airway hyperresponsiveness, but not airway remodeling, is attenuated during chronic pulmonary allergic responses to Aspergillus in $CCR4^{-/-}$ mice. FASEB J 2002;16(10):1313–1315.
20. Hogaboam CM, Carpenter KJ, Schuh JM, Buckland KF. Aspergillus and asthma–any link? Med Mycol 2005;43(Suppl 1):S197–202.
21. Carpenter KJ, Hogaboam CM. Immunosuppressive effects of CCL17 on pulmonary antifungal responses during pulmonary invasive aspergillosis. Infect Immun 2005;73(11):7198–7207.
22. Abi-Younes S, Si-Tahar M, Luster AD. The CC chemokines MDC and TARC induce platelet activation via CCR4. Thromb Res 2001;101(4):279–289.
23. Iellem A, Mariani M, Lang R, et al. Unique chemotactic response profile and specific expression of chemokine receptors CCR4 and CCR8 by CD4(+)CD25(+) regulatory T cells. J Exp Med 2001;194(6):847–853.

24. Yang ZZ, Novak AJ, Stenson MJ, Witzig TE, Ansell SM. Intratumoral $CD4^+CD25^+$ regulatory T-cell-mediated suppression of infiltrating $CD4^+$ T-cells in B-cell non-Hodgkin lymphoma. Blood 2006;107(9):3639–3646.
25. Amin K, Janson C, Harvima I, Venge P, Nilsson G. CC chemokine receptors CCR1 and CCR4 are expressed on airway mast cells in allergic asthma. J Allergy Clin Immunol 2005;116(6):1383–1386.
26. Krzysiek R, Lefevre EA, Bernard J, et al. Regulation of CCR6 chemokine receptor expression and responsiveness to macrophage inflammatory protein-3alpha/CCL20 in human B cells. Blood 2000;96(7):2338–2345.
27. Fitzhugh DJ, Naik S, Gonzalez E, Caughman SW, Hwang ST. CC chemokine receptor 6 (CCR6) is a marker for memory T cells that arrest on activated human dermal microvascular endothelium under shear stress. J Invest Dermatol 2000;115(2):332.
28. Kleinewietfeld M, Puentes F, Borsellino G, Battistini L, Rotzschke O, Falk K. CCR6 expression defines regulatory effector/memory-like cells within the $CD25^+CD4^+$ T-cell subset. Blood 2005;105(7):2877–2886.
29. Cook DN, Prosser DM, Forster R, et al. CCR6 mediates dendritic cell localization, lymphocyte homeostasis, and immune responses in mucosal tissue. Immunity 2000;12(5):495–503.
30. Dieu MC, Vanbervliet B, Vicari A, et al. Selective recruitment of immature and mature dendritic cells by distinct chemokines expressed in different anatomic sites. J Exp Med 1998;188(2):373–386.
31. Starner TD, Barker CK, Jia HP, Kang Y, McCray PB Jr. CCL20 is an inducible product of human airway epithelia with innate immune properties. Am J Respir Cell Mol Biol 2003;29(5):627–633.
32. Reibman J, Hsu Y, Chen LC, Bleck B, Gordon T. Airway epithelial cells release MIP-3alpha/CCL20 in response to cytokines and ambient particulate matter. Am J Respir Cell Mol Biol 2003;28(6):648–654.
33. Lukacs NW, Prosser DM, Wiekowski M, Lira SA, Cook DN. Requirement for the chemokine receptor CCR6 in allergic pulmonary inflammation. J Exp Med 2001;194(4):551–555.
34. Lundy SK, Lira SA, Smit JJ, Cook DN, Berlin AA, Lukacs NW. Attenuation of allergen-induced responses in $CCR6^{-/-}$ mice is dependent upon altered pulmonary T lymphocyte activation. J Immunol 2005;174(4):2054–2060.
35. Chung CD, Kuo F, Kumer J, et al. CCR8 is not essential for the development of inflammation in a mouse model of allergic airway disease. J Immunol 2003;170(1):581–587.
36. Chensue SW, Lukacs NW, Yang TY, et al. Aberrant in vivo T helper type 2 cell response and impaired eosinophil recruitment in CC chemokine receptor 8 knockout mice. J Exp Med 2001;193(5):573–584.
37. Goya I, Villares R, Zaballos A, et al. Absence of CCR8 does not impair the response to ovalbumin-induced allergic airway disease. J Immunol 2003;170(4):2138–2146.

38. Freeman CM, Chiu BC, Stolberg VR, et al. CCR8 is expressed by antigen-elicited, IL-10-producing CD4+CD25+ T cells, which regulate Th2-mediated granuloma formation in mice. J Immunol 2005;174(4):1962–1970.
39. Gombert M, Dieu-Nosjean MC, Winterberg F, et al. CCL1-CCR8 interactions: an axis mediating the recruitment of T cells and Langerhans-type dendritic cells to sites of atopic skin inflammation. J Immunol 2005;174(8):5082–5091.
40. Thomas MS, Kunkel SL, Lukacs NW. Regulation of cockroach antigen-induced allergic airway hyperreactivity by the CXCR3 ligand CXCL9. J Immunol 2004;173(1):615–623.
41. Thomas MS, Kunkel SL, Lukacs NW. Differential role of IFN-gamma-inducible protein 10 kDa in a cockroach antigen-induced model of allergic airway hyperreactivity: systemic versus local effects. J Immunol 2002;169(12):7045–7053.
42. Campbell JJ, Brightling CE, Symon FA, et al. Expression of chemokine receptors by lung T cells from normal and asthmatic subjects. J Immunol 2001;166(4):2842–2848.
43. Kurashima K, Fujimura M, Myou S, et al. Effects of oral steroids on blood CXCR3+ and CCR4+ T cells in patients with bronchial asthma. Am J Respir Crit Care Med 2001;164(5):754–758.
44. Liu LY, Jarjour NN, Busse WW, Kelly EA. Chemokine receptor expression on human eosinophils from peripheral blood and bronchoalveolar lavage fluid after segmental antigen challenge. J Allergy Clin Immunol 2003;112(3):556–562.
45. Aksoy MO, Yang Y, Ji R, et al. CXCR3 Surface expression in human airway epithelial cells: cell cycle dependence and effect on cell proliferation. Am J Physiol Lung Cell Mol Physiol 2005;290(5):L909–918.
46. Brightling CE, Kaur D, Berger P, Morgan AJ, Wardlaw AJ, Bradding P. Differential expression of CCR3 and CXCR3 by human lung and bone marrow-derived mast cells: implications for tissue mast cell migration. J Leukoc Biol 2005;77(5):759–766.
47. Brightling CE, Ammit AJ, Kaur D, et al. The CXCL10/CXCR3 axis mediates human lung mast cell migration to asthmatic airway smooth muscle. Am J Respir Crit Care Med 2005;171(10):1103–1108.
48. Cheong HS, Park CS, Kim LH, et al. CXCR3 polymorphisms associated with risk of asthma. Biochem Biophys Res Commun 2005;334(4):1219–1225.
49. Bandeira-Melo C, Bozza PT, Weller PF. The cellular biology of eosinophil eicosanoid formation and function. J Allergy Clin Immunol 2002;109(3):393–400.
50. Kampen GT, Stafford S, Adachi T, et al. Eotaxin induces degranulation and chemotaxis of eosinophils through the activation of ERK2 and p38 mitogen-activated protein kinases. Blood 2000;95(6):1911–1917.
51. Cui CH, Adachi T, Oyamada H, et al. The role of mitogen-activated protein kinases in eotaxin-induced cytokine production from bronchial epithelial cells. Am J Respir Cell Mol Biol 2002;27(3):329–335.
52. Ying S, Robinson DS, Meng Q, et al. Enhanced expression of eotaxin and CCR3 mRNA and protein in atopic asthma. Association with airway hyperresponsive-

ness and predominant co-localization of eotaxin mRNA to bronchial epithelial and endothelial cells. Eur J Immunol 1997;27(12):3507–3516.
53. Lilly CM, Woodruff PG, Camargo CA Jr, et al. Elevated plasma eotaxin levels in patients with acute asthma. J Allergy Clin Immunol 1999;104(4 Pt 1): 786–790.
54. Robinson DS, North J, Zeibecoglou K, et al. Eosinophil development and bone marrow and tissue eosinophils in atopic asthma. Int Arch Allergy Immunol 1999;118(2–4):98–100.
55. Zeibecoglou K, Ying S, Yamada T, et al. Increased mature and immature CCR3 messenger RNA+ eosinophils in bone marrow from patients with atopic asthma compared with atopic and nonatopic control subjects. J Allergy Clin Immunol 1999;103(1 Pt 1):99–106.
56. Wang CR, Chen PC, Liu MF. Increased circulating $CCR3^+$ type 2 helper T cells in house dust mite-sensitive Chinese patients with allergic diseases. Asian Pac J Allergy Immunol 2003;21(4):205–210.
57. Ravensberg AJ, Ricciardolo FL, van Schadewijk A, et al. Eotaxin-2 and eotaxin-3 expression is associated with persistent eosinophilic bronchial inflammation in patients with asthma after allergen challenge. J Allergy Clin Immunol 2005;115(4):779–785.
58. Humbles AA, Lu B, Friend DS, et al. The murine CCR3 receptor regulates both the role of eosinophils and mast cells in allergen-induced airway inflammation and hyperresponsiveness. Proc Natl Acad Sci U S A 2002;99(3):1479–1484.
59. Pope SM, Zimmermann N, Stringer KF, Karow ML, Rothenberg ME. The eotaxin chemokines and CCR3 are fundamental regulators of allergen-induced pulmonary eosinophilia. J Immunol 2005;175(8):5341–5350.
60. Leckie MJ, ten Brinke A, Khan J, et al. Effects of an interleukin-5 blocking monoclonal antibody on eosinophils, airway hyper-responsiveness, and the late asthmatic response. Lancet 2000;356(9248):2144–2148.
61. Flood-Page PT, Menzies-Gow AN, Kay AB, Robinson DS. Eosinophil's role remains uncertain as anti-interleukin-5 only partially depletes numbers in asthmatic airway. Am J Respir Crit Care Med 2003;167(2):199–204.
62. Buttner C, Lun A, Splettstoesser T, Kunkel G, Renz H. Monoclonal anti-interleukin-5 treatment suppresses eosinophil but not T-cell functions. Eur Respir J 2003;21(5):799–803.
63. Bertrand CP, Ponath PD. CCR3 blockade as a new therapy for asthma. Expert Opin Invest Drugs 2000;9(1):43–52.
64. Sabroe I, Peck MJ, Van Keulen BJ, et al. A small molecule antagonist of chemokine receptors CCR1 and CCR3. Potent inhibition of eosinophil function and CCR3-mediated HIV-1 entry. J Biol Chem 2000;275(34):25985–25992.
65. White JR, Lee JM, Dede K, et al. Identification of potent, selective non-peptide CC chemokine receptor-3 antagonist that inhibits eotaxin-, eotaxin-2-, and monocyte chemotactic protein-4-induced eosinophil migration. J Biol Chem 2000; 275(47):36626–36631.

66. Saeki T, Ohwaki K, Naya A, et al. Identification of a potent and nonpeptidyl ccr3 antagonist. Biochem Biophys Res Commun 2001;281(3):779–782.
67. Wacker DA, Santella JB 3rd, Gardner DS, et al. CCR3 antagonists: a potential new therapy for the treatment of asthma. Discovery and structure-activity relationships. Bioorg Med Chem Lett 2002;12(13):1785–1789.
68. Morokata T, Suzuki K, Masunaga Y, et al. A novel, selective, and orally available antagonist for CC chemokine receptor 3. J Pharmacol Exp Ther 2006;317(1):244–250.
69. Stellato C, Brummet ME, Plitt JR, et al. Expression of the C-C chemokine receptor CCR3 in human airway epithelial cells. J Immunol 2001;166(3):1457–1461.
70. Price KS, Friend DS, Mellor EA, De Jesus N, Watts GF, Boyce JA. CC chemokine receptor 3 mobilizes to the surface of human mast cells and potentiates immunoglobulin E-dependent generation of interleukin 13. Am J Respir Cell Mol Biol 2003;28(4):420–427.
71. Joubert P, Lajoie-Kadoch S, Labonte I, et al. CCR3 expression and function in asthmatic airway smooth muscle cells. J Immunol 2005;175(4):2702–2708.
72. Lukacs NW, Standiford TJ, Chensue SW, Kunkel RG, Strieter RM, Kunkel SL. C-C chemokine-induced eosinophil chemotaxis during allergic airway inflammation. J Leukoc Biol 1996;60(5):573–578.
73. Gonzalo JA, Lloyd CM, Wen D, et al. The coordinated action of CC chemokines in the lung orchestrates allergic inflammation and airway hyperresponsiveness. J Exp Med 1998;188(1):157–167.
74. Blease K, Mehrad B, Standiford TJ, et al. Airway remodeling is absent in $CCR1^{-/-}$ mice during chronic fungal allergic airway disease. J Immunol 2000; 165(3):1564–1572.
75. Carpenter KJ, Ewing JL, Schuh JM, et al. Therapeutic targeting of CCR1 attenuates established chronic fungal asthma in mice. Br J Pharmacol 2005;145(8):1160–1172.
76. Schaller M, Hogaboam CM, Lukacs N, Kunkel SL. Respiratory viral infections drive chemokine expression and exacerbate the asthmatic response. J Allergy Clin Immunol 2006;118(2):295–302.
77. John AE, Gerard CJ, Schaller M, et al. Respiratory syncytial virus-induced exaggeration of allergic airway disease is dependent upon CCR1-associated immune responses. Eur J Immunol 2005;35(1):108–116.
78. John AE, Berlin AA, Lukacs NW. Respiratory syncytial virus-induced CCL5/RANTES contributes to exacerbation of allergic airway inflammation. Eur J Immunol 2003;33(6):1677–1685.
79. Schaller MA, Lundy SK, Huffnagle GB, Lukacs NW. CD8+ T cell contributions to allergen induced pulmonary inflammation and airway hyperreactivity. Eur J Immunol 2005;35(7):2061–2070.
80. Mattoli S, Ackerman V, Vittori E, Marini M. Mast cell chemotactic activity of RANTES. Biochem Biophys Res Commun 1995;209(1):316–321.

81. Elsner J, Petering H, Hochstetter R, et al. The CC chemokine antagonist Met-RANTES inhibits eosinophil effector functions through the chemokine receptors CCR1 and CCR3. Eur J Immunol 1997;27(11):2892–2898.
82. Chvatchko Y, Proudfoot AE, Buser R, et al. Inhibition of airway inflammation by amino-terminally modified RANTES/CC chemokine ligand 5 analogues is not mediated through CCR3. J Immunol 2003;171(10):5498–5506.
83. Saeki T, Naya A. CCR1 chemokine receptor antagonist. Curr Pharm Des 2003;9(15):1201–1208.
84. Spagnolo P, Renzoni EA, Wells AU, et al. C-C chemokine receptor 2 and sarcoidosis: association with Lofgren's syndrome. Am J Respir Crit Care Med 2003;168(10):1162–1166.
85. Schuyler M, Gott K, Cherne A. Experimental hypersensitivity pneumonitis: role of MCP-1. J Lab Clin Med 2003;142(3):187–195.
86. Rose CE Jr, Sung SS, Fu SM. Significant involvement of CCL2 (MCP-1) in inflammatory disorders of the lung. Microcirculation 2003;10(3–4):273–288.
87. Hildebrandt GC, Duffner UA, Olkiewicz KM, et al. A critical role for CCR2/MCP-1 interactions in the development of idiopathic pneumonia syndrome after allogeneic bone marrow transplantation. Blood 2004;103(6):2417–2426.
88. Moore BB, Paine R 3rd, Christensen PJ, et al. Protection from pulmonary fibrosis in the absence of CCR2 signaling. J Immunol 2001;167(8):4368–4377.
89. Blease K, Mehrad B, Standiford TJ, et al. Enhanced pulmonary allergic responses to Aspergillus in $CCR2^{-/-}$ mice. J Immunol 2000;165(5):2603–2611.
90. Blease K, Mehrad B, Lukacs NW, Kunkel SL, Standiford TJ, Hogaboam CM. Antifungal and airway remodeling roles for murine monocyte chemoattractant protein-1/CCL2 during pulmonary exposure to Asperigillus fumigatus conidia. J Immunol 2001;166(3):1832–1842.
91. Koth LL, Rodriguez MW, Bernstein XL, et al. Aspergillus antigen induces robust Th2 cytokine production, inflammation, airway hyperreactivity and fibrosis in the absence of MCP-1 or CCR2. Respir Res 2004;5(1):12.
92. Kim Y, Sung S, Kuziel WA, Feldman S, Fu SM, Rose CE Jr. Enhanced airway Th2 response after allergen challenge in mice deficient in CC chemokine receptor-2 (CCR2). J Immunol 2001;166(8):5183–5192.
93. MacLean JA, De Sanctis GT, Ackerman KG, et al. CC chemokine receptor-2 is not essential for the development of antigen-induced pulmonary eosinophilia and airway hyperresponsiveness. J Immunol 2000;165(11):6568–6575.
94. Campbell EM, Charo IF, Kunkel SL, et al. Monocyte chemoattractant protein-1 mediates cockroach allergen-induced bronchial hyperreactivity in normal but not $CCR2^{-/-}$ mice: the role of mast cells. J Immunol 1999;163(4):2160–2167.
95. Nagase H, Kudo K, Izumi S, et al. Chemokine receptor expression profile of eosinophils at inflamed tissue sites: decreased CCR3 and increased CXCR4 expression by lung eosinophils. J Allergy Clin Immunol 2001;108(4):563–569.
96. Gonzalo JA, Lloyd CM, Peled A, Delaney T, Coyle AJ, Gutierrez-Ramos JC. Critical involvement of the chemotactic axis CXCR4/stromal cell-derived factor-1

alpha in the inflammatory component of allergic airway disease. J Immunol 2000;165(1):499–508.
97. Hatse S, Princen K, Bridger G, De Clercq E, Schols D. Chemokine receptor inhibition by AMD3100 is strictly confined to CXCR4. FEBS Lett 2002; 527(1–3):255–262.
98. Lukacs NW, Berlin A, Schols D, Skerlj RT, Bridger GJ. AMD3100, a CxCR4 antagonist, attenuates allergic lung inflammation and airway hyperreactivity. Am J Pathol 2002;160(4):1353–1360.
99. Matthys P, Hatse S, Vermeire K, et al. AMD3100, a potent and specific antagonist of the stromal cell-derived factor-1 chemokine receptor CXCR4, inhibits autoimmune joint inflammation in IFN-gamma receptor-deficient mice. J Immunol 2001;167(8):4686–4692.
100. Hogaboam CM, Carpenter KJ, Schuh JM, Proudfoot AA, Bridger G, Buckland KF. The therapeutic potential in targeting CCR5 and CXCR4 receptors in infectious and allergic pulmonary disease. Pharmacol Ther 2005;107(3):314–328.
101. Widdowson KL, Elliott JD, Veber DF, et al. Evaluation of potent and selective small-molecule antagonists for the CXCR2 chemokine receptor. J Med Chem 2004;47(6):1319–1321.
102. Pease JE, Sabroe I. The role of interleukin-8 and its receptors in inflammatory lung disease: implications for therapy. Am J Respir Med 2002;1(1):19–25.
103. Stemmler S, Arinir U, Klein W, et al. Association of interleukin-8 receptor alpha polymorphisms with chronic obstructive pulmonary disease and asthma. Genes Immun 2005;6(3):225–230.
104. Persson T, Monsef N, Andersson P, et al. Expression of the neutrophil-activating CXC chemokine ENA-78/CXCL5 by human eosinophils. Clin Exp Allergy 2003;33(4):531–537.
105. Persson-Dajotoy T, Andersson P, Bjartell A, Calafat J, Egesten A. Expression and production of the CXC chemokine growth-related oncogene-alpha by human eosinophils. J Immunol 2003;170(10):5309–5316.
106. Schuh JM, Blease K, Hogaboam CM. CXCR2 is necessary for the development and persistence of chronic fungal asthma in mice. J Immunol 2002;168(3): 1447–1456.
107. Miller AL, Strieter RM, Gruber AD, Ho SB, Lukacs NW. CXCR2 regulates respiratory syncytial virus-induced airway hyperreactivity and mucus overproduction. J Immunol 2003;170(6):3348–3356.
108. McKinley L, Kim J, Bolgos GL, Siddiqui J, Remick DG. CXC chemokines modulate IgE secretion and pulmonary inflammation in a model of allergic asthma. Cytokine 2005;32(3–4):178–185.
109. Umehara H, Goda S, Imai T, et al. Fractalkine, a CX3C-chemokine, functions predominantly as an adhesion molecule in monocytic cell line THP-1. Immunol Cell Biol 2001;79(3):298–302.
110. Haskell CA, Cleary MD, Charo IF. Molecular uncoupling of fractalkine-mediated cell adhesion and signal transduction. Rapid flow arrest of CX3CR1-expressing cells is independent of G-protein activation. J Biol Chem 1999;274(15):10053–10058.

111. Rimaniol AC, Till SJ, Garcia G, et al. The CX3C chemokine fractalkine in allergic asthma and rhinitis. J Allergy Clin Immunol 2003;112(6):1139–1146.
112. El-Shazly A, Berger P, Girodet PO, et al. Fraktalkine produced by airway smooth muscle cells contributes to mast cell recruitment in asthma. J Immunol 2006;176(3):1860–1868.
113. de Nadai P, Charbonnier AS, Chenivesse C, et al. Involvement of CCL18 in allergic asthma. J Immunol 2006;176(10):6286–6293.
114. Debes GF, Dahl ME, Mahiny AJ, et al. Chemotactic responses of IL-4-, IL-10-, and IFN-gamma-producing $CD4^+$ T cells depend on tissue origin and microbial stimulus. J Immunol 2006;176(1):557–566.
115. Fulkerson PC, Zimmermann N, Hassman LM, Finkelman FD, Rothenberg ME. Pulmonary chemokine expression is coordinately regulated by STAT1, STAT6, and IFN-gamma. J Immunol 2004;173(12):7565–7574.
116. Corry DB, Kiss A, Song LZ, et al. Overlapping and independent contributions of MMP2 and MMP9 to lung allergic inflammatory cell egression through decreased CC chemokines. FASEB J 2004;18(9):995–997.
117. Corry DB, Rishi K, Kanellis J, et al. Decreased allergic lung inflammatory cell egression and increased susceptibility to asphyxiation in MMP2-deficiency. Nat Immunol 2002;3(4):347–353.
118. McQuibban GA, Gong JH, Tam EM, McCulloch CA, Clark-Lewis I, Overall CM. Inflammation dampened by gelatinase A cleavage of monocyte chemoattractant protein-3. Science 2000;289(5482):1202–1206.
119. McQuibban GA, Butler GS, Gong JH, et al. Matrix metalloproteinase activity inactivates the CXC chemokine stromal cell-derived factor-1. J Biol Chem 2001;276(47):43503–43508.
120. McQuibban GA, Gong JH, Wong JP, Wallace JL, Clark-Lewis I, Overall CM. Matrix metalloproteinase processing of monocyte chemoattractant proteins generates CC chemokine receptor antagonists with anti-inflammatory properties in vivo. Blood 2002;100(4):1160–1167.

13

Chemokine Receptors and HIV/AIDS

Tzanko S. Stantchev and Christopher C. Broder

Summary

There have been tremendous advances made toward our understanding of chemokine receptor biology over the past decade. Much of the research conducted in this area was fueled by discoveries that certain chemokine receptor ligands (chemokines) could specifically block human immunodeficiency virus type 1 (HIV-1) infection and that certain chemokine receptors were the long-sought coreceptors that, together with CD4, were required for the productive entry of HIV-1, HIV-2, and simian immunodeficiency virus (SIV) isolates. In the current survey, along with the interactions of the HIV-1 envelope glycoprotein with chemokine receptors, we have focused on the relationship certain chemokine receptors have with particular cellular systems such as lipid rafts, chemokine receptor–mediated signaling, and the actin cytoskeleton; all of which appear to play roles in or influence the establishment of HIV-1 infection. We will also discuss the dichotomous effects that chemokines display at both entry and postentry levels of HIV-1 infection and their potential significance for HIV-1 pathogenesis and acquired immunodeficiency syndrome (AIDS).

Key Words: HIV; fusion; entry; infection; chemokine receptor; chemokine; signaling; gp120.

13.1. Introduction

The general consensus model of human immunodeficiency virus type 1 (HIV-1) entry describes a series of steps leading to the merger of the viral and host cell membranes that stem from the initial binding of the 120 kDa glycoprotein (gp120) subunit of the viral envelope glycoprotein (Env) to CD4, which in turn results in significant rearrangements and exposure of a conserved binding

From: *The Receptors: The Chemokine Receptors*
Edited by: J. K. Harrison and N. W. Lukacs © Humana Press Inc., Totowa, NJ

site on Env for a chemokine receptor coreceptor (reviewed in Refs. *1* and *2*). The engagement of the chemokine receptor triggers further conformational changes; exposing the hydrophobic amino-terminal fusion peptide of the 41 kDa Env glycoprotein (gp41), which inserts into the receptor-bearing host cell membrane followed by the subsequent formation of the trimer-of-hairpins structure or the six helix bundle *(3,4)*. Several studies have indicated that the process of gp41 folding and six-helix-bundle formation appears to create the driving force involved in the fusion pore formation but is certainly concomitant with membrane fusion *(5–7)*. This is one of the most-studied virus entry processes, and HIV-1 remains unique among viruses in its evolved entry process that utilizes two quite different cellular receptors for infection of its host cells.

The HIV-1 Env consists of a gp120 subunit which is noncovalently attached to a gp41 subunit that traverses the viral membrane. Because of their characteristic locations in the membrane, the gp120 and gp41 glycoproteins are also known as the surface (SU) and transmembrane (TM) subunits, respectively. Gp120 is heavily glycosylated and consists of five variable loops (V1–V5) interconnected by five more conserved regions (C1–C5), delineated by nine intrachain disulfide bonds *(8)*. Discontinuous epitopes from both the NH_2-terminus and the COOH-terminus of gp120 are involved in its noncovalent attachment to gp41. The ectodomain of gp41 encompasses a glycine-rich, hydrophobic N-terminus, followed by two hydrophobic regions with a characteristic coiled-coil structure, designated as heptad repeat 1 and 2 (HR1; HR2) *(2–4,9)*. The Env is synthesized in the endoplasmic reticulum as a 160 kDa polyprotein precursor (gp160), assembles into oligomers (trimers), and is subsequently cleaved or processed on its way to the plasma membrane through the Golgi/trans-Golgi network. This proteolytic processing is necessary for converting the Env precursor into a fusion competent form (reviewed in ref. *10*). The mature Env is expressed as a trimer of gp120-gp41 heterodimers on the surface of virions and/or infected cells. The crystal structure of gp120 has indicated that the molecule is organized into inner (facing the trimer axis) and outer domains. It also revealed a specific, antiparallel, four-stranded "bridging sheet" between the two *(9,11,12)*.

The vast majority of HIV-1 and HIV-2 strains predominately use either CCR5 or CXCR4, or both, in addition to CD4 for binding and entry, but a number of other chemokine receptors (CCR2b, CCR3, CCR5, CCR8, CCR9, CXCR4, CXCR6, CX3CR1) or chemokine receptor-like molecules [G protein-coupled receptor 1 (GPR1), GPR15, and angiotensin II receptor-like 1 (AGTRL1/APJ)] have been shown to support the entry of one or more particular HIV-1 strains. It is now generally recognized and widely accepted that CXCR4 and CCR5 are the principal coreceptors employed by HIV-1, HIV-2, and simian immunodeficiency virus (SIV) (reviewed in refs. *2, 13, 14*). Over the past decade, the nature of the receptors employed by HIV-1 coupled with the important roles CD4 and chemokine receptors play in immune system function offered researchers a myriad of new experimental approaches aimed at understanding the complexities of the interactions between virus and host including how HIV-1 may be modulating or influencing the host response. The current survey is not meant to be exhaustive, and we will focus on summarizing only the more recent findings on the roles the chemokine receptors play in HIV-1 infection and pathogenesis, which highlight several additional complexities that have emerged resulting from their use as HIV-1 receptors.

13.2. HIV-1 Tropism

Genotypic and phenotypic variation is a hallmark of HIV-1 infection. The ability of the virus to infect different cell types varies from one isolate to the next and is referred to as cellular tropism, or cytotropism. Initially, HIV-1 isolates were classified as M-tropic (able to infect primary human macrophages and $CD4^+$ lymphocytes), T-tropic (infecting primary $CD4^+$ T cells and T cell lines but not macrophages), and dual-tropic (capable of infecting all three cellular targets). In general, M-tropic isolates did not cause multinuclear giant cell formation in the lymphocytic MT-2 cell line and were also designated as non–syncytium inducing (NSI), in contrast with the syncytium-inducing (SI) T-tropic or dual-tropic strains. Subsequently, it was established that, in addition to CD4, M-tropic isolates predominately employed, but not exclusively, CCR5; T-tropic viruses principally exploited CXCR4; and dual-tropic strains utilized either CCR5 or CXCR4. The observation that MT-2 cell line expresses CXCR4, but not CCR5, further explained the differential effects of M-, T-, or dual-tropic strains when applied to these cells (reviewed in refs. *1, 2, 13, 15, 16*). Currently, it is considered more appropriate to classify HIV-1 isolates based on the coreceptor(s) they use instead of their cellular target preference, and this nomenclature has been universally adopted *(17)*.

In human cell lines, HIV-1 tropism largely coincides with the expression of CD4 and the appropriate coreceptor *(18)*. In primary cells, CD4 and coreceptor presence on the host cell surface does not always predict productive infection by HIV-1. Indeed, it was recently demonstrated that CCR5 using (R5) strains, isolated from asymptomatic patients, are usually replication competent in primary $CD4^+$ T lymphocytes, but interestingly most of them only inefficiently infected primary human macrophages (i.e., they did not fulfill the definition for an M-tropic virus) *(16)*. Primary macrophages fuse poorly with cells expressing HIV-1 Envs derived from laboratory-adapted T-tropic strains, even though the CXCR4 coreceptor is present on their cell surface. Further, it has been observed that CXCR4 on human macrophages can be used by some dual-tropic but not T-tropic laboratory-adapted strains for entry and infection *(19)*. The significantly reduced ability of most CXCR4 using (X4) laboratory-adapted Envs to mediate membrane fusion reactions with macrophages also implied that the restriction on virus infection was probably at the level of entry *(20–22)*. However, other studies supported the view that the inefficient infection of primary macrophages by laboratory-adapted X4 isolates was restricted at postentry levels of virus replication *(23–25)*. It was proposed that the weaker cell-signaling events in macrophages that are induced by these X4 Envs, compared with Envs derived from R5 strains, might be a factor in the reduced replication of T-tropic viral isolates *(26)*. More recently however, certain primary X4 isolates that could mediate fusion and entry and replicate to high levels in primary human macrophages have been identified and described *(22,27,28)*, and perhaps future detailed studies with Envs derived from these apparently unique viral strains may allow further clarification of the underlying mechanism(s) governing CXCR4 function as an HIV-1 coreceptor in these important cellular targets in the host.

It has also been established that both CCR5 *(29)* and CXCR4 *(30)* appear to display conformational heterogeneity, and such variations may influence their recognition and use as HIV-1 coreceptors. Several CXCR4 variants exhibiting alternative molecular weight (MW) species also differing in their ability to serve as coreceptors for HIV-1 infection have been observed and described *(31,32)*. For example, it was noted that peripheral $CD4^+$ T cells express several CXCR4 species with MW 47, 50, 62, and 98 kd *(33)*. Further, after phytohemagglutinin (PHA) and interleukin-2 (IL-2) stimulation, there was a decrease of the 47-kd and a concomitant increase in the 50-kd and 62-kd CXCR4 isoforms. Interestingly, the 62-kd CXCR4 predominately coimmunoprecipitated with CD4, and although the overall surface expression of CXCR4 did not change significantly after PHA/IL-2 treatment, the increased 62-kd fraction was associated with augmented HIV-1 Env-mediated fusion. Thus, although the discoveries of the principal coreceptors involved in HIV-1 entry and infection largely explains the cellular tropisms displayed by viruses, there continues to be new discoveries made that supply additional details on the nuances of coreceptor use depending on where and when they are expressed and utilized by HIV-1.

13.3. Transmission of HIV-1 Infection

It is the HIV-1 R5 strains that are predominant during the early and asymptomatic stages of infection and the strains that are also typically present throughout the symptomatic phase as well. Soon after the discovery of the important role CCR5 played as a principal coreceptor for most primary isolates of HIV-1, a particular genetic polymorphism in the CCR5 gene was noted among cohorts of multiply exposed yet uninfected individuals. This mutation was a 32-base-pair deletion (CCR5 Δ32), which essentially rendered cells homozygous for the defect CCR5-negative for cell surface expression. About 1% of Caucasians are homozygous CCR5 Δ32, which leaves them highly resistant to HIV-1 infection. The natural protection of these CCR5-negative individuals also provided compelling evidence that R5 viral strains were those primarily involved in HIV-1 transmission, and there have been only a few reported cases of CCR5 Δ32 homozygous individuals infected by X4 viral strains (reviewed in refs. *13, 18, 34*). Further, individuals heterozygous for the Δ32 deletion have reduced CCR5 levels compared with individuals with two wild-type CCR5 alleles. They are susceptible, yet probably with a slightly reduced risk of HIV-1 infection, but if infected they typically progress to AIDS at a slower rate *(18,35)*. Thus, there is not only clear evidence of a CCR5 role in the establishment of HIV-1 infection in humans but also in the pathogenic processes that ensue in those infected individuals.

The particular reasons or underlying mechanisms why X4 viral strains, which appear to be more pathogenic in vitro, are infrequently transmitted and fail to efficiently replicate during the early stages of infection have yet to be completely clarified (reviewed in refs. *13* and *36*). It has been suggested that mucosal epithelial cells, without becoming infected themselves, may preferentially transcytose R5 HIV-1 viruses in a CCR5- and galactosyl ceramide-dependent manner *(37,38)*. However, microabrasions during heterosexual or homosexual contacts would provide direct access of X4 viruses, if present, to the susceptible cells in lamina propria: such as immature Langerhans and dendritic cells (DCs), macrophages, as well as $CD4^+$ T lymphocytes. Although there is good evidence that CCR5 may be predominately expressed on these particular cellular targets (reviewed in ref. *13*), other studies have found significant and certainly sufficient levels of CXCR4 on these same cell types *(39–42)*.

In attempting to grasp the underlying biology of HIV-1 transmission and the early events of virus infection, it was hypothesized that that the constitutive production and secretion of stromal cell-derived factor 1 (SDF-1)/CXCL12, the natural ligand of CXCR4, at mucosal surfaces would cause downregulation of CXCR4 and hence a concomitant reduced level of an X4 HIV-1 strain's transmission success *(43)*. On the other hand, one could speculate that SDF-1 might attract $CD4^+$/$CXCR4^+$ lymphocytes susceptible to infection by X4 trains. Recently, gut-associated lymphoid tissue has emerged as an important site of both HIV-1 and SIV replication, where R5 viral strains may replicate more efficiently because of the high number of $CCR5^+$ memory T cells *(44–46)*. However, R5 viral strains are also preferentially transmitted via pathways that bypass these same mucosal surfaces, despite the fact that the majority of $CD4^+$ lymphocytes in the blood and lymph nodes express predominately CXCR4.

There are case reports describing the accidental inoculation of SI or mixture of SI and NSI strains in humans, which is followed by replication of SI viruses, but interestingly, there was a replacement of the SI with NSI isolates shortly thereafter (reviewed in ref. *13*). Subsequently, it was found that R5 viral strains evaded the cytotoxic T lymphocyte (CTL) responses more efficiently than X4 stains *(47)*. In a later study, it was demonstrated that rhesus macaques could be infected with an SIV that was pseudotyped with X4 or R5 HIV-1 Envs (SHIV), but the X4-SHIV replication was suppressed early in the infection by the specific $CD8^+$ T cells *(48)*. Therefore, although the transmission of X4 strains is still possible, it is likely that a combination of both host and viral factors gives advantage of R5 strains in the majority of cases during the initial stages of HIV-1 infection.

13.4. Exploitation of CCR5 and CXCR4 by HIV-1 During the Symptomatic Phase of Infection

Although X4 viruses are isolated rarely during the early, asymptomatic period of HIV-1 infection, it has been observed that they appear in approximately 40% to 50% of infected individuals after about 5 years of infection. The phenotypic switch of R5 (NSI) to X4 (SI) strains is usually associated with rapid $CD4^+$ lymphocyte decline and an accelerated progression of AIDS. An increased ability of X4 viral isolates to deplete $CD4^+$ cells has been attributed to the predominant expression of CXCR4 on lymphocyte precursors in the thymus and on $CD4^+$ lymphocytes in blood and lymph nodes (reviewed in refs. *13* and *49*). The precise mechanism(s) underlying the CCR5/CXCR4 coreceptor switch and why it appears so late during infection, despite the high viral turnover and mutation rates, has yet to be completely clarified (reviewed in ref. *50*). However, two major factors have emerged that may have roles in influencing the relatively late foothold that X4 HIV-1 strains experience: (i) the suppressive effect of the host immune system (discussed in the previous paragraph) and (ii) the findings of in vitro studies indicating that the CCR5-CXCR4 coreceptor switch occurs via viral intermediates that are characterized by reduced replication capacity and an enhanced sensitivity to both CCR5 and CXCR4 inhibitors *(51,52)*.

It is likely that initial deterioration of the antiviral immune response, combined with increased viral replication at later stages of infection, allows for the selection and propagation of X4 viruses, which in turn will lead to the accelerated decay of immune system function. Recent studies have in fact shown that the majority of patients that presented with AIDS were noted to harbor only R5 viral strains, and interestingly disease progression was associated with an enhanced ability of these R5 viruses to utilize low levels of CD4 and CCR5, infect cells of the monocyte/macrophage lineage, and cause $CD4^+$ T lymphocyte depletion *(49,53)*. Further, R5 viral strains isolated from symptomatic HIV-1 infected patients have been reported to display greater affinity and flexibility in CCR5 binding as well as a decreased sensitivity to certain β-chemokines that are natural ligands of CCR5 *(54)*, and these R5 viruses appeared to be more prone to neutralization by the anti-CD4 binding site monoclonal antibody (MAb) b12 *(16)*. Finally, and of importance to the possible future clinical application of HIV-1 entry inhibitors, results derived from in vitro studies have revealed that R5 viral strains could respond to CCR5 antagonists by either switching to CXCR4 usage or, in many cases, by developing decreased sensitivity to the suppressor without a change in the coreceptor used *(13,49,55)*. Thus, in light of these more recent observations, careful consideration should be given to the potential consequences that clinical therapy–induced selective pressure may have on HIV-1 coreceptor use during infection.

13.5. General Principles of HIV-1 Entry

13.5.1. The gp120 Chemokine Receptor Binding Sites

As discussed earlier, a major consequence of CD4 binding is the significant structural rearrangements in gp120 (repositioning of V2 and V3 loops and stabilization of the bridging sheet), resulting in the formation and exposure of the coreceptor binding site *(11,56–58)*. The V3 loop of gp120 has been established as a major factor involved in coreceptor binding and not surprisingly also in determining the specificity of chemokine receptor usage by varying HIV-1 strains *(8,59)*. Regardless that some of its segments are highly variable, a more detailed analysis reveals that the V3 loop is really a semiconservative region, having a relatively fixed size (30 to 39 amino acids) and containing a β-hairpin structure at its tip. Analysis by mutagenesis, coupled with the use of MAbs specific for CD4-induced epitopes (CD4i) of gp120 and data obtained from the crystal structure of gp120-sCD4-CD4i MAb complexes, has also revealed a second conserved coreceptor binding site that is formed by domains from both the bridging sheet, the base of the V3 loop, and certain neighboring residues *(36,60,61)*. In addition, it was reported that for some HIV-1 isolates, regions from V1 and/or V2 may also play a role in coreceptor binding *(36,62)*, and the apparent importance of V1 and especially V2 increases when cell-surface levels of CD4 are low on host cells *(63)*. A recent crystallographic study has proposed that CD4 binding results in positioning V3 in such a way that its coreceptor binding tip is presented as a protrusion that is some 30 Å from the gp120 core toward the target cell membrane *(61)*. In addition to receptor-induced structural alterations in HIV-1 Env, it has also been suggested, at least for some HIV-1 isolates, that reduction of certain disulfide bonds by protein disulfide isomerase (PDI) (reviewed in ref. *64*) and/or thioredoxin *(65)* is required for the structural rearrangements of gp120 during the exposure/formation of the coreceptor binding site.

In addition, the external surfaces of CXCR4 have been noted to be more electronegative compared with CCR5, and in general the V3 loop of X4 viral strains has a more positive charge than the V3 loop of CCR5 using isolates. Basic amino acids at position 11 or 25 of the V3 loop are usually associated with X4 isolates *(66)*. However, changes in V3 during the process of CCR5 to CXCR4 coreceptor switching may result in the generation of less-fit viral intermediates, and it has been shown that gain-of-fitness mutations in or near V1/V2 were able to compensate for the deleterious effects of V3 loop alterations on virus viability, although they did not confer CXCR4 use *(51,52)*, and this suggests that the appearance and establishment of functional X4 variants during the course of infection may be a multistep process (reviewed in ref. *50*).

Clinically relevant HIV-1 strains critically depend on CD4 for fusion and infection, however some HIV-1 and HIV-2 isolates may become CD4-independent after being cultured in vitro under selective pressure, and many primary SIV strains are indeed capable of CD4-independent infection via direct coreceptor (CCR5) interactions. Taken together, these observations have led to the speculation that the chemokine receptors are the probable obligatory primordial receptors with CD4 being a more recent adaptation in HIV evolution. One likely advantage of CD4 usage as an HIV-1 receptor may be to reduce to a minimum the exposure of the conserved chemokine receptor binding sites and hence shield the virus from the immune system *(1,3)*. Indeed, it has been demonstrated that CD4-independent HIV-1 and HIV-2 strains are more sensitive to antibody neutralization by having their chemokine receptor binding sites constitutively exposed *(67–71)*. For different HIV-1 isolates, mutations outside *(72)*, within *(73)*, or both outside and within the coreceptor binding domains *(67,70)* have been shown to be responsible for acquisition of CD4-independence. The observations that mutations in the transmembrane *(72)* or the cytoplasmic domain of gp41 *(67,70,74)* affect the exposure of CD4i epitopes further demonstrate the importance of viewing the formation of the multidomain chemokine receptor binding site in the context of the structurally interconnected HIV-1 Env.

13.5.2. Chemokine Receptor Domains Involved in gp120 Binding

Understanding the nature of the domains or elements on the chemokine receptor coreceptors is also of significant importance, especially with regard to the development of coreceptor-blocking therapeutics. HIV-1 gp120 interacts with multiple extracellular domains of the chemokine receptors to mediate binding, and the N-terminus and extracellular loop (ECL) number 2 of CCR5 or predominately ECL2 of CXCR4 appears to play the major roles (reviewed in ref. *18*). Although HIV-1 Envs and chemokines may interact with overlapping binding domains on one and the same chemokine receptor(s), the chemokine and gp120 binding patterns are not identical. CCR5 variants with mutated extracellular cysteines were unable to bind and respond to macrophage inflammatory protein 1β (MIP-1β)/CCL4 but still support, albeit at reduced levels, HIV-1 binding and entry *(75)*. By using a panel of MAbs specific for different epitopes of CCR5, it was observed that N-terminal MAbs blocked gp120 binding more efficiently than those directed against ECL2, whereas an inverse correlation was found for Regulated upon Activation Normal T-cell Expressed and Secreted (RANTES)/CCL5 and MIP-1β/CCL4 binding. Unexpectedly, anti-ECL2 MAbs were more potent inhibitors of HIV-1 infection, indicating that the ability to suppress gp120 binding did not always reflect the potency to block infection *(29)*. This phenomenon was further supported by the finding that P438A mutation in the β21 strand of gp120 significantly reduced the binding of the YU-2 HIV-1 isolate to CCR5 but had little effect on the Env-induced fusion and virus infectivity *(76)*.

It has also been demonstrated that tyrosine sulfation of the CCR5 N-terminus was required for both β-chemokine *(77)* and gp120 binding *(78)*. In a recent model, based on the crystal structure of the gp120 core-sCD4-CD4i Fab complex, it was proposed that the CCR5 N-terminus extends up and binds to the base of the V3 loop and probably some additional domains of the gp120 core, while the V3 loop tip, exposed toward the plasma membrane after CD4 binding, interacts with ECL2 *(61)*. Subsequently, de Parseval et al. *(79)* observed that the conserved Arg_{298} at the base of the V3 loop governed its binding to the sulfated N-terminus of CCR5 as well as to sulfated moieties of cell surface syndecans. Interestingly, variants of the R5 strain JR-CSF, created in vitro under selective pressure to become independent of this N-terminal sulfated region, showed enhanced fusogenicity, and some of these variants surprisingly developed an increased affinity for the ECL2 rather than for the expected adaptation to the mutated N-terminus *(80)*. Related to these observations, another study observed that R5 viruses isolated during late stages of symptomatic infection, in contrast with the viruses from the asymptomatic phase, displayed higher infectivity, reduced sensitivity to β-chemokines, and were in fact also able to exploit CCR5 chimeric coreceptors for binding and entry, where the CCR5 N-terminus or N-terminus and the ECL1 had been replaced with the corresponding regions of CXCR4 *(54)* (i.e., the binding of these isolates to CCR5 resembled that of the way X4 strains interact with CXCR4) via ECL2.

To induce intracellular signaling, SDF-1/CXCL12 interacts with the N-terminus of CXCR4 and subsequently binds to ECL2 and ECL3. Binding of X4 HIV-1 isolate Envs depends predominately on ECL2 of CXCR4, and several

CXCR4 mutants with abrogated SDF-1 binding and signaling can in fact still support HIV-1 infection. Results obtained from alanine-scanning mutagenesis experiments revealed that several residues outside the ECL2 may also contribute to the coreceptor function of CXCR4. Interestingly, some CXCR4 mutations that do not bind soluble gp120 could still function as coreceptors *(81–84)*. Tyrosine sulfation of CXCR4 N-terminus was also required for SDF-1 binding *(85)*, however, it was important only for infection by some HIV-1 CXCR4 using isolates *(86)* and played only a minor role for others *(85)*.

Although the majority of well-characterized viruses exclusively employ only one of the principal coreceptors, some HIV-1 strains are able to make use of either CCR5 or CXCR4 in the fusion process (R5X4 or dual-tropic stains), but the preference for either coreceptor may vary depending on the virus isolate or the target cell type *(22,27,87)*. R5 isolates displayed flexibility in using the N-terminus or ECL2 for their binding to CCR5, although neither of the two coreceptor domains was shown to be completely dispensable during the R5 Env-CCR5 interactions. However, the ability of R5X4 isolates to employ CCR5 during virus entry was considerably more dependent on the binding of their Envs with the CCR5 N-terminus. On the other hand, the interactions with ECL2 emerged to be the major factor in CXCR4 usage by R5X4 strains. Based on the above findings, it has been proposed that R5 isolates evolve into R5X4 variants by acquiring the ability to interact with the ECL2 of CXCR4 while retaining an ability to bind to the CCR5 N-terminus *(88,89)*. Interestingly however, it has been demonstrated that removal of the single N-linked glycosylation site located in the N-terminus of CXCR4 allowed it to serve as an universal coreceptor for both X4 and R5 HIV-1 strains, suggesting the existence of conserved elements between CXCR4 and CCR5 that are involved in the binding of HIV-1 Env; thus, acquisition of alternative coreceptor capacity through Env mutation may not be the sole mechanism that governs coreceptor use, and posttranslational modification of coreceptors might also be of influence *(90)*.

13.5.3. Cooperation and Interactions of Multiple Receptors During HIV-1 Entry Process

Clustering of a certain number of CD4 and coreceptor molecules is presumed to be necessary for the efficient HIV-1 Env-mediated fusion pore formation. It has been proposed that four to six CCR5 molecules *(91)* and three CD4 binding events are needed to induce fusion between the viral and host cell membranes *(92)*. Both CD4 *(93)* and chemokine receptors can form functional dimers *(94)* in the plasma membrane. It was proposed that formation of CD4 dimers, mediated by a disulfide bond between the cysteine residues of the D2 domain, might enhance HIV-1 entry and infection *(95,96)*. In contrast, others have provided evidence that chemokine receptor dimerization may be inhibitory in the HIV-1 entry process *(97)*. Further, this dimerization phenomenon has also been suggested as a potential mechanism for resistance to HIV-1 infection in certain genetically determined cases: (i) formation of dimers between wild-type and Δ32 CCR5 molecules prevents surface expression of wild-type CCR5, and (ii) heterodimer formation between the CCR2V64I mutant and CCR5 or CXCR4 impairs their ability to support HIV-1 entry (reviewed in ref. *94*).

Not only formation of homooligomers of CD4 or coreceptors appears capable of influencing HIV-1 Env binding and virus entry, but also CD4 and coreceptor interactions might be at play. Observations from coimmunoprecipitation experiments have shown that CD4 and CXCR4 or CCR5 physically coassociate with one another *(31,98)*. Further, CD4-CCR5 but not CD4-CXCR4 complexes have been demonstrated in primary macrophages and could be correlated with the ability of macrophages to support fusion with cells expressing R5 but not laboratory-adapted HIV-1 Envs, respectively *(20)*. CD4 but not CXCR4 overexpression resulted in a significant increase in the X4 Env-mediated fusion, suggesting that different chemokine receptors may compete for association with CD4 during virus entry particularly in the human macrophage target *(20,99)*.

Additional evidence of a physical association of CD4 and CCR5 has come from observations demonstrating receptor cross-talk with allosteric CD4-dependent regulation of the signaling properties of CCR5 *(100)*, and another report suggested that the CD4 domains involved in this interaction may be located primarily in D1 and D2 *(101)*. Indeed, it was noted earlier that reciprocal receptor cross-desensitization between CD4 and CCR5 (and not CCR 1, 2, or 3) existed and that a selective relationship between these two receptors was at play *(102)*. It has now been shown that although IL-16 (CD4's ligand) does not bind CCR5 alone, the presence of CCR5 significantly increases IL-16 binding to CD4, providing additional functional correlates to the CD4-CCR5 interaction with important functional consequences such as the CCR5 influence of IL-16–induced migration of $CD4^+$ T cells *(103)*. The existence of a primary X4 HIV-1 strain that efficiently entered human macrophages *(22,27,28)* also implies that the CD4-CCR5 association is reversible and perhaps under certain circumstances is modulated by gp120 binding.

The pattern of CD4-coreceptor association is also likely to be cell type specific, because in lymphocytes CD4 coimmunoprecipitated preferentially with the 62-kd CXCR4 species discussed earlier. Recently, using CXCR4 and CD4 molecules tagged with enhanced cyan fluorescent protein (ECFP) or enhanced yellow fluorescent protein (EYFP), Toth et al. *(104)* observed CXCR4-CXCR4, CD4-CD4, and CXCR4-CD4 homo- and heterodimers in a fluorescence resonance energy transfer (FRET) study. Also, as one might predict, the addition of gp120 (IIIB X4 isolate) further increased CD4-CD4 dimerization and colocalization of labeled CXCR4 molecules if CD4 was also expressed on the cell surface *(104)*.

13.5.4. The Significance of the Virological/Infectious Synapse in HIV-1 Infection

It is either the fusion of free virions to susceptible host cells or Env-mediated fusion between infected cells with partner cells expressing CD4 and an appropriate coreceptor that facilitates HIV-1 transmission. Recently, a specialized intercellular contact structure, termed the *virological/infectious synapse* because of its resemblance to the already well described immunological synapse, has been shown to greatly enhance HIV-1 transmission between effector and target cells. The virological synapse consists of multiple, ordered intercellular junctions that provide a stable environment for the directed secretion and transfer of virus particles through the synaptic space (reviewed in refs. *105* and *106*). It was also demonstrated that virological synapse formation between HIV-1 infected Jurkat (effector) cells and $CD4^+/CXCR4^+$ primary T lymphocytes (target cells) followed the clustering of CD4, CXCR4, lymphocyte function-associated antigen 1 (LFA-1), and talin at the target cell interface and HIV-1 Env and gag proteins in effector cells. The CD4, CXCR4, LFA-1 clustering and virological synapse formation was also dependent on actin cytoskeleton rearrangements in target but not effector cells *(107)*. However, the integrity of lipid rafts in effector cells was shown to be necessary for the virological synapse assembly *(108)*. It was proposed that the virological synapse formation between virus carrying DCs and $CD4^+$ T cells is particularly important for HIV-1 dissemination *(105,106,109)*. DCs bind and endocytose HIV-1, predominately via a C-type lectin receptor [DC-specific intracellular adhesion molecule-3 (ICAM-3)-grabbing nonintegrin; DC-SIGN], and subsequently deliver the virus from a specific endocytic compartment to the area of contact with susceptible T cells *(109–111)*, as the virological synapse significantly increases the efficiency of HIV-1 transmission of both R5 and X4 strains *(107,109,110,112)*. Infection of DCs by HIV-1 is not considered very efficient, but if it occurs it is likely that the newly synthesized virus particles bud into certain part(s) of the endocytic compartment and are directed from there to the virological synapse during DC interaction with T lymphocytes *(109,113)*. In vivo, HIV-1 faces the challenge of selecting cells susceptible to productive infection. Because apoptotic cells were shown to be incapable of efficient actin cytoskeleton reorganization, it was hypothesized that the actin-mediated formation of the synapse and/or active CD4-coreceptor clustering in general might also be a mechanism by which HIV-1 would avoid entering cells, considered to be a "dead end" for viral replication *(114)*.

13.6. Significance of HIV-1 gp120-Induced Chemokine Receptor Signaling

An extensive coverage of the chemokine receptors signaling pathways is beyond the scope of the current survey, but some details important for the subsequent discussion will be briefly highlighted. Chemokine receptors are often given as an example of G protein–coupled receptors (GPCRs) that associate exclusively with pertussis toxin (PT)-sensitive $G_{\alpha i}$ proteins; however, they may also couple with other subclasses of G_{α} molecules, such as $G_{\alpha q}$ or $G_{\alpha s}$. The mode of these interactions is complex and depends on the type of chemokine receptor molecule and the cell type or even the activation state of the cell. In addition to modulation of intracellular cAMP levels, chemokine receptors are also able to activate various nonreceptor tyrosine kinases: the Janus kinases (JAKs), proline-rich tyrosine kinase Pyk2 (also known as related focal adhesion kinase; RAFTK), and members of the Src or Syk kinase subfamilies in both PT-sensitive and PT-insensitive modes (reviewed in refs. *115* and *116*). Activation of certain tyrosine phosphatases by chemokines has been described as well *(117–119)*. Chemokine receptors are also able to stimulate the extracellular regulated kinases (ERKs) *(120,121)* and stress-activated protein kinases (SAPKs) *(120,122)* branches of the mitogen-activated protein kinases (MAPKs) pathway (reviewed in refs. *123* and *124*). The known signaling patterns of CD4 are far less complex and have been primarily associated with activation of the CD4 cytoplasmic tail-associated Src-related tyrosine kinase, $p56^{lck}$ *(125)*. Interestingly, a more recent study has shown that CD4 is able to induce Ca^{2+} influx, tyrosine phosphorylation, and activation of phospholipase C gamma (PLC-γ), phosphatidylinositol 3-kinase (PI-3K), and upregulation of SAPK in promonocytic cell lines, which like primary monocytes/macrophages do not express $p56^{lck}$ *(126)*.

It is now well established that during its interactions with CD4 and/or chemokine receptors, HIV-1 gp120 elicits various intracellular signaling events both in primary cells and in cell lines (reviewed in refs. *115, 116, 127*) and, although similar, gp120-induced signaling is not identical to that caused by chemokines *(115,128)*. The ability of gp120 to induce calcium signaling and activation of certain tyrosine kinases (pyk2, ZAP-70, Lyn, focal adhesion kinase) was established in several cell types, and some of the effects were shown to be PT insensitive *(26,115,129–132)*. In macrophages, the gp120-caused elevation of intracellular Ca^{2+} was markedly higher and more sustained in response to HIV-1 Envs from R5 strains that efficiently replicate in these cells compared with Envs from laboratory-adapted X4 isolates *(26,115,131)*. Also, stimulation of PI-3K, c-Jun, and p38 MAPKs in primary macrophages, after gp120 stimulation, have been described as events that are likely downstream effects of Ca^{2+} and/or tyrosine kinase signaling *(115,133,134)*. This gp120-mediated signaling may influence HIV-1 pathogenesis in several ways: increased secretion of proinflammatory cytokines *(115,132,135)*, attraction of cells to the sites of viral replication *(136–138)*, induction of apoptosis in $CD4^+$ and/or coreceptor-expressing bystander cells *(139–141)*, and augmentation of HIV-1 replication in already infected cells *(142–144)*.

At present, the prevailing view, based on studies using PT *(59,145)* and mutated CD4 *(146)* or chemokine receptors *(147–150)*, is that signaling is dispensable for HIV-1 Env-mediated fusion. Nevertheless, it is now evident that signaling via both CD4 and, more importantly, the chemokine coreceptor molecules is far more complex than was appreciated just a few years ago, and in the light of these new findings, it is unknown whether the mutant receptors or PT treatment employed previously were able to completely suppress the gp120-induced signaling, especially via possible $G\alpha_i$ alternative pathways. Further, an examination of HIV-1 entry and infection in which both CD4 and the coreceptor molecules were altered in such a manner has not yet been explored. Two early studies demonstrated that laboratory-adapted X4 Env-mediated syncytia formation was sensitive to tyrosine kinase inhibitors *(151,152)*. It has also been observed that in primary lymphocytes, PT and especially its B oligomer were able to suppress HIV-1 infection at both entry and postentry levels *(153–157)*. Because the B oligomer does not inhibit G_i but is able to bind to certain cell surface glycoproteins and induce intracellular signaling *(158,159)*, its effect on HIV-1 entry was attributed to heterologous desensitization of CCR5 (reviewed in ref. *160*).

Gp120-mediated signaling via CD4 and/or coreceptors most likely involves certain tyrosine kinases or other $G_{\alpha i}$-independent signaling pathways and may affect the rearrangements of lipid rafts and actin cytoskeleton—events that appear to be important for virus entry and infection. Indeed, there is already evidence from studies on the immunological synapse formation that tyrosine kinase signaling, lipid raft coalescence, and actin cytoskeleton reorganization are closely interconnected events *(161,162)*, and the remodeling and coalescence of lipid rafts may be required for HIV-1 entry (reviewed in ref. *163*). Rafts differentially associate a variety of signaling molecules, including tyrosine kinases *(164)*, and certain stimuli, including tyrosine kinase signaling, may be able to induce patching of lipid rafts and changes in their composition and size *(164,165)*.

The role of tyrosine kinase signaling in actin cytoskeleton regulation is well established *(166–171)*, and certain steps in the activation pathways of both Rac1 and RhoA, two molecules known to modulate actin function, may also require tyrosine kinase activity *(172,173)*. In fact, two recent reports have also indicated that the Rho family GTPases Rac1 *(174)* or RhoA *(175)* play a substantial role in HIV-1 Env-induced fusion. These actin cytoskeleton rearrangements may be necessary at several steps in the process of productive HIV-1 infection: (i) remodeling of cellular microvilli *(176,177)* where CD4, CCR5, and CXCR4 are preferentially localized *(178)*; (ii) clustering of a sufficient number of CD4 and coreceptor molecules in a timely manner for fusion pore formation *(91,116,177,179)*; (iii) assembly of the newly described virological synapse, which enhances HIV-1 spread and infection *(107)*; (iv) passage of the viral nucleocapsid through the cortical actin *(180–182)*; and (v) the formation and transport of the viral reverse transcription complex *(183,184)*. It has been shown that HIV-1 Env-mediated cell fusion is inhibited by reagents that interfere with actin function *(174,177,185)*, and the sensitivity to this inhibition was inversely correlated with the levels of CD4 and coreceptors on the cell surface *(177)*. Inhibition of CD4/CXCR4 co-capping and HIV-1 infection by the actin-disrupting agent cytochalasin D was observed in primary peripheral blood mononuclear cells (PBMCs) *(179)* but not in human osteosarcoma (HOS) CD4/CXCR4 positive cell line *(186)*. The inconsistency between these studies may reflect variations in CD4/CXCR4 expression and/or differences in actin cytoskeleton regulation between primary cells and cell lines. Indeed cancer cells, respectively cell lines, may differentially express particular actin binding proteins, may have increased activity of certain molecules involved in actin cytoskeleton regulation, and may show a different phosphoprotein profile compared with primary cells and hence be less dependent on gp120 signaling for HIV-1 entry and infection *(187–191)*.

Finally, during budding, HIV-1 may also incorporate into its membrane envelope a variety of different molecules, including proteins that may subsequently interact with their counterparts on the host cell membrane (reviewed in ref. *192*), resulting in intracellular signaling and facilitation of virus fusion *(193,194)*. However, the incorporation of cell membrane–derived molecules does not appear to be an absolute requirement for virus entry *(195)*, indicating the leading role of CD4 and the coreceptor for any such mechanism. However, this phenomenon may account for lower levels of inhibition when the effects of mutant CD4 and/or chemokine receptor are studied.

13.7. Chemokines in HIV-1 Infection

Currently, it is well established that the natural ligands of CCR5 (β-chemokines RANTES/CCL5, MIP-1β/CCL4, and MIP-1α/CCL3) or CXCR4 (SDF-1/CXCL12) are able to suppress HIV-1 infection in vitro, both in cell lines [β-chemokines (196–198); SDF-1/CXCL12 *(199–202)*] and in primary cells [β-chemokines *(196,203–208)*; SDF-1/CXCL12 *(199,208–211)*]. Among the

CCR5 natural ligands, the nonallelic isoform of MIP-1α/CCL3, LD78β/CCL3L1 *(212)*, has been reported as the most effective HIV-1 inhibitor in both PBMCs *(213)* and macrophages *(214)*. The preponderance of experimental data imply that the inhibitory effect of chemokines at the level of virus entry is achieved via three possible and not mutually excluded mechanisms: steric hindrance by chemokine binding of the gp120/coreceptor interaction sites, coreceptor dimerization, and/or coreceptor endocytosis and surface downregulation (reviewed in ref. *94*).

However, chemokines may also enhance HIV-1 infection at postentry levels. It has been noted that β-chemokines increased replication of X4 isolates, and SDF-1/CXCL-12 had the same effect on R5 strains. Further, observations of β-chemokines (RANTES, MIP-1α, or MIP-1β) effects on human macrophages revealed dichotomous effects on HIV-1 replication depending on the time of their application: an increase in viral replication when cells were only pretreated with β-chemokines 48 hours prior to virus infection; whereas they were inhibitory if they were added to the cells during or immediately after a 4-hour infection period. Maximal suppression was achieved when β-chemokines were present both during and after infection (reviewed in refs. *116, 215–217*). It is also specific for RANTES to aggregate at high concentrations (≥500 ng/mL), interact with cell surface glycosaminoglycans (GAGs), and induce PT-independent/tyrosine kinase–dependent signaling that augments HIV-1 replication (reviewed in ref. *218*). These dichotomous effects of β-chemokines on HIV-1 entry and/or postentry levels are probably the underlying reasons for the seemingly inconsistent results reported in the literature on their role in HIV-1 infection in vivo.

The overproduction of RANTES, MIP-1α, or MIP-1β has also been linked to the HIV-1 resistance observed in some multiply-exposed but uninfected individuals (EU) expressing wild-type CCR5 molecules. In addition, an increased β-chemokine secretion by $CD4^+$ or $CD8^+$ T cells or T cell–derived clones has been associated with an HIV-1–infected asymptomatic state and delayed $CD4^+$ T-cell decline (reviewed in ref. *219*). However, other studies have found no correlation between β-chemokines levels and disease progression (reviewed in ref. *220*). Additional correlations between chemokines and their effects on HIV-1 infection or disease progression have been made to genetic variants observed in certain chemokine alleles. Individuals with genetic variants that result in decreased LD78β/CCL3L1 expression have a markedly enhanced susceptibility to HIV-1/AIDS *(212)*. Mutations leading to increased or reduced expression of RANTES/CCL5 have been correlated with delayed or accelerated disease progression, respectively (reviewed in ref. *221*), although such correlation has not been found to be always statistically significant *(222)*. Paradoxically, one of these mutations (403A) in the RANTES gene promoter region, resulting in elevated chemokine production and a more benign disease course, was also associated with an almost doubled risk of acquiring HIV-1 infection. Here, it was speculated that RANTES might enhance transmission by inducing inflammation at mucosal surfaces, but acting later during infection predominately as a virus-entry inhibitor *(221)*. In a related phenomenon, the increased secretion of β-chemokines observed in vitro in response to the HIV-1 glycoprotein gp120 or the virus infection itself was also proposed to be a way to attract susceptible cells to sites of virus replication *(133,216)*.

Interestingly, during HIV-1 infection, certain chemokines themselves may be modified in ways that enhance viral replication. CD26 is a serine protease that can cleave a variety of proteins, including chemokines that possess a proline or alanine residue in the penultimate position. The CD26-mediated removal of the two N-terminal amino acids of RANTES results in reduced signaling through CCR1 and CCR3, whereas signaling via CCR5 remains unaffected. But perhaps of greater importance is that truncated RANTES (RANTES 3-68) inhibited infection of mononuclear cells by an R5 HIV-1 isolate fivefold more efficiently than the intact molecule (reviewed in refs. *223–225*). Adding further complexity to these relationships is the observation that in vitro, HIV-1 Tat inhibits the peptidase activity of CD26, and CD26 is hypersialylated during HIV-1 infection, which may further enhance its interaction with Tat under physiologic conditions. Therefore, it is almost certain that during HIV-1 infection, extracellular Tat will shift the ratio between intact and cleaved RANTES 3-68 in a direction that will facilitate entry and increase viral replication. This hypothetical model is indeed supported by the observed inverse correlation between the dipeptidyl peptidase enzymatic activity of plasma CD26 and HIV-1 RNA in infected individuals (reviewed in ref. *116*). Several β-chemokines with an altered N-terminus have already been created and tested as an efficient HIV-1 inhibitor (reviewed in ref. *14*). It was recently observed *(226)* that RANTES can be converted by a still unidentified enzyme(s) to a variant lacking the first tree N-terminal residues (4-69 RANTES), which exhibited reduced chemotactic and anti-HIV suppressive activities. However, if the cleavage of RANTES to a 4-69 amino acid molecule, which does not appear to depend on cellular activation *(226)*, has any significance for HIV-1 pathogenesis in vivo, it is currently unknown. In contrast with RANTES, both the chemotactic and anti-HIV-1 activities of SDF-1 are abolished after CD26-mediated cleavage *(223)*.

In addition, HIV-1 infection also induces the secretion of certain metalloproteinases (MMP-2, MMP-9) that are capable of cleaving the N-terminus of SDF-1 but not that of the β-chemokines, and this decreases SDF-1 affinity for CXCR4 binding *(227–229)*. Proteolytic modification of SDF-1 by a serum factor(s) that is different from CD26 and MMPs, which results in reduced anti-HIV activity, has been described as well *(230)*. How SDF-1 proteolytic processing changes during HIV-1 infection is unknown, but it is possible that initially the Tat-induced suppression of CD26 enzymatic activity may increase the fraction of uncleaved SDF-1 that will favor the propagation of R5 strains. Later, during HIV-1 infection, SDF-1 production probably decreases as a result of the progressive degradation of the lymph nodes architecture *(231)*, and SDF-1 cleavage by MMPs and reduction of its anti-HIV-1 properties may further facilitate the appearance of X4 strains. It was also suggested that the HIV-1 Tat protein may interact with CXCR4 and inhibit X4 viral strains *(232,233)*. There is also evidence that Tat may influence CCR5 and to a lesser extent CXCR4 expression *(234)*, but this observation was not confirmed in a later study *(232)*. Tat also suppresses the secretion of MIP-1α and MIP-1β *(235)* but may increase the production of MCP-1 *(236)* and certain proinflammatory cytokines, thus enhancing HIV-1 replication *(237,238)*.

Finally, complex formation between β-chemokines and soluble GAG (an event occurring naturally in $CD8^+$ cytolytic granules) increases β-chemokine anti-HIV-1 effects while reducing their signaling efficiency *(239,240)*. Cell surface proteoglycans are also important for the initial attachment of β-chemokines *(198)* and SDF-1α *(241)* and can potentiate their ability to suppress HIV-1 entry. It is probable that the HIV-1 induced metalloproteinases mentioned above may also affect the chemokine-GAG interactions *(229)*, but this possibility has not been investigated. The fact that chemokines typically act in the presence of other proinflammatory cytokines that may augment virus replication should also be taken into consideration when evaluating their effect on HIV-1 infection, especially in vivo *(215,242,243)*.

Surprisingly, the influence of HIV-1 on CCR5 and CXCR4 expression during the course of infection is still not well defined. A summary of several in vitro HIV-1 infection studies suggests that CCR5 expression is decreased, whereas CXCR4 levels are not affected or may increase (reviewed in ref. *244*). However, an analysis of $CD4^+$ cells from HIV-1 infected patients revealed higher levels of CCR5 and reduced CXCR4 levels in comparison with uninfected control individuals *(245)*. In different groups of HIV-1 infected people, the upregulation of CCR5 expression on $CD4^+$ lymphocytes was confirmed while no diminution of CXCR4 expression was observed *(246)*. Two recent studies on HIV-1–positive individuals with concomitant tuberculosis infection showed increased levels of β-chemokines but with either increased *(247)* or reduced *(248)* CCR5 and CXCR4 expression levels. Elevated β-chemokine levels and/or gp120 may cause CCR5 downregulation, and this effect may be prevalent in vitro *(244)*. However, the activation of the immune system may lead to enhanced CCR5 expression during HIV-1 infection *(13,245)*. Taken together, these data particularly highlight the point that observations made in vitro may not always be predictive of findings observed in HIV-1 infected individuals.

13.8. Conclusions

The predilection for mutation that HIV-1 possesses along with its ability to create long-lasting and varied viral reservoirs have so far been insurmountable obstacles for devising a highly effective prevention or eradication modality for HIV-1 infection. The ever-expanding quantity of more detailed knowledge concerning viral Env structures and their interaction with receptors has the potential for facilitating the identification of new and effective vaccine immunogens and/or entry inhibitors. Yet, a better understanding of the more conserved host cellular factors involved in HIV-1 infection and pathogenesis should also be beneficial in the development of new antiviral therapies in the future.

Acknowledgments

The views expressed in this manuscript are solely those of the authors, and they do not represent official views or opinions of the U.S. Department of Defense or The Uniformed Services University of the Health Sciences.

References

1. Dimitrov DS, Broder CC. HIV and Membrane Receptors. Austin, TX: Landes Bioscience; 1997.
2. Berger EA, Murphy PM, Farber JM. Chemokine receptors as HIV-1 coreceptors: roles in viral entry, tropism, and disease. Annu Rev Immunol 1999;17: 657–700.
3. Wyatt R, Sodroski J. The HIV-1 envelope glycoproteins: fusogens, antigens, and immunogens. Science 1998;280(5371):1884–1888.
4. Chan DC, Kim PS. HIV entry and its inhibition. Cell 1998;93(5):681–684.
5. Melikyan GB, Markosyan RM, Hemmati H, Delmedico MK, Lambert DM, Cohen FS. Evidence that the transition of HIV-1 gp41 into a six-helix bundle, not the bundle configuration, induces membrane fusion. J Cell Biol 2000;151(2): 413–424.
6. Markosyan RM, Cohen FS, Melikyan GB. HIV-1 envelope proteins complete their folding into six-helix bundles immediately after fusion pore formation. Mol Biol Cell 2003;14(3):926–938.
7. Abrahamyan LG, Mkrtchyan SR, Binley J, Lu M, Melikyan GB, Cohen FS. The cytoplasmic tail slows the folding of human immunodeficiency virus type 1 Env from a late prebundle configuration into the six-helix bundle. J Virol 2005;79(1):106–115.
8. Hunter E. Viral entry and receptors. In: Coffin SH, Hughes SH, Varmus HE, eds. Retroviruses. New York: Cold Spring Harbor Laboratory Press; 1997:71–119.
9. Kwong PD, Wyatt R, Robinson J, Sweet RW, Sodroski J, Hendrickson WA. Structure of an HIV gp120 envelope glycoprotein in complex with the CD4 receptor and a neutralizing human antibody. Nature 1998;393(6686):648–659.

10. Moulard M, Decroly E. Maturation of HIV envelope glycoprotein precursors by cellular endoproteases. Biochim Biophys Acta 2000;1469(3):121–132.
11. Rizzuto CD, Wyatt R, Hernandez-Ramos N, et al. A conserved HIV gp120 glycoprotein structure involved in chemokine receptor binding [see comments]. Science 1998;280(5371):1949–1953.
12. Xiang SH, Doka N, Choudhary RK, Sodroski J, Robinson JE. Characterization of CD4-induced epitopes on the HIV type 1 gp120 envelope glycoprotein recognized by neutralizing human monoclonal antibodies. AIDS Res Hum Retroviruses 2002;18(16):1207–1217.
13. Moore JP, Kitchen SG, Pugach P, Zack JA. The CCR5 and CXCR4 coreceptors—central to understanding the transmission and pathogenesis of human immunodeficiency virus type 1 infection. AIDS Res Hum Retroviruses 2004; 20(1):111–126.
14. Princen K, Schols D. HIV chemokine receptor inhibitors as novel anti-HIV drugs. Cytokine Growth Factor Rev 2005;16(6):659–677.
15. Unutmaz D, Littman DR. Expression pattern of HIV-1 coreceptors on T cells: implications for viral transmission and lymphocyte homing. Proc Natl Acad Sci U S A 1997;94:1615–1618.
16. Gray L, Sterjovski J, Churchill M, et al. Uncoupling coreceptor usage of human immunodeficiency virus type 1 (HIV-1) from macrophage tropism reveals biological properties of CCR5-restricted HIV-1 isolates from patients with acquired immunodeficiency syndrome. Virology 2005;337(2):384–398.
17. Berger EA, Doms RW, Fenyo EM, et al. A new classification for HIV-1. Nature 1998;391(6664):240.
18. Reeves J, Doms R. The role of chemokine receptors in HIV infection of host cells. In: Bradshaw R, Dennis E, eds. Handbook of Cell Signaling: New York: Academic Press; 2004:191–196.
19. Yi Y, Rana S, Turner JD, Gaddis N, Collman RG. CXCR-4 is expressed by primary macrophages and supports CCR5- independent infection by dual-tropic but not T-tropic isolates of human immunodeficiency virus type 1. J Virol 1998;72(1):772–777.
20. Dimitrov DS, Norwood D, Stantchev TS, Feng Y, Xiao X, Broder CC. A mechanism of resistance to HIV-1 entry: inefficient interactions of CXCR4 with CD4 and gp120 in macrophages. Virology 1999;259(1):1–6.
21. Lapham CK, Zaitseva MB, Lee S, Romanstseva T, Golding H. Fusion of monocytes and macrophages with HIV-1 correlates with biochemical properties of CXCR4 and CCR5. Nat Med 1999;5(3):303–308.
22. Collman RG, Yi Y, Liu QH, Freedman BD. Chemokine signaling and HIV-1 fusion mediated by macrophage CXCR4: implications for target cell tropism. J Leukoc Biol 2000;68(3):318–323.
23. Schmidtmayerova H, Alfano M, Nuovo G, Bukrinsky M. Human immunodeficiency virus type 1 T-lymphotropic strains enter macrophages via a CD4- and CXCR4-mediated pathway: replication is restricted at a postentry level. J Virol 1998;72(6):4633–4642.

24. Neil S, Martin F, Ikeda Y, Collins M. Postentry restriction to human immunodeficiency virus-based vector transduction in human monocytes. J Virol 2001;75(12):5448–5456.
25. Eisert V, Kreutz M, Becker K, et al. Analysis of cellular factors influencing the replication of human immunodeficiency virus type I in human macrophages derived from blood of different healthy donors. Virology 2001;286(1):31–44.
26. Weissman D, Rabin RL, Arthos J, et al. Macrophage-tropic HIV and SIV envelope proteins induce a signal through the CCR5 chemokine receptor. Nature 1997;389(6654):981–985.
27. Singh A, Yi Y, Isaacs SN, Kolson DL, Collman RG. Concordant utilization of macrophage entry coreceptors by related variants within an HIV type 1 primary isolate viral swarm. AIDS Res Hum Retroviruses 2001;17(10):957–963.
28. Singh A, Besson G, Mobasher A, Collman RG. Patterns of chemokine receptor fusion cofactor utilization by human immunodeficiency virus type 1 variants from the lungs and blood. J Virol 1999;73(8):6680–6690.
29. Lee B, Sharron M, Blanpain C, et al. Epitope mapping of CCR5 reveals multiple conformational states and distinct but overlapping structures involved in chemokine and coreceptor function. J Biol Chem 1999;274(14):9617–9626.
30. Baribaud F, Edwards TG, Sharron M, et al. Antigenically distinct conformations of CXCR4. J Virol 2001;75(19):8957–8967.
31. Lapham CK, Zaitseva MB, Lee S, Romanstseva T, Golding H. Fusion of monocytes and macrophages with HIV-1 correlates with biochemical properties of CXCR4 and CCR5 [published erratum appears in Nat Med 1999;5(5):590]. Nat Med 1999;5(3):303–308.
32. Sloane AJ, Raso V, Dimitrov DS, et al. Marked structural and functional heterogeneity in CXCR4: separation of HIV-1 and SDF-1alpha responses. Immunol Cell Biol 2005;83(2):129–143.
33. Zaitseva M, Romantseva T, Manischewitz J, Wang J, Goucher D, Golding H. Increased CXCR4-dependent HIV-1 fusion in activated T cells: role of CD4/CXCR4 association. J Leukoc Biol 2005;78(6):1306–1317.
34. O'Brien SJ, Moore JP. The effect of genetic variation in chemokines and their receptors on HIV transmission and progression to AIDS. Immunol Rev 2000;177:99–111.
35. Clapham PR, McKnight A. Cell surface receptors, virus entry and tropism of primate lentiviruses. J Gen Virol 2002;83(Pt 8):1809–1829.
36. Hartley O, Klasse PJ, Sattentau QJ, Moore JP. V3: HIV's switch-hitter. AIDS Res Hum Retroviruses 2005;21(2):171–189.
37. Bomsel M, David V. Mucosal gatekeepers: selecting HIV viruses for early infection. Nat Med 2002;8(2):114–116.
38. Meng G, Wei X, Wu X, et al. Primary intestinal epithelial cells selectively transfer R5 HIV-1 to $CCR5^+$ cells. Nat Med 2002;8(2):150–156.
39. Smith PD, Meng G, Salazar-Gonzalez JF, Shaw GM. Macrophage HIV-1 infection and the gastrointestinal tract reservoir. J Leukoc Biol 2003;74(5):642–649.

40. Prakash M, Patterson S, Gotch F, Kapembwa MS. Ex vivo analysis of HIV-1 co-receptors at the endocervical mucosa of women using oral contraceptives. BJOG 2004;111(12):1468–1470.
41. Frank I, Pope M. The enigma of dendritic cell-immunodeficiency virus interplay. Curr Mol Med 2002;2(3):229–248.
42. Teleshova N, Frank I, Pope M. Immunodeficiency virus exploitation of dendritic cells in the early steps of infection. J Leukoc Biol 2003;74(5):683–690.
43. Agace WW, Amara A, Roberts AI, et al. Constitutive expression of stromal derived factor-1 by mucosal epithelia and its role in HIV transmission and propagation. Curr Biol 2000;10(6):325–328.
44. Harouse JM, Gettie A, Tan RC, Blanchard J, Cheng-Mayer C. Distinct pathogenic sequela in rhesus macaques infected with CCR5 or CXCR4 utilizing SHIVs. Science 1999;284(5415):816–819.
45. Veazey RS, Lackner AA. HIV swiftly guts the immune system. Nat Med 2005;11(5):469–470.
46. Brenchley JM, Schacker TW, Ruff LE, et al. $CD4^+$ T cell depletion during all stages of HIV disease occurs predominantly in the gastrointestinal tract. J Exp Med 2004;200(6):749–759.
47. Schutten M, van Baalen CA, Guillon C, et al. Macrophage tropism of human immunodeficiency virus type 1 facilitates in vivo escape from cytotoxic T-lymphocyte pressure. J Virol 2001;75(6):2706–2709.
48. Harouse JM, Buckner C, Gettie A, et al. $CD8^+$ T cell-mediated CXC chemokine receptor 4-simian/human immunodeficiency virus suppression in dually infected rhesus macaques. Proc Natl Acad Sci U S A 2003;100(19):10977–10982.
49. Gorry PR, Churchill M, Crowe SM, Cunningham AL, Gabuzda D. Pathogenesis of macrophage tropic HIV-1. Curr HIV Res 2005;3(1):53–60.
50. Regoes RR, Bonhoeffer S. The HIV coreceptor switch: a population dynamical perspective. Trends Microbiol 2005;13(6):269–277.
51. Pastore C, Ramos A, Mosier DE. Intrinsic obstacles to human immunodeficiency virus type 1 coreceptor switching. J Virol 2004;78(14):7565–7574.
52. Pastore C, Nedellec R, Ramos A, Pontow S, Ratner L, Mosier DE. Human immunodeficiency virus type 1 coreceptor switching: V1/V2 gain-of-fitness mutations compensate for V3 loss-of-fitness mutations. J Virol 2006;80(2):750–758.
53. Karlsson I, Grivel JC, Chen SS, et al. Differential pathogenesis of primary CCR5-using human immunodeficiency virus type 1 isolates in ex vivo human lymphoid tissue. J Virol 2005;79(17):11151–11160.
54. Karlsson I, Antonsson L, Shi Y, et al. Coevolution of RANTES sensitivity and mode of CCR5 receptor use by human immunodeficiency virus type 1 of the R5 phenotype. J Virol 2004;78(21):11807–11815.
55. Marozsan AJ, Kuhmann SE, Morgan T, et al. Generation and properties of a human immunodeficiency virus type 1 isolate resistant to the small molecule CCR5 inhibitor, SCH-417690 (SCH-D). Virology 2005;338(1):182–199.
56. Kwong PD. Human immunodeficiency virus: refolding the envelope. Nature 2005;433(7028):815–816.

57. Chen B, Vogan EM, Gong H, Skehel JJ, Wiley DC, Harrison SC. Structure of an unliganded simian immunodeficiency virus gp120 core. Nature 2005;433(7028): 834–841.
58. Pan Y, Ma B, Nussinov R. CD4 binding partially locks the bridging sheet in gp120 but leaves the beta2/3 strands flexible. J Mol Biol 2005;350(3):514–527.
59. Cocchi F, DeVico AL, Garzino-Demo A, Cara A, Gallo RC, Lusso P. The V3 domain of the HIV-1 gp120 envelope glycoprotein is critical for chemokine-mediated blockade of infection. Nat Med 1996;2(11):1244–1247.
60. Zolla-Pazner S. Identifying epitopes of HIV-1 that induce protective antibodies. Nat Rev Immunol 2004;4(3):199–210.
61. Huang CC, Tang M, Zhang MY, et al. Structure of a V3-containing HIV-1 gp120 core. Science 2005;310(5750):1025–1028.
62. Saunders CJ, McCaffrey RA, Zharkikh I, et al. The V1, V2, and V3 regions of the human immunodeficiency virus type 1 envelope differentially affect the viral phenotype in an isolate-dependent manner. J Virol 2005;79(14):9069–9080.
63. Walter BL, Wehrly K, Swanstrom R, Platt E, Kabat D, Chesebro B. Role of low CD4 levels in the influence of human immunodeficiency virus type 1 envelope V1 and V2 regions on entry and spread in macrophages. J Virol 2005;79(8):4828–4837.
64. Ryser HJ, Fluckiger R. Progress in targeting HIV-1 entry. Drug Discov Today 2005;10(16):1085–1094.
65. Ou W, Silver J. Role of protein disulfide isomerase and other thiol-reactive proteins in HIV-1 envelope protein-mediated fusion. Virology 2006;350(2):406–417.
66. Resch W, Hoffman N, Swanstrom R. Improved success of phenotype prediction of the human immunodeficiency virus type 1 from envelope variable loop 3 sequence using neural networks. Virology 2001;288(1):51–62.
67. Edwards TG, Hoffman TL, Baribaud F, et al. Relationships between CD4 independence, neutralization sensitivity, and exposure of a CD4-induced epitope in a human immunodeficiency virus type 1 envelope protein. J Virol 2001;75(11): 5230–5239.
68. Hoffman TL, LaBranche CC, Zhang W, et al. Stable exposure of the coreceptor-binding site in a CD4-independent HIV-1 envelope protein. Proc Natl Acad Sci U S A 1999;96(11):6359–6364.
69. Kolchinsky P, Kiprilov E, Sodroski J. Increased neutralization sensitivity of CD4-independent human immunodeficiency virus variants. J Virol 2001;75(5):2041–2050.
70. Edwards TG, Wyss S, Reeves JD, et al. Truncation of the cytoplasmic domain induces exposure of conserved regions in the ectodomain of human immunodeficiency virus type 1 envelope protein. J Virol 2002;76(6):2683–2691.
71. Thomas ER, Shotton C, Weiss RA, Clapham PR, McKnight A. CD4-dependent and CD4-independent HIV-2: consequences for neutralization. AIDS 2003;17(3): 291–300.

72. LaBranche CC, Hoffman TL, Romano J, et al. Determinants of CD4 independence for a human immunodeficiency virus type 1 variant map outside regions required for coreceptor specificity J Virol 1999;73(12):10310–10319.
73. Zhang PF, Bouma P, Park EJ, et al. A variable region 3 (V3) mutation determines a global neutralization phenotype and CD4-independent infectivity of a human immunodeficiency virus type 1 envelope associated with a broadly cross-reactive, primary virus-neutralizing antibody response. J Virol 2002;76(2):644–655.
74. Wyss S, Dimitrov AS, Baribaud F, Edwards TG, Blumenthal R, Hoxie JA. Regulation of human immunodeficiency virus type 1 envelope glycoprotein fusion by a membrane-interactive domain in the gp41 cytoplasmic tail. J Virol 2005; 79(19):12231–12241.
75. Blanpain C, Lee B, Vakili J, et al. Extracellular cysteines of CCR5 are required for chemokine binding, but dispensable for HIV-1 coreceptor activity. J Biol Chem 1999;274(27):18902–18908.
76. Reeves JD, Miamidian JL, Biscone MJ, et al. Impact of mutations in the coreceptor binding site on human immunodeficiency virus type 1 fusion, infection, and entry inhibitor sensitivity. J Virol 2004;78(10):5476–5485.
77. Bannert N, Craig S, Farzan M, et al. Sialylated O-glycans and sulfated tyrosines in the NH2-terminal domain of CC chemokine receptor 5 contribute to high affinity binding of chemokines. J Exp Med 2001;194(11):1661–1673.
78. Farzan M, Mirzabekov T, Kolchinsky P, et al. Tyrosine sulfation of the amino terminus of CCR5 facilitates HIV-1 entry. Cell 1999;96(5):667–676.
79. de Parseval A, Bobardt MD, Chatterji A, et al. A highly conserved arginine in gp120 governs HIV-1 binding to both syndecans and CCR5 via sulfated motifs. J Biol Chem 2005;280(47):39493–39504.
80. Platt EJ, Kuhmann SE, Rose PP, Kabat D. Adaptive mutations in the V3 loop of gp120 enhance fusogenicity of human immunodeficiency virus type 1 and enable use of a CCR5 coreceptor that lacks the amino-terminal sulfated region. J Virol 2001;75(24):12266–12278.
81. Crump MP, Gong JH, Loetscher P, et al. Solution structure and basis for functional activity of stromal cell-derived factor-1; dissociation of CXCR4 activation from binding and inhibition of HIV-1. Embo J 1997;16(23):6996–7007.
82. Heveker N, Montes M, Germeroth L, et al. Dissociation of the signalling and antiviral properties of SDF-1-derived small peptides. Curr Biol 1998;8(7):369–376.
83. Doranz BJ, Orsini MJ, Turner JD, et al. Identification of CXCR4 domains that support coreceptor and chemokine receptor functions. J Virol 1999;73(4):2752–2761.
84. Chabot DJ, Zhang PF, Quinnan GV, Broder CC. Mutagenesis of CXCR4 identifies important domains for human immunodeficiency virus type 1 X4 isolate envelope-mediated membrane fusion and virus entry and reveals cryptic coreceptor activity for R5 isolates. J Virol 1999;73(8):6598–6609.
85. Farzan M, Babcock GJ, Vasilieva N, et al. The role of post-translational modifications of the CXCR4 amino terminus in stromal-derived factor 1 alpha association and HIV-1 entry. J Biol Chem 2002;277(33):29484–29489.

86. Brelot A, Heveker N, Montes M, Alizon M. Identification of residues of CXCR4 critical for human immunodeficiency virus coreceptor and chemokine receptor activities. J Biol Chem 2000;275(31):23736–23744.
87. Yi Y, Shaheen F, Collman RG. Preferential use of CXCR4 by R5X4 human immunodeficiency virus type 1 isolates for infection of primary lymphocytes. J Virol 2005;79(3):1480–1486.
88. Lu Z, Berson JF, Chen Y, et al. Evolution of HIV-1 coreceptor usage through interactions with distinct CCR5 and CXCR4 domains. Proc Natl Acad Sci U S A 1997;94(12):6426–6431.
89. Yi Y, Singh A, Shaheen F, Louden A, Lee C, Collman RG. Contrasting use of CCR5 structural determinants by R5 and R5X4 variants within a human immunodeficiency virus type 1 primary isolate quasispecies. J Virol 2003;77(22): 12057–12066.
90. Chabot DJ, Chen H, Dimitrov DS, Broder CC. N-Linked glycosylation of CXCR4 Masks coreceptor function for CCR5-dependent human immunodeficiency virus type 1 isolates. J Virol 2000;74(9):4404–4413.
91. Kuhmann SE, Platt EJ, Kozak SL, Kabat D. Cooperation of multiple CCR5 coreceptors is required for infections by human immunodeficiency virus type 1. J Virol 2000;74(15):7005–7015.
92. Doms RW. Beyond receptor expression: the influence of receptor conformation, density, and affinity in HIV-1 infection. Virology 2000;276(2):229–237.
93. Moldovan MC, Yachou A, Levesque K, et al. CD4 dimers constitute the functional component required for T cell activation. J Immunol 2002;169(11): 6261–6268.
94. Mellado M, Rodriguez-Frade JM, Manes S, Martinez AC. Chemokine signaling and functional responses: the role of receptor dimerization and TK pathway activation. Annu Rev Immunol 2001;19:397–421.
95. Goldsmith MA, Doms RW. HIV entry: are all receptors created equal? Nat Immunol 2002;3(8):709–710.
96. Matthias LJ, Yam PT, Jiang XM, et al. Disulfide exchange in domain 2 of CD4 is required for entry of HIV-1. Nat Immunol 2002;3(8):727–732.
97. Vila-Coro AJ, Mellado M, Martin de Ana A, et al. HIV-1 infection through the CCR5 receptor is blocked by receptor dimerization. Proc Natl Acad Sci U S A 2000;97(7):3388–3393.
98. Xiao X, Wu L, Stantchev TS, et al. Constitutive cell surface association between CD4 and CCR5. Proc Natl Acad Sci U S A 1999;96(13):7496–7501.
99. Lee S, Lapham CK, Chen H, et al. Coreceptor competition for association with CD4 may change the susceptibility of human cells to infection with T-tropic and macrophagetropic isolates of human immunodeficiency virus type 1. J Virol 2000;74(11):5016–5023.
100. Staudinger R, Phogat SK, Xiao X, Wang X, Dimitrov DS, Zolla-Pazner S. Evidence for CD4-enchanced signaling through the chemokine receptor CCR5. J Biol Chem 2003;278(12):10389–10392.
101. Wang X, Staudinger R. Interaction of soluble CD4 with the chemokine receptor CCR5. Biochem Biophys Res Commun 2003;307(4):1066–1069.

102. Mashikian MV, Ryan TC, Seman A, Brazer W, Center DM, Cruikshank WW. Reciprocal desensitization of CCR5 and CD4 is mediated by IL-16 and macrophage-inflammatory protein-1 beta, respectively. J Immunol 1999;163(6):3123–3130.
103. Lynch EA, Heijens CA, Horst NF, Center DM, Cruikshank WW. Cutting edge: IL-16/CD4 preferentially induces Th1 cell migration: requirement of CCR5. J Immunol 2003;171(10):4965–4968.
104. Toth PT, Ren D, Miller RJ. Regulation of CXCR4 receptor dimerization by the chemokine SDF-1alpha and the HIV-1 coat protein gp120: a fluorescence resonance energy transfer (FRET) study. J Pharmacol Exp Ther 2004;310(1): 8–17.
105. Piguet V, Sattentau Q. Dangerous liaisons at the virological synapse. J Clin Invest 2004;114(5):605–610.
106. Jolly C, Sattentau QJ. Retroviral spread by induction of virological synapses. Traffic 2004;5(9):643–650.
107. Jolly C, Kashefi K, Hollinshead M, Sattentau QJ. HIV-1 cell to cell transfer across an Env-induced, actin-dependent synapse. J Exp Med 2004;199(2):283–293.
108. Jolly C, Sattentau QJ. Human immunodeficiency virus type 1 virological synapse formation in T cells requires lipid raft integrity. J Virol 2005;79(18): 12088–12094.
109. Lekkerkerker AN, van Kooyk Y, Geijtenbeek TB. Viral piracy: HIV-1 targets dendritic cells for transmission. Curr HIV Res 2006;4(2):169–176.
110. Geijtenbeek TB, Kwon DS, Torensma R, et al. DC-SIGN, a dendritic cell-specific HIV-1-binding protein that enhances trans-infection of T cells. Cell 2000;100(5): 587–597.
111. Garcia E, Pion M, Pelchen-Matthews A, et al. HIV-1 trafficking to the dendritic cell-T-cell infectious synapse uses a pathway of tetraspanin sorting to the immunological synapse. Traffic 2005;6(6):488–501.
112. Arrighi JF, Pion M, Garcia E, et al. DC-SIGN-mediated infectious synapse formation enhances X4 HIV-1 transmission from dendritic cells to T cells. J Exp Med 2004;200(10):1279–1288.
113. Marsh M, Helenius A. Virus entry: open sesame. Cell 2006;124(4):729–40.
114. Iyengar S, Schwartz DH. How do cell-free HIV virions avoid infecting dead-end host cells and cell fragments? AIDS Rev 2004;6(3):155–160.
115. Freedman BD, Liu QH, Del Corno M, Collman RG. HIV-1 gp120 chemokine receptor-mediated signaling in human macrophages. Immunol Res 2003; 27(2–3):261–276.
116. Stantchev TS, Broder CC. Human immunodeficiency virus type-1 and chemokines: beyond competition for common cellular receptors. Cytokine Growth Factor Rev 2001;12(2–3):219–243.
117. Kim CH, Qu CK, Hangoc G, et al. Abnormal chemokine-induced responses of immature and mature hematopoietic cells from motheaten mice implicate the protein tyrosine phosphatase SHP-1 in chemokine responses. J Exp Med 1999; 190(5):681–690.

118. Vila-Coro AJ, Rodriguez-Frade JM, Martin De Ana A, Moreno-Ortiz MC, Martinez AC, Mellado M. The chemokine SDF-1alpha triggers CXCR4 receptor dimerization and activates the JAK/STAT pathway. FASEB J 1999;13(13): 1699–1710.
119. Ganju RK, Brubaker SA, Chernock RD, Avraham S, Groopman JE. Beta-chemokine receptor CCR5 signals through SHP1, SHP2, and Syk. J Biol Chem 2000;275(23):17263–17268.
120. Ganju RK, Brubaker SA, Meyer J, et al. The alpha-chemokine, stromal cell-derived factor-1alpha, binds to the transmembrane G-protein-coupled CXCR-4 receptor and activates multiple signal transduction pathways. J Biol Chem 1998;273(36):23169–23175.
121. Sotsios Y, Whittaker GC, Westwick J, Ward SG. The CXC chemokine stromal cell-derived factor activates a Gi-coupled phosphoinositide 3-kinase in T lymphocytes. J Immunol 1999;163(11):5954–5963.
122. Knall C, Worthen GS, Johnson GL. Interleukin 8-stimulated phosphatidylinositol-3-kinase activity regulates the migration of human neutrophils independent of extracellular signal-regulated kinase and p38 mitogen-activated protein kinases. Proc Natl Acad Sci U S A 1997;94(7):3052–3057.
123. Pierce KL, Premont RT, Lefkowitz RJ. Seven-transmembrane receptors. Nat Rev Mol Cell Biol 2002;3(9):639–650.
124. Hall RA, Premont RT, Lefkowitz RJ. Heptahelical receptor signaling: beyond the G protein paradigm. J Cell Biol 1999;145(5):927–932.
125. Veillette A, Bookman MA, Horak EM, Samelson LE, Bolen JB. Signal transduction through the CD4 receptor involves the activation of the internal membrane tyrosine-protein kinase p56lck. Nature 1989;338(6212):257–259.
126. Graziani-Bowering G, Filion LG, Thibault P, Kozlowski M. CD4 is active as a signaling molecule on the human monocytic cell line Thp-1. Exp Cell Res 2002;279(1):141–152.
127. Popik W, Pitha PM. Exploitation of cellular signaling by HIV-1: unwelcome guests with master keys that signal their entry. Virology 2000;276(1): 1–6.
128. Liu QH, Williams DA, McManus C, et al. HIV-1 gp120 and chemokines activate ion channels in primary macrophages through CCR5 and CXCR4 stimulation. Proc Natl Acad Sci U S A 2000;97(9):4832–4837.
129. Cicala C, Arthos J, Ruiz M, et al. Induction of phosphorylation and intracellular association of CC chemokine receptor 5 and focal adhesion kinase in primary human $CD4^+$ T cells by macrophage-tropic HIV envelope. J Immunol 1999;163(1):420–426.
130. Davis CB, Dikic I, Unutmaz D, et al. Signal transduction due to HIV-1 envelope interactions with chemokine receptors CXCR4 or CCR5. J Exp Med 1997; 186(10):1793–1798.
131. Arthos J, Rubbert A, Rabin RL, et al. CCR5 signal transduction in macrophages by human immunodeficiency virus and simian immunodeficiency virus envelopes. J Virol 2000;74(14):6418–6424.

132. Tomkowicz B, Lee C, Ravyn V, Cheung R, Ptasznik A, Collman R. The Src kinase Lyn is required for CCR5 signaling in response to MIP-1β and R5 HIV-1 gp120 in human macrophages. Blood 2006;108(4):1145–1150.
133. Del Corno M, Liu QH, Schols D, et al. HIV-1 gp120 and chemokine activation of Pyk2 and mitogen-activated protein kinases in primary macrophages mediated by calcium-dependent, pertussis toxin-insensitive chemokine receptor signaling. Blood 2001;98(10):2909–2916.
134. Lee C, Liu QH, Tomkowicz B, Yi Y, Freedman BD, Collman RG. Macrophage activation through CCR5- and CXCR4-mediated gp120-elicited signaling pathways. J Leukoc Biol 2003;74(5):676–682.
135. Lee C, Tomkowicz B, Freedman BD, Collman RG. HIV-1 gp120-induced TNF-α production by primary human macrophages is mediated by phosphatidylinositol-3 (PI-3) kinase and mitogen-activated protein (MAP) kinase pathways. J Leukoc Biol 2005;78(4):1016–1023.
136. Lin CL, Sewell AK, Gao GF, Whelan KT, Phillips RE, Austyn JM. Macrophage-tropic HIV induces and exploits dendritic cell chemotaxis. J Exp Med 2000;192(4):587–594.
137. Iyengar S, Schwartz DH, Hildreth JE. T cell-tropic HIV gp120 mediates CD4 and CD8 cell chemotaxis through CXCR4 independent of CD4: implications for HIV pathogenesis. J Immunol 1999;162(10):6263–6267.
138. Misse D, Cerutti M, Noraz N, et al. A CD4-independent interaction of human immunodeficiency virus-1 gp120 with CXCR4 induces their cointernalization, cell signaling, and T-cell chemotaxis. Blood 1999;93(8):2454–2462.
139. Cicala C, Arthos J, Rubbert A, et al. HIV-1 envelope induces activation of caspase-3 and cleavage of focal adhesion kinase in primary human CD4(+) T cells. Proc Natl Acad Sci U S A 2000;97(3):1178–1183.
140. Corasaniti MT, Bagetta G, Rotiroti D, Nistico G. The HIV envelope protein gp120 in the nervous system: interactions with nitric oxide, interleukin-1β and nerve growth factor signalling, with pathological implications in vivo and in vitro. Biochem Pharmacol 1998;56(2):153–156.
141. Russo R, Navarra M, Rotiroti D, Di Renzo G. Evidence for a role of protein tyrosine kinases in cell death induced by gp120 in CHP100 neuroblastoma cells. Toxicol Lett 2003;139(2–3):207–211.
142. Cicala C, Arthos J, Selig SM, et al. HIV envelope induces a cascade of cell signals in non-proliferating target cells that favor virus replication. Proc Natl Acad Sci U S A 2002;99(14):9380–9385.
143. Arthos J, Cicala C, Selig SM, et al. The role of the CD4 receptor versus HIV coreceptors in envelope-mediated apoptosis in peripheral blood mononuclear cells. Virology 2002;292(1):98–106.
144. Kinter AL, Umscheid CA, Arthos J, et al. HIV envelope induces virus expression from resting $CD4^+$ T cells isolated from HIV-infected individuals in the absence of markers of cellular activation or apoptosis. J Immunol 2003;170(5):2449–2455.

145. Alkhatib G, Locati M, Kennedy PE, Murphy PM, Berger EA. HIV-1 coreceptor activity of CCR5 and its inhibition by chemokines: independence from G protein signaling and importance of coreceptor downmodulation. Virology 1997; 234(2):340–348.
146. Bedinger P, Moriarty A, II RCvB, Donovan NJ, Steimer KS, Littman DR. Internalization of the human immunodeficiency virus does not require the cytoplasmic domain of CD4. Nature 1988;334:162–165.
147. Atchison RE, Gosling J, Monteclaro FS, et al. Multiple extracellular elements of CCR5 and HIV-1 entry: dissociation from response to chemokines. Science 1996;274:1924–1926.
148. Amara A, Vidy A, Boulla G, et al. G protein-dependent CCR5 signaling is not required for efficient infection of primary T lymphocytes and macrophages by R5 human immunodeficiency virus type 1 isolates. J Virol 2003;77(4): 2550–2558.
149. Gosling J, Monteclaro FS, Atchison RE, et al. Molecular uncoupling of C-C chemokine receptor 5-induced chemotaxis and signal transduction from HIV-1 coreceptor activity. Proc Natl Acad Sci U S A 1997;94(10):5061–5066.
150. Farzan M, Choe H, Martin KA, et al. HIV-1 entry and macrophage inflammatory protein-1beta-mediated signaling are independent functions of the chemokine receptor CCR5. J Biol Chem 1997;272(11):6854–6857.
151. Cohen DI, Tani Y, Tian H, Boone E, Samelson LE, Lane HC. Participation of tyrosine phosphorylation in the cytopathic effect of human immunodeficiency virus-1. Science 1992;256(5056):542–545.
152. Yoshida H, Koga Y, Moroi Y, Kimura G, Nomoto K. The effect of p56lck, a lymphocyte specific protein tyrosine kinase, on the syncytium formation induced by human immunodeficiency virus envelope glycoprotein. Int Immunol 1992;4(2):233–242.
153. Alfano M, Schmidtmayerova H, Amella CA, Pushkarsky T, Bukrinsky M. The B-oligomer of pertussis toxin deactivates CC chemokine receptor 5 and blocks entry of M-tropic HIV-1 strains. J Exp Med 1999;190(5):597–605.
154. Alfano M, Vallanti G, Biswas P, et al. The binding subunit of pertussis toxin inhibits HIV replication in human macrophages and virus expression in chronically infected promonocytic U1 cells. J Immunol 2001;166(3):1863–1870.
155. Guntermann C, Murphy BJ, Zheng R, Qureshi A, Eagles PA, Nye KE. Human immunodeficiency virus-1 infection requires pertussis toxin sensitive G-protein-coupled signalling and mediates cAMP downregulation. Biochem Biophys Res Commun 1999;256(2):429–435.
156. Lapenta C, Spada M, Santini SM, et al. Pertussis toxin B-oligomer inhibits HIV infection and replication in hu-PBL-SCID mice. Int Immunol 2005;17(4):469–475.
157. Alfano M, Grivel JC, Ghezzi S, et al. Pertussis toxin B-oligomer dissociates T cell activation and HIV replication in CD4 T cells released from infected lymphoid tissue. AIDS 2005;19(10):1007–1014.

158. Wong WS, Rosoff PM. Pharmacology of pertussis toxin B-oligomer. Can J Physiol Pharmacol 1996;74(5):559–564.
159. Li H, Wong WS. Pertussis toxin activates tyrosine kinase signaling cascade in myelomonocytic cells: a mechanism for cell adhesion. Biochem Biophys Res Commun 2001;283(5):1077–1082.
160. Wang JM, Oppenheim JJ. Interference with the signaling capacity of CC chemokine receptor 5 can compromise its role as an HIV-1 entry coreceptor in primary T lymphocytes. J Exp Med 1999;190(5):591–595.
161. Delon J, Bercovici N, Liblau R, Trautmann A. Imaging antigen recognition by naive $CD4^+$ T cells: compulsory cytoskeletal alterations for the triggering of an intracellular calcium response. Eur J Immunol 1998;28(2):716–729.
162. Trautmann A, Randriamampita C. Initiation of TCR signalling revisited. Trends Immunol 2003;24(8):425–428.
163. Manes S, del Real G, Martinez AC. Pathogens: raft hijackers. Nat Rev Immunol 2003;3(7):557–568.
164. Simons K, Toomre D. Lipid rafts and signal transduction. Nat Rev 2000; 1:31–39.
165. Viola A, Schroeder S, Sakakibara Y, Lanzavecchia A. T lymphocyte costimulation mediated by reorganization of membrane microdomains. Science 1999; 283(5402):680–682.
166. Fujio Y, Yamada F, Takahashi K, Shibata N. Responses of smooth muscle cells to platelet-derived growth factor are inhibited by herbimycin-A tyrosine kinase inhibitor. Biochem Biophys Res Commun 1993;195(1):79–83.
167. Chrzanowska-Wodnicka M, Burridge K. Tyrosine phosphorylation is involved in reorganization of the actin cytoskeleton in response to serum or LPA stimulation. J Cell Sci 1994;107(Pt 12):3643–3654.
168. Suter DM, Forscher P. Transmission of growth cone traction force through apCAM-cytoskeletal linkages is regulated by Src family tyrosine kinase activity. J Cell Biol 2001;155(3):427–438.
169. Hirshman CA, Zhu D, Panettieri RA, Emala CW. Actin depolymerization via the beta-adrenoceptor in airway smooth muscle cells: a novel PKA-independent pathway. Am J Physiol Cell Physiol 2001;281(5):C1468–1476.
170. Ridley AJ. Rho GTPases and cell migration. J Cell Sci 2001;114(Pt 15): 2713–2722.
171. Kuwahara M. Involvement of Rho and tyrosine kinase in angiotensin II-induced actin reorganization in mesothelial cells. Eur J Pharmacol 2002;436(1–2): 15–21.
172. Murasawa S, Matsubara H, Mori Y, et al. Angiotensin II initiates tyrosine kinase Pyk2-dependent signalings leading to activation of Rac1-mediated c-Jun NH2-terminal kinase. J Biol Chem 2000;275(35):26856–26863.
173. Sakurada S, Okamoto H, Takuwa N, Sugimoto N, Takuwa Y. Rho activation in excitatory agonist-stimulated vascular smooth muscle. Am J Physiol Cell Physiol 2001;281(2):C571–578.

174. Pontow SE, Heyden NV, Wei S, Ratner L. Actin cytoskeletal reorganizations and coreceptor-mediated activation of rac during human immunodeficiency virus-induced cell fusion. J Virol 2004;78(13):7138–7147.
175. del Real G, Jimenez-Baranda S, Mira E, et al. Statins inhibit HIV-1 infection by down-regulating Rho activity. J Exp Med 2004;200(4):541–547.
176. Popik W, Alce TM. CD4 receptor localized to non-raft membrane microdomains supports HIV-1 entry. Identification of a novel raft localization marker in CD4. J Biol Chem 2004;279(1):704–712.
177. Viard M, Parolini I, Sargiacomo M, et al. Role of cholesterol in human immunodeficiency virus type 1 envelope protein-mediated fusion with host cells. J Virol 2002;76(22):11584–11595.
178. Singer, II, Scott S, Kawka DW, et al. CCR5, CXCR4, and CD4 are clustered and closely apposed on microvilli of human macrophages and T cells. J Virol 2001;75(8):3779–3790.
179. Iyengar S, Hildreth JE, Schwartz DH. Actin-dependent receptor colocalization required for human immunodeficiency virus entry into host cells. J Virol 1998;72(6):5251–5255.
180. Marsh M, Bron R. SFV infection in CHO cells: cell-type specific restrictions to productive virus entry at the cell surface. J Cell Sci 1997;110 (Pt 1):95–103.
181. Pelkmans L, Helenius A. Insider information: what viruses tell us about endocytosis. Curr Opin Cell Biol 2003;15(4):414–422.
182. Komano J, Miyauchi K, Matsuda Z, Yamamoto N. Inhibiting the Arp2/3 complex limits infection of both intracellular mature vaccinia virus and primate lentiviruses. Mol Biol Cell 2004;15(12):5197–5207.
183. Stevenson M. Portals of entry: uncovering HIV nuclear transport pathways. Trends Cell Biol 1996;6(1):9–15.
184. Bukrinskaya A, Brichacek B, Mann A, Stevenson M. Establishment of a functional human immunodeficiency virus type 1 (HIV-1) reverse transcription complex involves the cytoskeleton. J Exp Med 1998;188(11):2113–2125.
185. Rawat SS, Eaton J, Gallo SA, et al. Functional expression of CD4, CXCR4, and CCR5 in glycosphingolipid-deficient mouse melanoma GM95 cells and susceptibility to HIV-1 envelope glycoprotein-triggered membrane fusion. Virology 2004;318(1):55–65.
186. Campbell EM, Nunez R, Hope TJ. Disruption of the actin cytoskeleton can complement the ability of Nef to enhance human immunodeficiency virus type 1 infectivity. J Virol 2004;78(11):5745–5755.
187. Irish JM, Hovland R, Krutzik PO, et al. Single cell profiling of potentiated phospho-protein networks in cancer cells. Cell 2004;118(2):217–228.
188. Pendaries C, Tronchere H, Plantavid M, Payrastre B. Phosphoinositide signaling disorders in human diseases. FEBS Lett 2003;546(1):25–31.
189. Kallergi G, Tsapara A, Kampa M, et al. Distinct signaling pathways regulate differential opioid effects on actin cytoskeleton in malignant MCF7 and nonmalignant MCF12A human breast epithelial cells. Exp Cell Res 2003;288(1): 94–109.

190. Rao J, Li N. Microfilament actin remodeling as a potential target for cancer drug development. Curr Cancer Drug Targets 2004;4(4):345–354.
191. Kwiatkowski DJ. Functions of gelsolin: motility, signaling, apoptosis, cancer. Curr Opin Cell Biol 1999;11(1):103–108.
192. Tremblay MJ, Fortin JF, Cantin R. The acquisition of host-encoded proteins by nascent HIV-1. Immunol Today 1998;19(8):346–351.
193. Tardif MR, Tremblay MJ. Presence of host ICAM-1 in human immunodeficiency virus type 1 virions increases productive infection of $CD4^+$ T lymphocytes by favoring cytosolic delivery of viral material. J Virol 2003;77(22):12299–12309.
194. Liao Z, Roos JW, Hildreth JE. Increased infectivity of HIV type 1 particles bound to cell surface and solid-phase ICAM-1 and VCAM-1 through acquired adhesion molecules LFA-1 and VLA-4. AIDS Res Hum Retroviruses 2000;16(4): 355–366.
195. Tardif MR, Tremblay MJ. LFA-1 is a key determinant for preferential infection of memory $CD4^+$ T cells by human immunodeficiency virus type 1. J Virol 2005;79(21):13714–13724.
196. Cocchi F, DeVico AL, Garzino-Demo A, Arya SK, Gallo RC, Lusso P. Identification of RANTES, MIP-1 alpha, and MIP-1 beta as the major HIV-suppressive factors produced by $CD8^+$ T cells. Science 1995;270(5243):1811–1815.
197. Deng H, Liu R, Ellmeier W, et al. Identification of a major co-receptor for primary isolates of HIV-1 [see comments]. Nature 1996;381(6584):661–666.
198. Oravecz T, Pall M, Wang J, Roderiquez G, Ditto M, Norcross MA. Regulation of anti-HIV-1 activity of RANTES by heparan sulfate proteoglycans. J Immunol 1997;159(9):4587–4592.
199. Bleul CC, Farzan M, Choe H, et al. The lymphocyte chemoattractant SDF-1 is a ligand for LESTR/fusin and blocks HIV-1 entry. Nature 1996;382(6594): 829–833.
200. Oberlin E, Amara A, Bachelerie F, et al. The CXC chemokine SDF-1 is the ligand for LESTR/fusin and prevents infection by T-cell-line-adapted HIV-1. Nature 1996;382(6594):833–835.
201. Amara A, Gall SL, Schwartz O, et al. HIV coreceptor downregulation as antiviral principle: SDF-1alpha-dependent internalization of the chemokine receptor CXCR4 contributes to inhibition of HIV replication. J Exp Med 1997;186(1):139–146.
202. Lacey SF, McDanal CB, Horuk R, Greenberg ML. The CXC chemokine stromal cell-derived factor 1 is not responsible for CD8+ T cell suppression of syncytia-inducing strains of HIV-1. Proc Natl Acad Sci U S A 1997;94(18):9842–9847.
203. Scarlatti G, Tresoldi E, Bjorndal A, et al. In vivo evolution of HIV-1 co-receptor usage and sensitivity to chemokine-mediated suppression. Nat Med 1997;3(11): 1259–1265.
204. Trkola A, Paxton WA, Monard SP, et al. Genetic subtype-independent inhibition of human immunodeficiency virus type 1 replication by CC and CXC chemokines. J Virol 1998;72(1):396–404.
205. Kinter AL, Ostrowski M, Goletti D, et al. HIV replication in $CD4^+$ T cells of HIV-infected individuals is regulated by a balance between the viral suppressive

effects of endogenous beta-chemokines and the viral inductive effects of other endogenous cytokines. Proc Natl Acad Sci U S A 1996;93(24):14076–14081.
206. Granelli-Piperno A, Moser B, Pope M, et al. Efficient interaction of HIV-1 with purified dendritic cells via multiple chemokine coreceptors. J Exp Med 1996;184:2433–2438.
207. Verani A, Scarlatti G, Comar M, et al. C-C chemokines released by lipopolysaccharide (LPS)-stimulated human macrophages suppress HIV-1 infection in both macrophages and T cells. J Exp Med 1997;185(5):805–816.
208. Stantchev TS, Broder CC. Consistent and significant inhibition of human immunodeficiency virus type 1 envelope-mediated membrane fusion by beta-chemokines (RANTES) in primary human macrophages. J Infect Dis 2000;182(1):68–78.
209. Ayehunie S, Garcia-Zepeda EA, Hoxie JA, et al. Human immunodeficiency virus-1 entry into purified blood dendritic cells through CC and CXC chemokine coreceptors. Blood 1997;90(4):1379–1386.
210. Zaitseva MB, Lee S, Rabin RL, et al. CXCR4 and CCR5 on human thymocytes: biological function and role in HIV-1 infection. J Immunol 1998;161(6):3103–3113.
211. Tchou I, Misery L, Sabido O, et al. Functional HIV CXCR4 coreceptor on human epithelial Langerhans cells and infection by HIV strain X4. J Leukoc Biol 2001;70(2):313–321.
212. Gonzalez E, Kulkarni H, Bolivar H, et al. The influence of CCL3L1 gene-containing segmental duplications on HIV-1/AIDS susceptibility. Science 2005;307(5714):1434–1440.
213. Menten P, Struyf S, Schutyser E, et al. The LD78beta isoform of MIP-1alpha is the most potent CCR5 agonist and HIV-1-inhibiting chemokine. J Clin Invest 1999;104(4):R1–5.
214. Aquaro S, Menten P, Struyf S, et al. The LD78beta isoform of MIP-1alpha is the most potent CC-chemokine in inhibiting CCR5-dependent human immunodeficiency virus type 1 replication in human macrophages. J Virol 2001;75(9):4402–4406.
215. Kinter A, Arthos J, Cicala C, Fauci AS. Chemokines, cytokines and HIV: a complex network of interactions that influence HIV pathogenesis. Immunol Rev 2000;177:88–98.
216. Fantuzzi L, Belardelli F, Gessani S. Monocyte/macrophage-derived CC chemokines and their modulation by HIV-1 and cytokines: a complex network of interactions influencing viral replication and AIDS pathogenesis. J Leukoc Biol 2003;74(5):719–725.
217. Grainger DJ, Lever AM. Blockade of chemokine-induced signalling inhibits CCR5-dependent HIV infection in vitro without blocking gp120/CCR5 interaction. Retrovirology 2005;2(1):23.
218. Appay V, Rowland-Jones SL. RANTES: a versatile and controversial chemokine. Trends Immunol 2001;22(2):83–87.
219. Verani A, Lusso P. Chemokines as natural HIV antagonists. Curr Mol Med 2002;2(8):691–702.

220. Hogan CM, Hammer SM. Host determinants in HIV infection and disease. Part 2: genetic factors and implications for antiretroviral therapeutics. Ann Intern Med 2001;134(10):978–996.
221. Anastassopoulou CG, Kostrikis LG. The impact of human allelic variation on HIV-1 disease. Curr HIV Res 2003;1(2):185–203.
222. Vidal F, Peraire J, Domingo P, et al. Polymorphism of RANTES chemokine gene promoter is not associated with long-term nonprogressive HIV-1 infection of more than 16 years. J Acquir Immune Defic Syndr 2006;41(1):17–22.
223. Ohtsuki T, Tsuda H, Morimoto C. Good or evil: CD26 and HIV infection. J Dermatol Sci 2000;22(3):152–160.
224. De Meester I, Korom S, Van Damme J, Scharpe S. CD26, let it cut or cut it down. Immunol Today 1999;20(8):367–375.
225. Augustyns K, Bal G, Thonus G, et al. The unique properties of dipeptidyl-peptidase IV (DPP IV / CD26) and the therapeutic potential of DPP IV inhibitors. Curr Med Chem 1999;6(4):311–327.
226. Lim JK, Burns JM, Lu W, DeVico AL. Multiple pathways of amino terminal processing produce two truncated variants of RANTES/CCL5. J Leukoc Biol 2005;78(2):442–452.
227. Elkington PT, O'Kane CM, Friedland JS. The paradox of matrix metalloproteinases in infectious disease. Clin Exp Immunol 2005;142(1):12–20.
228. Van Damme J, Struyf S, Opdenakker G. Chemokine-protease interactions in cancer. Semin Cancer Biol 2004;14(3):201–208.
229. Comerford I, Nibbs RJ. Post-translational control of chemokines: a role for decoy receptors? Immunol Lett 2005;96(2):163–174.
230. Villalba S, Salvucci O, Aoki Y, et al. Serum inactivation contributes to the failure of stromal-derived factor-1 to block HIV-I infection in vivo. J Leukoc Biol 2003;74(5):880–888.
231. Michael NL, Moore JP. HIV-1 entry inhibitors: evading the issue [news]. Nat Med 1999;5(7):740–742.
232. Xiao H, Neuveut C, Tiffany HL, et al. Selective CXCR4 antagonism by tat: implications for in vivo expansion of coreceptor use by HIV-1. Proc Natl Acad Sci U S A 2000;97(21):11466–11471.
233. Ghezzi S, Noonan DM, Aluigi MG, et al. Inhibition of CXCR4-dependent HIV-1 infection by extracellular HIV-1 Tat. Biochem Biophys Res Commun 2000;270(3): 992–996.
234. Huang L, Bosch I, Hofmann W, Sodroski J, Pardee AB. Tat protein induces human immunodeficiency virus type 1 (HIV-1) coreceptors and promotes infection with both macrophage-tropic and T-lymphotropic HIV-1 strains. J Virol 1998;72(11):8952–8960.
235. Zagury D, Lachgar A, Chams V, et al. Interferon alpha and Tat involvement in the immunosuppression of uninfected T cells and C-C chemokine decline in AIDS. Proc Natl Acad Sci U S A 1998;95(7):3851–3856.
236. Conant K, Garzino-Demo A, Nath A, et al. Induction of monocyte chemoattractant protein-1 in HIV-1 Tat-stimulated astrocytes and elevation in AIDS dementia. Proc Natl Acad Sci U S A 1998;95(6):3117–3121.

237. Vicenzi E, Alfano M, Ghezzi S, et al. Divergent regulation of HIV-1 replication in PBMC of infected individuals by CC chemokines: suppression by RANTES, MIP-1alpha, and MCP-3, and enhancement by MCP-1. J Leukoc Biol 2000; 68(3):405–412.
238. Lafrenie RM, Wahl LM, Epstein JS, Yamada KM, Dhawan S. Activation of monocytes by HIV-Tat treatment is mediated by cytokine expression. J Immunol 1997;159(8):4077–4083.
239. Burns JM, Lewis GK, DeVico AL. Soluble complexes of regulated upon activation, normal T cells expressed and secreted (RANTES) and glycosaminoglycans suppress HIV-1 infection but do not induce Ca^{2+} signaling. Proc Natl Acad Sci U S A 1999;96(25):14499–14504.
240. Wagner L, Yang OO, Garcia-Zepeda EA, et al. Beta-chemokines are released from HIV-1-specific cytolytic T-cell granules complexed to proteoglycans. Nature 1998;391(6670):908–911.
241. Valenzuela-Fernandez A, Palanche T, Amara A, et al. Optimal inhibition of X4 HIV isolates by the CXC chemokine stromal cell-derived factor 1 alpha requires interaction with cell surface heparan sulfate proteoglycans. J Biol Chem 2001;276(28):26550–26558.
242. Fauci AS. Host factors and the pathogenesis of HIV-induced disease. Nature 1996;384:529–534.
243. Copeland KF. Modulation of HIV-1 transcription by cytokines and chemokines. Mini Rev Med Chem 2005;5(12):1093–1101.
244. Ruibal-Ares BH, Belmonte L, Bare PC, Parodi CM, Massud I, de Bracco MM. HIV-1 infection and chemokine receptor modulation. Curr HIV Res 2004;2(1): 39–50.
245. Ostrowski MA, Justement SJ, Catanzaro A, et al. Expression of chemokine receptors CXCR4 and CCR5 in HIV-1-infected and uninfected individuals. J Immunol 1998;161(6):3195–3201.
246. Oishi K, Hayano M, Yoshimine H, et al. Expression of chemokine receptors on $CD4^+$ T cells in peripheral blood from HIV-infected individuals in Uganda. J Interferon Cytokine Res 2000;20(6):597–602.
247. Wolday D, Tegbaru B, Kassu A, et al. Expression of chemokine receptors CCR5 and CXCR4 on $CD4^+$ T cells and plasma chemokine levels during treatment of active tuberculosis in HIV-1-coinfected patients. J Acquir Immune Defic Syndr 2005;39(3):265–271.
248. Hanna LE, Bose JC, Nayak K, Subramanyam S, Swaminathan S. Short communication: influence of active tuberculosis on chemokine and chemokine receptor expression in HIV-infected persons. AIDS Res Hum Retroviruses 2005;21(12): 997–1002.

14

Chemokines and Their Receptors in Fibrosis

Glenda Trujillo and Cory M. Hogaboam

Summary

Tissue fibrosis, which results in the destruction of normal organ function, is a leading cause of morbidity and mortality. Current strategies for treating fibrosis have been unsuccessful, largely because of the difficulty in distinguishing whether inflammatory or fibrogenic events sustain the progression of the disease. The causes of fibrosis are diverse regardless of the tissue involved, and the common features include the sequential recruitment of inflammatory cells, overproliferation of matrix-producing cells, and the overproduction of extracellular matrix. An excessive wound-healing response presumably represents disruption in this sequence thereby leading to a disturbance in the balance between tissue remodeling, matrix degradation, and permanent scarring. The mechanisms involved in pulmonary fibrosis also represent three sequential events characterized by an initial insult, inflammation, and tissue repair. Central to the progression of these events is the balance between a T helper 1 (Th1) and a T helper 2 (Th2) environment, in which Th2-specific signals have been shown to be immunomodulatory and profibrotic. However, the release of these Th2-specific cytokines and chemokines by both inflammatory and resident cells maintains the fibrotic response, consequently leading to fibrotic disease. Evidence from animal models and human studies have identified a number of Th1/Th2-associated chemokines and chemokine receptors as profibrotic or antifibrotic. Therefore, investigating the chemokines, chemokine receptors, and the cells that they impact is an attractive approach to identifying therapeutic targets in fibrosis.

Key Words: Fibrosis; chemokines; chemokine receptors; inflammation; wound healing; Th1; Th2.

From: *The Receptors: The Chemokine Receptors*
Edited by: J. K. Harrison and N. W. Lukacs © Humana Press Inc., Totowa, NJ

14.1. Introduction

Fibrosis is a progressive, pathologic consequence of a wound-healing process gone awry. It stems from a tightly regulated sequence of events, initially orchestrated to resolve damaged tissue, through the deposition of new extracellular matrix (ECM). Normally, there is a balance between matrix formation and degradation *(1,2)*. However, this balance is altered in fibrosis, where a gradual expansion of the scar tissue (fibrotic mass) leads to the eventual impairment of the involved tissue and organs *(1–3)*. The pathogenesis of a fibrotic disorder is similar regardless of the tissue and involves recruitment and proliferation of matrix-producing and inflammatory cells, with subsequent overproduction and accumulation of ECM *(1,3–5)*. Although the exact molecular mechanism leading to fibrosis is not yet clear, evidence from both in vitro and in vivo studies strongly implicates chemokines and their cognate receptors in the propagation of this process. This chapter will focus on the role of chemokines (released by inflammatory cells and resident tissue cells) and the differential expression of chemokine receptors that affect recruitment, activation, and proliferation of inflammatory cells in fibrotic disease. Specifically, we concentrate on the role of chemokines in pulmonary fibrosis; however, the mechanisms can just as readily be applied to the development of hepatic, renal, dermal, or myocardial fibrosis.

14.2. Fibrosis: Is It Linked to Inflammation?

The primary causes of fibrosis are diverse and include toxic vapors, inorganic dusts, drugs, and radiation *(4,6)*. Physical or chemical injuries and immunologic disorders can lead to cutaneous fibrosis such as keloids, hypertrophic scars, and scleroderma (systemic sclerosis) *(4)*. Alcohol and viral infections are major causes of hepatic fibrosis, and glomerulonephritis, diabetic mellitus, and hypertension are major causes of renal scarring *(4,6)*. Diffuse cardiac fibrosis is one of the major complications of hypertension usually associated with progressive heart failure *(7)*. A number of drugs such as bleomycin, cisplatin, cyclosporine, and gentamicin can also induce fibrosis of the lung and kidney and have been instrumental in inducing fibrosis in certain experimental models *(7,8)*.

Regardless of the cause of tissue injury, an inflammatory response immediately ensues that is followed by a complex, highly regulated, dynamic wound-healing process. However, certain types of injury can lead to a dysregulated inflammatory and wound-healing response resulting in formation of permanent scar tissue. The fibrotic process involves the replacement of functional tissue with excessive fibrous tissue in an attempt to maintain tissue integrity after repeated cycles of injury, inflammation, and repair *(4,8,9)*. Unlike normal wound healing, no true resolution occurs in fibrosis. Instead, fibroblast proliferation and matrix synthesis persist, thereby resulting in progressive destruction of the normal parenchyma *(4,6)*. In some cases, fibrosis ultimately results in organ failure and death. Fibrotic disease is therefore often defined as a failed wound-healing process after chronic, sustained injury *(4,8–10)*.

Inflammatory disorders usually do not result in fibrosis, but fibrotic responses are thought to almost always be preceded by, and potentially propagated by, chronic inflammation *(4,10,11)*. Chronic inflammation is defined as a reaction that persists for several weeks or months where inflammation, tissue destruction, and repair processes may all occur simultaneously *(4,6)*. This form of inflammation is characterized by a large infiltrate of mononuclear cells, which includes macrophages, lymphocytes, plasma cells, and sometimes eosinophils *(4,12)*. These cells are often recruited by chemokines generated at the site of inflammation. Furthermore, the release of specific profibrotic cytokines and chemokines can also recruit T cells that in turn can activate fibroblasts to produce collagen *(4,13–15)*. Thus, when repeated injury occurs, chronic inflammation and repair can cause an excessive accumulation of ECM components and lead to the formation of a permanent scar, a hallmark of fibrotic disease *(2,4,16,17)*. This process is a consequence of complex interactions among the fibroblasts, cytokines, growth factors, proteases, and ECM proteins, which function to amplify the process. However, the exact mechanism underlying the fibrotic process is not completely understood.

Recently, the notion that the chronicity of inflammation may not actually drive the fibrogenic process has been widely appreciated (Tables 1, 2, and 3). Some propose that it is indeed the alteration of the mesenchymal cell phenotypes that disrupts the balance between collagen synthesis and degradation in the wound-healing process, highlighted by clinical evidence that shows unsuccessful treatment of fibrosis with anti-inflammatory or immunosuppressive drugs *(18,19)*. One scenario is that mesenchymal cells (myofibroblasts and fibroblasts) are phenotypically altered and thus do not undergo apoptosis after resolution.

Table 1
Hypothesis for the Pathogenesis of Fibrotic Disease

1. Chronic inflammation is a leading component and contributing event to the pathogenesis of fibrotic disease, where a normal inflammatory response to injury becomes a chronic, pathologic wound-healing response.
2. Fibrotic disease results from epithelial injury and abnormal wound repair in the absence of preceding inflammation.
3. Inflammation is a normal response to injury and, with persistent antigen or injury, results in the polarization of a profibrotic microenvironment where resident cells and inflammatory cells promote matrix deposition and not degradation.

Table 2
Chemokine Receptors and Ligands in Fibrosis

Chemokine receptor	Chemokine ligand(s)	Cell expression	Profibrotic	Antifibrotic
CCR2	CCL2/MCP-1	Macrophage, immature DC, basophil, fibroblast, T, NK, endothelial	X	
CCR3	CCL11/eotaxin	Eosinophil, basophil, platelet, mast, Th2, airway epithelial	X	
CCR4	CCL17/TARC CCL22 CKLF1	DC, basophil Th2, Treg, platelet	X	
CCR5	CCL3/MIP-1α CCL8/MCP-2 CCL5/RANTES CCL7	DC, macrophage, Th1, NK		X
CCR6	CCL20	Immature DC, T, B	X	
CCR7	CCL21 SLC	DC, fibroblast, T, B	X	
CXCR3	CXCL10/IP-10 CXCL9/MIG CXCL11/I-TAC	Th1, B, smooth muscle, mesangial, microglia		X
CXCR4	CXCL12/SDF-1	DC, platelet, neutrophil, T, B, macrophage, astrocyte	X	

DC, dendritic cell; NK, natural killer; Th1, T helper 1; Th2, T helper 2; Treg, T regulatory.

Table 3
Th1 and Th2 Cytokines in Fibrosis

Th1 antifibrotic	Th2 antifibrotic
IFN-γ	IL-4
IL-2	IL-5
IL-12	IL-10
IL-18	IL-13
TNF-β	CCL2/MCP-1

Increased ECM production and reduced ECM degradation persists and drives fibrotic disease. A "profibrogenic microenvironment" is maintained by the release of profibrogenic factors (TGF-β, platelet-derived growth factor (PDGF), IL-1) by the altered mesenchymal cells and resident epithelial cells *(18,20)*. Altered structural cells and ECM are readily capable of supplying growth factors to sustain the fibrogenic process, independent of inflammatory cells *(18)*. This dichotomy is illustrated in tendonitis ("tennis elbow"), which is now more commonly known as tendinosis. Histopathologic studies demonstrate that patients with tendinosis have small numbers of inflammatory cells in the presence of large populations of mesenchymal cells (fibroblasts and endothelial cells), in addition to increased hyperplasia, and disorganized collagen deposition *(21)*. Interestingly, although tendinosis occurs as a result of repetitive microtrauma to the tendon, evidence of chronic inflammation is undetectable *(21)*. Essentially, this example illustrates that the inflammation commonly observed in fibrosis possibly arises from conditions defined by the injured tissue itself and the subsequent change in the tissue microenvironment *(18)*.

Moreover, a genetic predisposition to fibrosis is also a possible cause, further questioning the role of chronic inflammation in fibrotic disease. Profibrotic, phenotypic alterations observed in fibroblasts suggest that abnormal wound repair may occur in the absence of preceding inflammation *(11,22)*. Overexpression of active TGF-β in lung epithelial cells causes progressive fibrosis without apparent inflammation *(23)*. Only after fibrosis has been established (by other mechanisms) for a prolonged duration is there this accumulation of inflammatory cells (mast cells), possibly through the expression of transmembrane stem cells factor (SCF) by myofibroblasts, release of chemokines (monocyte chemoattractant protein-1 (MCP-1)/CCL2) by resident fibroblasts, or expression of IL-15 in the fibrotic tissue *(24–26)*.

Recently, the pulmonary effects of the proinflammatory cytokine IL-1β were investigated to further delineate the relation between inflammation and fibrosis *(27)*. In contrast with TNF-α, overexpression of IL-β caused severe acute inflammation, which was associated with tissue destruction and subsequent progressive lung fibrosis *(27)*. The fibrotic remodeling that was observed was associated with a persistent upregulation of endogenous TGF-β, strongly suggesting that progressive fibrosis is potentially more related to an impairment of the repair process and less to chronic inflammation *(27)*. The *Smad* signaling pathway is believed to be the major signaling mechanism through which active TGF-β stimulates the induction of profibrotic genes *(27,28)*. Interestingly, *Smad3*-deficient mice are resistant to development of lung fibrosis but not to inflammation caused by bleomycin *(27)*. This evidence demonstrates that IL-1β promotes pulmonary fibrosis not through the inflammatory component but rather through induction of TGF-β *(27,28)*.

Regardless of the exact role of inflammation in the process, wound healing in fibrosis (or fibroplasia) consists of two events: recruitment and proliferation of fibroblasts at the site of injury and deposition of excessive ECM by these cells. The resulting scar tissue is observed in varying degrees in essentially every type of chronic inflammatory disease. Several cellular pathways, including immune cell activation, have been identified as the major avenues for the generation of the matrix-producing cells in diseased conditions. Although severe acute (nonrepetitive) injuries can cause substantial tissue remodeling, fibrosis that is associated with chronic (repetitive) injury is unique in that the adaptive immune response is thought to have an important role *(4,14,15)*.

14.2.1. Cells Involved in the Fibrotic Process

Although inflammation typically precedes fibrosis, the amount of scarring is not necessarily linked with the severity of the inflammatory response, suggesting that the mechanisms that regulate fibrogenesis are distinct from those that regulate inflammation *(1,4,29)*. Most of the fibrotic diseases have in common a persistent inflammatory stimulus and lymphocyte-monocyte interactions that sustain the production of growth factors, proteolytic enzymes, and fibrogenic cytokines and chemokines, which together promote the deposition of connective tissue elements *(1,29)*. Fibrosis is a characteristic feature in the pathogenesis of a wide spectrum of diseases and is a leading cause of morbidity and mortality in numerous disorders *(2,4,17)*. Therefore, in order to develop successful treatment for fibrotic disease, it is imperative to understand the upstream mediators involved in the regulation of the fibrotic mechanism.

A relatively newly described immune cell, the fibrocyte, has been implicated in fibrotic disease, as well as in normal wound repair, granuloma formation, and antigen presentation *(3,5,30–33)*. Evidence from patients with hypertrophic scars such as keloids, and those affected by scleroderma and other fibrosing disorders, have demonstrated fibrocytes in their lesions *(3,30)*. Recently, a newly appreciated disease, nephrogenic fibrosing dermopathy (NFD), has been described in humans. NFD is a rare fibrotic skin condition that affects patients with renal disease, and the fibrocyte may play an important role in disease development *(34,35)*. Initially identified as a blood-borne cell, fibrocytes also have been localized to various tissues in both normal and pathologic conditions *(3,5,32,33)*. Morphologically, they are distinguishable from fibroblasts and leukocytes. Fibrocytes can further be characterized by their expression patterns of extracellular markers, cytokines, chemokines, and growth factors *(3,5,30–33)*. The precise origin of the fibrocyte is still unclear, and a fibrocyte-specific cell surface marker has yet to be identified *(3,30)*. However, fibrocytes are known to express CD45, CD34, CD11b, and/or collagen 1 and to produce matrix proteins such as vimentin, collagens I and III, and are known to participate in the remodeling response by secreting matrix metalloproteinases (MMPs) *(3,5,30–33)*. They are a rich source of inflammatory cytokines, growth factors, and chemokines that provide important intercellular markers typical of an antigen-presenting cell *(5,30,33)*. Moreover, fibrocytes express several chemokine receptors, such as CCR3, CCR5, CCR7, and CXCR4 *(5)*. Secondary lymphoid tissue chemokine (SLC)/CCL21, has been reported to be a key mediator of fibrocyte trafficking, as intradermal instillation of recombinant SLC results in recruitment of fibrocytes to the injection site *(5,30)*.

The link between fibrocytes and myofibroblasts is not clear, although some believe that fibrocytes can differentiate into myofibroblasts *(19,30,33,36,37)*. Early evidence demonstrates that within the wound, fibrocytes differentiate in response to TGF-β and collagen and are signaled to express increased smooth muscle actin (αSMA), a marker of myofibroblasts, possibly through the IL-6 signaling *(30,33,38,39)*. In this context, it is proposed that fibrocytes may represent a systemic source of contractile myofibroblasts that appears in many fibrotic lesions, such as those seen in chronic asthma, and various forms of pulmonary fibrosis *(33,38)*. Indeed, they do appear to play a role in the proliferative phase of wound repair and because they appear early in the wound site, it is suggestive of their role in the inflammatory phase of wound healing *(5)*.

Currently, it is widely accepted that peripheral blood fibrocytes are recruited (by chemokinetic signals) to sites of injury after fibrosis is initiated *(3,5,30–33,40)*. It is proposed that the fibrocyte is derived from the bone marrow and is recruited to the site of fibrosis, via the vasculature, only after a signal has occurred (such as in pulmonary fibrosis) *(36,40)*. This point is strengthened by the finding that bleomycin (BLM) treatment significantly increases chemokines (CXCR4 ligand and SLC/CCL21) specific for the chemokines receptors (CXCR4 and CCR7) expressed by peripheral blood fibrocytes *(36)*. Using a murine BLM-induced model of pulmonary fibrosis, Moore et al. demonstrated that the transition of fibrocyte to fibroblast in vitro is associated with the loss of expression of the chemokine receptor, CCR2, and expression and enhanced production of collagen 1 *(40)*. The origin of the fibrocyte, however, and its role in mediating fibrosis is particularly controversial. Recently, using a murine model of BLM-induced pulmonary fibrosis, Hashimoto et al. reported that the collagen-expressing myofibroblasts found in the fibrotic lung appear to be derived from local, quiescent fibroblasts rather than from the bone marrow *(36)*. These findings, which indicate a phenotypic difference between resident lung fibroblasts and those derived from bone marrow progenitor cells, challenge the theory that blood-borne fibrocytes are precursors for the myofibroblasts found in fibrotic tissue *(30,33,36)*.

Whereas the study by Moore et al. implicates the fibrocyte as having a central, profibrotic (profibrogenic) role in the fibrotic disease, other studies suggest otherwise. Ortiz et al. show that not all bone marrow–derived cells are damaging during tissue repair *(19)*. They demonstrate that murine mesenchymal stem cells migrate to the fibrotic lung, adopt epithelial-like phenotype, and reduce inflammation and collagen deposition in BLM-challenged mice *(19)*. Possibly, this study can support a bone-marrow origin for fibrocytes, if they indeed are part of the milieu of cells derived from the bone marrow engraftment, while also supporting an anti-fibrotic role for fibrocytes during tissue repair *(19)*. A recent study by Choi et al. demonstrates an increase in CCR7 expression in fibrotic lung biopsies that, interestingly, did not correlate with the presence of fibrocytes or myofibroblasts in the focal areas *(37)*. Although the study mainly provides evidence of a pivotal role for CCR7 in idiopathic interstitial pneumonia (IIP), it also challenges the profibrotic role of the fibrocyte proposed previously. The study emphasizes that it is unlikely that the cells expressing elevated levels of CCR7 in IIP are indeed infiltrating fibrocytes or myofibroblasts, affirming that it is the activation of *resident* pulmonary fibroblasts that are profibrotic *(37)*. Notably, CCR7 has been investigated in the context of differentiation of memory T cells, dendritic cell chemotaxis during homeostatic and inflammatory conditions, and trafficking of T cells from blood to lymph nodes *(41–44)*. Thus, identifying signals that either *promote* fibrogenesis or *prevent* an excessive wound-healing response is central in understanding fibrotic disease.

14.3. Chemokines and Their Role in Fibrotic Disease

Chemokines and their receptors mediate numerous effects on both the innate and adaptive arms of the immune system and are among the many factors that regulate the fibrotic process. Various functions have been attributed to chemokines, including recruitment of inflammatory cells (leukocytes), integrin activation, and granulocyte degranulation *(45–48)*. In addition, chemokines are known to modulate cytokine production, adhesion molecule expression, and mononuclear cell proliferation *(13–15)*. Specifically, chemokines play a major role in adaptive immunity by regulating T-cell localization to areas of inflammation and their subsequent participation in the recruitment and activation of other effector cells *(46,48)*. In contrast with neutrophils and eosinophils, which represent a rather nonspecific inflammatory response, T cells are equipped with the capacity to initiate, amplify, and terminate antigen-specific immune responses *(48)*. These functions are carefully synchronized by chemokines and chemokine receptors, expressed by both T cells and other effector cells at sites of fibrosis. Although this chapter is focused on chemokines and chemokine receptors in pulmonary fibrosis, we contend that chemokines are intricately linked to the cytokine environment within several organs including the lung. Several studies, including those presented herein, suggest that chemokines and cytokines cross-regulate each other in fibrotic disease.

In fibrosis, the balance between T helper 1 (Th1)- and T helper 2 (Th2)-specific signals dictates the progression of the disease, in which a Th2 environment has been shown to be profibrotic *(49)*. Data supporting this concept (Fig. 1; see color plate) stems from both in vitro and in vivo studies. The Th1/Th2 balance is controlled by the release and regulation of cytokines (by inflammatory and structural cells at sites of injury) specifically aimed at favoring either a Th1 response or a Th2 response *(4,14)*. Each of the main Th2 cytokines (IL-4, IL-5, IL-13) has a specialized role in the regulation of tissue remodeling and fibrosis *(4,50)*. For example, Th2 cytokines have been shown to regulate MCP-1 (CCL2) activity to favor fibrosis in a variety of tissues, including lung, liver, kidney, and skin *(47,51)*. Specifically, IL-4 has been observed to drive TGF-β synthesis by pulmonary fibroblasts, promoting collagen deposition within the interstitial spaces of the fibrotic lung *(22,52)*.

Findings from studies of schistosomiasis-induced liver fibrosis, as well as other models of pulmonary, kidney, and liver fibrosis, strongly support the role of $CD4^+$ Th2 cells in the progression of fibrosis *(4)*. In this regard, analyses of gene and protein expression after stimulation by Th1 (vs. Th2) cytokines indicates that IL-4 is found at increased concentrations in the bronchoalveolar lavage (BAL) fluid of patients with idiopathic pulmonary fibrosis, as well as in the peripheral blood mononuclear cells of those afflicted with periportal fibrosis *(10,53–56)*.

Moreover, development of fibrotic disease in models using irradiation-induced fibrosis is also associated with increased concentrations of IL-4 *(57)*. Similar effects are observed with IL-13, where overexpression of IL-13 in the lungs of mice induces considerable subepithelial airway fibrosis *(58)*. In addition, induced IL-13–specific antibodies have been found to attenuate collagen deposition in the lungs of animals challenged with *Aspergillus fumigatus* conida or BLM *(59–61)*. Recent evidence has also shown that BLM-induced fibrosis is exacerbated in transgenic mice that overexpress IL-5, another Th2 cytokine that is notable for its effect on eosinophil function *(62–64)*. Thus, it is clear that chemokines drive leukocyte recruitment and activation and cooperate with Th2 (profibrotic) cytokines in the development of fibrosis.

14.3.1. Pulmonary Fibrosis

Pulmonary fibrosis is a potentially fatal condition that is the end result of persistent lung inflammation and can occur in a variety of clinical settings *(1,8,65)*. It can be a manifestation of environmental or occupational exposure, asthma, sarcoidosis, or a consequence of an autoimmune disease such as scleroderma (systemic sclerosis), rheumatoid arthritis, or even polymyositis *(1,7,19,29,65–67)*. Most commonly, however, progressive pulmonary fibrosis can occur without an evident cause or systemic disease and is referred to as idiopathic pulmonary fibrosis (IPF) *(68,69)*. It is the most common of the interstitial pneumonias of unknown etiology and the most aggressive interstitial lung disease *(29)*. The hallmark lesions of IPF are the fibroblast foci, representing focal areas of active fibrogenesis, in which vigorous fibroblast proliferation and ECM deposition leads to restriction of the distal air space *(29,36)*. The fibrotic mass continues to gradually expand, destroying the air sacs surrounding the lung tissues and capillaries, and causing permanent loss of function (i.e., oxygen transport) *(7)*. The level of disability experienced by a person is directly related to the extent of tissue scarring *(7)*. Fibroblasts, fibrocytes, and myofibroblasts are the major sources of the excessive accumulation of matrix proteins that forms the fibrotic mass (scar) *(3,5,22,29,32,37,70)*. These cells are therefore responsible for the derangement of alveolar epithelium, loss of elasticity, and development of a rigid lung *(5,7,22)*. In IPF, the damaged alveolar epithelial cells are believed to initiate the pathologic process by expressing specific cytokines (TGF-β) and growth factors (TNF-α) to promote fibroblast migration, activation, and proliferation *(29)*. This principle has led to a greater focus on the interplay between structural cells and inflammatory cells in the fibrotic pathway *(14,22,29,32,37,67,70,71)*.

By and large, the process and progression of pulmonary fibrosis is composed of three events: initial triggering events, inflammatory events, and fibrotic events *(7)*. Accumulation of neutrophils, eosinophils, lymphocytes, mast cells, monocytes, and alveolar macrophages is a consistent histologic feature of the early stage of various pulmonary fibrotic diseases *(7)*. The release of specific chemokines and cytokines by these inflammatory cells, as well as by the activated resident cells (alveolar epithelial cells), may intensify the inflammatory and fibrotic events in the lung *(72,73)*. The contributions of these chemokines and their cognate receptors, as determined by various experimental studies, will be discussed herein.

14.3.2. Chemokines and Chemokine Receptors in Animal Models of Pulmonary Fibrosis

Murine models have been instrumental in representing human disease in vivo. Transgenic approaches have produced mice with genetic defects that promote susceptibility to fibrosis, including abnormal expression of chemokines and/or chemokine receptors. These models have elucidated the factors that are conducive to fibrosis and have thus revealed the central roles of certain chemokines and chemokine receptors. Conventional methods of inducing experimental pulmonary fibrosis in mice include direct pulmonary instillation of a fibrogenic agent, such as BLM or fluorescein isothiocyanate (FITC), or irradiation of the thoracic cavity *(7,18,74,75)*. The BLM model of pulmonary fibrosis has been the leading model for dissecting the regulation of pulmonary fibrogenesis, particularly because of its remarkable reproducibility as a scaled-down model of human pulmonary fibrosis *(7)*. Induction of experimental pulmonary fibrosis results in the accumulation of inflammatory cells within the interstitial area of the lungs prior to the development of the fibrotic process, both of which resemble the human pathology *(76)*. Evidence from these studies confirms that the transmigration of these cells from the blood into the inflammatory sites is tightly regulated by a variety of chemokines and the differential expression of the associated receptors.

Murine models of pulmonary fibrosis strongly implicate a role for CCL2 (MCP-1) and its receptor, CCR2, in the pathogenesis of the disease *(22,77,78)*. Widely known to be a chemoattractant specific for monocytes, CCL2 has also been shown to contribute to de novo synthesis of type 1 procollagen by signaling lung fibroblasts to synthesize TGF-β, a known inducer of collagen production *(22,69)*. The profibrotic effects of CCL2 have been widely appreciated in fibrosis formation in a variety of organs, such as the lung, kidney, liver, pancreas, and, most recently, the intestine *(77–82)*. CCR2 is expressed on monocytes, as well as on activated T cells, B cells, NK cells, fibroblasts, and mast cells *(22,83)*. Moore et al. have demonstrated previously that profibrotic signaling is diminished in $CCR2^{-/-}$ mice after FITC or BLM treatment *(84)*. These findings suggest that CCR2 signaling is part of a generalized pathway that leads to pulmonary fibrosis *(84)*. Interestingly, this study also demonstrates that similar numbers and types of leukocytes are recruited to the lung in both $CCR2^{+/+}$ and $CCR2^{-/-}$ mice, suggesting that CCR2 may play a significant role in the fibrotic rather than inflammatory phases of the injury. These data support a role for CCR2 beyond that of a chemotactic receptor and suggest that signal transduction via CCR2 plays an important role in cellular activation in the context of fibrosis. In addition, another study has investigated the expression of chemokines and chemokine receptors during the fibrotic phase of injury after thoracic irradiation in fibrosis-sensitive and fibrosis-resistant mice *(75)*. It was reported that chronic expression of specific chemokine and chemokine receptors induced by thoracic irradiation (to induce fibrosis) plays a critical role in the chronic inflammatory response leading to the development of fibrosis *(75)*. Specifically, elevated levels of CCR1, CCR2, CCR5, and CCR6 mRNA were found in fibrosis-sensitive mice and were correlated with the development of disease *(75)*.

The role of CCL2 and CCR2 in the pathogenesis of BLM-induced pulmonary fibrosis is also supported by the work of Okuma et al. using $CCR2^{-/-}$ mice, with a focus on MMP-2 and MMP-9 activity *(85)*. MMPs are necessary for the degradation of ECM components including collagen, gelatin, and proteoglycans and may limit the progression to fibrosis. The basement membrane of the lung epithelium is composed largely of type IV collagen and is digested primarily by MMP-2 and MMP-9 *(20,86,87)*. The major source of MMP-2 and MMP-9 in the lung is the macrophage, although fibroblasts, bronchial epithelial cells, and neutrophils also can act as sources of these proteases *(20,86,87)*. The findings by Okuma et al. demonstrate that pulmonary inflammation and fibrosis are attenuated in $CCR2^{-/-}$ mice, which may correlate with reduced macrophage infiltration and macrophage-derived MMP-2 and MMP-9 production *(85)*. They propose that CCR2 deficiency may thus protect hosts from severe lung injury and respiratory dysfunction in BLM-induced pulmonary fibrosis *(85)*. The evidence emphasizes the balance between matrix deposition (fibrogenesis) and matrix degradation (resolution) may become skewed toward a profibrotic environment and that this regulation is possibly influenced by the expression of chemokines and their receptors.

Another method to evaluate the role of CCR2 in murine disease is to employ a mutant CCL2 that functions as a receptor antagonist. The NH_2-terminal deletion mutant of CCL2 (7ND) binds to CCR2 with comparable affinity and blocks the binding of wild-type CCL2, inhibiting its ability to mediate monocyte chemotaxis *(78,88)*. Transfection of the mutant CCL2 gene into skeletal muscles of mice resulted in attenuation of BLM-induced pulmonary fibrosis *(78)*. Inoshima et al. demonstrate that overexpression of the mutant gene 10 to 14 days after intratracheal instillation of BLM results in decreased pulmonary fibrosis at day 14 *(78)*. Therefore, these findings suggest that the interaction of CCL2 with CCR2 may have an important role in the development of fibrogenesis but not in the development of early lung inflammation because inflammatory cell infiltration begins 3 days after BLM treatment in both the control and transfected animals *(78)*.

CXC chemokines and receptors also have been implicated in the pathogenesis of pulmonary fibrosis. CXCR3 is the receptor for CXC chemokines IFN-γ–induced protein 10-kd (IP-10, CXCL10), monokine induced by IFN-γ (MIG, CXCL9), and IFN-inducible T-cell α-chemoattractant (I-TAC, CXC11). CXCR3 and its ligands act primarily on activated T and NK cells and have been implicated in mediating the effects of IFN-γ. Both CXCR3 and CCR5 are preferentially expressed on Th1 cells, whereas Th2 cells favor the expression of CCR3 and CCR4. CXCR3 ligands that attract Th1 cells can concomitantly block the migration of Th2 cells in response to CCR3 ligands, thus enhancing the polarization of effector T-cell recruitment.

Th1 cells producing IFN-γ and IL-12 have been demonstrated to limit the development of fibrosis, whereas Th2 cells producing IL-4 and IL-13 have been shown to exacerbate lung fibrosis after BLM-induced injury *(4,48,65,89)*. Thus, as discussed previously, polarized T-cell responses have been implicated in the development of tissue fibrosis, where a Th2 environment promotes fibrosis *(22,66)*. By inducing pulmonary fibrosis with BLM in CXCR3-deficient mice, Jiang et al. demonstrate that CXCR3 plays a major, nonredundant role in limiting lung fibrosis in response to noninfectious tissue injury by promoting endogenous production of IFN-γ *(66)*. Moreover, that an immune response dominated by IFN-γ and other Th1 cytokines (IL-12, IL-18) prevents overt fibroblast activation and subsequent pulmonary fibrosis has been widely established *(89–92)*. Recently, IFN-γ–producing natural killer T (NKT) cells have been shown to play an antifibrotic role in the development of pulmonary fibrosis by regulating TGF-β production *(93)*.

Belperio et al., demonstrate elevated levels of the Th2 chemokines CCL17 thymus and activation-regulated chemokine (TARC) and CCL22 and their shared receptor CCR4 (which is highly expressed on macrophages) in the BLM model of pulmonary fibrosis *(94)*. These findings support a central role for the Th2 chemokines and the macrophage in the pathogenesis of pulmonary fibrosis *(94)*. Collectively, this evidence reinforces the concept that Th2 cells are prominent mediators of the chronic inflammation that precedes pulmonary fibrosis. Furthermore, results from these studies also support the profibrotic roles of Th2 cytokines and an antifibrotic role for CXCR3 in pulmonary injury, inflammation, and fibrosis in vivo.

The natural history and sequence of events that dictate the pathogenesis of pulmonary fibrosis is not well characterized, but insight into the chemokine receptor expression patterns may help to elucidate the mechanism. Phillips et al. show that circulating fibrocytes contribute to the pathogenesis of pulmonary fibrosis *(32)*. Using the murine model of BLM-induced pulmonary fibrosis, this group observed a peak period of fibrocyte infiltration 8 days after BLM exposure *(32)*. In addition, they observed a CXCL12 gradient between the lungs and plasma of BLM-treated mice, which was inhibited by the addition of neutralizing anti-CXCL12 *(32)*. Interestingly, complete obliteration of lung fibrosis with treatment of anti-CXCL12 antibodies was not reported, suggesting that other mechanisms may be involved in fibrocyte recruitment during lung fibrosis.

Eosinophil recruitment is often encountered in many fibrotic lesions and may contribute to disease pathogenesis. Cellular sources of the eosinophil chemotactic factor CCL11 (eotaxin-1) are bronchial epithelial cells, T cells, macrophages, and eosinophils. Its biological effects are mediated by binding to its receptor CCR3, which is highly expressed by eosinophils as well as other immune cells *(62,95–97)*. Recently, Huaux et al. have examined the pulmonary expression of eotaxin-1/CC11 and CCR3 during BLM-induced lung injury and determined their importance in the recruitment of inflammatory cells and the development of lung fibrosis *(62)*. They find that BLM treatment induced increased pulmonary expression of CCL11 and CCR3 and that CCR3 is highly expressed by both eosinophils and neutrophils *(62)*. Furthermore, overexpression of CCL11 significantly enhances BLM-induced lung fibrosis and production of profibrotic cytokines with an increase in lung infiltrates of eosinophils and neutrophils, suggesting that CCL11 and CCR3 are important in the pulmonary recruitment of granulocytes and may play a significant role in BLM-induced lung fibrosis *(62)*.

14.4. Evidence from Human Studies

Experimental animal models have been informative to understand the mechanism of fibrosis and appear to replicate some patterns observed in human disease. Data from human subjects relating to chemokines and chemokine receptor expression in fibrosis correlates with that from animal studies. Human lung epithelial cells from patients with idiopathic pulmonary fibrosis strongly express CCL2 mRNA and its protein product, CCL2, in contrast with those cells from healthy patients *(98)*. Schmidt et al. correlate observations from human and animal studies within the same study and show that both models support a role for fibrocytes in pulmonary fibrosis in the context of bronchial asthma *(33)*. In patients with allergen-induced asthma, endobronchial biopsies (that were performed after antigen challenge) show airway fibrosis is associated with the presence of fibrocytes *(33)*. Similar to murine fibroblasts, human fibrocytes isolated from human peripheral blood express CD45, collagen I, and CXCR4 and migrate to the fibrotic lung in response to CXCL12 *(32)*.

Examination of BAL fluid from patients affected by idiopathic pulmonary fibrosis demonstrates increased levels of CCL2, CCL3, and CCL4, whereas CCR5 expression by lymphocytes is significantly reduced *(99)*. Both CCL3 and CCL5 bind CCR5 in activated T cells and alveolar macrophages *(70,99–101)*. IL-4 is a Th2 cytokine that stimulates the increased expression of CCR by fibroblasts. IFN-γ, on the other hand, is a Th1 cytokine that downregulates collagen production by fibroblasts. Because the activation of CCR5 induces an increased production of IFN-γ, the downregulation of this receptor (as observed in BAL fluid from IPF patients) could be correlated with a decrease in IFN-γ thereby promoting a profibrotic environment *(99,101)*. The downregulation of CCR5 expression is observed in lymphocytes and macrophages between the first and third stages of fibrosis, coupled with an increase of IL-4 levels and a decrease in IFN-γ in the BAL fluid of patients with advanced disease (compared with first and second stages) *(99)*. Interestingly, Choi et al. observed increased expression of CCR5 by cultured fibroblasts from surgical biopsies of usual interstitial pneumonia (UIP) but did not evaluate alveolar macrophages and lymphocytes *(71)*. In the same study, increased levels of CCL7 were observed in UIP. The presence of high levels of CCL7 could contribute to the reduced activity of CCR5 and, consequently, to the decreased production of IFN-γ *(55,71,99,100)*.

Recent evidence has proposed a nonredundant role for CXCR3 as an antifibrotic mediator in the BLM mouse model of IPF *(66)*, as well as an important role for its agonists ligands, CXCL11 *(74)* and CXCL10 *(102)*. Accordingly, Pignatti et al. evaluate the effects of CXCR3 and CCR4 on T cells derived from BAL fluid and peripheral blood of patients with IPF and other interstitial lung diseases *(103)*. They report decreased expression of CXCR3 on BAL fluid $CD4^+$ cells from IPF patients, in agreement with the murine studies that demonstrate $CXCR3^{-/-}$ mice show increased mortality with progressive interstitial fibrosis (compared with wild type after BLM-induced fibrosis) *(66,103)*. In addition, they report increased CCR4 expression by $CD4^+$ lymphocytes, in accordance with recent evidence that CCL17 and CCL22 (the agonists of CCR4 receptor) favor the development of pulmonary fibrosis in BLM-treated mice *(94,103)*. Interestingly, a phase II study of biomarker expression in IPF-treated patients by Streiter et al. demonstrates increased expression of CXCL11 but not CXCL9 or CXCL10 in the BAL fluid and plasma of these patients *(104)*. Taken together, these findings suggest that CXCR3 expression could be protective toward the development of fibrosis *(103)*.

14.5. Current Therapy for Treatment

Corticosteroids or immunosuppressive agents have traditionally been used to treat patients with various forms of fibrotic disease, especially those affected by IPF. As previously discussed, the link between chronic inflammation and fibrosis remains controversial, which has driven the exploration for new therapeutic approaches. Recently, innovative attempts have been made to restore Th2 cytokine balance with respect to inflammation. The Th1 cytokine IFN-γ has more recently been widely used therapeutically in fibrotic disease, as it has been found to reduce fibroblast proliferation and collagen deposition in vitro while alleviating progression of fibrosis (such as lung and kidney) in vivo *(49,105,106)*. Strieter et al. have recently reported that IPF patients treated with IFN-γ-1b display a marked reduction in the synthesis of profibrotic molecules, such as elastin, procollagen I and III, and PDGF, as well as a decrease in the proangiogenic chemokine CXCL5 *(104)*.

Inhibition of IFN-γ entails inhibition of fibroblast proliferation and differentiation, subsequent collagen synthesis, and increased expression of MMP-1 to promote degradation of matrix *(105)*. IFN-γ also triggers robust T-lymphocyte recruitment to sites of inflammation by upregulating I-TAC/CXCL11 activity, which has been recently shown to act as a natural antagonist for CCR5 *(107)*. I-TAC dramatically reduces CCR5-mediated activities (chemotaxis) by blocking the binding of MIP-1β and potentially other promiscuous ligands of CCR5 *(107)*.

Therapy that targets inflammation, although marginally successful in humans with various fibrotic diseases, has been promising in experimental models *(68)*. Recently, a novel IκB kinase-β inhibitor, IMD-0354, has been shown to attenuate BLM-induced pulmonary fibrosis in mice *(108)*. IκB kinase-β functions as a critical regulator in the activation of NFκB, a transcription factor activated in response to inflammatory cytokines such as IL-1β and TNF-α *(108)*. Thus, inhibiting a downstream mediator of inflammation appears to be an attractive approach to ameliorating the progression of fibrosis. Moreover, other strategies for reducing the progression of fibrosis focus on targeting resident cells within the afflicted tissue. With respect to IPF, the alveolar epithelial cells demonstrate that they can be manipulated to regenerate in response to exogenous epithelial mitogens (such as keratinocyte growth factor; KGF) *(105)*. Approaches to regenerate and repair the alveolar epithelium in this case promotes resolution instead of fibrosis. As previously discussed, fibroblasts and myofibroblasts are important regulators of wound healing and thus attractive targets for slowing the progression of fibrosis. In this context, strategies to inhibit fibroblast recruitment, activation, and proliferation and induce apoptosis of myofibroblasts (which is altered is fibrotic disease) have already been widely addressed experimentally and clinically *(105)*. Therefore, chemokines and chemokine receptors are promising therapeutic targets, because they appear to regulate cell function on a variety of levels.

14.6. Conclusions

Fibrosis is a dynamic progression of dysregulated wound healing that results from chronic inflammation. It is a common pathology regardless of the tissue involved, and therefore, the mechanisms that progress to fibrosis can be widely applied. The recruitment, activation, and proliferation of inflammatory cells and their cooperation with resident cells appears to rely on the action of chemokines and the differential expression of the chemokine receptors by these cells. Thus, chemokine receptors make particularly attractive therapeutic targets.

References

1. Razzaque MS, Taguchi T. Pulmonary fibrosis: Cellular and molecular events. Pathol Int 2003;53(3):133–145.

2. Mutsaers SE, Bishop JE, McGrouther G, Laurent GJ. Mechanisms of tissue repair: from wound healing to fibrosis. Int J Biochem Cell Biol 1997;29(1): 5–17.
3. Metz CN. Fibrocytes: a unique cell population implicated in wound healing. Cell Mol Life Sci 2003;60(7):1342–1350.
4. Wynn TA. Fibrotic disease and the T(H)1/T(H)2 paradigm. Nat Rev Immunol 2004;4(8):583–594.
5. Quan TE, Cowper S, Wu SP, Bockenstedt LK, Bucala R. Circulating fibrocytes: collagen-secreting cells of the peripheral blood. Int J Biochem Cell Biol 2004;36(4):598–606.
6. Cotran RS, Kumar V, Collins T. Robbins Pathologic Basis of Disease. 6th ed. Philadelphia: W.B. Saunders; 1999.
7. Chua F, Gauldie J, Laurent GJ. Pulmonary fibrosis: searching for model answers. Am J Respir Cell Mol Biol 2005;33(1):9–13.
8. Liu Y. Renal fibrosis: New insights into the pathogenesis and therapeutics. Kidney Int 2006;69(2):213–217.
9. Macdonald TT. Decoy receptor springs to life and eases fibrosis. Nat Med 2006;12(1):13–14.
10. Kaviratne M, Hesse M, Leusink M, et al. IL-13 activates a mechanism of tissue fibrosis that is completely TGF-beta independent. J Immunol 2004;173(6):4020–4029.
11. Buckley CD, Pilling D, Lord JM, Akbar AN, Scheel-Toellner D, Salmon M. Fibroblasts regulate the switch from acute resolving to chronic persistent inflammation. Trends Immunol 2001;22(4):199–204.
12. Sakai N, Wada T, Furuichi K, et al. MCP-1/CCR2-dependent loop for fibrogenesis in human peripheral CD14-positive monocytes. J Leukoc Biol 2006;79(3):555–563.
13. White FA, Bhangoo SK, Miller RJ. Chemokines: integrators of pain and inflammation. Nat Rev Drug Discov 2005;4(10):834–844.
14. Coelho AL, Hogaboam CM, Kunkel SL. Chemokines provide the sustained inflammatory bridge between innate and acquired immunity. Cytokine Growth Factor Rev 2005;16(6):553–560.
15. Esche C, Stellato C, Beck LA. Chemokines: key players in innate and adaptive immunity. J Invest Dermatol 2005;125(4):615–628.
16. Strutz F, Neilson EG. New insights into mechanisms of fibrosis in immune renal injury. Springer Semin Immunopathol 2003;24(4):459–476.
17. Mutsaers SE, Prele CM, Brody AR, Idell S. Pathogenesis of pleural fibrosis. Respirology 2004;9(4):428–440.
18. Gauldie J. Inflammatory mechanisms are a minor component of the pathogenesis of idiopathic pulmonary fibrosis. Am J Respir Crit Care Med 2002;165(9): 1205–1206.
19. Ortiz LA, Gambelli F, McBride C, et al. Mesenchymal stem cell engraftment in lung is enhanced in response to bleomycin exposure and ameliorates its fibrotic effects. Proc Natl Acad Sci U S A 2003;100(14):8407–8411.

20. Corbel M, Belleguic C, Boichot E, Lagente V. Involvement of gelatinases (MMP-2 and MMP-9) in the development of airway inflammation and pulmonary fibrosis. Cell Biol Toxicol 2002;18(1):51–61.
21. Kraushaar BS, Nirschl RP. Tendinosis of the elbow (tennis elbow). Clinical features and findings of histological, immunohistochemical, and electron microscopy studies. J Bone Joint Surg Am 1999;81(2):259–278.
22. Hogaboam CM, Bone-Larson CL, Lipinski S, et al. Differential monocyte chemoattractant protein-1 and chemokine receptor 2 expression by murine lung fibroblasts derived from Th1- and Th2-type pulmonary granuloma models. J Immunol 1999;163(4):2193–2201.
23. Sime PJ, Xing Z, Graham FL, Csaky KG, Gauldie J. Adenovector-mediated gene transfer of active transforming growth factor-beta1 induces prolonged severe fibrosis in rat lung. J Clin Invest 1997;100(4):768–776.
24. Hogaboam C, Kunkel SL, Strieter RM, et al. Novel role of transmembrane SCF for mast cell activation and eotaxin production in mast cell-fibroblast interactions. J Immunol 1998;160(12):6166–6171.
25. Yamamoto T, Hartmann K, Eckes B, Krieg T. Role of stem cell factor and monocyte chemoattractant protein-1 in the interaction between fibroblasts and mast cells in fibrosis. J Dermatol Sci 2001;26(2):106–111.
26. Wang HW, Tedla N, Hunt JE, Wakefield D, McNeil HP. Mast cell accumulation and cytokine expression in the tight skin mouse model of scleroderma. Exp Dermatol 2005;14(4):295–302.
27. Bonniaud P, Margetts PJ, Ask K, Flanders K, Gauldie J, Kolb M. TGF-beta and Smad3 signaling link inflammation to chronic fibrogenesis. J Immunol 2005;175(8):5390–5395.
28. Hagood JS, Prabhakaran P, Kumbla P, et al. Loss of fibroblast Thy-1 expression correlates with lung fibrogenesis. Am J Pathol 2005;167(2):365–379.
29. Harari S, Caminati A. Idiopathic pulmonary fibrosis. Allergy 2005;60(4):421–435.
30. Abe R, Donnelly SC, Peng T, Bucala R, Metz CN. Peripheral blood fibrocytes: differentiation pathway and migration to wound sites. J Immunol 2001;166(12):7556–7562.
31. Bucala R, Spiegel LA, Chesney J, Hogan M, Cerami A. Circulating fibrocytes define a new leukocyte subpopulation that mediates tissue repair. Mol Med 1994;1(1):71–81.
32. Phillips RJ, Burdick MD, Hong K, et al. Circulating fibrocytes traffic to the lungs in response to CXCL12 and mediate fibrosis. J Clin Invest 2004;114(3):438–446.
33. Schmidt M, Sun G, Stacey MA, Mori L, Mattoli S. Identification of circulating fibrocytes as precursors of bronchial myofibroblasts in asthma. J Immunol 2003;171(1):380–389.
34. Dupont A, Majithia V, Ahmad S, McMurray R. Nephrogenic fibrosing dermopathy, a new mimicker of systemic sclerosis. Am J Med Sci 2005;330(4):192–194.

35. Daram SR, Cortese CM, Bastani B. Nephrogenic fibrosing dermopathy/nephrogenic systemic fibrosis: report of a new case with literature review. Am J Kidney Dis 2005;46(4):754–759.
36. Hashimoto N, Jin H, Liu T, Chensue SW, Phan SH. Bone marrow-derived progenitor cells in pulmonary fibrosis. J Clin Invest 2004;113(2):243–252.
37. Choi ES, Pierce EM, Jakubzick C, et al. Focal interstitial CC chemokine receptor 7 (CCR7) expression in idiopathic interstitial pneumonia. J Clin Pathol 2006;59(1):28–39.
38. Zhang K, Rekhter MD, Gordon D, Phan SH. Myofibroblasts and their role in lung collagen gene expression during pulmonary fibrosis. A combined immunohistochemical and in situ hybridization study. Am J Pathol 1994;145(1):114–125.
39. Gallucci RM, Lee EG, Tomasek JJ. IL-6 modulates alpha-smooth muscle actin expression in dermal fibroblasts from IL-6-deficient mice. J Invest Dermatol 2006;126(3):561–568.
40. Moore BB, Kolodsick JE, Thannickal VJ, et al. CCR2-mediated recruitment of fibrocytes to the alveolar space after fibrotic injury. Am J Pathol 2005;166(3):675–684.
41. Lanzavecchia A, Sallusto F. Understanding the generation and function of memory T cell subsets. Curr Opin Immunol 2005;17(3):326–332.
42. Humrich JY, Humrich JH, Averbeck M, et al. Mature monocyte-derived dendritic cells respond more strongly to CCL19 than to CXCL12: consequences for directional migration. Immunology 2006;117(2):238–247.
43. Bromley SK, Thomas SY, Luster AD. Chemokine receptor CCR7 guides T cell exit from peripheral tissues and entry into afferent lymphatics. Nat Immunol 2005;6(9):895–901.
44. Jang MH, Sougawa N, Tanaka T, et al. CCR7 is critically important for migration of dendritic cells in intestinal lamina propria to mesenteric lymph nodes. J Immunol 2006;176(2):803–810.
45. Stein JV, Nombela-Arrieta C. Chemokine control of lymphocyte trafficking: a general overview. Immunology 2005;116(1):1–12.
46. Moser B, Willimann K. Chemokines: role in inflammation and immune surveillance. Ann Rheum Dis 2004;63(Suppl 2):ii84–ii9.
47. Bendall L. Chemokines and their receptors in disease. Histol Histopathol 2005;20(3):907–926.
48. D'Ambrosio D, Mariani M, Panina-Bordignon P, Sinigaglia F. Chemokines and their receptors guiding T lymphocyte recruitment in lung inflammation. Am J Respir Crit Care Med 2001;164(7):1266–1275.
49. Sime PJ, O'Reilly KM. Fibrosis of the lung and other tissues: new concepts in pathogenesis and treatment. Clin Immunol 2001;99(3):308–319.
50. Ansel KM, Djuretic I, Tanasa B, Rao A. Regulation of Th2 differentiation and Il4 locus accessibility. Annu Rev Immunol 2006;24:607–656.
51. Wada T, Furuichi K, Sakai N, et al. Gene therapy via blockade of monocyte chemoattractant protein-1 for renal fibrosis. J Am Soc Nephrol 2004;15(4):940–948.

52. Gharaee-Kermani M, Denholm EM, Phan SH. Costimulation of fibroblast collagen and transforming growth factor beta1 gene expression by monocyte chemoattractant protein-1 via specific receptors. J Biol Chem 1996;271(30):17779–17784.
53. Booth M, Mwatha JK, Joseph S, et al. Periportal fibrosis in human Schistosoma mansoni infection is associated with low IL-10, low IFN-gamma, high TNF-alpha, or low RANTES, depending on age and gender. J Immunol 2004;172(2):1295–1303.
54. Alves Oliveira LF, Moreno EC, Gazzinelli G, et al. Cytokine production associated with periportal fibrosis during chronic schistosomiasis mansoni in humans. Infect Immun 2006;74(2):1215–1221.
55. Keane MP. Chemokine profiling in pulmonary fibrosis: ready for prime time? Am J Respir Crit Care Med 2004;170(5):475–476.
56. Marra F. Chemokines in liver inflammation and fibrosis. Front Biosci 2002;7:d1899–d1914.
57. Buttner C, Skupin A, Reimann T, et al. Local production of interleukin-4 during radiation-induced pneumonitis and pulmonary fibrosis in rats: macrophages as a prominent source of interleukin-4. Am J Respir Cell Mol Biol 1997;17(3):315–325.
58. Zhu Z, Homer RJ, Wang Z, et al. Pulmonary expression of interleukin-13 causes inflammation, mucus hypersecretion, subepithelial fibrosis, physiologic abnormalities, and eotaxin production. J Clin Invest 1999;103(6):779–788.
59. Belperio JA, Dy M, Burdick MD, et al. Interaction of IL-13 and C10 in the pathogenesis of bleomycin-induced pulmonary fibrosis. Am J Respir Cell Mol Biol 2002;27(4):419–427.
60. Blease K, Jakubzick C, Westwick J, Lukacs N, Kunkel SL, Hogaboam CM. Therapeutic effect of IL-13 immunoneutralization during chronic experimental fungal asthma. J Immunol 2001;166(8):5219–5224.
61. Hogaboam CM, Blease K, Schuh JM. Cytokines and chemokines in allergic bronchopulmonary aspergillosis (ABPA) and experimental Aspergillus-induced allergic airway or asthmatic disease. Front Biosci 2003;8:e147–e156.
62. Huaux F, Gharaee-Kermani M, Liu T, et al. Role of eotaxin-1 (CCL11) and CC chemokine receptor 3 (CCR3) in bleomycin-induced lung injury and fibrosis. Am J Pathol 2005;167(6):1485–1496.
63. Huaux F, Liu T, McGarry B, Ullenbruch M, Xing Z, Phan SH. Eosinophils and T lymphocytes possess distinct roles in bleomycin-induced lung injury and fibrosis. J Immunol 2003;171(10):5470–5481.
64. Rothenberg ME, Hogan SP. The eosinophil. Annu Rev Immunol 2006;24:147–174.
65. Green FH. Overview of pulmonary fibrosis. Chest 2002;122(6 Suppl):334S–339S.
66. Jiang D, Liang J, Fan J, et al. Regulation of lung injury and repair by Toll-like receptors and hyaluronan. Nat Med 2005;11(11):1173–1179.
67. Coker RK, Laurent GJ. Pulmonary fibrosis: cytokines in the balance. Eur Respir J 1998;11(6):1218–1221.

68. Abdelaziz MM, Samman YS, Wali SO, Hamad MM. Treatment of idiopathic pulmonary fibrosis: is there anything new? Respirology 2005;10(3):284–289.
69. Gharaee-Kermani M, Phan SH. Molecular mechanisms of and possible treatment strategies for idiopathic pulmonary fibrosis. Curr Pharm Des 2005;11(30):3943–3971.
70. Hogaboam CM, Carpenter KJ, Schuh JM, Proudfoot AA, Bridger G, Buckland KF. The therapeutic potential in targeting CCR5 and CXCR4 receptors in infectious and allergic pulmonary disease. Pharmacol Ther 2005;107(3): 314–328.
71. Choi ES, Jakubzick C, Carpenter KJ, et al. Enhanced monocyte chemoattractant protein-3/CC chemokine ligand-7 in usual interstitial pneumonia. Am J Respir Crit Care Med 2004;170(5):508–515.
72. Fehrenbach H. Alveolar epithelial type II cell: defender of the alveolus revisited. Respir Res 2001;2(1):33–46.
73. Williams MC. Alveolar type I cells: molecular phenotype and development. Annu Rev Physiol 2003;65:669–695.
74. Burdick MD, Murray LA, Keane MP, et al. CXCL11 attenuates bleomycin-induced pulmonary fibrosis via inhibition of vascular remodeling. Am J Respir Crit Care Med 2005;171(3):261–268.
75. Johnston CJ, Williams JP, Okunieff P, Finkelstein JN. Radiation-induced pulmonary fibrosis: examination of chemokine and chemokine receptor families. Radiat Res 2002;157(3):256–265.
76. Yara S, Kawakami K, Kudeken N, et al. FTS reduces bleomycin-induced cytokine and chemokine production and inhibits pulmonary fibrosis in mice. Clin Exp Immunol 2001;124(1):77–85.
77. Lloyd CM, Minto AW, Dorf ME, et al. RANTES and monocyte chemoattractant protein-1 (MCP-1) play an important role in the inflammatory phase of crescentic nephritis, but only MCP-1 is involved in crescent formation and interstitial fibrosis. J Exp Med 1997;185(7):1371–1380.
78. Inoshima I, Kuwano K, Hamada N, et al. Anti-monocyte chemoattractant protein-1 gene therapy attenuates pulmonary fibrosis in mice. Am J Physiol Lung Cell Mol Physiol 2004;286(5):L1038–L1044.
79. Motomura Y, Khan WI, El-Sharkawy RT, et al. Induction of a fibrogenic response in mouse colon by overexpression of monocyte chemoattractant protein 1. Gut 2006;55(5):662–670.
80. Iyonaga K, Takeya M, Saita N, et al. Monocyte chemoattractant protein-1 in idiopathic pulmonary fibrosis and other interstitial lung diseases. Hum Pathol 1994;25(5):455–463.
81. Kanno K, Tazuma S, Nishioka T, Hyogo H, Chayama K. Angiotensin II participates in hepatic inflammation and fibrosis through MCP-1 expression. Dig Dis Sci 2005;50(5):942–948.
82. Inoue M, Ino Y, Gibo J, et al. The role of monocyte chemoattractant protein-1 in experimental chronic pancreatitis model induced by dibutyltin dichloride in rats. Pancreas 2002;25(4):e64–e70.

83. Frade JM, Mellado M, del Real G, Gutierrez-Ramos JC, Lind P, Martinez AC. Characterization of the CCR2 chemokine receptor: functional CCR2 receptor expression in B cells. J Immunol 1997;159(11):5576–5584.
84. Moore BB, Paine R, 3rd, Christensen PJ, et al. Protection from pulmonary fibrosis in the absence of CCR2 signaling. J Immunol 2001;167(8):4368–4377.
85. Okuma T, Terasaki Y, Kaikita K, et al. C-C chemokine receptor 2 (CCR2) deficiency improves bleomycin-induced pulmonary fibrosis by attenuation of both macrophage infiltration and production of macrophage-derived matrix metalloproteinases. J Pathol 2004;204(5):594–604.
86. Fukuda Y, Ishizaki M, Kudoh S, Kitaichi M, Yamanaka N. Localization of matrix metalloproteinases-1, -2, and -9 and tissue inhibitor of metalloproteinase-2 in interstitial lung diseases. Lab Invest 1998;78(6):687–698.
87. Corbel M, Caulet-Maugendre S, Germain N, Molet S, Lagente V, Boichot E. Inhibition of bleomycin-induced pulmonary fibrosis in mice by the matrix metalloproteinase inhibitor batimastat. J Pathol 2001;193(4):538–545.
88. Zhang Y, Ernst CA, Rollins BJ. MCP-1: structure/activity analysis. Methods 1996;10(1):93–103.
89. Izbicki G, Or R, Christensen TG, et al. Bleomycin-induced lung fibrosis in IL-4-overexpressing and knockout mice. Am J Physiol Lung Cell Mol Physiol 2002;283(5):L1110–L1116.
90. Keane MP, Belperio JA, Burdick MD, Strieter RM. IL-12 attenuates bleomycin-induced pulmonary fibrosis. Am J Physiol Lung Cell Mol Physiol 2001;281(1): L92–L97.
91. Kitasato Y, Hoshino T, Okamoto M, et al. Enhanced expression of interleukin-18 and its receptor in idiopathic pulmonary fibrosis. Am J Respir Cell Mol Biol 2004;31(6):619–625.
92. Segel MJ, Izbicki G, Cohen PY, et al. Role of interferon-gamma in the evolution of murine bleomycin lung fibrosis. Am J Physiol Lung Cell Mol Physiol 2003;285(6):L1255–L1262.
93. Kim JH, Kim HY, Kim S, Chung JH, Park WS, Chung DH. Natural killer T (NKT) cells attenuate bleomycin-induced pulmonary fibrosis by producing interferon-gamma. Am J Pathol 2005;167(5):1231–1241.
94. Belperio JA, Dy M, Murray L, et al. The role of the Th2 CC chemokine ligand CCL17 in pulmonary fibrosis. J Immunol 2004;173(7):4692–4698.
95. Garcia G, Godot V, Humbert M. New chemokine targets for asthma therapy. Curr Allergy Asthma Rep 2005;5(2):155–160.
96. Joubert P, Lajoie-Kadoch S, Labonte I, et al. CCR3 expression and function in asthmatic airway smooth muscle cells. J Immunol 2005;175(4):2702–2708.
97. Pope SM, Zimmermann N, Stringer KF, Karow ML, Rothenberg ME. The eotaxin chemokines and CCR3 are fundamental regulators of allergen-induced pulmonary eosinophilia. J Immunol 2005;175(8):5341–5350.
98. Antoniades HN, Neville-Golden J, Galanopoulos T, Kradin RL, Valente AJ, Graves DT. Expression of monocyte chemoattractant protein 1 mRNA in human

idiopathic pulmonary fibrosis. Proc Natl Acad Sci U S A 1992;89(12):5371–5375.

99. Capelli A, Di Stefano A, Gnemmi I, Donner CF. CCR5 expression and CC chemokine levels in idiopathic pulmonary fibrosis. Eur Respir J 2005;25(4):701–707.
100. Blanpain C, Migeotte I, Lee B, et al. CCR5 binds multiple CC-chemokines: MCP-3 acts as a natural antagonist. Blood 1999;94(6):1899–1905.
101. Loetscher P, Uguccioni M, Bordoli L, et al. CCR5 is characteristic of Th1 lymphocytes. Nature 1998;391(6665):344–345.
102. Tager AM, Kradin RL, LaCamera P, et al. Inhibition of pulmonary fibrosis by the chemokine IP-10/CXCL10. Am J Respir Cell Mol Biol 2004;31(4):395–404.
103. Pignatti P, Brunetti G, Moretto D, et al. Role of the chemokine receptors CXCR3 and CCR4 in human pulmonary fibrosis. Am J Respir Crit Care Med 2006;173(3):310–317.
104. Strieter RM, Starko KM, Enelow RI, Noth I, Valentine VG. Effects of interferon-gamma 1b on biomarker expression in patients with idiopathic pulmonary fibrosis. Am J Respir Crit Care Med 2004;170(2):133–140.
105. Selman M, Thannickal VJ, Pardo A, Zisman DA, Martinez FJ, Lynch JP, 3rd. Idiopathic pulmonary fibrosis: pathogenesis and therapeutic approaches. Drugs 2004;64(4):405–430.
106. Antoniou KM, Ferdoutsis E, Bouros D. Interferons and their application in the diseases of the lung. Chest 2003;123(1):209–216.
107. Petkovic V, Moghini C, Paoletti S, Uguccioni M, Gerber B. I-TAC/CXCL11 is a natural antagonist for CCR5. J Leukoc Biol 2004;76(3):701–708.
108. Inayama M, Nishioka Y, Azuma M, et al. A novel IkappaB kinase-beta inhibitor ameliorates bleomycin-induced pulmonary fibrosis in mice. Am J Respir Crit Care Med 2006;173(9):1016–1022.

15

Chemokines and Angiogenesis

Michael P. Keane, John A. Belperio, and Robert M. Strieter

Summary

Angiogenesis is the process of new blood vessel growth and is a critical biological process under both physiologic and pathologic conditions. Angiogenesis can occur under physiologic conditions that include embryogenesis and the ovarian/menstrual cycle. In contrast, pathologic angiogenesis is associated with chronic inflammation/chronic fibroproliferative disorders and tumorigenesis of cancer. Similarly, aberrant angiogenesis associated with chronic inflammation/fibroproliferative disorders is analogous to neovascularization of tumorigenesis of cancer. Net angiogenesis is determined by a balance in the expression of angiogenic compared with angiostatic factors. CXC chemokines are heparin-binding proteins that display unique disparate roles in the regulation of angiogenesis. Based on their structure, CXC chemokines can be divided into two groups that either promote or inhibit angiogenesis, and they are therefore uniquely placed to regulate net angiogenesis in both physiologic and pathologic conditions.

Key Words: Chemokine; chemokine receptor; angiogenesis; inflammation; fibrosis; cancer.

15.1. Introduction

Angiogenesis is the process of new blood vessel growth and is a critical biological process under both physiologic and pathologic conditions. Angiogenesis can occur under physiologic conditions that include embryogenesis and the ovarian/menstrual cycle. In contrast, pathologic angiogenesis is associated with chronic inflammation/chronic fibroproliferative disorders and tumorigenesis of cancer. For purposes of this chapter, *angiogenesis* and

From: *The Receptors: The Chemokine Receptors*
Edited by: J. K. Harrison and N. W. Lukacs © Humana Press Inc., Totowa, NJ

neovascularization will be used interchangeably in the context of pathologic or aberrant angiogenesis.

Inflammation and angiogenesis, though distinct and separable processes, are closely related events and often temporally overlap *(1)*. Chronic inflammation associated with chronic fibroproliferation histologically appears as granulation-like tissue, a prominent feature of which is neovascularization. The metabolic demands of tissue undergoing proliferative, reparative, and hyperplastic changes are extremely high and require a proportionally greater capillary blood supply compared with normal tissue. Thus, aberrant angiogenesis associated with chronic inflammation/fibroproliferative disorders is analogous to neovascularization of tumorigenesis of cancer and provides a therapeutic target for novel intervention in these difficult-to-treat disorders.

A variety of factors have been described that either promote or inhibit angiogenesis. In the local microenvironment, net angiogenesis is determined by a balance in the expression of angiogenic compared with angiostatic factors. CXC chemokines are heparin-binding proteins that display unique disparate roles in the regulation of angiogenesis. The family has four highly conserved cysteine amino acid residues, with the first two cysteines separated by a nonconserved amino acid residue *(2,3)*. A second structural domain dictates their functional activity. The NH_2-terminus of several CXC chemokines contains three amino acid residues (Glu-Leu-Arg; ELR motif), which immediately precedes the first cysteine amino acid residue *(2,3)*. The CXC chemokines with the ELR motif (ELR^+) promote angiogenesis (Tables 1 and 2) *(2)*. In contrast, CXC chemo-

Table 1
The CXC Chemokines That Display Disparate Angiogenic Activity

Angiogenic CXC chemokines containing the ELR motif	
CXCL1	Growth-related oncogene alpha (GRO-α)
CXCL2	Growth-related oncogene beta (GRO-β)
CXCL3	Growth-related oncogene gamma (GRO-γ)
CXCL5	Epithelial neutrophil activating protein-78 (ENA-78)
CXCL6	Granulocyte chemotactic protein-2 (GCP-2)
CXCL7	Neutrophil activating protein-2 (NAP-2)
CXCL8	Interleukin-8 (IL-8)
Angiostatic CXC chemokines that lack the ELR motif	
CXCL4	Platelet factor-4 (PF-4)
CXCL9	Monokine induced by interferon-γ (MIG)
CXCL10	Interferon-γ–inducible protein (IP-10)
CXCL11	Interferon-inducible T cell alpha chemoattractant (I-TAC)
CXCL12	Stromal cell–derived factor-1 (SDF-1)
CXCL14	Breast and kidney–expressed chemokine (BRAK)

Table 2
CXC Chemokine Ligands and Receptors That Have Been Implicated in Angiogenesis

Receptor	Ligand
CXCR2	CXCL1, CXCL2, CXCL3, CXCL5, CXCL6, CXCL7, CXCL8
CXCR4	CXCL12

kines that are, in general, interferon-inducible and lack the ELR motif (ELR$^-$) inhibit angiogenesis (Tables 1 and 3) *(2)*. The dissimilarity in structure dictates the use of different CXC chemokine receptors on endothelial cells, which ultimately leads to signal coupling and either promotion or inhibition of angiogenesis. CXCR2 mediates the angiogenic signals of the ELR$^+$ CXC chemokines, whereas the ELR$^-$ CXC chemokines mediate their angiostatic actions through CXCR3. Furthermore, it has recently been suggested that angiostatic signals are specifically mediated through CXCR3B, whereas CXCR3A may mediate angiogenic signals *(4,5)*.

15.2. The CXC Chemokines

The CXC chemokines can be divided into two groups on the basis of a structure/function domain consisting of the presence or absence of three amino acid residues (Glu-Leu-Arg; ELR motif) that precedes the first cysteine amino acid residue in the primary structure of these cytokines. The ELR$^+$ CXC chemokines are chemoattractants for neutrophils and act as potent angiogenic factors *(6)*. In contrast, the ELR$^-$ CXC chemokines are chemoattractants for mononuclear cells and are potent inhibitors of angiogenesis (Table 1) *(6)*.

Based on the structural/functional difference, the members of the CXC chemokine family are unique cytokines in their ability to behave in a disparate manner in the regulation of angiogenesis. The angiogenic members include

Table 3
CXC Chemokine Ligands and Receptors That Have Been Implicated in Angiostasis

Receptor	Ligand
CXCR3B	CXCL4, CXCL9, CXCL10, CXCL11
Unknown/non–receptor mediated	CXCL4, CXCL14

Note: CXCL4 may act through CXCR3B or non–receptor-mediated mechanisms.

CXCL 1, 2, 3, 5, 6, 7, and 8. CXCL 1, 2, and 3 are closely related CXC chemokines, with CXCL1 originally described for its melanoma growth stimulatory activity (Table 1). CXCL5, CXCL6, and CXCL8 were all initially identified on the basis of neutrophil activation and chemotaxis. The angiostatic (ELR$^-$) members of the CXC chemokine family include CXCL4, which was originally described for its ability to bind heparin and inactivate heparin's anticoagulation function. Other angiostatic ELR$^-$ CXC chemokines include CXCL9, CXCL10, CXCL11, and CXCL14 (Table 1).

15.2.1. CXCR2 Is the Receptor for Angiogenic ELR$^+$ CXC Chemokine-Mediated Angiogenesis

The fact that all ELR$^+$ CXC chemokines mediate angiogenesis highlights the importance of identifying a common receptor that mediates their biological function in promoting angiogenesis. Although the candidate CXC chemokine receptors are CXCR1 and/or CXCR2, only CXCL8 and CXCL6 specifically bind to CXCR1, whereas all ELR$^+$ CXC chemokines bind to CXCR2 *(7)*. The ability of all ELR$^+$ CXC chemokine ligands to bind to CXCR2 supports the notion that this receptor mediates the angiogenic activity of ELR$^+$ CXC chemokines.

Whereas CXCR1 and CXCR2 are detected in endothelial cells *(7–9)*, the expression of CXCR2 but not CXCR1 has been found to be the primary functional chemokine receptor in mediating endothelial cell chemotaxis *(7,8)*. Heidemann et al. *(10)* have further confirmed the importance of CXCR2 in mediating the effects of angiogenesis in human microvascular endothelial cells. They found that endothelial cells respond to CXCL8 with rapid stress fiber assembly, chemotaxis, enhanced proliferation, and phosphorylation of extracellular signal-regulated protein kinase 1/2 (ERK 1/2) related to activation of CXCR2 *(10)*. Blocking the function of CXCR2 by either specific neutralizing antibodies or inhibiting downstream signaling using specific inhibitors of ERK1/2 and phosphoinositide 3-kinase (PI3kinase) impaired CXCL8-induced stress fiber assembly, chemotaxis, and endothelial tube formation in endothelial cells *(10)*.

What is the fate of CXCR2 after activation by ELR$^+$ CXC chemokines? CXCR2 activation leads to receptor internalization, recycling of the receptor back to the cell membrane, or targets CXCR2 for degradation. ELR$^+$ CXC chemokine activation of CXCR2 under conditions in which the receptor is transiently exposed or stimulated with less than saturable concentrations results in movement of CXCR2 into clathrin-coated pits, movement into the early endosome, the sorting endosome, and on to the recycling endosome with trafficking back to the plasma membrane compartment and reexpression on the cell surface *(11)*. However, if CXCR2 is exposed to prolonged saturating concentrations of ELR^+ CXC chemokines, a significant proportion of CXCR2 will move into the late endosome and on to the lysosome for degradation *(11)*. Interestingly, CXCR2 internalization is necessary for generating a chemotactic response. Mutation of the CXCR2, which impairs receptor internalization by altering the binding of adaptor proteins AP-2 (activating protein-2) or beta-arrestin to the receptor, results in a marked reduction in the chemotactic response *(11)*.

The importance of CXCR2 in mediating ELR^+ CXC chemokine-induced angiogenesis has been shown in vivo using the cornea micropocket assay of angiogenesis in $CXCR2^{+/+}$ and $CXCR2^{-/-}$ animals. ELR^+ CXC chemokine-mediated angiogenesis was inhibited in the corneas of $CXCR2^{-/-}$ mice and in the presence of neutralizing antibodies to CXCR2 in the rat corneal micropocket assay *(7)*. These studies have been further substantiated using $CXCR2^{-/-}$ mice in a wound-repair model system *(12)*. Devalaraja et al. *(12)* have examined the significance of CXC chemokines in wound healing. In this study, full excisional wounds were created on $CXCR2^{+/+}$, heterozygous +/-, or $CXCR2^{-/-}$ mice. Significant delays in wound-healing parameters were found in $CXCR2^{-/-}$ mice, including decreased neovascularization *(12)*. These studies have now been extended to a lung cancer syngeneic tumor model system in $CXCR2^{-/-}$ compared with $CXCR2^{+/+}$ mice. Lung cancer in $CXCR2^{-/-}$ demonstrate reduced growth, increased tumor-associated necrosis, inhibited tumor-associated angiogenesis, and metastatic potential *(13)*. These in vitro and in vivo studies establish that CXCR2 is an important receptor that mediates ELR^+ CXC chemokine-dependent angiogenic activity.

CXCR2 is the putative receptor that mediates the angiogenic activity ELR^+ CXC chemokines. Therefore, the aberrant expression of this receptor or a homologue of this receptor may be expected to be associated with cellular transformation relevant to pre-neoplastic to neoplastic transformation. The pathogenesis of lung cancer is a multistep process that involves sequential morphologic and molecular changes that precede invasive lung cancer; these events are not too dissimilar from other neoplasms. For example, the human Kaposi sarcoma herpes virus (KSHV) that mediates the pathogenesis of Kaposi sarcoma (KS) encodes a seven-transmembrane G protein–coupled receptor (7TM-GPCR) (KSHV-GPCR) that is homologous to CXCR2 *(14,15)*. 7TM-GPCR activation leads to dissociation of the heterotrimeric protein complex ($G_{\alpha\beta\gamma}$) to α and $\beta\gamma$ subunits that mediate downstream regulation of several intracellular signaling pathways [i.e., cAMP/protein kinase A (PKA), protein kinase C (PKC), phospholipase C (PLC), PI3kinase/AKT/mTOR, Ras/Raf/MEK/JNK/p38/ERK1/ERK2] and activates NFκB pathways *(16–19)*. Some of these signaling pathways are identical to signal transduction by receptor protein tyrosine kinases that are important for cellular proliferation, migration, and regulation of apoptosis *(16–18)*. These findings support the notion that

7TM-GPCRs, like CXCR2, may be involved in pre-neoplastic to neoplastic transformation and the development of cancer.

The KSHV-GPCR has been determined to constitutively signal-couple, and signal-coupling of this receptor can be further augmented with CXC chemokine ligand binding (i.e., CXCL8 and/or CXCL1) *(20–23)*. To ascertain the relevance of KSHV-GPCR in promoting the pathogenesis of KS, transgenic mice overexpressing KSHV-GPCR have been found to constitutively produce tumors similar to KS *(24,25)*. The mice develop angioproliferative lesions in multiple organs that morphologically resemble KS lesion *(24,25)*. These findings suggest that the expression of only one viral chemokine receptor-like gene can lead to the histopathologic recapitulation of KS with cellular transformation and the development of a lesion that resembles an angiosarcoma. This supports the notion that a CXCR2-like receptor facilitates pre-neoplastic to neoplastic cellular transformation. The KSHV-GPCR has been found to signal-couple a number of signaling pathways relevant to cellular transformation.

In further support of this contention is the finding that a point mutation of CXCR2, but not CXCR1, results in constitutive signaling of the receptor and cellular transformation of transfected cells in a similar manner as KSHV-GPCR *(14)*. Furthermore, the persistent activation of CXCR2 by specific CXC chemokine ligands can lead to a similar cellular transformation as seen with either the point mutation of CXCR2 or KSHV-GPCR *(14)*. CXCR2 has been further identified in the cellular transformation of melanocytes into melanoma *(26,27)*. Therefore, the expression of CXCR2 on certain cells in the presence of persistent autocrine and paracrine stimulation with specific CXC chemokine ligands has important implications in promoting cellular transformation that may be relevant to the pre-neoplastic to neoplastic transformation of lung cancer.

Are there decoy receptors for ELR^+ CXC chemokines that prevent their angiogenic activity? The Duffy antigen receptor for chemokines (DARC) is known to be a promiscuous chemokine receptor that binds chemokines in the absence of any detectable signal transduction events *(28)*. Within the ELR^+ CXC chemokines, DARC binds the angiogenic CXC chemokines including CXCL8, CXCL1, and CXCL5, all of which have previously been shown to be important for promoting tumor growth in a variety of tumors, including non–small cell lung carcinoma (NSCLC) tumor growth *(29–31)*. Addison et al. *(28)* demonstrated that stable transfection and overexpression of DARC in a NSCLC tumor cell line resulted in the binding of the angiogenic ELR^+ CXC chemokines by the tumor cells. The binding of tumor cell–derived ELR^+ CXC chemokines to the tumor cells themselves interfered with the local tumor paracrine microenvironment of tumor cell interaction with host responding endothelial cells and prevented the ability of these angiogenic factors to stimulate endothelial cells and promote tumor-associated angiogenesis *(28)*. NSCLC tumor cells that constitutively expressed DARC in vitro were similar in their growth characteristics compared with control-transfected cells. However, they found that tumors derived from DARC-expressing cells were significantly larger in size than tumors derived from control-transfected cells. Interestingly, upon histologic examination, DARC-expressing tumors had significantly more necrosis and decreased tumor cellularity compared with control tumors. Expression of DARC by NSCLC cells was also associated with a marked decrease in tumor-associated vasculature and a reduction in metastatic potential. The finding of this study suggested that competitive binding of ELR^+ CXC chemokines by tumor cells expressing a decoy receptor could prevent paracrine activation of endothelial cells in the tumor microenvironment and reduce tumor-associated angiogenesis.

15.2.2. CXCR3 Appears to Be the Major Receptor for CXC Chemokines That Inhibit Angiogenesis

The major receptor that has been identified for angiostatic CXC chemokines is CXCR3, and it exists as alternative splice forms (i.e., CXCR3A, CXCR3B, and CXCR3-alt) that are involved in mediating recruitment of T helper 1 (Th1) cells and acts as the receptor for inhibition of angiogenesis *(32–36)*. CXCR3A is the major chemokine receptor found on Th1 effector T cells, cytotoxic CD8 T cells, activated B cells, and natural killer (NK) cells *(37,38)*. IL-2 is a major agonist for the expression of CXCR3A on these cells *(38)*. CXCR3 was originally identified on murine endothelial cells *(39);* subsequent studies demonstrated that CXCR3 ligands could block both human microvascular endothelial cell migration and proliferation in response to a variety of angiogenic factors *(9,40)*. Further clarification of the role of CXCR3 in mediating angiostatic activity has come from the discovery that CXCR3 exists as two alternative splice forms *(4)*. These variants have been termed CXCR3A and CXCR3B *(4)*. CXCR3A mediates the CXCR3 ligand-dependent chemotactic activity of mononuclear cells *(4)*. CXCR3B mediates the angiostatic activity of CXCL4, CXCL9, CXCL10, and CXCL11 on human microvascular endothelial cells (4). Moreover, specific antibodies to CXCR3B immunolocalize to endothelial cells within neoplastic tissues *(4)*. This supports the notion that if CXCR3 ligands can be spatially expressed within the tumor, then CXCR3B activation can inhibit tumor-associated angiogenesis *(4)*. To add to the complexity of CXCR3 biology, a variant of human CXCR3 has been recently identified, which is generated by posttranscriptional exon skipping referred to as CXCR3-alt *(36)*. This receptor is expressed and responds to CXCL11 greater than CXCL9 and CXCL10 *(36)*. These findings support the notion that augmenting CXCR3/CXCR3 ligand biology will be a therapeutic strategy to enhance angiostasis within the tumor.

Although the above studies have supported CXCR3 as the putative receptor for CXCL4, CXCL9, CXCL10, and CXCL11, it has remained unclear in vivo whether these CXCR3 ligands use CXCR3 on endothelium to mediate their angiostatic effect. Yang and Richmond *(41)* have recently demonstrated that CXCL10 mediates its angiostatic activity in vivo by binding to CXCR3 and not via binding to glycosaminoglycans. To clarify this issue, they created expression constructs for mutants of CXCL10 that exhibit partial or total loss of binding to CXCR3 or loss of binding to glycosaminoglycans. They transfected a human melanoma cell line with these expression vectors, and stable clones were selected and inoculated into immunodeficient mice *(41)*. Tumor cells expressing wild-type CXCL10 showed remarkable reduction in tumor growth compared with control vector-transfected tumor cells. Surprisingly, mutation of CXCL10 resulting in partial loss of receptor binding (IP-10C), or loss of glycosaminoglycans binding (IP-10H), did not significantly alter the ability to inhibit tumor growth. The reduction in tumor growth was associated with a reduction in tumor-associated angiogenesis, leading to the observed increase in both tumor cell apoptosis and necrosis *(41)*. In contrast, expression of the CXCL10 mutant that fails to bind to CXCR3 failed to inhibit tumor growth *(41)*. The above study has been confirmed in another in vivo angiogenesis-dependent model. Burdick et al. *(42)* have found that CXCL11 in a CXCR3-dependent manner inhibits angiogenesis in a murine model of pulmonary fibrosis. These data suggest that CXCR3 receptor binding, but not glycosaminoglycan binding, is essential for the tumor angiostatic activity of CXCR3 ligands.

15.2.3. The Role of CXCR4 in Angiogenesis

CXCL12 has been shown to have an important role in metastasis of cancer *(43)*. However, CXCL12 is also a non-ELR$^+$ CXC chemokine that via CXCR4 has been implicated in promoting angiogenesis *(44–47)*. This has led to the speculation that the predominant function of this ligand/receptor pair in tumorigenesis is due to its angiogenic effect, not necessarily due to its potential of mediating organ-specific metastases. However, in order for the biological axis of CXCL12/CXCR4 to mediate tumor-associated angiogenesis, then both the ligand and receptor should be temporally and spatially present within the tumor. Schrader et al. *(48)* demonstrated in both renal cell carcinoma cell lines and actual patient specimens that CXCR4 is expressed predominately by the tumor cells, and its ligand CXCL12 is essentially absent within the tumor. These findings have been further substantiated in human breast cancer and NSCLC tumor specimens, in which CXCR4 was found expressed on the tumor cells and does not mediate tumor-associated angiogenesis in vivo *(43,49)*. The studies demonstrated that when animals bearing breast or NSCLC tumors were treated with either neutralizing anti-CXCL12 or anti-CXCR4 antibodies, there was no change in the size of the primary tumor nor was there any evidence for a decline in primary tumor-associated angiogenesis *(43);* however, there was a marked attenuation of tumor metastases in an organ-specific manner *(43,49)*. These studies support the notion that CXCL12/CXCR4 biology mediates metastases of the tumor cells in an angiogenesis-independent manner.

An explanation for the disparity of the tumor studies in vivo from in vitro studies of CXCL12/CXCR4 mediated angiogenesis is that tumor cells expressing CXCR4 are themselves able to "out-compete" endothelial cells for CXCL12 if present. In support of this contention, classical angiogenic factors are elevated in human tumors, whereas CXCL12 is not *(29,43,48,50–52)*. Moreover, the depletion of classical angiogenic factors in vivo results in a net reduction of angiogenesis and a consequent reduction in primary tumor size and metastatic potential *(29,50–52)*. These findings suggest a dichotomy in the function for CXCL12 versus classical angiogenic factors; such that angiogenic factors promote metastasis through their effect in mediating angiogenesis, whereas CXCL12 promotes metastasis in an angiogenesis-independent manner via CXCR4-dependent tumor cell migration. This concept supports the notion that expression of CXCR4 on tumor cells may represent a critical biomarker for their propensity to metastasize. Therefore, further understanding the molecular mechanisms that are involved in the regulation of CXCR4 expression on tumor cells could lead to targets to modify expression of this receptor that impacts on metastases.

15.2.4. Possible Non–Receptor-Mediated Inhibition of Angiogenesis

Platelet factor-4 (PF-4)/CXCL4 was the first chemokine described to inhibit neovascularization *(53)*. Although this angiostatic chemokine was the subject of extensive research as a candidate anticancer drug *(54)*, its nonallelic gene variant PF-4_{alt}/PF-4_{var1}/ SCYB4V1 (accession numbers P10720 and M26167 at Swiss-Prot and Genbank databases for the CXCL4L1 protein and gene, respectively) has not been previously investigated *(55,56)*. The product of the nonallelic variant gene of CXCL4, PF-4var1/PF-4alt, designated CXCL4L1, was recently isolated from thrombin-stimulated human platelets and purified to homogeneity *(57)*. Although secreted CXCL4 and CXCL4L1 differ in only three amino acid residues, CXCL4L1 is more potent for inhibiting angiogenesis in response to angiogenic factors in both in vitro and in vivo models of angiogenesis *(57)*. By contrast, CXCL9, CXCL10, and CXCL11 are induced by both type I and II interferons *(32–34,58,59)*.

The molecular mechanism for CXCL4 angiostatic function is still a matter of debate. Brandt et al. *(60)* suggested that CXCL4 is a unique chemokine that does not bind to a GPCR, however it activates cells (i.e., neutrophils) through binding to cell surface glycosaminoglycans (GAGs). Yet, it is not clear whether CXCL4 binding to GAG sites alone is both necessary and sufficient to trigger endothelial cell signaling. For instance, CXCL4 is reported to prevent activation of the extracellular signal-regulated kinase by basic fibroblast growth factor (bFGF) and to inhibit downregulation of the cyclin-dependent kinase inhibitor p21 *(61,62)*. Furthermore, CXCL4 function is not abrogated in heparan sulfate–deficient cells, and CXCL4 mutants or peptides lacking heparin affinity are capable of inhibiting angiogenesis *(54,63)*. Recently, Lasagni et al. *(4)* have identified a splice variant of CXCR3, designated CXCR3B, and have found that this GPCR binds CXCL4 and mediates its angiostatic activity. Finally, other studies have reported that the inhibitory effect of CXCL4 is mediated through complex formation with bFGF or CXCL8 *(63,64)*. These findings suggest that the mechanisms involved in CXCL4L1-mediated attenuation of angiogenesis are complex. Furthermore, the important discovery of a variant of CXCL4 that is more efficacious for inhibiting angiogenesis than authentic CXCL4 has significant implications for the use of this angiostatic factor as a therapeutic tool to inhibit aberrant angiogenesis in a variety of diseases.

Breast and kidney–expressed chemokine (BRAK)/CXCL14 is another non-ELR$^+$ CXC chemokine that has recently been identified to inhibit angiogenesis *(65)*. CXCL14 was first identified by differential display of normal oral epithelial cells and head and neck squamous cell carcinoma *(66)*. CXCL14 was downregulated in tumor specimens compared with normal adjacent tissue *(66)*. The biological significance of the absence of CXCL14 in these tumors remained to be elucidated until Shellenberger et al. *(65)* discovered that CXCL14 inhibited microvascular endothelial cell chemotaxis in vitro in response to CXCL8, bFGF, and vascular endothelial growth factor (VEGF) and inhibited neovascularization in vivo in response to the same angiogenic agonists. Schwarze et al. *(67)* have found that CXCL14 expression is observed in normal and tumor prostate epithelium and focally in stromal cells adjacent to prostate cancer. Interestingly, CXCL14 was found to be significantly upregulated in localized prostate cancer and positively correlated with Gleason score *(67)*. In contrast, CXCL14 levels were unchanged in benign prostate hypertrophy (BPH) specimens *(67)*. Using a model of human prostate cancer in immunodeficient mice, prostate cancer cells transfected with CXCL14 were found to have a 43% reduction of tumor growth compared with controls *(67)*. The above studies support the notion that the loss or inadequate expression of CXCL14 is associated with the transformation of normal epithelial cells to cancer and the promotion of a proangiogenic microenvironment suitable for tumor growth. The receptor that mediates the actions of CXCL14 remains to be determined.

15.3. Conclusions

Although CXC chemokine biology was originally thought to be restricted to recruitment of subpopulations of leukocytes, it has become increasingly clear that these cytokines can display pleiotropic effects in mediating biology that go beyond their originally described function. CXC chemokines are a unique cytokine family that exhibits on the basis of structure/function and receptor binding/activation either angiogenic or angiostatic biological activity in the regulation of angiogenesis. CXC chemokines appear to be important in the regulation of angiogenesis associated with tumorigenesis relevant to cancer and also to chronic fibroproliferative disorders. These findings support the notion that therapy directed at either inhibition of angiogenic or augmentation of angiostatic CXC chemokines may be a novel approach in the treatment of a variety of cancers and chronic fibroproliferative diseases.

Acknowledgments

This work was supported in part by National Institutes of Health grants P50HL67665 (M.P.K), HL04493 (J.A.B.), P50HL67665, HL66027, and CA87879 (R.M.S.).

References

1. Jackson JR, Seed MP, Kircher CH, Willoughby DA, Winkler JD. The codependence of angiogenesis and chronic inflammation. FASEB J 1997;11(6):457–465.
2. Strieter RM, Polverini PJ, Kunkel SL, et al. The functional role of the ELR motif in CXC chemokine-mediated angiogenesis. J Biol Chem 1995;270(45):27348–27357.
3. Belperio JA, Keane MP, Arenberg DA, et al. CXC chemokines in angiogenesis. J Leukoc Biol 2000;68(1):1–8.
4. Lasagni L, Francalanci M, Annunziato F, et al. An alternatively spliced variant of CXCR3 mediates the inhibition of endothelial cell growth induced by IP-10, Mig, and I-TAC, and acts as functional receptor for platelet factor 4. J Exp Med 2003;197(11):1537–1549.
5. Boulday G, Haskova Z, Reinders ME, Pal S, Briscoe DM. Vascular endothelial growth factor-induced signaling pathways in endothelial cells that mediate overexpression of the chemokine IFN-(gamma)-inducible protein of 10 kDa in vitro and in vivo. J Immunol 2006;176(5):3098–3107.
6. Strieter RM, Polverini PJ, Kunkel SL, et al. The functional role of the "ELR" motif in CXC chemokine-mediated angiogenesis. J Biol Chem 1995;270(45):27348–27357.

7. Addison CL, Daniel TO, Burdick MD, et al. The CXC chemokine receptor 2, CXCR2, is the putative receptor for ELR(+) CXC chemokine-induced angiogenic activity. J Immunol 2000;165(9):5269–5277.
8. Murdoch C, Monk PN, Finn A. CXC chemokine receptor expression on human endothelial cells. Cytokine 1999;11(9):704–712.
9. Salcedo R, Resau JH, Halverson D, et al. Differential expression and responsiveness of chemokine receptors (CXCR1-3) by human microvascular endothelial cells and umbilical vein endothelial cells. FASEB J 2000;14(13):2055–2064.
10. Heidemann J, Ogawa H, Dwinell MB, et al. Angiogenic effects of interleukin 8 (CXCL8) in human intestinal microvascular endothelial cells are mediated by CXCR2. J Biol Chem 2003;278(10):8508–8515.
11. Richmond A, Fan GH, Dhawan P, Yang J. How do chemokine/chemokine receptor activations affect tumorigenesis? Novartis Found Symp 2004;256:74–89.
12. Devalaraja RM, Nanney LB, Qian Q, et al. Delayed wound healing in CXCR2 knockout mice. J Invest Dermatol 2000;115(2):234–244.
13. Keane MP, Belperio JA, Xue YY, Burdick MD, Strieter RM. Depletion of CXCR2 inhibits tumor growth and angiogenesis in a murine model of lung cancer. J Immunol 2004;172(5):2853–2860.
14. Burger M, Burger JA, Hoch RC, Oades Z, Takamori H, Schraufstatter IU. Point mutation causing constitutive signaling of CXCR2 leads to transforming activity similar to Kaposi's sarcoma herpesvirus-G protein-coupled receptor. J Immunol 1999;163(4):2017–2022.
15. Gershengorn MC, Geras-Raaka E, Varma A, Clark-Lewis I. Chemokines activate Kaposi's sarcoma-associated herpesvirus G protein-coupled receptor in mammalian cells in culture. J Clin Invest 1998;102(8):1469–1472.
16. Sugden PH, Clerk A. Regulation of the ERK subgroup of MAP kinase cascades through G protein-coupled receptors. Cell Signal 1997;9(5):337–351.
17. Pawson T, Scott JD. Signaling through scaffold, anchoring, and adaptor proteins. Science 1997;278(5346):2075–2080.
18. Shyamala V, Khoja H. Interleukin-8 receptors R1 and R2 activate mitogen-activated protein kinases and induce c-fos, independent of Ras and Raf-1 in Chinese hamster ovary cells. Biochemistry 1998;37(45):15918–15924.
19. Couty JP, Gershengorn MC. Insights into the viral G protein-coupled receptor encoded by human herpesvirus type 8 (HHV-8). Biol Cell 2004;96(5):349–354.
20. Arvanitakis L, Geras-Raaka E, Varma A, Gershengorn MC, Cesarman E. Human herpesvirus KSHV encodes a constitutively active G-protein-coupled receptor linked to cell proliferation. Nature 1997;385(6614):347–350.
21. Bais C, Santomasso B, Coso O, et al. G-protein-coupled receptor of Kaposi's sarcoma-associated herpesvirus is a viral oncogene and angiogenesis activator. Nature 1998;391(6662):86–89.
22. Geras-Raaka E, Arvanitakis L, Bais C, Cesarman E, Mesri EA, Gershengorn MC. Inhibition of constitutive signaling of Kaposi's sarcoma-associated herpesvirus G protein-coupled receptor by protein kinases in mammalian cells in culture. J Exp Med 1998;187(5):801–806.

23. Geras-Raaka E, Varma A, Ho H, Clark-Lewis I, Gershengorn MC. Human interferon-gamma-inducible protein 10 (IP-10) inhibits constitutive signaling of Kaposi's sarcoma-associated herpesvirus G protein-coupled receptor. J Exp Med 1998;188(2):405–408.
24. Yang TY, Chen SC, Leach MW, et al. Transgenic expression of the chemokine receptor encoded by human herpesvirus 8 induces an angioproliferative disease resembling Kaposi's sarcoma. J Exp Med 2000;191(3):445–454.
25. Guo HG, Sadowska M, Reid W, Tschachler E, Hayward G, Reitz M. Kaposi's sarcoma-like tumors in a human herpesvirus 8 ORF74 transgenic mouse. J Virol 2003;77(4):2631–2639.
26. Luan J, Shattuck-Brandt R, Haghnegahdar H, et al. Mechanism and biological significance of constitutive expression of MGSA/GRO chemokines in malignant melanoma tumor progression. J Leukoc Biol 1997;62(5):588–597.
27. Owen JD, Strieter R, Burdick M, et al. Enhanced tumor-forming capacity for immortalized melanocytes expressing melanoma growth stimulatory activity/growth-regulated cytokine beta and gamma proteins. Int J Cancer 1997;73(1):94–103.
28. Addison CL, Belperio JA, Burdick MD, Strieter RM. Overexpression of the duffy antigen receptor for chemokines (DARC) by NSCLC tumor cells results in increased tumor necrosis. BMC Cancer 2004;4(1):28.
29. Arenberg DA, Kunkel SL, Polverini PJ, Glass M, Burdick MD, Strieter RM. Inhibition of interleukin-8 reduces tumorigenesis of human non-small cell lung cancer in SCID mice. J Clin Invest 1996;97(12):2792–2802.
30. Arenberg DA, Keane MP, DiGiovine B, et al. Epithelial-neutrophil activating peptide (ENA-78) is an important angiogenic factor in non-small cell lung cancer. J Clin Invest 1998;102(3):465–472.
31. Moore BB, Arenberg DA, Stoy K, et al. Distinct CXC chemokines mediate tumorigenicity of prostate cancer cells. Am J Pathol 1999;154(5):1503–1512.
32. Luster AD. Chemokines-chemotactic cytokines that mediate inflammation. N Engl J Med 1998;338(7):436–445.
33. Rollins BJ. Chemokines. Blood 1997;90(3):909–928.
34. Balkwill F. The molecular and cellular biology of the chemokines. J Viral Hepat 1998;5(1):1–14.
35. Loetscher M, Loetscher P, Brass N, Meese E, Moser B. Lymphocyte-specific chemokine receptor CXCR3: regulation, chemokine binding and gene localization. Eur J Immunol 1998;28(11):3696–3705.
36. Ehlert JE, Addison CA, Burdick MD, Kunkel SL, Strieter RM. Identification and partial characterization of a variant of human CXCR3 generated by posttranscriptional exon skipping. J Immunol 2004;173(10):6234–6240.
37. Moser B, Loetscher P. Lymphocyte traffic control by chemokines. Nat Immunol 2001;2(2):123–128.
38. Beider K, Nagler A, Wald O, et al. Involvement of CXCR4 and IL-2 in the homing and retention of human NK and NK T cells to the bone marrow and spleen of NOD/SCID mice. Blood 2003;102(6):1951–1958.
39. Soto H, Wang W, Strieter RM, et al. The CC chemokine 6Ckine binds the CXC chemokine receptor CXCR3. Proc Natl Acad Sci U S A 1998;95(14):8205–8210.

40. Romagnani P, Annunziato F, Lasagni L, et al. Cell cycle-dependent expression of CXC chemokine receptor 3 by endothelial cells mediates angiostatic activity. J Clin Invest 2001;107(1):53–63.
41. Yang J, Richmond A. The angiostatic activity of interferon-inducible protein-10/CXCL10 in human melanoma depends on binding to CXCR3 but not to glycosaminoglycan. Mol Ther 2004;9(6):846–855.
42. Burdick MD, Murray LA, Keane MP, et al. CXCL11 attenuates bleomycin-induced pulmonary fibrosis via inhibition of vascular remodeling. Am J Respir Crit Care Med 2005;171(3):261–268.
43. Phillips RJ, Burdick MD, Lutz M, Belperio JA, Keane MP, Strieter RM. The stromal derived factor-1/CXCL12-CXC chemokine receptor 4 biological axis in non-small cell lung cancer metastases. Am J Respir Crit Care Med 2003;167(12):1676–1686.
44. Bachelder RE, Wendt MA, Mercurio AM. Vascular endothelial growth factor promotes breast carcinoma invasion in an autocrine manner by regulating the chemokine receptor CXCR4. Cancer Res 2002;62(24):7203–7206.
45. Salcedo R, Oppenheim JJ. Role of chemokines in angiogenesis: CXCL12/SDF-1 and CXCR4 interaction, a key regulator of endothelial cell responses. Microcirculation 2003;10(3–4):359–370.
46. Kijowski J, Baj-Krzyworzeka M, Majka M, et al. The SDF-1-CXCR4 axis stimulates VEGF secretion and activates integrins but does not affect proliferation and survival in lymphohematopoietic cells. Stem Cells 2001;19(5):453–466.
47. Salcedo R, Wasserman K, Young HA, et al. Vascular endothelial growth factor and basic fibroblast growth factor induce expression of CXCR4 on human endothelial cells: in vivo neovascularization induced by stromal-derived factor-1alpha. Am J Pathol 1999;154(4):1125–1135.
48. Schrader AJ, Lechner O, Templin M, et al. CXCR4/CXCL12 expression and signalling in kidney cancer. Br J Cancer 2002;86(8):1250–1256.
49. Muller A, Homey B, Soto H, et al. Involvement of chemokine receptors in breast cancer metastasis. Nature 2001;410(6824):50–56.
50. Smith DR, Polverini PJ, Kunkel SL, et al. Inhibition of interleukin 8 attenuates angiogenesis in bronchogenic carcinoma. J Exp Med 1994;179(5):1409–1415.
51. Belperio JA, Keane MP, Arenberg DA, et al. CXC chemokines in angiogenesis. J Leukoc Biol 2000;68(1):1–8.
52. Arenberg DA, Kunkel SL, Polverini PJ, et al. Interferon-gamma-inducible protein 10 (IP-10) is an angiostatic factor that inhibits human non-small cell lung cancer (NSCLC) tumorigenesis and spontaneous metastases. J Exp Med 1996;184(3):981–992.
53. Maione TE, Gray GS, Petro J, et al. Inhibition of angiogenesis by recombinant human platelet factor-4 and related peptides. Science 1990;247(4938):77–79.
54. Bikfalvi A, Gimenez-Gallego G. The control of angiogenesis and tumor invasion by platelet factor-4 and platelet factor-4-derived molecules. Semin Thromb Hemost 2004;30(1):137–144.

55. Eisman R, Surrey S, Ramachandran B, Schwartz E, Poncz M. Structural and functional comparison of the genes for human platelet factor 4 and PF4alt. Blood 1990;76(2):336–344.
56. Green CJ, Charles RS, Edwards BF, Johnson PH. Identification and characterization of PF4varl, a human gene variant of platelet factor 4. Mol Cell Biol 1989;9(4):1445–1451.
57. Struyf S, Burdick MD, Proost P, Van Damme J, Strieter RM. Platelets release CXCL4L1, a nonallelic variant of the chemokine platelet factor-4/CXCL4 and potent inhibitor of angiogenesis. Circ Res 2004;95(9):855–857.
58. Strieter RM, Belperio JA, Phillips RJ, Keane MP. CXC chemokines in angiogenesis of cancer. Semin Cancer Biol 2004;14(3):195–200.
59. Strieter RM, Belperio JA, Phillips RJ, Keane MP. Chemokines: angiogenesis and metastases in lung cancer. Novartis Found Symp 2004;256:173–84.
60. Brandt E, Petersen F, Ludwig A, Ehlert JE, Bock L, Flad HD. The beta-thromboglobulins and platelet factor 4: blood platelet-derived CXC chemokines with divergent roles in early neutrophil regulation. J Leukoc Biol 2000;67(4):471–478.
61. Gentilini G, Kirschbaum NE, Augustine JA, Aster RH, Visentin GP. Inhibition of human umbilical vein endothelial cell proliferation by the CXC chemokine, platelet factor 4 (PF4), is associated with impaired downregulation of p21(Cip1/WAF1). Blood 1999;93(1):25–33.
62. Sulpice E, Bryckaert M, Lacour J, Contreres JO, Tobelem G. Platelet factor 4 inhibits FGF2-induced endothelial cell proliferation via the extracellular signal-regulated kinase pathway but not by the phosphatidylinositol 3-kinase pathway. Blood 2002;100(9):3087–3094.
63. Perollet C, Han ZC, Savona C, Caen JP, Bikfalvi A. Platelet factor 4 modulates fibroblast growth factor 2 (FGF-2) activity and inhibits FGF-2 dimerization. Blood 1998;91(9):3289–3299.
64. Dudek AZ, Nesmelova I, Mayo K, Verfaillie CM, Pitchford S, Slungaard A. Platelet factor 4 promotes adhesion of hematopoietic progenitor cells and binds IL-8: novel mechanisms for modulation of hematopoiesis. Blood 2003;101(12):4687–4694.
65. Shellenberger TD, Wang M, Gujrati M, et al. BRAK/CXCL14 is a potent inhibitor of angiogenesis and is a chemotactic factor for immature dendritic cells. Cancer Res 2004;64:8262–8270.
66. Frederick MJ, Henderson Y, Xu X, et al. In vivo expression of the novel CXC chemokine BRAK in normal and cancerous human tissue. Am J Pathol 2000;156(6):1937–1950.
67. Schwarze SR, Luo J, Isaacs WB, Jarrard DF. Modulation of CXCL14 (BRAK) expression in prostate cancer. Prostate 2005;13:13.

16

Chemokine Receptors in Cancer: Pathobiology and Potential Therapeutic Targets

Tonya C. Walser and Amy M. Fulton

Summary

Many chemokine receptors have now been detected in a variety of malignancies. Considerable data now exist that two receptors, CXCR4 and CCR7, are widely expressed in epithelial cancers and contribute to the ability of some tumors to metastasize. Preclinical studies support a role for these receptors in mediating tumor cell migration, but other functions have also been identified including support of tumor cell proliferation in response to ligand stimulation. Less is known about the function of other chemokine receptors expressed in malignant cells, but it is likely that at least some of these also contribute to aggressive behavior. Although the specific role of each receptor is complex, it is becoming clear that these receptors may mediate metastasis to some, but not all, metastatic sites characteristic of each histologic type. Some of the site specificity is due to expression of the cognate ligand at the secondary site; however, other factors to be defined also contribute to chemokine receptor–determined metastatic patterns.

Key Words: Chemokine; receptor; cancer; metastasis; migration.

16.1. Introduction

Chemokines are a superfamily of low-molecular-weight chemotactic cytokines that exert their effects through seven transmembrane domain G protein–coupled receptors. Although some chemokines are constitutively expressed in certain settings, most are induced by proinflammatory mediators, such as IFN-γ and TNF-α. Upon binding to the appropriate receptor, chemokines initiate a

From: *The Receptors: The Chemokine Receptors*
Edited by: J. K. Harrison and N. W. Lukacs

signaling cascade that leads to the production of molecules critical to events such as leukocyte development, maturation, and trafficking. There are more than 40 human chemokines and 19 corresponding chemokine receptors, as well as nonsignaling molecules (e.g., D6 and Duffy; see Chapters 2 and 3 for details). Although chemokine receptors were first described on leukocytes, functional chemokine receptors are now known to be expressed on endothelial cells *(1)*, epithelial cells, and some malignant cells *(2,3)*. Some host cells infiltrating tumors express chemokine receptors, and there is growing evidence that malignancies of diverse histologic types can also express a wide range of chemokine receptors. This chapter will focus on several receptors expressed by malignant cells for which there is significant evidence in support of their role in the control of tumor behavior.

16.2. CXC Chemokine Receptors

16.2.1. CXCR4

Muller et al. performed a comprehensive examination of chemokine receptor expression on a series of breast cancer and melanoma cell lines *(2)*. Using quantitative RT-PCR and specific probes for CXCR1-5, CCR1-10, CX3CR1, and XCR1, seven breast cancer cell lines expressed mRNA primarily for CXCR4, CXCR2, and CCR7. In comparison with normal mammary epithelial cultures, CXCR4 and CCR7 were consistently elevated in malignant cell lines. Like breast cancer cell lines, melanoma cell lines also expressed high levels of CXCR4 and CCR7, but in addition, these cells expressed high levels of CCR10 compared with normal primary melanocytes. Examination of pleural effusions from breast cancer patients, primary invasive lobular carcinomas, or ductal carcinomas also revealed heightened expression of both CXCR4 and CCR7 mRNA in malignant tissues compared with normal mammary gland. CXCR4 expression in primary tumors, axillary lymph nodes, and pulmonary and hepatic metastases was confirmed by immunohistochemistry. Using in vitro migration and invasion assays as measures of metastatic potential, breast cancer cells were responsive to CXCL12 and CCL21, respectively the ligands for CXCR4 and CCR7. Chemotaxis and invasion induced by CXCL12 were both blocked by neutralizing CXCR4 and CXCL12 antibodies.

These studies led the authors to hypothesize that, like hematopoietic cells, malignant cells employ chemokine receptors to migrate toward chemokine gradients. Using a panel of human tissues, CXCL12 mRNA was detected primarily in lymph nodes, lungs, liver, and bone marrow, which are preferred sites of breast cancer metastasis. Finally, treatment with antibody to CXCR4 inhibited metastasis in severe combined immunodeficient (SCID) mice bearing xenografts of human breast cancer cell line MDA-MB-231.

A global gene expression study also detected increased CXCR4 mRNA in invasive breast cancers *(4)*. Likewise, invasive ductal carcinomas metastasizing to lymph nodes, but not other sites, expressed more CXCR4 *(5)*. Several laboratories have shown that inhibition of CXCR4 signaling using either small interfering RNA (siRNA) *(6)*, small-molecular-weight peptides *(7)*, or neutralizing antibody *(2)* blocked migration of breast cancer cell lines to CXCL12. Conversely, forced overexpression of CXCR4 in B16 melanoma cells increased metastatic spread to the lungs *(8)*. Interestingly, in melanoma, CXCR4 expression mediated metastasis to lungs, but not to the lymph nodes. Taken together, these findings are consistent with a mechanism whereby breast tumor cells expressing CXCR4 migrate in response to chemokine gradients toward lymph nodes expressing higher levels of the CXCR4-specific ligand, CXCL12, whereas melanoma cells migrate toward lungs. CXCR4-CXCL12 interactions may direct chemotaxis to only particular sites, and this organ specificity may differ depending on the cancer type.

CXCR4 receptor function has also been characterized in ovarian cancer. Unlike breast cancer, in which several receptors are detected, CXCR4 appears to be the only chemokine receptor expressed in ovarian cancers *(3)*. Expression of CXCR4 and migration to CXCL12 has also been documented in many other cancer types, including neuroblastoma, glioblastoma, non–Hodgkin lymphoma, small cell lung carcinoma, renal cell carcinoma, and pancreatic carcinoma *(9–14)*, among others, suggesting that this may be an important mechanism common to many cancers.

In addition to mediating tumor cell migration and invasion, CXCR4 may play a role in other critical functions that affect metastatic success, and the mechanisms are beginning to be elucidated. Proliferation of ovarian carcinoma cells was stimulated in vitro by recombinant CXCL12 *(15)*, an effect associated with Akt activation. The effects of CXCL12 on proliferation were blocked with AMD3100, a specific inhibitor of CXCR4. Likewise, neutralizing antibody to CXCR4 inhibited proliferation and induced apoptosis of non–Hodgkin lymphoma cells *(10)*. Studies in vivo have confirmed that CXCR4 can play a role in local tumor growth. Non–Hodgkin lymphoma tumor-bearing mice treated with antibody to CXCR4 experienced a reduction in tumor weight and increased survival time. Knockdown of CXCR4 with siRNA also limited growth of orthotopically transplanted 4T1 mammary tumor cells *(16)*. Mice transplanted with control tumor cells died from pulmonary metastases, whereas those treated with CXCR4-silenced tumor cells survived without developing macroscopic metastases, confirming a role for CXCR4 in promoting metastasis.

Although it seems clear that CXCR4 is often highly expressed in malignancies, the mechanism responsible for upregulation of CXCR4 have not been completely elucidated. CXCR4 on breast cancer cell lines MDA-MB-361 and SKBR3 has been implicated in CXCL12-mediated transactivation of HER2-neu, an effect associated with Src kinase activation *(17)*. Inhibitors of CXCR4, epidermal growth factor receptor (EGFR)/HER2-neu tyrosine kinase, or Src kinase all blocked CXCL12-induced HER2-neu phosphorylation and tumor cell migration. Hypoxia has also been implicated in increased CXCR4 expression by tumor cells *(18,19)*. Tumors rapidly outgrow the local vasculature and become hypoxic. In order to continue expanding, tumors must induce new blood vessel formation. Vascular endothelial growth factor (VEGF) is suspected to regulate CXCR4 expression on MDA-MB-231 breast cancer cells, because VEGF antisense oligodeoxynucleotides blocked migration of the cells to exogenous CXCL12 *(20)*. This finding was confirmed in ovarian cancer, where VEGF and CXCL12 synergistically induced in vivo angiogenesis *(18)*. Specifically, VEGF upregulated CXCR4 expression on vascular endothelial cells, facilitating tumor cell migration, and CXCR4 upregulated VEGF on vascular endothelial cells, facilitating tumor cell growth and escape from oxygen starvation–induced apoptosis. Hypoxia was identified as the trigger for this angiogenic synergism. Constitutively active nuclear factor kappa B (NFκB), the extracellular signal-activated transcription factor, is a common feature of malignancies, and it was also examined as a potential regulator of CXCR4 expression *(21)*. NFκB p65 and p50 were reported to bind and activate the CXCR4 promoter in breast cancer cells. Other studies have shown that the tumor suppressor von Hippel–Lindau gene, which is linked to the development of several tumor types, negatively regulated expression of CXCR4 *(22,23)*. This may represent an additional regulatory mechanism for those cancers in which the von Hippel–Lindau gene is inactivated.

Thus, expression of CXCR4 on tumor cells, which is common to many breast cancers as well as other cancer types, may contribute to malignant behavior in several ways. CXCR4 expression by tumor cells may interact with CXCL12 to facilitate tumor cell growth and escape from oxygen starvation–induced apoptosis and as a mechanism to home (metastasize) to secondary sites.

16.2.2. CXCR3

Like CXCR4, CXCR3 has been detected on several malignant cell lines. Chronic lymphocytic leukemia cells express CXCR3 *(24)*, as do human melanomas and neuroblastomas *(25,26)*. In the neuroblastoma study, CXCR3 was expressed on six malignant cell lines, and CXCL10, one of three CXCR3 ligands, was secreted by bone marrow stromal cell lines *(27)*, suggesting a possible mechanism by which tumor cells would home to the bone marrow. In this study, CXCL10 and CXCL11 ligands induced Erk phosphorylation in the neuroblastoma cell lines in a $G\alpha_i$-dependent manner.

CXCR3 expression on human breast cancer cell lines MDA-MB-231, MCF-7, and T-47D has also been reported *(25)*. Increased expression of CXCR3 was observed when the tumor cells were serum-starved. The authors of this study observed that CXCR3 was upregulated, instead of downregulated, after exposure to high concentrations of the cognate ligand CXCL10, and the tumor cells themselves secreted CXCL10. This finding was somewhat unexpected, because ligand-induced receptor internalization is the norm. These data support a mechanism by which tumor-produced CXCL10 regulates tumor behavior via CXCR3 in an autocrine loop. We have shown that, like human breast cancer cells, murine mammary tumor cells express CXCR3 that is functionally coupled to migration and calcium mobilization *(28)*. Using a small-molecular-weight pharmacologic antagonist of CXCR3, tumor metastasis in vivo was inhibited.

B16.F10 melanoma cells also express functional CXCR3, as determined by ligand-induced actin polymerization, migration, invasion, and cell survival *(29)*. When CXCR3 gene expression by melanoma cells was silenced using antisense RNA, syngeneic mice developed fewer lymph node metastases than mice injected with control tumor cells. Mice treated with Freund's adjuvant to increase levels of CXCL9 and CXCL10 in the draining lymph node experienced a slight increase in lymph node metastasis of the melanoma cells. The extent of pulmonary metastases was not affected by CXCR3 gene silencing, indicating that CXCR3 mediates organ-specific metastasis.

Like CXCR4, CXCR3 mediates metastasis of breast and melanoma cells; however, the same receptor directs migration to different secondary sites. Breast cancers employ CXCR3 to migrate to the lungs, whereas melanoma CXCR3 mediates lymph node metastasis. Conversely, B16 melanoma employs CXCR4 to home to the lungs *(8)*. Whether the same pattern of divergent receptor function applies in other malignancies remains to be determined.

16.3. CC Chemokine Receptors

16.3.1. CCR7

CCR7 was highly expressed in breast, melanoma, gastric, non–small cell lung, esophageal, head and neck, and colorectal cancer, as well as in chronic lymphocytic leukemia *(2,30–36)*. Higher expression of CCR7 in each of these malignancies was associated with lymph node metastasis and disease progression. These findings were confirmed in a murine model of melanoma, where forced overexpression of CCR7 in B16 melanoma cells resulted in increased homing to lymph nodes *(36)*. In colorectal carcinoma, focal CCR7 expression at the tumor invasion front, as opposed to disseminated membrane expression, was linked to decreased survival *(31)*. Because both major ligands for CCR7,

CCL19 and CCL21, are expressed constitutively in lymph nodes, interaction of CCR7 with these ligands is predicted to promote metastasis to lymph nodes in a manner analogous to CXCR4/CXCL12.

The function of CCR7 on malignant cells of squamous cell carcinoma of the head and neck was dependent upon phosphoinositide-3 kinase and treatment with the ligand CCL19 *(37)*. Expression of both CCR7 and CXCR4 on melanoma cells was epigenetically regulated. Expression of functional receptors was enhanced by treatment with the histone deacetylase inhibitor trichostatin A and the demethylating agent, 5-aza-2-deoxycytidine *(38)*. Therefore, chemotherapy with histone deacetylase inhibitors, currently being evaluated in clinical trials, may have the undesirable side effect of increasing CCR7 and CXCR4 expression on malignant cells.

16.3.2. CCR5

Aggressive natural killer (NK) cell leukemia is associated with a poor outcome, likely because of the multiorgan infiltration by the malignant NK cells *(39)*. Whereas normal human NK cells and those from less aggressive lymphoproliferative diseases were negative for CCR5, unless activated with IL-2 and IL-15, malignant NK cells from the aggressive leukemia were simultaneously positive for CXCR1 and CCR5 and showed enhanced chemotaxis toward CXCR1 and CCR5 ligands. These data suggest that chemokine receptor expression may play a significant role in the characteristic multiorgan involvement in leukemia. Additionally, expression of CCR5 may represent a marker of aggressive NK cell leukemia.

Both surface and intracellular CCR5 were detected in prostate cancer, however, intracellular expression of CCR5 was much more abundant by flow cytometry *(40)*. CCL5 stimulation of the cancer cells induced invasion, and treatment with the CCR5 antagonist TAK-779 blocked this invasion. CCL5, therefore, acts directly on prostate cancer cells through CCR5, affecting their growth and survival in an autocrine manner.

One of the earliest attempts at discerning the mechanism by which CCR5 influences cancer progression was in a breast cancer xenograft model *(41)*. Stimulation of MCF-7 cells with CCL5, a CCR5 ligand, resulted in activation of the tumor suppressor p53 through a mechanism that depended on CCR5 expression on MCF-7 and G_i, Janus kinase-2 (JAK2), and p38-mitogen activated protein kinase (MAPK) activation. CCR5 and p53 together regulated in vivo proliferation of MCF-7 and MDA-MB-231 tumor cells, where reduced expression of CCR5 resulted in increased proliferation of tumor cells bearing wild-type p53. This finding was confirmed in biopsy specimens from breast cancer patients, where breast tumors bearing wild-type p53 grew faster and the patient relapsed sooner, if the patient possessed a CCR5Δ32 allele, which yields nonfunctional CCR5.

16.4. Other CXC Chemokine Receptors

16.4.1. CXCR1 and CXCR2

Elevated CXCR1 and CXCR2 expression was found in pancreatic cancer, colon cancer, non–small cell lung cancer, small cell lung cancer, melanoma, and murine squamous cell carcinoma. The CXCR1 and CXCR2 ligand, CXCL8, was more prevalent in malignant cells versus normal tissues *(42–48)*. Expression of receptor and ligand by the same malignant cells suggests autocrine chemokine stimulation, leading to increased proliferation, survival, tumor cell migration, and disease progression. In fact, levels of receptor and ligand expression were linked to both the proliferative and invasive potential of melanoma *(48)*. In pancreatic cancer, receptor and ligand expression were localized to the cytoplasm of malignant cells but still correlated positively with advanced stage, proliferation, and angiogenesis *(47)*.

Androgen independence is a common feature of prostate cancer progression. CXCR1 expression and the emergence of androgen independence were examined in prostate cancer specimens from patients undergoing hormonal treatment *(49)*. CXCR1, CXCR2, and CXCL8 promoted androgen-independent growth of prostate cancer cells. CXCL8 produced by the neuroendocrine tumor cells acted on CXCR1 in non–neuroendocrine tumor cells to promote androgen-independent growth. Conversely, CXCL8 acted on the closely related receptor, CXCR2, in neuroendocrine tumor cells to regulate their differentiation and function.

Strieter et al. determined that a particular amino acid motif in CXC chemokines, glu-leu-arg (ELR), conferred angiogenic potential to chemokines *(50)*. This angiogenic activity was mediated by the binding of ELR^+ CXC chemokines to CXCR2 *(51)* (see Chapter 15 in this volume). Depletion of CXCR2 using knockout mice or neutralizing antibodies to CXCR2 inhibited tumor growth and angiogenesis in a murine model of lung cancer *(52)*. Likewise, the absence of CXCR2 in knockout mice resulted in decreased angiogenesis, increased tumor necrosis, and diminished metastasis of renal cell carcinoma to lungs *(53)*.

16.4.2. CXCR5

Primary central nervous system lymphoma is a rare form of non–Hodgkin lymphoma that rarely spreads beyond the central nervous system. As stated previously, malignant cells of classical Hodgkin lymphoma express CXCR4, CXCR5, and CCR5 that may direct their metastasis to lymph nodes *(34)*. Expression of these receptors on malignant B cells of primary central nervous system lymphomas was confirmed, but expression was restricted to the cytoplasm *(54)*. Primary central nervous system lymphoma may be a malignant variant of differentiated B cells that physiologically downregulate CXCR4,

CXCR5, and CCR5 during normal B-cell differentiation. The authors propose that the cytoplasmic location of the receptors interferes with migration, thereby accounting for diminished dissemination characteristic of this histologic type.

16.5. Other CC Chemokine Receptors

16.5.1. CCR1

CCR1 and CXCR4 expression was observed on multiple myeloma cells *(55)*. The CCR1 and CXCR4 ligands, CCL3 and CXCL12, were expressed at high levels in the bone marrow of patients with multiple myeloma, suggesting that CCR1 and CXCR4 expressed on multiple myeloma cells may mediate their migration to and retention in the bone marrow.

CCR1 was the only chemokine receptor found in hepatitis C–induced hepatocellular carcinoma *(56)*. CCL3 is also expressed and may interact in an autocrine fashion with CCR1 to promote the growth and survival of hepatoma cells. CCR1 and CCL3 were also abundantly expressed in *N*-nitrosodiethylamine–induced hepatocellular carcinoma tissues and hepatitis B virus surface antigen transgenic mice. In both $CCR1^{-/-}$ and $CCL3^{-/-}$ mice treated with *N*-nitrosodiethylamine, the number and size of tumor foci were reduced compared with tumor foci in wild-type mice treated with the carcinogen. In contrast, tumor incidence was higher in the knockout mice. Both tumor angiogenesis and the number of intratumoral Kupffer cells, a source of growth factors and matrix metalloproteinases (MMPs), were diminished in *N*-nitrosodiethylamine–treated $CCR1^{-/-}$ and $CCL3^{-/-}$ mice. MMPs are proteolytic enzymes implicated in metastatic invasion of the extracellular matrix. Expression of MMP-9 and MMP-13 was augmented in the livers of wild-type mice treated with *N*-nitrosodiethylamine, but MMP-9 gene induction by carcinogen was reduced in $CCR1^{-/-}$ and $CCL3^{-/-}$ mice.

16.5.2. CCR2

Functional CCR2 was expressed on multiple myeloma, and the CCR2 ligands, CCL2, CCL7, and CCL8, were produced by stromal cells cultured from normal and malignant myeloma bone marrow samples *(57)*. Ligand expression likely facilitated the migration of multiple myeloma cells to the bone marrow.

CCR2 expression was also examined in liver tumor formation induced by intraportal injection of a murine colon adenocarcinoma cell line *(58)*. Tumor growth occurred at similar rates in wild-type and $CCR2^{-/-}$ mice, but the number and size of the tumor foci were significantly inhibited in the knockout mice. Inhibition of angiogenesis and MMP-2 expression was observed in the CCR2 knockout mice, along with a reduction in the accumulation of macrophages and hepatic stellate cells. The authors observed that MMP-2 was expressed in hepatic stellate cells, not in macrophages, and they hypothesized CCR2 mediates trafficking of hepatic stellate cells, a main source of MMP-2, thereby facilitating angiogenesis during liver tumor formation.

The CCR2-64I, a single nucleotide polymorphism, was considered in association with the risk of developing invasive cervical cancer from squamous intraepithelial lesions *(59)*. In an examination of 326 cases, the mutated CCR2 allele had a protective role in evolution from high-grade squamous intraepithelial lesions to invasive cervical cancer. The CCR2-64I polymorphism was also considered in relation to breast, bladder, and nonmelanoma skin cancers *(60)*. In this study, the mutated CCR2 allele was found to confer protection from breast cancer only. Taken together, these data suggest that CCR2 signaling normally contributes to the progression of certain tumor types.

16.5.3. CCR4

CCR4 was expressed in adult T-cell leukemia/lymphoma and was associated with skin involvement, and the CCR4 ligand, CCL17, was expressed on normal and inflamed cutaneous endothelia *(61)*. CCR4 expression also characterized other subtypes of T-cell and NK-cell lymphomas. In peripheral T-cell lymphoma, unspecified, CCR4 expression was an independent unfavorable prognostic factor. Interestingly, in this setting, CCR4 mRNA expression positively correlated with FoxP3, a putative marker of T regulatory (Treg) cell activity that suppresses antitumor immunity.

Because CCR4 is a selective marker for adult T-cell leukemia/lymphoma, therapy with antibody directed to CCR4 was examined. In this study, a defucosylated antibody against CCR4 was compared with a conventional CCR4 antibody, and the former was found to possess superior antitumor activity and antibody-dependent cellular cytotoxicity *(62)*. The defucosylated antibody inhibited CCR4 expression, inhibited Treg activity, and enhanced suppression of tumor growth *(63)*.

16.5.4. CCR6

Mouse plasmacytoma grows rapidly in situ and quickly metastasizes to the lymph nodes and bone marrow and later to the liver and lungs. Functional CCR6 and CCR7 were expressed at high levels in malignant cells found within metastatic nodules in the liver compared with normal liver tissue *(64)*. Likewise, functional CCR6 and CCR7 were expressed in areas of liver metastases arising from colon, thyroid, and ovarian carcinomas. Because the human liver constitutively expresses the CCR6 ligand, CCL20, and attracts malignant cells expressing CCR6, metastasis of several cancer types to the liver may be mediated by the CCR6-CCL20 axis.

In a study of colorectal cancer, CCR6 was expressed in primary tumors and metastases, whereas CCL20 expression was downregulated in metastases compared with normal intestinal epithelial cells *(65)*. Tumor cell proliferation and metastasis were stimulated by treatment with CCL20. Likewise, CCR6 and CCL20 were also expressed in pancreatic cancer *(66)*. Increasing concentrations of the ligand promoted a dose-dependent increase in pancreatic tumor cell invasion of type IV collagen, an effect inhibited by incubation with antibody to CCR6. These data again indicate that the metastasis of certain malignant cell types may be mediated by CCR6 expression.

16.5.5. CCR9

Epithelial tumor metastasis to the small intestine is rare. CCR9 was highly expressed in all melanoma cells derived from metastatic lesions in the small intestine *(65)*. CCL25 was also expressed in the small intestine and in the thymus. Only malignant cells derived from metastatic lesions in the small intestine responded to CCL25, as determined by receptor downregulation and actin polymerization after binding. This suggests that the CCR9-CCL25 axis guides the metastasis of malignant cell types to the small intestine. Similarly, CCR9 expression was elevated in prostate cancer compared with normal prostate tissue, and CCL25 was expressed in common sites of prostate cancer metastasis *(67)*. Cells derived from cancerous prostate tissue were more responsive to CCL25 than cells from normal tissue, again indicating that the CCR9-CCL25 axis may drive the metastasis of prostate cancer.

16.6. Conclusions

The earliest studies of chemokines and their receptors in cancer were surveys of chemokine expression in tumors and their metastases. In an early report, Muller et al. described expression of chemokine receptors on malignant cells *(2)*. The authors suggested that tumor cells bearing a particular chemokine receptor travel out of the local tumor toward distant sites of metastasis where their ligand is expressed at a higher level. For example, CXCR4 is expressed by breast cancer cells, but not by normal breast tissue, and CXCL12, the ligand for CXCR4, is expressed in the lung, but not in the tumor. The gradient favors migration of the CXCR4-expressing breast cancer cells to the lung where CXCL12 is expressed. The literature is now replete with reports of chemokine receptor expression in malignant cell lines and clinical samples and supports their usefulness as markers of disease progression and prognosis.

For CXCR4/CXCL12, considerable data exist to suggest that this receptor/ligand axis is an important determinant of metastatic capacity. Less is known about the function of other chemokine receptors, however, it is highly likely that at least some additional receptors play a similar role in promoting metastasis. Evidence for organ specificity of different receptors is emerging, but the mechanisms of the specificity remain to be determined. Preclinical studies in which several receptors have been targeted support the continued examination of strategies to inhibit chemokine receptor function to treat malignant disease.

Acknowledgments

Work in the authors' laboratory was supported by National Institutes of Health and Susan G. Komen Breast Cancer Foundation grants.

References

1. Gupta SK, Lysko PG, Pillarisetti K, Ohlstein E, Stadel JM. Chemokine receptors in human endothelial cells. J Biol Chem 1998;273:4282–4287.
2. Muller A, Homey B, Soto H, et al. Involvement of chemokine receptors in breast cancer metastasis. Nature 2001;410:50–56.
3. Scotton CJ, Wilson JL, Milliken D, Stamp G, Balkwill FR. Epithelial cancer cell migration: a role for chemokine receptors? Cancer Res 2001;61:4961–4965.
4. Porter DA, Krop IE, Nasser S, et al. A SAGE (serial analysis of gene expression) view of breast tumor progression. Cancer Res 2001;61:5697–5702.
5. Kato M, Kitayama J, Kazama S, Nagawa H. Expression pattern of CXC chemokine receptor-4 is correlated with lymph node metastasis in human invasive ductal carcinoma. Breast Cancer Res 2003;5:144–150.
6. Chen Y, Stamatoyannopoulos G, Song C-Z. Down-regulation of CXCR4 by inducible small interfering RNA inhibits breast cancer cell invasion *in vitro*. Cancer Res 2003;63:4801–4804.
7. Tamamura H, Hori A, Kanzaki N, et al. T140 analogs as CXCR4 antagonists identified as anti-metastatic agents in the treatment of breast cancer. FEBS Lett 2003;550:79–83.
8. Murakami T, Maki W, Cardones AR, et al. Expression of CXC chemokine receptor-4 enhances the pulmonary metastatic potential of murine B16 melanoma cells. Cancer Res 2002;62:7328–7334.
9. Barbero S, Bonavia R, Bajetto A, et al. Stromal cell-derived factor 1α stimulates human glioblastoma cell growth through the activation of both extracellular signal-regulated kinases 1/2 and AKT. Cancer Res 2003;63:1969–1974.
10. Bertolini F, Dell'Agnola C, Mancuso P, et al. CXCR4 neutralization, a novel therapeutic approach for non-Hodgkin's lymphoma. Cancer Res 2002;62:3106–3112.

11. Germinder H, Sagi-Assif O, Goldberg L, et al. A possible role for CXCR4 and its ligand, the CXC chemokine stromal cell-derived factor-1, in the development of bone marrow metastases in neuroblastoma. J Immunol 2001;167:4747–4757.
12. Kijima T, Maulik G, Ma PC, et al. Regulation of cellular proliferation, cytoskeletal function and signal transduction through CXCR4 and c-kit in small cell lung cancer cells. Cancer Res 2002:62:6304–6311.
13. Marchesi F, Monti P, Leone BE, et al. Increased survival, proliferation, and migration in metastatic human pancreatic tumor cells expressing functional CXCR4. Cancer Res 2004;64:8420–8427.
14. Schrader AJ, Lechner O, Templin M, et al. CXCR4/CXCL12 expression and signaling in kidney cancer. Br J Cancer 2002;86:1250–1256.
15. Scotton CJ, Wilson JL, Scott K, et al. Multiple actions of the chemokine CXCL12 on epithelial tumor cells in human ovarian cancer. Cancer Res 2002;62:5930–5938.
16. Smith MCP, Luker KE, Gargow JR, et al. CXCR4 regulates growth of both primary and metastatic breast cancer. Cancer Res 2004;64:8604–8612.
17. Cabioglu N, Summy J, Miller C, et al. CXCL-12/stromal cell-derived factor-1α transactivates HER2-neu in breast cancer cells by a novel pathway involving src kinase activation. Cancer Res 2005;65:6493–6497.
18. Kryczek I, Lange A, Mottram P, et al. CXCL12 and vascular endothelial growth factor synergistically induce neoangiogenesis in human ovarian cancer. Cancer Res 2005;65:465–472.
19. Schioppa T, Uranchimeg B, Saccani A, et al. Regulation of the chemokine receptor CXCR4 by hypoxia. J Exp Med 2003;198:1391–1402.
20. Bachelder RE, Wendt MA, Mercurio AM. Vascular endothelial growth factor promotes breast carcinoma invasion in an autocrine manner by regulating the chemokine receptor CXCR4. Cancer Res 2002;62:7203–7206.
21. Helbig G, Christopherson KW, Bhat-Nakshatri P, et al. NF-κB promotes breast cancer cell migration and metastasis by inducing the expression of the chemokine receptor CXCR4. J Biol Chem 2003;278:21631–21638.
22. Staller P, Sulitkova J, Lisztwan J, et al. Chemokine receptor CXCR4 downregulated by von Hippel-Lindau tumor suppressor pVHL. Nature 2003;425:307–311.
23. Zagzag D, Krishnamachary B, Yee H, et al. Stromal cell-derived factor-1α and CXCR4 expression in hemangioblastoma and clear cell-renal cell carcinoma: von Hippel-Lindau loss-of-function induces expression of a ligand and its receptor. Cancer Res 2005;65:6178–6188.
24. Jones D, Benjamin RJ, Shahsafaei A, Dorfman DM. The chemokine receptor CXCR3 is expressed in a subset of B-cell lymphomas and is a marker of B-cell chronic lymphocytic leukemia. Blood 2000;95:627–632.
25. Goldberg-Bittman L, Neumark E, Sagi-Assif O, et al. The expression of the chemokine receptor CXCR3 and its ligand, CXCL10, in human breast adenocarcinoma cell lines. Immunol Lett 2004;92:171–178.

26. Robledo MM, Bartolome RA, Longo N, et al. Expression of functional chemokine receptors CXCR3 and CXCR4 on human melanoma cells. J Biol Chem 2001; 48:45098–45105.
27. Goldberg-Bittman L, Sagi-Assif O, Meshel T, et al. Cellular characteristics of neuroblastoma cells: regulation by the ELR$^-$-CXC chemokine CXCL10 and expression of a CXCR3-like receptor. Cytokine 2005;29:105–117.
28. Walser TC, Rifat S, Ma X, et al. Antagonism of CXCR3 inhibits lung metastasis in a murine model of metastatic breast cancer. Cancer Res 2006;66:7701–7707.
29. Kawada K, Sonoshita M, Sakashita H, et al. Pivotal role of CXCR3 in melanoma cell metastasis to lymph nodes. Cancer Res 2004;64:4010–4017.
30. Ding Y, Shimada Y, Maeda M, et al. Association of CC chemokine receptor 7 with lymph node metastasis of esophageal squamous cell carcinoma. Clin Cancer Res 2003;9:3406–3412.
31. Gunther K, Leier J, Henning G, et al. Prediction of lymph node metastasis in colorectal carcinoma by expression of chemokine receptor CCR7. Int J Oncol 2005;116:726–733.
32. Mashino K, Sadanaga N, Yamaguchi H, et al. Expression of chemokine receptor CCR7 is associated with lymph node metastasis of gastric carcinoma. Cancer Res 2002;62:2937–2941.
33. Takanami, I. Overexpression of CCR7 mRNA in nonsmall cell lung cancer: correlation with lymph node metastasis. Int J Cancer 2003;105:186–189.
34. Till KJ, Lin K, Zuzel M, Cawley JC. The chemokine receptor CCR7 and α4 integrin are important for migration of chronic lymphocytic leukemia cells into lymph nodes. Neoplasia 2002;99:2977–2984.
35. Wang J, Hunt JL, Gooding W, et al. Expression pattern of chemokine receptor 6 (CCR6) and CCR7 in squamous cell carcinoma of the head and neck identifies a novel metastatic phenotype. Cancer Res 2004;64:1861–1866.
36. Wiley HE, Gonzalez EB, Maki W, Wu M-T, Hwang ST . Expression of CC chemokine receptor-7 and regional lymph node metastasis of B16 murine melanoma. J Nat Cancer Inst 2001;93:1638–1643.
37. Wang J, Zhang X, Thomas SM, et al. Chemokine receptor 7 activates phosphoinositide-3-kinase-mediated invasive and prosurvival pathways in head and neck cancer cells independent of EGFR. Oncogene 2005;24:5897–5904.
38. Mori T, Kim J, Yamano T, et al. Epigenetic up-regulation of C-C chemokine receptor 7 and C-X-C chemokine receptor 4 expression in melanoma cells. Cancer Res 2005;65:180–187.
39. Makishima H, Ito T, Asano N, et al. Significance of chemokine receptor expression in aggressive NK cell leukemia. Leukemia 2005;19:1169–1174.
40. Vaday GG, Peehl DM, Kadam PA, Lawrence DM. Expression of CCL5 (RANTES) and CCR5 in prostate cancer. Prostate 2006;66:124–134.
41. Manes S, Mira E, Colomer R, et al. CCR5 expression influences the progression of human breast cancer in a p53-dependent manner. J Exp Med 2003;198:1381–1389.

42. Kuwada Y, Sasaki T, Morinaka K, Kitadai Y, Mukaida N, Chayama K. Potential involvement of IL-8 and its receptors in the invasiveness of pancreatic cancer cells. Int J Oncol 2003;22:765–771.
43. Li A, Varney ML, Singh RK. Constitutive expression of growth regulated oncogene (gro) in human colon carcinoma cells with different metastatic potential and its role in regulating their metastatic potential. Clin Exp Metastasis 2004; 21:571–579.
44. Loukinova E, Dong G, Enamorado-Ayalya I, et al. Growth regulated oncogene-α expression by murine squamous cell carcinoma promotes tumor growth, metastasis, leukocyte infiltration and angiogenesis by a host CXC Receptor-2 dependent mechanism. Oncogene 2000;19:3477–3486.
45. Ramjeesingh R, Leung R, Siu C-H. Interleukin-8 secreted by endothelial cell induces chemotaxis of melanoma cells through the chemokine receptor CXCR1. FASEB J 2003;17:1292–1294.
46. Zhu YM, Webster SJ, Flower D, Woll PJ. Interleukin-8/CXCL8 is a growth factor for human lung cancer cells. Br J Cancer 2004;91:1970–1976.
47. Murphy C, McGurk M, Pettigrew J, et al. Nonapical and cytoplasmic expression of interleukin-8, CXCR1, CXCR2 correlates with cell proliferation and microvessel density in prostate cancer. Clin Cancer Res 2005;11:4117–4127.
48. Varney ML, Li A, Dave BJ, Bucana CD, Johansson SL, Singh RK. Expression of CXCR1 and CXCR2 receptors in malignant melanoma with different metastatic potential and their role in interleukin-8 (CXCL-8)-mediated modulation of metastatic phenotype. Clin Exp Metastasis 2003;20:723–731.
49. Huang J, Yao JL, Zhang L, et al. Differential expression of interleukin-8 and its receptors in the neuroendocrine and non-neuroendocrine compartments of prostate cancer. Am J Pathol 2005;166:1807–1815.
50. Strieter RM, Polverini PJ, Kunkel SL, et al. The functional role of the ELR motif in CXC chemokine-mediated angiogenesis. J Biol Chem 1995;270:27348–27357.
51. Addison CL, Daniel TO, Burdick MD, et al. The CXC chemokine receptor 2, CXCR2, is the putative receptor for ELR^+ CXC chemokine-induced angiogenic activity. J Immunol 2000;165:5269–5277.
52. Keane MP, Belperio JA, Xue YY, Burdick MD, Strieter RM. Depletion of CXCR2 inhibits tumor growth and angiogenesis in a murine model of lung cancer. J Immunol 2004;172:2853–2860.
53. Mestas J, Burdick MD, Reckamp K, Pantuck A, Figlin RA, Strieter RM. The role of CXCR2/CXCR2 ligand biological axis in renal cell carcinoma. J Immunol 2005;175:5352–5357.
54. Jahnke K, Coupland SE, Na IK, et al. Expression of the chemokine receptors CXCR4, CXCR5, and CCR7 in primary central nervous system lymphoma. Blood 2005;106:384–5.
55. Moller C, Stromberg T, Juremalm M, Nilsson K, Nisson G. Expression and function of chemokine receptors in human multiple myeloma. Leukemia 2003;17: 203–210.

56. Yang X, Lu P, Fujii C, et al. Essential contribution of a chemokine, CCL3, and its receptor, CCR1, to hepatocellular carcinoma progression. Int J Cancer 2006;118:1869–1876.
57. Vande Broek I, Asosingh K, Vanderkerken K, et al. Chemokine receptor CCR2 is expressed by human multiple myeloma cells and mediates migration to bone marrow stromal cell-produced monocyte chemotactic proteins MCP-1, -2, and -3. Br J Cancer 2003;88:855–862.
58. Yang X, Lu P, Ishida Y, Kuziel WA, Fujii C, Mukaida N. Attenuated liver tumor formation in the absence of CCR2 with a concomitant reduction in the accumulation of hepatic stellate cells, macrophages and neovascularization. Int J Cancer 2006;118:335–345.
59. Coelho A, Matos A, Catarino R, et al. Protective role of the polymorphism CCR2–64I in the progression from squamous intraepithelial lesions to invasive cervical carcinoma. Gynecol Oncol 2005;96:760–764.
60. Zafiropoulos A, Crikas N, Passam AM, Spandidos DA. Significant involvement of CCR2-64I and CCL12-3a in the development of sporadic breast cancer. J Med Genet 2004;41:e59.
61. Ishida T, Inagaki H, Utsunomiya A, et al. CXC chemokine receptor 3 and CC chemokine receptor 4 expression in T-cell and NK-cell lymphomas with special reference to clinicopathological significance for peripheral T-cell lymphoma, unspecified. Clin Cancer Res 2004;10:5494–5500.
62. Niwa R, Shoji-Hosaka E, Sakurada M, et al. Defucosylated chimeric anti-CC chemokine receptor 4 IgG1 with enhanced antibody-dependent cellular cytotoxicity shows potent therapeutic activity to T-cell leukemia and lymphoma. Cancer Res 2004;64:2127–2133.
63. Ishida T, Iida S, Akatsuka Y, et al. The CC chemokine receptor 4 as a novel specific molecular target for immunotherapy in adult T-cell leukemia/lymphoma. Clin Cancer Res 2004;10:7529–7539.
64. Dellacasagrande J, Schreurs OJ, Hofgaard O, et al. Liver metastasis of cancer facilitated by chemokine receptor CCR6. Scand J Immunol 2003;57:534–544.
65. Letsch A, Keilholz U, Schadendorf D, et al. Functional CCR9 expression is associated with small intestinal metastasis. J Invest Derm 2004;122:685–690.
66. Kimsey TF, Campbell AS, Albo D, Wang TN. Co-localization of macrophage inflammatory protein-3α (Mip-3α) and its receptor, CCR6, promotes pancreatic cancer cell invasion. Cancer J 2004;10:374–380.
67. Singh S, Singh UP, Stiles JK, Grizzle WE, Lillard JW Jr. Expression and functional role of CCR9 in prostate cancer cell migration and invasion. Clin Cancer Res 2004;10:8743–8750.

17

Chemokine Receptors in Neuroinflammation

Astrid E. Cardona and Richard M. Ransohoff

Summary

Actions of chemokines and the interaction with specific receptors within the central nervous system (CNS) surpass their original defined role of leukocyte recruitment to inflamed tissues. Chemokine receptor expression by resident CNS cells is crucial for normal brain development and architectural organization, neuronal protection during inflammatory and neurotoxic challenges, and, among many others, protective mechanisms during inflammatory conditions such as multiple sclerosis. The chemokine/chemokine receptor systems involved in such significant functions include CXCR4/CXCL12, CXCR2/CXCL1, and CX3CR1/CX3CL1. In this chapter, we discuss how these receptors might contribute to modulate communication within the CNS and with peripheral elements, and we also suggest potential mechanisms of action of fractalkine and the translation of these into the understanding of microglial function during neuroinflammatory conditions.

Key Words: Chemokine receptors; CNS; multiple sclerosis; microglia; NK cells; EAE; fractalkine; neuroinflammation.

17.1. Introduction

Chemokines, a superfamily of small (8 to 14 kd) chemoattractant cytokines, were originally identified by their capacity to mediate gradient-dependent cell migration, accompanied by Ca^{2+} flux, in vitro. Sequence homology of mammal, bird, and fish chemokines indicates that this family of proteins is highly conserved throughout evolution. Chemokines are multifunctional molecules with a vast repertoire of specialized functions in different organs. Broadly, chemokine actions are associated with cell adhesion, cytokine secretion, cellular

From: *The Receptors: The Chemokine Receptors*
Edited by: J. K. Harrison and N. W. Lukacs © Humana Press Inc., Totowa, NJ

proliferation, phagocytosis, apoptosis, angiogenesis, atherosclerosis, autoimmunity, viral pathogenesis, and neurodegeneration *(1,2)*. Most recently, it was hypothesized that the CNS interaction of chemokines and chemokine receptors acts in concert with the neuropeptide and neurotransmitter system to regulate brain function *(3,4)*, adding a novel mechanistic role for chemokines during homeostatic and inflammatory conditions.

Because of the inefficient transport of antigens to lymphoid organs, and to the unique and selective functions of the blood-brain barrier, the CNS was often regarded as an immunoprivileged site. However, it has been demonstrated that the CNS is an immunologically competent organ, with the potential of developing efficient adaptive and regulatory immune responses. Intense focus in recent years has been concentrated on understanding the physiologic and molecular pathways involved in the recruitment of peripheral leukocytes into the brain. Leukocyte attraction and migration into the CNS is a highly regulated multistep process that involves coordinated actions of chemokines, chemokine receptors, and adhesion molecules. This topic of leukocyte migration through the blood-brain barrier has been recently reviewed *(5)*.

Importantly, over the past decade it has been demonstrated that chemokines and chemokine receptors are not restricted to leukocytes. In the brain, chemokine receptors are not only found in microglia and astrocytes but also in oligodendrocytes, neurons, and along the brain microvasculature. In this chapter, we will focus on the functions of chemokine receptors expressed by resident CNS cells during physiologic and inflammatory conditions.

17.2. Chemokines and Chemokine Receptors

To date, more than 40 chemokines have been identified in humans, and they are further divided into four subfamilies (C, CC, CXC, and CX3C) according to the number and spacing of conserved cysteine residues near the N-terminus *(6,7)*. Chemokines exert their biological functions by binding to high-affinity G protein–coupled receptors; a superfamily of seven-transmembrane spanning molecules involved in transducing signals through heterotrimeric GTP-binding proteins. Generally, the G proteins activated by chemokine receptors belong to the $G\alpha_i$ family and are sensitive to pertussis toxin treatment. The intracellular signals elicited by chemokine receptor binding of ligand involve reduction of cAMP and transient increases in intracellular calcium. Downstream activation of mitogen-activated protein kinases (MAPKs), phosphatidylinositol-3′-kinase (PI3K), and small GTP-binding proteins such as RAC, RhoA, and CDC42H with roles in cytoskeletal organization required for cellular migration, have been described.

17.2.1. Ligand and Receptor Relationship

Chemokines exhibit complex ligand-receptor relationships. Often, individual chemokines can bind productively to more than one receptor, and many chemokine receptors can signal in response to more than one chemokine. Usually, these chemokine receptors bind multiple chemokines within the same subfamily. The best documented example of chemokine receptor promiscuity across subfamilies involves CXCR3 binding to CCL21.

17.2.2. Nonsignaling Receptors

Three receptor-like molecules are known and include (i) the Duffy antigen receptor for chemokines (DARC), (ii) CCX-CKR, and (iii) D6 (*see* Table 1 in Chapter 3 for details). Binding of chemokines to these receptors do not evoke prototypical cellular responses such as chemotaxis or activation *(8)*. There is evidence that endothelial cells are active in transcytosing chemokines to their luminal surfaces, where they are presented to leukocytes *(9)*. In this aspect, DARC has been attributed the role of transferring chemokines across the endothelium, and because of its abundance on erythrocytes, DARC is seen as a receptor involved in clearance of chemokines present at high levels in the blood *(10)*. D6 is a seven-transmembrane spanning receptor that binds 12 inflammatory CC chemokines such as CCL2 and CCL3, but it does not bind constitutively expressed chemokines such as CCL19 and CCL21. After ligand binding, D6 internalizes the bound chemokine and targets it for degradation, indicating that D6 might act as a decoy receptor, clearing away residual CC chemokines in inflamed skin *(11)*, and therefore resolves inflammation and decreases pathologic accumulation of inflammatory cells *(12)*. Direct evidence of the scavenging actions of D6 comes from analyses of D6-deficient mice, which exhibited markedly increased cutaneous cellular responses in two distinct models of skin inflammation induced by complete Fraund's adjuvant (CFA) *(11)* and phorbol esters *(12)*. However, D6-deficient mice were resistant to induction of autoimmune inflammation (experimental autoimmune encephalomyelitis) by immunization with myelin oligodendroglial glycoprotein (MOG_{35-55}), due to impaired encephalitogenic responses. D6-deficient mice exhibited reduced spinal cord inflammation and demyelination and a lower incidence and severity of experimental autoimmune encephalomyelitis (EAE) attacks compared with wild-type littermates. In adoptive transfer studies, MOG-primed $D6^{-/-}$ T cells transferred disease poorly to $D6^{+/-}$ mice. These findings demonstrate that D6 is required for the generation of adaptive immune responses *(13)*.

The redundancy and cross talk between chemokines and receptors in vivo creates a complex immunologic and regulatory communication system. As proposed by Rot and von Andrian *(14)*, our challenge is therefore to understand the language that underlies this complex communication system during normal and pathologic conditions.

17.3. Expression of Chemokine Receptors in Cells Intrinsic to the Central Nervous System

Increasing evidence highlights the prominence of chemokines in a variety of physiologic and pathologic processes in the CNS. In particular, chemokines have been shown to be critical determinants in the positioning of cellular population in the development of CNS inflammation due to autoimmune reactions or infectious diseases *(2,15)*. Several lines of evidence indicate that all resident cells of the CNS express functional chemokine receptors in the intact human brain and in the CNS of rodent and macaques as experimental models. Astrocytes and microglia express most of the chemokine receptors including CCR3 *(16–18)*, CCR5 *(17)*, CXCR3 *(19–21)*, and CXCR4 *(18,22,23)*. Functional expression of CCR2 by fetal human astrocytes (24) and by reactive microglia in multiple sclerosis (MS) lesions *(25)* has been documented. Confined exclusively to microglia in vivo is the expression of CX3CR1 *(26,27)*. Neurons exhibit expression of CCR1 *(28)*, CXCR1 *(29)*, CXCR2 *(29)*, and CXCR4. Neuronal CCR1 expression, however, is not a generalized marker during neurodegeneration but is proposed to be part of the neuroimmune response to Abeta42-positive neuritic plaques in Alzheimer disease *(30)*. Furthermore, constitutive expression of the CXC chemokine receptors, CXCR1, CXCR2 and CXCR3, on oligodendrocytes in normal adult human CNS tissue has been demonstrated *(31)*. In addition to chemokine receptor expression by resident brain cells, chemokine receptors are also expressed along the brain microvasculature at the intersection of the CNS-peripheral system boundaries and include CCR1 *(32)*, CCR2 *(33,34)*, CCR3 *(35)*, CCR4 *(35)*, and CCR5 *(17)*. It has been demonstrated that endothelial cell expression of CCR2 is involved in the passage of CCL2 from circulation to brain parenchyma across the brain microvessels *(33,34)*; a mechanism of definitive relevance for initiation of cellular responses within the CNS.

Contrasting the large number of chemokines described in the periphery, relatively few chemokines are present in the brain under pathologic conditions. Among the most commonly found chemokines in the challenged CNS are CCL2, CCL3, CCL4, CCL5, CCL8, CCL21, CXCL8, CXCL10, and CXCL12 *(36)*. Recent studies have shown that glial chemokines are crucial mediators of accumulation of infiltrating cells into the CNS. In order to understand the mechanisms involved in CNS infiltration of leukocytes and the contribution of systemic and local brain cells to the pathogenesis of disease, it is necessary to dissect the effects of constitutive versus inducible regulation of chemokine receptor expression.

From the large list of chemokines and chemokine receptors and their complex interactions, three chemokine/chemokine-receptor pairs constitute interesting examples due to their intrinsic CNS expression (Fig. 1; see color plate). These systems include the chemokine receptors and ligands CXCR4/CXCL12, CXCR2/CXCL1, and CX3CR1/CX3CL1, which will be further discussed below. We will also describe the inducible system CXCR3/CCL21 and its proposed role in cellular communication within the CNS.

17.4. CXCR4/CXCL12 Linked to CNS Development, and Much More

As mentioned previously, CXCR4 is expressed in glial cells (astrocytes and microglia but not oligodendrocytes) as well as in neurons. In the adult brain, neuronal expression of CXCR4 is localized to cerebral cortex, caudate putamen, globus pallidus, substantia innominata, supraoptic and paraventricular hypothalamic nuclei, ventromedial thalamic nucleus, dentate gyrus, cerebellum, entorhinal cortex, and substantia nigra. Astrocytes represent the major cellular source of the CXCR4 ligand CXCL12 in the brain. In vitro studies have demonstrated that CXCL12 is a potent chemoattractant for several types of neural cells, including neuronal precursor cells from the external germinal layer, cortical neuronal progenitors, cerebellar granule neurons, and dentate gyrus granular neurons *(37,38)*. Disruption of either CXCR4 or CXCL12 in gene-targeted mice gives similar but not identical pathophysiologic outcomes; mice died soon after birth and exhibited severe abnormalities in cerebellar development *(39)*. Defects included distorted laminar structure and abnormal migration of cells from the proliferative external granule cells in the developing cerebellum, with establishment of proliferating aggregates and absence of foliation. CXCR4-deficient mice also showed abnormal development of hippocampal dentate gyrus, which compared with wild-type mice appeared small and immature. This abnormality was attributed to deficits in granule cell migration and a reduction in the number of dividing neuronal precursors. In addition, the distribution of interneurons in the neocortex of CXCR4- and CXCL12-deficient mice was severely perturbed. Additionally, it is critical for normal brain and spinal cord development to receive CXCL12/CXCR4 signals for axonal responsiveness to known guidance cues *(40,41)*.

CXCR4 actions, however, go beyond physiologic cellular positioning of neuron progenitors in the developing CNS *(42)*. Further, activation of CXCR4 in astrocytes causes release of TNF-α and subsequent downmodulation of critically important astrocytic transporters (responsible for removing extracellular glutamate) and promoting neuronal apoptosis via excitotoxicity. During CNS inflammation, microglia enhance this neurotoxic mechanism by producing TNF-α and triggering release of glutamate *(43)*. This complex signaling pathway by which CXCL12 evokes glutamate release and modulation of neuronal function has been termed the glutamate-chemokine connection *(43,44)*. Another important role of CXCR4 is highlighted by various in vivo studies showing that blockade of CXCR4 by antibodies or small antagonists inhibits tumor cell migration *(45)*. Therefore in cancer biology, CXCR4 is envisioned as a target for novel strategies in the treatment of various malignancies, including gliomas and medulloblastomas *(46)*. Such new approaches include blockade of mechanisms involved in tumor cell growth and metastatic spreading *(45,47–49)*. Another important role is evident by the finding of *Cxcr4* mutations that are associated with a rare inherited human immunologic disorder named WHIM (for warts, hypogammaglobulinemia, infections, and myelokathexis) *(50)*. This human immunodeficiency appears to rise from aberrant signaling of the mutant receptor and is clinically manifested by a prolonged retention of neutrophils in the bone marrow compartment, chronic leucopenia, and extensive human papillomavirus infection *(50–52)*. In addition to their instrumental role in architectural organization, brain morphogenesis, and axonal guidance, CXCL12/CXCR4 is also involved in neuronal function, tumor development, and immune and stem cell trafficking *(53,54)*.

17.5. CXCR2/CXCL1: Role in Oligodendrocyte Development

CXCR2 is expressed at high levels by specific populations of neurons in multiple regions of the brain (including hippocampus, dentate nucleus of the cerebellum, locus caeruleus, pontine nuclei in the brain stem) and spinal cord. CXCL1 has been shown to be expressed by spinal astrocytes and to be a potent promoter of oligodendrocyte precursor proliferation. In vitro, the proliferative response of immature spinal cord precursors to their major mitogen, platelet-derived growth factor (PDGF), was dramatically enhanced by CXCL1 and was shown to be dependent on the concentration of both ligands and the maturation state of the cells. Furthermore, signaling through CXCR2, which has been shown to be highly expressed on CNS oligodendrocyte precursors in rats, mice, and humans, plays an important role in patterning of the developing CNS. CXCL1 signaling to its receptor CXCR2 inhibited oligodendrocyte precursor migration *(55)*. The migration arrest was rapid, reversible, and concentration dependent. In the developing spinal cords of CXCR2-deficient mice, there were reduced numbers of oligodendrocytes, abnormally concentrated at the periphery. Therefore, in contrast with the originally described functions of chemokines and their receptors in migration, CXCR2/CXCL1 interactions control the positioning of oligodendrocyte precursors in the developing spinal cord by arresting their migration.

Studies on MS and its animal model, experimental autoimmune encephalomyelitis (EAE), have suggested that chemokines may be involved in the establishment of disease. The accumulation of inflammatory monocytes in the afflicted MS and EAE CNS involves actions of chemokine receptors such as CCR2. Chemokine receptors and the interaction with their ligands might also provide beneficial mechanisms during the neuroinflammatory processes. The expression of the chemokine receptors CXCR1, CXCR2, and CXCR3 in human oligodendrocytes *(31)* and production of their respective ligands (CXCL8, CXCL1, and CXCL10) by hypertrophic astrocytes suggest novel functional roles of these molecules in oligodendrocyte recruitment and remyelination. These later studies have demonstrated that IL-1β produced by local and infiltrating cells regulates CXCL1 production by astrocytes, and prominent expression of CXCR2 was detected on regions of proliferating oligodendrocytes around active MS lesions. Association of CXCR2-expressing oligodendrocytes with CXCL1-positive astrocytes suggests functional interactions between these two cell types in the MS brain. A mechanism linking proinflammatory cytokine secretion during early stages of MS and the subsequent upregulation of CXCL1 by astrocytes is proposed *(31,56)*. The concurrence of CXCR2 on oligodendrocytes and induced CXCL1 expression by astrocytes in the MS brain provides a proliferative signal and/or prevents migration of CXCR2-bearing oligodendrocytes, both of which could potentially lead to increase in cell number and provide a mechanism for lesion repair.

17.6. CX3CR1/CX3CL1: A Complex and Enigmatic System

Fractalkine (CX3CL1) has a number of unique properties that mediate various biological activities. Structurally, the chemokine is characterized by three amino acids intervening between the first two conserved cysteines. Fractalkine is synthesized as a type I transmembrane molecule with the chemokine domain tethered by a 241-amino-acid glycosylated stalk, a 19-amino-acid transmembrane region, and a 37-amino-acid intracellular tail *(57)*. A soluble form of the protein comprising the chemokine domain and most of the stalk region is generated by cleavage at a region proximal to the cell membrane. Its localization pattern in the CNS characterizes fractalkine as a unique chemokine present on neuronal membranes and capable of being released as a soluble protein by constitutive or stress-activated ADAM (A disintegrin and metalloproteinase)-family protease activity *(58–61)*. Distribution studies in mouse, rat, and human have shown that, unlike other chemokines that are expressed mostly by cells of the immune system, fractalkine is localized predominately in the brain, where its receptor is also highly expressed.

In addition to CX3CL1, a number of diverse proteins exist both in full-length, membrane-bound forms and in soluble forms generated by posttranslational cleavage. These include TNF-α, which is involved in inflammatory responses; L-selectin, a cell adhesion molecule *(62)*; β-amyloid precursor protein, involved in Alzheimer disease; and angiotensin-converting enzyme, involved in blood pressure regulation *(63)*. An appropriate balance between the soluble and full-length forms of these proteins appears necessary for homeostasis. The mechanism for regulating CX3CL1 cleavage is unknown and may have implications in the understanding of the roles of soluble and membrane-bound forms.

Fractalkine exerts its functions by binding to its only identified receptor CX3CR1. The cellular distribution of the fractalkine receptor in the brain is still controversial. Clearly, CX3CR1 can be expressed in cultured neurons *(64)*, although CX3CR1 mRNA is not detected under various conditions of stimulation in neuronal populations *(65)*. In vitro stimulated astrocytes also upregulate CX3CR1, although in vivo this observation has not been reproduced. Most recently, neurons, astrocytes, and navy glial antigen 2 ($NG2^+$) did not reveal expression of the CX3CR1 transcription unit in vivo *(66)*.

Mice lacking fractalkine or its receptor exhibit heterogeneous differences from wild-type mice in peripheral challenges. None of the phenotypes of CX3CR1 deficiency was obvious; the first report of *Cx3cr1*$^{-/-}$ mice appeared in the "Mammalian Genetic Models with Minimal Phenotypes" section of *Molecular and Cellular Biology (27)*. In one productive genetic-engineering project, CX3CR1-deficient mice were generated by disrupting the CX3CR1 locus through insertion of enhanced green fluorescent protein (GFP). Because of the fact that GFP labels all CX3CR1-expressing microglia in heterozygous and null mice, these mice have been extremely valuable to monitor and understand microglial physiologic changes in vivo. *Cx3cr1*$^{-/-}$ mice show impaired replacement of tissue macrophages in noninflamed peripheral organs *(67)*. It was also shown that lamina propria dendritic cells are dependent on CX3CR1 to form transepithelial dendrites, which enable the cells to directly sample luminal antigens *(68)*. *Cx3cr1*$^{-/-}$ mice also fail to recruit natural killer (NK) cells to cardiac allografts *(69,70)* or to the inflamed CNS of mice with EAE *(71)*. Important clues about fractalkine/CX3CR1 function came from genetic studies in humans, where two single nucleotide polymorphisms result in amino acid substitutions, 249 (V→I) and 280 (T→M), in substantial linkage disequilibrium *(72)*. The variant $CX3CR1^{I249/M280}$ receptor is present at an allele frequency of 25% to 30% and exerts a dominant effect that leads to reduced adhesive properties *(73,74)*. $CX3CR1^{I249/M280}$ is associated with reduced risk of atherosclerosis or restenosis *(75–77)* while the isoform containing T^{280} confers increased risk of stenosing inflammatory bowel disease *(78)*.

Although they are mainly produced in the CNS, CX3CL1 and CX3CR1 also have a distinctive peripheral pattern of expression. CX3CL1 is found at low levels in endothelial and some epithelial cells of selected peripheral (not CNS) tissues such as kidney, lung, prostate, and heart, but not spleen or liver *(57)*. In the periphery, monocytes and NK cells express CX3CR1 *(27)*. Interaction of membrane-tethered fractalkine with CX3CR1 mediates high-affinity adhesion. The adhesive event is equivalent in strength to integrin/cell adhesion molecule (CAM)-mediated adhesion and can arrest cells under physiologic shear forces. In vitro studies suggested that soluble fractalkine mediates microglial chemoattraction but, surprisingly, inhibits microglial activation *(65,79,80)* and protects microglia from apoptosis *(81,82)*. However, the role of CX3CL1/CX3CR1 during neuroinflammation in vivo has not been well understood, and the consequences of neuron-microglial adhesion, mediated by CX3CL1/CX3CR1, have not been addressed. The precise role of fractalkine receptor expression in selective peripheral- and CNS-cell types (NK cells and microglia respectively) during neuroinflammation is a topic of current investigation as described below.

17.6.1. CX3CR1 and NK Cells

It was established that depletion of NK cells by in vivo treatment with monoclonal antibodies against NK1.1 resulted in increased severity and relapsing pattern of EAE disease in C57BL/6 mice *(83,84)*. Importantly, MS patients tend to have lower numbers and activities of NK cells than healthy individuals and patients with other neurologic diseases *(85)*. It was also observed that new and enlarging or recurring MS lesions developed after periods of low NK cell activities. Although NK cells play a role in the regulation of EAE and MS, the mechanisms of NK cell accumulation in the inflamed CNS have not been clear. It was recently demonstrated that in the absence of CX3CR1, progression of EAE revealed a very interesting phenotype *(71)*. CX3CR1-null mice developed severe clinical disease with higher EAE scores and increased mortality when compared with wild-type littermates. In addition, inflammatory lesions in CX3CR1-deficient mice were accompanied by spinal cord hemorrhages. Of significance was the observation that EAE brains of CX3CR1-deficient mice selectively lacked $CD3^-$ NK cells *(71)*. CX3CR1 therefore is crucial for the recruitment of regulatory and neuroprotective NK cells into the CNS. These important findings encourage us to define the exact role and the mechanism of action of CX3CR1 in NK cells and determine their precise functions during autoimmune inflammation.

17.6.2. CX3CR1 and Microglial Cells

Microglia are hematopoietic in origin and populate the CNS early during development to form a regularly spaced network of highly ramified cells. Resident microglial cells in the healthy brain are thought to rest in a quiescent state, and activation is associated with rapid structural changes, such as motile branches or migration of somata. It was recently shown that in vivo, microglial cells display a territorial organization. Time-lapse imaging experiments of up to 10 hours showed that somata of microglial cells remained fixed with slight signs of migration *(86,87)*. In contrast, microglial processes were remarkably motile, continuously undergoing cycles of projection and retractions. These dynamics enable microglia to sample the extracellular space in a random fashion. Remarkably, it is estimated that during health, the brain parenchyma is completely screened by microglia once every few hours.

Microglial cells represent the resident immune phagocytes of the CNS and have the capacity to become rapidly activated in most pathologic conditions of the CNS. Microglia are critically involved in brain damage, neurodegenerative diseases, stroke, and brain tumor *(88,89)*. For this, microglia are best described as the primary sensors of pathologic damage of the CNS *(90)*. Although microglial activation has been described extensively in many CNS diseases, the impact of microglial responses on disease pathogenesis remains still to be clearly defined. Microglia are also often regarded as double-edge swords, because of the fact that they can mediate beneficial and detrimental functions in the CNS. In autoimmune diseases such as MS, most data point to a detrimental role of microglia, for example by producing neurotoxic molecules, proinflammatory cytokines, chemokines, or by presenting self-antigens. Close association of activated microglia to dying oligodendrocytes in MS lesions raises the possibility that microglial activation might be detrimental to oligodendrocyte survival *(91)*. However, microglia activation may also counteract damage and promote repair by providing neurotrophic or immunosuppressive factors.

Whereas the importance of autoreactive $CD4^+$ T cells in EAE has been extensively documented, particularly by their ability to initiate disease, the effector mechanisms of microglia and their contribution to inflammation and demyelination within the CNS have only recently been directly addressed. During EAE, it was elegantly shown that microglial paralysis results in substantial amelioration of the clinical signs and in strong reduction of CNS inflammation *(92)*. These studies used a pharmacogenetically induced in vivo model of microglial paralysis by generating transgenic mice in which the thymidine kinase of herpes simplex virus (HSVTK) is driven by the CD11b promoter. By virtue of the ability of ganciclovir to be converted to a toxic metabolite by HSVTK, treatment of $HSVTK^+$ microglia with ganciclovir induces cellular toxicity and ablation of cellular function. This model of microglial paralysis therefore established a crucial role of microglial cells in the development of EAE. Furthermore, recent studies using bone marrow chimera mice to distinguish between activated microglial cells and peripheral macrophages allow the study of the kinetics of microglial activation during EAE in vivo *(93)*. It was found that activation of microglia takes place before the entrance of peripheral monocytes/macrophages into the CNS during EAE, and their localization within the inflammatory lesions suggest that microglial cells play a fundamental role in the onset and progression of CNS autoimmunity.

Using a rat microglial cell culture model *(80)*, it was demonstrated that lipopolysaccharide (LPS) induced a marked increased in the release of the proinflammatory cytokine TNF-α. LPS-activated microglia were found to be neurotoxic when added to neuronal hippocampal culture, and treatment of microglial cells with fractalkine reduced TNF-α production. This study postulated that expression of fractalkine in the normal and healthy brain might serve as a mechanism to maintain microglia in a state of hyporesponsiveness. Fractalkine was also shown to increase microglial cell survival and to block Fas-induced programmed cell death of primary cells *(81)*. Fractalkine was found to be rapidly cleaved from cultured neurons in response to an excitotoxic stimulus. More specifically, fractalkine cleavage preceded actual neuronal death and represented the principal chemokine released from the neurons into the culture medium upon an excitotoxic stimulus to promote chemotaxis of primary microglial and monocytic cells (94).

17.6.3. A Neuronal-Microglial and CNS-Peripheral Immune System Communication System

An interesting and puzzling aspect in the biology of CX3CL1 is the fact that it is a chemokine mostly produced by neurons and peripheral endothelial cells; hence its receptor in vivo is present in microglia and subpopulations of T cells, monocytes, and NK cells. Microglial cells during the natural process of immune surveillance contact astrocytes, neuronal cell bodies, and vessels suggesting that in the healthy CNS, microglia dynamically interact with various cortical elements. It is of particular interest to understand how microglia sense changes in the brain environment through engagement of constitutive receptors such as CX3CR1, whose ligand is also present at all times in the CNS extracellular milieu. As reinforced by in vitro studies, one might hypothesize that this pattern of cellular distribution of CX3CR1 and its ligand CX3CL1 is linked to a neuronal-microglial communication system and also to a discrete communication mechanism between the CNS expressing CX3CL1 and the peripheral immune system bearing $CX3CR1^+$ leukocytes.

Direct evidence of a neuronal-microglia communication mechanism through the fractalkine system came from in vivo experiments using CX3CR1-deficient mice showing that neuronal-derived fractalkine and its interaction with fractalkine receptor plays a critical role in tonic inhibition of microglial mediated neurotoxicity *(66)*. In three clinically relevant models—CNS response to systemic inflammation, the l-methyl-4 phenyl-1,2,3,6 tetra-hydropyridine (MPTP) model of Parkinson disease, and the SOD1^{G93A} model of amyotrophic lateral sclerosis—absence of the fractalkine receptor was associated with intense microglial activation and increased neuronal damage. Using an adoptive transfer protocol to examine neurotoxic mechanisms in LPS-injected *Cx3cr1*$^{-/-}$ mice, IL-1 was identified as a downstream mediator of microglial neurotoxicity after systemic inflammation. Previous studies support the notion that the roles of CX3CR1 in microglial neurotoxicity will differ, depending on the nature and chronicity of the activating stimulus. As examples, *Cx3cr1*$^{-/-}$ mice displayed a normal level of microglial activation after facial nerve axotomy *(27)* and after laser-induced injury *(86)*. Furthermore, *Cx3cl1*$^{-/-}$ mice showed relative protection from cerebral ischemia *(95)*, and intrathecal injection of CX3CL1 enhanced nociception *(96)* possibly by activating microglia.

17.7. CXCR3/CCL21 Inducible System of Microglial/ Neuronal Communication

CCL21 is a chemokine constitutively expressed in secondary lymphoid organs (lymph nodes and Peyer patches). CCL21 is involved in homing of mature dendritic cells and naïve T cells into lymph nodes and is crucial for the formation of secondary lymphoid tissue acting via CCR7. It was recently shown that CCL21 in addition to being part of the peripheral immunosurveillance mechanism is also present in the CNS, with neurons being its primary source *(97)*. In particular, CCL21 is specifically induced and released from neurons under various conditions of neurotoxicity *(97)*. Therefore, CCL21 is envisioned as a chemokine produced by endangered neurons. Interestingly, actions of CCL21 in the brain do not primarily involve CCR7. The alternative CCL21 receptor, CXCR3, is expressed by astrocytes and microglia in an inducible manner *(98,99)*. The functional properties of CCL21 on microglia were demonstrated by the induction of morphologic microglial activation and chemotaxis. In the absence of CXCR3, entorhinal cortex lesion failed to evoke microglial activation, whereas astrogliosis was indistinguishable from the wild-type situation. Neuronal-derived CCL21 might act as a potent chemoattractant for T cells, and in addition CCL21/CXCR3 thus represents a potential neuronal microglia communication system that operates under conditions that challenge the CNS either by neurodegeneration or inflammation. As suggested by De Jong et al. *(97)*, CCL21 serves to activate microglial cells at distance from the primary neuronal damaged site.

17.8. Conclusions

One of the biggest challenges is to be able to clinically improve therapeutic treatments for patients with various neurodegenerative and neuroinflammatory conditions, such as multiple sclerosis, Parkinson disease, Alzheimer disease, and motor neuron disease, as well as various viral and bacterial encephalopathies. One major focus addresses the molecular bases of CXCR2 action and seeks to define the precise role of this system in remyelination, using experimental models of EAE and chemically induced demyelination such as in cuprizone treatment. A number of relevant hypotheses are shown schematically in Fig. 2 (see color plate). Understanding the mechanisms that underlie the process of microglia activation and how microglia responses are translated into beneficial or detrimental actions will have a great impact in the modulation of various CNS inflammatory conditions. We hypothesize that interaction of neuronal membrane-bound fractalkine and microglial CX3CR1 causes deleterious adhesion, leading to neuronal damage, whereas soluble fractalkine is neuroprotective (Fig. 2; see color plate). We hope that in the near future, we will see delineation of the actions of the fractalkine receptor of NK cells and microglia in experimental models of MS and definition of microglia regulation by the soluble and membrane-bound forms of fractalkine. With these goals, hopefully chemokine receptor research will be more easily translated to clinically relevant approaches.

References

1. Mackay CR. Chemokines: immunology's high impact factors. Nat Immunol 2001;2:95–101.
2. Charo IF, Ransohoff RM. The many roles of chemokines and chemokine receptors in inflammation. N Engl J Med 2006;354:610–621.
3. Adler MW, Rogers TJ. Are chemokines the third major system in the brain? J Leukoc Biol 2005;78:1204–1209.
4. Adler MW, Geller EB, Chen X, et al. Viewing chemokines as a third major system of communication in the brain. AAPS J 2005;7:E865–E870.
5. Ransohoff RM, Kivisakk P, Kidd G. Three or more routes for leukocyte migration into the central nervous system. Nat Rev Immunol 2003;3:569–581.
6. Murphy PM, Baggiolini M, Charo IF, et al. International union of pharmacology. XXII. Nomenclature for chemokine receptors. Pharmacol Rev 2000;52:145–176.
7. Zlotnik A, Yoshie O. Chemokines: a new classification system and their role in immunity. 2000;12:121–127.
8. Locati M, Torre YM, Galliera E, et al. Silent chemoattractant receptors: D6 as a decoy and scavenger receptor for inflammatory CC chemokines. Cytokine Growth Factor Rev 2005;16:679–686.
9. Middleton J, Patterson AM, Gardner L, et al. Leukocyte extravasation: chemokine transport and presentation by the endothelium. Blood 2002;100:3853–3860.
10. Rot A. Contribution of Duffy antigen to chemokine function. Cytokine Growth Factor Rev 2005;16:687–694.
11. Martinez dlT, Locati M, Buracchi C, et al. Increased inflammation in mice deficient for the chemokine decoy receptor D6. Eur J Immunol 2005;35:1342–1346.

12. Jamieson T, Cook DN, Nibbs RJ, et al. The chemokine receptor D6 limits the inflammatory response in vivo. Nat Immunol 2005;6:403–411.
13. Liu L, Graham G, Hu T, et al. The silent chemokine receptor D6 is required for generating T cell responses that mediate experimental autoimmune encephalomyelitis. J Immunol 2006;177(1):17–21.
14. Rot A, von Andrian UH. Chemokines in innate and adaptive host defense: basic chemokinese grammar for immune cells. Ann Rev Immunol 2004;22:891–928.
15. Ubogu EE, Cossoy MB, Ransohoff RM. The expression and function of chemokines involved in CNS inflammation. Trends Pharmacol Sci 2006;27:48–55.
16. van der Meer P, Goldberg SH, Fung KM, et al. Expression pattern of CXCR3, CXCR4, and CCR3 chemokine receptors in the developing human brain. J Neuropathol Exp Neurol 2001;60:25–32.
17. van der Meer P, Ulrich AM, alez-Scarano F, et al. Immunohistochemical analysis of CCR2, CCR3, CCR5, and CXCR4 in the human brain: potential mechanisms for HIV dementia. Exp Mol Pathol 2000;69:192–201.
18. Westmoreland SV, Rottman JB, Williams KC, et al. Chemokine receptor expression on resident and inflammatory cells in the brain of macaques with simian immunodeficiency virus encephalitis. Am J Pathol 1998;152:659–665.
19. Goldberg SH, van der Meer P, Hesselgesser J, et al. CXCR3 expression in human central nervous system diseases. Neuropathol Appl Neurobiol 2001;27:127–138.
20. Biber K, Dijkstra I, Trebst C, et al. Functional expression of CXCR3 in cultured mouse and human astrocytes and microglia. Neuroscience 2002;112:487–497.
21. Biber, K. Microglial chemokines and chemokine receptors. In: Universes in delicate balance: chemokines and the nervous system. In: Ransohoff RM, Suzuki K, Proudfoot AEI, et al., eds. Amsterdam: Elsevier; 2002:289–300.
22. Boutet A, Salim H, Leclerc P, et al. Cellular expression of functional chemokine receptor CCR5 and CXCR4 in human embryonic neurons. Neurosci Lett 2001;311: 105–108.
23. Westmoreland SV, Alvarez X, deBakker C, et al. Developmental expression patterns of CCR5 and CXCR4 in the rhesus macaque brain. J Neuroimmunol 2002;122:146–158.
24. Andjelkovic AV, Song L, Dzenko KA, et al. Functional expression of CCR2 by human fetal astrocytes. J Neurosci Res 2002;70:219–231.
25. Simpson J, Rezaie P, Newcombe J, et al. Expression of the beta-chemokine receptors CCR2, CCR3 and CCR5 in multiple sclerosis central nervous system tissue. J Neuroimmunol 2000;108:192–200.
26. Hughes PM, Botham MS, Frentzel S, et al. Expression of fractalkine (CX3CL1) and its receptor, CX3CR1, during acute and chronic inflammation in the rodent CNS. Glia 2002;37:314–327.
27. Jung S, Aliberti J, Graemmel P, et al. Analysis of fractalkine receptor CX(3)CR1 function by targeted deletion and green fluorescent protein reporter gene insertion. Mol Cell Biol 2000;20:4106–4114.
28. Cowell RM, Silverstein FS. Developmental changes in the expression of chemokine receptor CCR1 in the rat cerebellum. J Comp Neurol 2003;457:7–23.

29. Danik M, Puma C, Quirion R, et al. Widely expressed transcripts for chemokine receptor CXCR1 in identified glutamatergic, gamma-aminobutyric acidergic, and cholinergic neurons and astrocytes of the rat brain: a single-cell reverse transcription-multiplex polymerase chain reaction study. J Neurosci Res 2003; 74:286–295.
30. Halks-Miller M, Schroeder ML, Haroutunian V, et al. CCR1 is an early and specific marker of Alzheimer's disease. Ann Neurol 2003;54:638–646.
31. Omari KM, John G, Lango R, et al. Role for CXCR2 and CXCL1 on glia in multiple sclerosis. Glia 2006;53(1):24–31.
32. Sanders VJ, Pittman CA, White MG, et al. Chemokines and receptors in HIV encephalitis. AIDS 1998;12:1021–1026.
33. Dzenko KA, Andjelkovic AV, Kuziel WA, et al. The chemokine receptor CCR2 mediates the binding and internalization of monocyte chemoattractant protein-1 along brain microvessels. J Neurosci 2001;21:9214–9223.
34. Dzenko KA, Song L, Ge S, et al. CCR2 expression by brain microvascular endothelial cells is critical for macrophage transendothelial migration in response to CCL2. Microvasc Res 2005;90(1–2):53–64.
35. Berger O, Gan X, Gujuluva C, et al. CXC and CC chemokine receptors on coronary and brain endothelia. Mol Med 1999;5:795–805.
36. Biber K, Zuurman MW, Dijkstra IM, et al. Chemokines in the brain: neuroimmunology and beyond. Curr Opin Pharmacol 2002;2:63–68.
37. Dziembowska M, Tham TN, Lau P, et al. A role for CXCR4 signaling in survival and migration of neural and oligodendrocyte precursors. Glia 2005;50:258–269.
38. Banisadr G, Skrzydelski D, Kitabgi P, et al. Highly regionalized distribution of stromal cell-derived factor-1/CXCL12 in adult rat brain: constitutive expression in cholinergic, dopaminergic and vasopressinergic neurons. Eur J Neurosci 2003;18: 1593–1606.
39. Vilz TO, Moepps B, Engele J, et al. The SDF-1/CXCR4 pathway and the development of the cerebellar system. Eur J Neurosci 2005;22:1831–1839.
40. Xiang Y, Li Y, Zhang Z, et al. Nerve growth cone guidance mediated by G protein-coupled receptors. Nat Neurosci 2002;5:843–848.
41. Chalasani SH, Sabelko KA, Sunshine MJ, et al. A chemokine, SDF-1, reduces the effectiveness of multiple axonal repellents and is required for normal axon pathfinding. J Neurosci 2003;23:1360–1371.
42. Belmadani A, Tran PB, Ren D, et al. The chemokine stromal cell-derived factor-1 regulates the migration of sensory neuron progenitors. J Neurosci 2005;25: 3995–4003.
43. Allen NJ, Attwell D. A chemokine-glutamate connection. Nat Neurosci 2001;4:676–678.
44. Bezzi P, Domercq M, Brambilla L, et al. CXCR4-activated astrocyte glutamate release via TNFalpha: amplification by microglia triggers neurotoxicity. Nat Neurosci 2001;4:702–710.
45. Ehtesham M, Winston JA, Kabos P, et al. CXCR4 expression mediates glioma cell invasiveness. Oncogene 2006;25(19):2801–2806.

46. Bajetto A, Barbieri F, Dorcaratto A, et al. Expression of CXC chemokine receptors 1-5 and their ligands in human glioma tissues: role of CXCR4 and SDF1 in glioma cell proliferation and migration. Neurochem Int 2006;49:423–432.
47. Woerner BM, Warrington NM, Kung AL, et al. Widespread CXCR4 activation in astrocytomas revealed by phospho-CXCR4-specific antibodies. Cancer Res 2005;65:11392–11399.
48. Airoldi I, Raffaghello L, Piovan E, et al. CXCL12 does not attract $CXCR4^+$ human metastatic neuroblastoma cells: clinical implications. Clin Cancer Res 2006;12:77–82.
49. Kucia M, Reca R, Miekus K, et al. Trafficking of normal stem cells and metastasis of cancer stem cells involve similar mechanisms: pivotal role of the SDF-1-CXCR4 axis. Stem Cells 2005;23:879–894.
50. Diaz GA. CXCR4 mutations in WHIM syndrome: a misguided immune system? Immunol Rev 2005;203:235–243.
51. Diaz GA, Gulino AV. WHIM syndrome: a defect in CXCR4 signaling. Curr Allergy Asthma Rep 2005;5:350–355.
52. Kawai T, Choi U, Whiting-Theobald NL, et al. Enhanced function with decreased internalization of carboxy-terminus truncated CXCR4 responsible for WHIM syndrome. Exp Hematol 2005;33:460–468.
53. Ruiz DA, Luttun A, Carmeliet P. An SDF-1 trap for myeloid cells stimulates angiogenesis. Cell 2006;124:18–21.
54. Lapidot T, Dar A, Kollet O. How do stem cells find their way home? Blood 2005;106:1901–1910.
55. Tsai HH, Frost E, To V, et al. The chemokine receptor CXCR2 controls positioning of oligodendrocyte precursors in developing spinal cord by arresting their migration. Cell 2002;110:373–383.
56. Omari KM, John GR, Sealfon SC, et al. CXC chemokine receptors on human oligodendrocytes: implications for multiple sclerosis. Brain 2005;128:1003–1015.
57. Lucas AD, Chadwick N, Warren BF, et al. The transmembrane form of the CX3CL1 chemokine fractalkine is expressed predominantly by epithelial cells in vivo. Am J Pathol 2001;158:855–866.
58. Garton KJ, Gough PJ, Blobel CP, et al. Tumor necrosis factor-alpha-converting enzyme (ADAM17) mediates the cleavage and shedding of fractalkine (CX3CL1). J Biol Chem 2001;276:37993–38001.
59. Tsou CL, Haskell CA, Charo IF. Tumor necrosis factor-alpha-converting enzyme mediates the inducible cleavage of fractalkine. J Biol Chem 2001;276:44622–44626.
60. Ludwig A, Schiemann F, Mentlein R, et al. Dipeptidyl peptidase IV (CD26) on T cells cleaves the CXC chemokine CXCL11 (I-TAC) and abolishes the stimulating but not the desensitizing potential of the chemokine. J Leukoc Biol 2002;72: 183–191.
61. Hundhausen C, Misztela D, Berkhout TA, et al. The disintegrin-like metalloproteinase ADAM10 is involved in constitutive cleavage of CX3CL1 (fractalkine) and regulates CX3CL1-mediated cell-cell adhesion. Blood 2003;102:1186–1195.

62. Smalley DM, Ley K. L-selectin: mechanisms and physiological significance of ectodomain cleavage. J Cell Mol Med 2005;9:255–266.
63. Ludwig A, Hundhausen C, Lambert MH, et al. Metalloproteinase inhibitors for the disintegrin-like metalloproteinases ADAM10 and ADAM17 that differentially block constitutive and phorbol ester-inducible shedding of cell surface molecules. Comb Chem High Throughput Screen 2005;8:161–171.
64. Meucci O, Fatatis A, Simen AA, et al. Expression of CX3CR1 chemokine receptors on neurons and their role in neuronal survival. Proc Natl Acad Sci U S A 2000; 97:8075–8080.
65. Maciejewski-Lenoir D, Chen S, Feng L, et al. Characterization of fractalkine in rat brain cells: migratory and activation signals for CX3CR-1-expressing microglia. J Immunol 1999;163:1628–1635.
66. Cardona A, Pioro EP, Sasse ME, et al. Control of microglial neurotoxicity by the fractalkine receptor. Nat Neurosci 2006;9:917–924.
67. Geissmann F, Jung S, Littman DR. Blood monocytes consist of two principal subsets with distinct migratory properties. Immunity 2003;19:71–82.
68. Niess JH, Brand S, Gu X, et al. CX3CR1-mediated dendritic cell access to the intestinal lumen and bacterial clearance. Science 2005;307:254–258.
69. Haskell CA, Hancock WW, Salant DJ, et al. Targeted deletion of CX(3)CR1 reveals a role for fractalkine in cardiac allograft rejection. J Clin Invest 2001; 108:679–688.
70. Robinson LA, Nataraj C, Thomas DW, et al. A role for fractalkine and its receptor (CX3CR1) in cardiac allograft rejection. J Immunol 2000;165:6067–6072.
71. Huang D, Shi FD, Jung S, et al. The neuronal chemokine CX3CL1/ fractalkine selectively recruits NK cells that modify experimental autoimmune encephalomyelitis within the central nervous system. FASEB J 2006;20:896–905.
72. Moatti D, Faure S, Fumeron F, et al. Polymorphism in the fractalkine receptor CX3CR1 as a genetic risk factor for coronary artery disease. Blood 2001;97: 1925–1928.
73. McDermott DH, Halcox JP, Schenke WH, et al. Association between polymorphism in the chemokine receptor CX3CR1 and coronary vascular endothelial dysfunction and atherosclerosis. Circ Res 2001;89:401–407.
74. Daoudi M, Lavergne E, Garin A, et al. Enhanced adhesive capacities of the naturally occurring Ile249-Met280 variant of the chemokine receptor CX3CR1. J Biol Chem 2004;279:19649–19657.
75. Apostolakis S, Baritaki S, Kochiadakis GE, et al. Effects of polymorphisms in chemokine ligands and receptors on susceptibility to coronary artery disease. Thromb Res 2007;119:63–71.
76. Cybulsky MI, Hegele RA. The fractalkine receptor CX3CR1 is a key mediator of atherogenesis. J Clin Invest 2003;111:1118–1120.
77. Damas JK, Boullier A, Waehre T, et al. Expression of fractalkine (CX3CL1) and its receptor, CX3CR1, is elevated in coronary artery disease and is reduced during statin therapy. Arterioscler Thromb Vasc Biol 2005;25:2567–2572.

78. Lavergne E, Labreuche J, Daoudi M, et al. Adverse associations between CX3CR1 polymorphisms and risk of cardiovascular or cerebrovascular disease. Arterioscler Thromb Vasc Biol 2005;25:847–853.
79. Zujovic V, Benavides J, Vige X, et al. Fractalkine modulates TNF-alpha secretion and neurotoxicity induced by microglial activation. Glia 2000;29:305–315.
80. Zujovic V, Taupin V. Use of cocultured cell systems to elucidate chemokine-dependent neuronal/microglial interactions: control of microglial activation. Methods 2003;29:345–350.
81. Boehme SA, Lio FM, Maciejewski-Lenoir D, et al. The chemokine fractalkine inhibits Fas-mediated cell death of brain microglia. J Immunol 2000;165:397–403.
82. Mizuno T, Kawanokuchi J, Numata K, et al. Production and neuroprotective functions of fractalkine in the central nervous system. Brain Res 2003;979:65–70.
83. Xu W, Fazekas G, Hara H, et al. Mechanism of natural killer (NK) cell regulatory role in experimental autoimmune encephalomyelitis. J Neuroimmunol 2005;163: 24–30.
84. Jahng AW, Maricic I, Pedersen B, et al. Activation of natural killer T cells potentiates or prevents experimental autoimmune encephalomyelitis. J Exp Med 2001; 194:1789–1799.
85. Infante-Duarte C, Weber A, Kratzschmar J, et al. Frequency of blood CX3CR1-positive natural killer cells correlates with disease activity in multiple sclerosis patients. FASEB J 2005;19:1902–1904.
86. Davalos D, Grutzendler J, Yang G, et al. ATP mediates rapid microglial response to local brain injury in vivo. Nat Neurosci 2005;8:752–758.
87. Nimmerjahn A, Kirchhoff F, Helmchen F. Resting microglial cells are highly dynamic surveillants of brain parenchyma in vivo. Science 2005;308:1314–1318.
88. Hickey WF. Basic principles of immunological surveillance of the normal central nervous system. Glia 2001;36:118–124.
89. Benveniste EN. Role of macrophages/microglia in multiple sclerosis and experimental allergic encephalomyelitis. J Mol Med 1997;75:165–173.
90. Kreutzberg GW. Microglia:a sensor for pathological evets in the CNS. Trends Neurosci 1996;19:312–318.
91. Peterson JW, Bo L, Mork S, et al. VCAM-1-positive microglia target oligodendrocytes at the border of multiple sclerosis lesions. J Neuropathol Exp Neurol 2002;61:539–546.
92. Heppner FL, Greter M, Marino D, et al. Experimental autoimmune encephalomyelitis repressed by microglial paralysis. Nat Med 2005;11:146–152.
93. Ponomarev ED, Shriver LP, Maresz K, et al. Microglial cell activation and proliferation precedes the onset of CNS autoimmunity. J Neurosci Res 2005;81: 374–389.
94. Chapman GA, Moores K, Harrison D, et al. Fractalkine cleavage from neuronal membranes represents an acute event in the inflammatory response to excitotoxic brain damage. J Neurosci 2000;20:RC87.

95. Soriano SG, Amaravadi LS, Wang YF, et al. Mice deficient in fractalkine are less susceptible to cerebral ischemia-reperfusion injury. J Neuroimmunol 2002;125: 59–65.
96. Milligan ED, Zapata V, Chacur M, et al. Evidence that exogenous and endogenous fractalkine can induce spinal nociceptive facilitation in rats. Eur J Neurosci 2004;20:2294–2302.
97. De Jong EK, Dijkstra IM, Hensens M, et al. Vesicle-mediated transport and release of CCL21 in endangered neurons: a possible explanation for microglia activation remote from a primary lesion. J Neurosci 2005;25:7548–7557.
98. Biber K, Sauter A, Brouwer N, et al. Ischemia-induced neuronal expression of the microglia attracting chemokine secondary lymphoid-tissue chemokine (SLC). Glia 2001;34:121–133.
99. Rappert A, Biber K, Nolte C, et al. Secondary lymphoid tissue chemokine (CCL21) activates CXCR3 to trigger a Cl-current and chemotaxis in murine microglia. J Immunol 2002;168:3221–3226.

18

Pharmaceutical Targeting of Chemokine Receptors

Sofia Ribeiro and Richard Horuk

Summary

In this chapter, we will give a perspective of in vitro assays used in drug discovery when targeting chemokine receptors. We outline the complexity of the chemokine system and give a historical perspective on the in vitro assays and types of assays used at different stages of discovery, followed by several examples of successes and failures in clinical trials. Finally, we discuss the rationale for continuing, after several failures, to target chemokine receptors and how screening may change with the increasing evidence of dimerization of chemokine receptors.

Key Words: Chemokine receptor; drug discovery; in vitro assay.

18.1. Introduction

The great Russian biologist Elie Metchnikoff laid one of the major foundations of modern immunology when he showed that the motile cells of starfish larvae can migrate around and engulf a rose thorn placed in their midst *(1)*. From these simple experiments came the concept of phagocytosis and cellular immunity. Just like the primitive starfish cells, leukocytes, which represent a major line of defense against invasion by pathogenic organisms, can migrate toward and overpower harmful cellular invaders. In the late 1980s, several groups of scientists isolated the signaling molecules that allowed immune cells to communicate with one another and to seek out and destroy invading pathogens *(2–4)*. These chemotactic (from chemo-chemical taxis) molecules are now collectively known as chemokines.

The chemokines are small, basic proteins that can be divided into two major (CXC and CC) and two minor (C and CX3C) branches based on the position

From: *The Receptors: The Chemokine Receptors*
Edited by: J. K. Harrison and N. W. Lukacs © Humana Press Inc., Totowa, NJ

of the first two cysteines *(5)*. The chemokine family is continuously expanding and currently numbers more than 40 members that include CXC chemokines like IL-8 (CXCL8) and melanoma growth stimulating activity (CXCL1) and CC chemokines like RANTES (CCL5), monocyte chemotactic protein-1 (CCL2), eotaxin, and macrophage inflammatory protein (CCL3). Each of these proteins recognizes and induces the chemotaxis of a particular subset of leukocytes. For example, CXC chemokines, like CXCL8 and CXCL1, preferentially attract neutrophils and induce their activation by producing changes in neutrophil shape, transient increases in cellular calcium concentration, and the upregulation of surface adhesion proteins. In contrast, CC chemokines like CCL2, CCL5, and eotaxin (CCL11) have little effect on neutrophils but chemoattract other leukocytes: CCL11 for example is a potent eosinophil chemoattractant and induces their degranulation, and CCL5 and CCL2 are chemoattractants for T cells and monocytes, respectively.

Chemokines mediate their biological effects by binding to cell surface receptors. Chemokine receptors belong to a superfamily of serpentine proteins that signal through coupled heterotrimeric G proteins. At the latest count, well over 850 members of this G protein–coupled receptor (GPCR) superfamily have been identified and classified into families *(6)*. Nineteen chemokine receptors have been cloned so far including 7 CXC, 10 CC, 1 CX3C, and 1 C receptor *(6)*. Although each of these receptors binds only a single class of chemokines, they can bind several members of the same class with high affinity. Only the promiscuous chemokine binding protein duffy antigen receptor for chemokines (DARC) *(7)* and the viral receptor M3 *(8)* have been shown to bind both CC and CXC chemokines with equal affinity.

Chemokine receptor binding initiates a cascade of intracellular events that culminate in the expression of biological effects *(5)*. The first step in this complex process is the ligation of the receptor by its high-affinity ligand. This induces a conformational change that leads to a dissociation of the receptor-associated heterotrimeric G proteins into α and $\beta\gamma$ subunits. Considerable evidence exists for a role for both α and $\beta\gamma$ subunits as second messengers in receptor signaling. These G-protein subunits can then activate various effector enzymes, including phospholipases, which leads to inositol phosphate production and an increase in intracellular Ca^{2+} and activation of protein kinases. This signal transduction cascade leads to the activation not only of chemotaxis by modulating actin-dependent cellular processes and upregulating adhesion proteins but also of a wide range of functions in different leukocytes such as an increase in the respiratory burst, degranulation, phagocytosis, and lipid mediator synthesis *(5)*.

18.2. Promiscuity of Chemokines and Their Receptors

This system of chemokine action is under very tight control, and this stringent regulation includes a number of interesting control points. For example, it has been shown that different patterns of chemokine receptor expression are obtained on immune cells depending on the cytokine stimulus that is applied. Expression of CCR1 and CCR2 in T cells is strongly induced by the cytokines IL-2 and IL-12 but not by IFN-γ, and both cellular migration and receptor expression are rapidly downregulated when IL-2 is withdrawn but are fully restored when IL-2 is added again *(9)*. Another control point that has been highlighted recently is proteolytic processing of chemokines by enzymes such as dipeptidyl peptidase (CD26), which can influence both the target cell and its receptor profile *(10)*. For example, when CCL11 is cleaved by CD26, the N-terminally truncated molecule has reduced chemotactic activity for eosinophils and desensitized calcium signaling and inhibited chemotaxis relative to intact CCL11. In contrast, both the truncated and full-length CCL11 inhibit HIV-2 infection of CCR3 to a similar extent. Interestingly, chemokines can even act as agonists on one cell and antagonists on another, setting up further physiologic checks and balances during the host defense response *(11)*.

Chemokine biology is further complicated by the fact that there are 19 receptors and more than 40 different chemokines *(12)*. Some receptors such as CCR1 are highly promiscuous and can respond physiologically to as many as eight different chemokine agonists, whereas other receptors such as CCR9 are monogamous and respond to only one agonist *(12)*. The converse is also true, and some chemokines such as CCL3 are promiscuous and activate several receptors (CCR1, CCR3, and CCR5), whereas fractalkine (CX3CL1) is faithful to its partner CX3CR1.

The consequences of a breakdown in this stringent regulation of immune cell mobilization leads to autoimmunity including diseases such as rheumatoid arthritis, diabetes, and multiple sclerosis *(12)*. The realization that chemokines are involved in the pathogenesis of these diseases and that their receptors belong to one of the most pharmacologically exploited families of proteins (GPCRs) has made them the focus of intense interest by pharmaceutical companies. A variety of antagonists have been described, and the most advanced of these is currently in phase III trials as a CCR5 inhibitor for the treatment of AIDS *(13)*.

18.3. Targeting Chemokine Receptors

Screening of chemokine receptors has focused mostly on ligand displacement assays. These assays use a radiolabeled ligand and cells or membranes expressing the receptor of interest. A small molecule/ligand with affinity for the receptor inhibits the binding of the natural radiolabeled ligand (^{125}I). This method allows for the identification of competitive agonists, antagonists, or inverse agonists. Most of the development candidates currently in phase II and above *(13)* and many small laboratories still report heterogenous binding assays, where a separation step by either centrifugation or several washing steps of a filter plate allow for the recovery and detection of a receptor-radiolabeled ligand complex. With the advent of scintillation proximity assays (SPAs) in the 1990s, these assays became homogenous and amenable to high throughput, automation, and miniaturization to 384-well and even 1536-well formats. The SPA system is based on one of two types of beads: polyvinyltoluene (PVT) or yttrium silicate (Ysi). The outer surface of the beads is modified by coating of the surface with a wide variety of materials. For chemokine receptor binding, PVT coated with wheat germ agglutinin (WGA) or polyethyleneimine-treated WGA are usually the best choices. During early assay development testing, a wide range of beads against the radiolabeled ligand allows the identification of the best bead type for a low nonspecific binding for the particular ligand in the absence of receptors (Fig. 1). The aim is to have less than 1% of the total

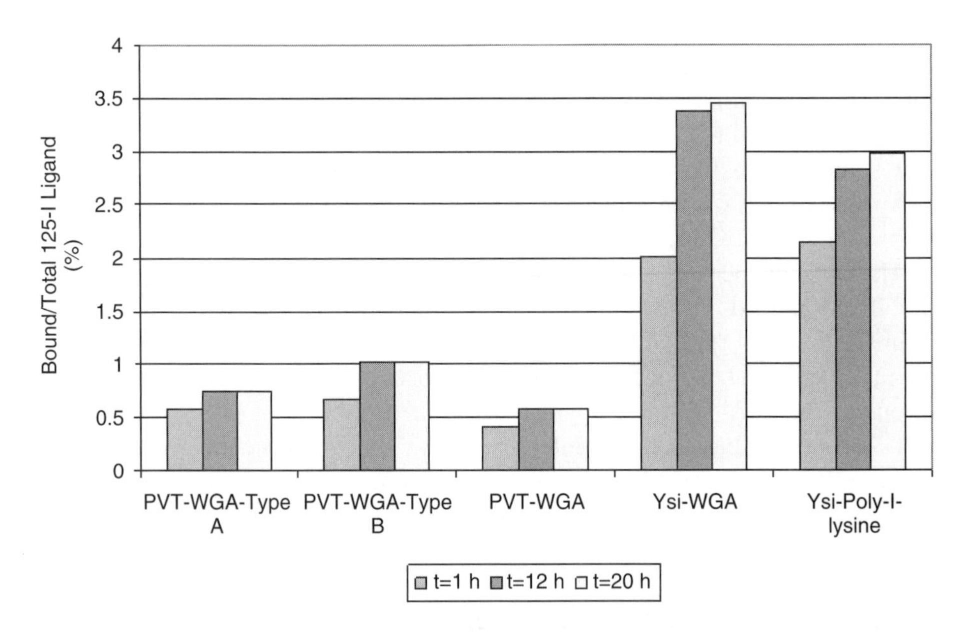

Fig. 1. Competition binding screens: Bead/ligand compatibility test in early assay development. The beads PVT-WGA, Type A and Type B are based on polyvinyltoluene (PVT) not polyethyleneimine (PEI). PVT, polyvinyltoluene; WGA, wheat germ agglutinin; Ysi, yttrium silicate.

activity added being bound to the SPA beads. Bead selection is the first of several factors during assay development: production of high receptor expressor cells allows for low number of cells/well and easier miniaturization to 384-well format; radiolabeled ligand is usually labeled with ^{125}I, although tritium is also suitable with most ligands readily available in catalogues. The use of whole live cells, frozen cells, or membrane preparations is also an important factor. Screening with whole live cells presents some major advantages: the system is intact, ready to use, and no extra membrane preparation is required. On the other hand, over time one may find decreased receptor expression at the cell surface. This is mainly due to two factors: confluency and passage number. Preparation of membranes in large batches allows for, in most cases, uniform assay quality, although different membrane preparation techniques can result in loss of receptor activity. The use of frozen cells, which are prepared in large culture batches and stored in liquid N_2, is a practical solution for eliminating these issues. The SPA technology is also well characterized with respect to buffers and is compatible with a variety of commonly used buffer systems (HEPES, PBS, Hank's buffer, and Tris at neutral pH). Most described assays vary mostly in their salt (NaCl) and bovine serum albumin content, which confers specificity to the interaction observed. Depending on receptor expression and characteristics of the radiolabeled ligand, assay development is achieved quickly in most instances. These types of assays are still widely in use in industry. Figure 2 illustrates representative results of a series of competition assays using CCR8-expressing cells. However, these assays are not always possible, depending on the unique characteristics of some ligand/receptor pairs, which can give rise to quite unacceptable assay variability. Moreover, the inherent safety and environmental risks and high cost of radioactivity has led to an increased focus on alternative technologies mainly using fluorescent-labeled ligands to detect binding. These are very successful but not universal, because labeling of the ligands can be rather difficult, and the ligand affinity can be reduced or abolished upon labeling. Alternatively, the development of relevant functional assays is in many cases preferred because one can gain the ability to distinguish between agonists and antagonists and to detect an increasingly important class of compounds—allosteric modulators and noncompetitive antagonists and agonists.

18.3.1. Functional Screening Assays

The first functional event upon ligand activation of the GPCR is the exchange of GDP for GTP on the α-subunit of the heterotrimeric G protein. The use of a labeled (^{35}S or fluorescent) nonhydrolyzable GTP analogue, GTPγS, allows for the accumulation of the Gα*GTPγS complex, which is directly proportional to the activation of the GPCR. The major advantage is the proximity to the

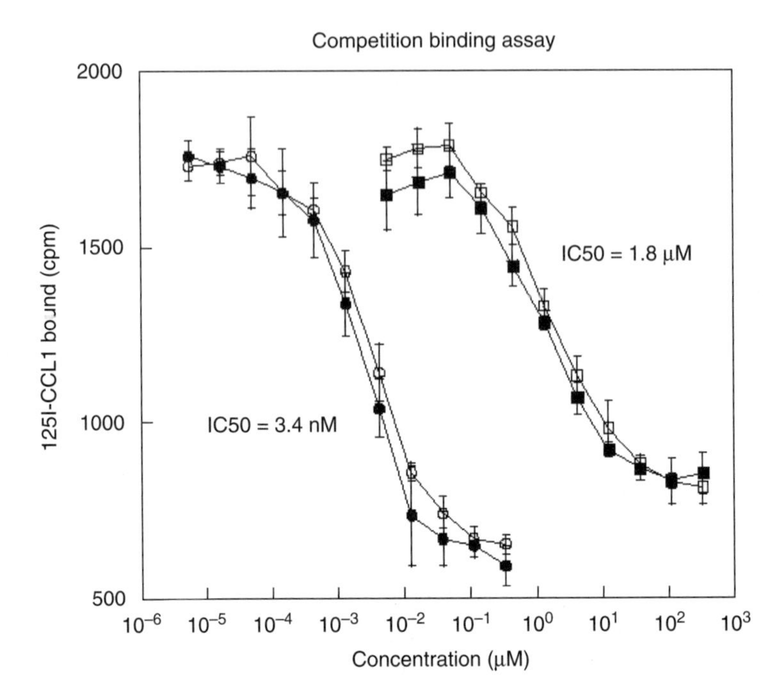

Fig. 2. Competition binding screens: Examples of concentration-response curves for ^{125}I-CCL1 binding to human CCR8. U87 cells expressing hCCR8 were incubated with trace amounts of ^{125}I-CCL1 (100 pM) and increasing concentrations of cold CCL1 (circles) or compound A (squares).

main event, activation or inactivation of the receptor, allowing for the determination of pharmacologic characteristics (affinity, potency, and efficacy) in a cell-free environment. The radiolabeled analogue allows for a homogenous SPA assay using membranes (crucial for a good signal-to-background), but this radiolabeled method usually yields low signal-to-background ratios and has the disadvantages described above relating to the use of radioactivity. There are several alternative labels, for instance, the use of europium-conjugated GTPγS for time-resolved fluorescence resonance energy transfer (FRET), which has two major advantages: it is a nonradioactive and higher signal-to-background detection method. However, nonradioactive GTP binding assays require separation of the free GTP from the bound (Gα*GTP) form, confining these assays to a much lower throughput.

In contrast, functional assays that look at post–receptor events, such as cAMP stimulation (G_s), cAMP inhibition (G_i), inositol triphosphate (IP3)/monophosphate (IP1) increase (G_q), or intracellular calcium mobilization (G_q), are homogeneous, for the most part nonradioactive, and easy to automate (with some exceptions). Measurement of intracellular calcium mobilization (Ca flux) is one of the most widely used functional assays not only to monitor G_q-coupled receptors but also cell lines transfected with promiscuous or chimeric G proteins (Gα16, $G_{qi}5$), which redirect the normal signal pathway through phospholipase Cβ and Ca^{2+}-mobilization. Calcium flux is a transient response, usually not more than 10 to 30 seconds, and although a widely used method, it requires specialized instrumentation to detect these subtle kinetic cellular changes. Two main detection technologies are available: fluorescence using calcium-sensitive dyes and luminescence using aequorin (photoprotein isolated from jellyfish). Assay development can be challenging depending on the type of cell culture (adherent or suspension), receptor and G-protein levels, and sensitivity and stability over time of cell line to the fluorescence dyes. The aequorin system has the advantage of a robust signal-to-background, with no toxicity issues or leakage of dye (common when using fluorescence dyes). However, it requires the coexpression of yet another protein.

Detection of cAMP levels is one of the preferred functional high throughput screen (HTS) assays when dealing with receptor signaling via G_s and if looking for receptor agonists via G_i. cAMP detection technologies have also evolved from traditional radioactive methods to homogeneous, mix and read assays *(15)*. Table 1 summarizes commonly used technologies for cAMP detection; these technologies are robust with Z values from 0.6 to 0.9.

Detection of IP3 is also possible although difficult to automate and implement due to the short life of IP3 (~20 seconds). An alternative approach to screen G_q-coupled receptors relies on the accumulation of IP1, by addition of 50 mM LiCl, after activation of the receptor with an agonist. This assay is based on the homogeneous time resolved fluorescence (HTRF) technology from CisBio (Cisbio, Bedford, MA), which developed specific antibodies for IP1. Assay development is in general straightforward, but agonists sometimes appear to have lower affinities than suggested by other assays, as an apparent decrease of affinity can sometimes be observed (Fig. 3); in the example shown, the agonist had an EC_{50} of 0.42 nM in the calcium mobilization assay compared with 120 nM in the IP1 assay.

The above described assays, both binding and functional, are usually part of primary screens in HTS format or used as secondary screens for a rapid understanding of the pharmacology of a hit, defined as a small molecule, peptide-like molecule, or antibody that binds to and functionally inhibits or stimulates the chemokine receptor activity. These assays are "household" items in most laboratories involved in GPCR screening and adapted on a case-by-case basis for the chemokine receptor of interest. The most frequent problem for chemokine receptor screening is the inherent ability of chemokines to bind nonspecifically to common labware materials. Some chemokines are also problematic due to their ability to self-aggregate giving rise to receptor-binding artifacts,

Table 1
Common Technologies Available for cAMP Detection

Parameters	Radiometric proximity assays (SPA, Flashplate)*	Fluorescence polarization (FP)	Time-resolved fluorescence (HTRF)	Amplified luminescence (ALPHAScreen)*	Enzyme (β-galactosidase) complementation	Electrochemiluminescence
cAMP label	^{125}I	Fluorescein	D2 acceptor	Biotin	Enzyme donor	Ruthenium derivative
Anti-cAMP modification	Bead/plate	None	Cryptate	Streptavidin	Enzyme acceptor	Plate
Number of additions	3	3	2–3	3	2–6	4
Detection limit (fmol/well)	200	<200	0.6	~7.5	~10	~10

*Flashplate and AlphaScreen assays (Perkin Elmer Life and Analytical Sciences Walthan, MA).

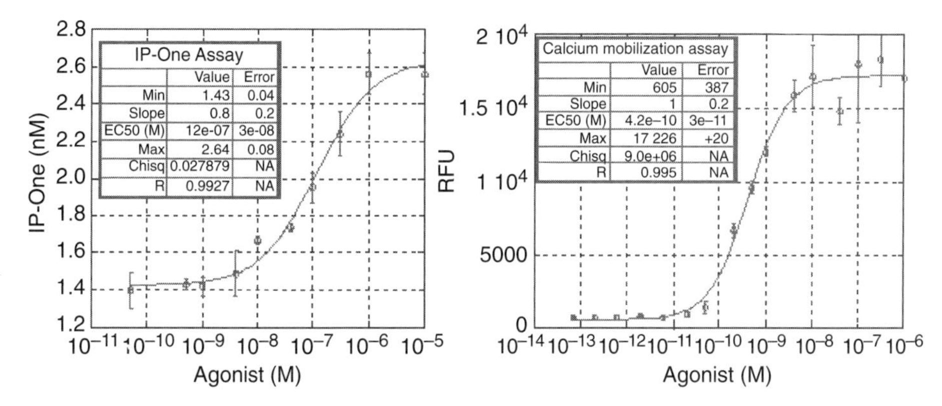

Fig. 3. Comparison between IP-One kit versus calcium mobilization assay (384-well format). Human embryonic kidney (HEK) 293 cells expressing a chemokine receptor were evaluated on HTRF IP-One kit (CisBio, Bedford, MA) and fluroescent imaging plate reader (FLIPR) with Calcium 3 kit (Molecular Devices, Mountain View, CA).

necessitating heterologous displacement rather than homologous displacement in binding assays to avoid this. Some chemokines are notorious for their problematic behavior in assays, and although this is well appreciated in the scientific community, publications are scarce, and these are usually discussed as "war stories" during HTS conventions at poster sessions. There are several examples of difficult chemokines, for example, CCL5, CXCL11, and CXCL10. Here we give an example using CCL5 where the goal of the experiment was to determine the B_{max} of a human CCR1-expressing cell line. For this determination, it is possible to use ^{125}I-labeled CCL5 or CCL3, as both ligands are available and bind to CCR1; determination of B_{max} is however very difficult (extremely dependent on buffer, cell line, number of receptors, and origin of the ligand) when using ^{125}I-labeled CCL5, because addition of cold CCL3 will actually increase the total binding signal.

The gold-standard assay used for all chemokine receptor inhibitors that reach clinical-phase trials is the chemotaxis functional assay. This assay relies on the ability of chemokines to recruit cells expressing their respective receptor to areas of inflammation. In vitro, this assay was first described in detail by Taub et al. *(16)* for 24/48-well plates; currently, this can be achieved by using 96-well plates. Cells are incubated in the upper chamber with an antagonist for a particular receptor (at different concentrations or with buffer) and challenged to migrate to the lower chamber, which has the relevant chemokine. After 2 to 4 hours of incubation at 37°C, the upper chamber inlet is removed and the cells in the lower chamber quantified by fluorescence with, for example, Calcein AM (Invitrogen, Carlsbad, CA).

18.4. How Successful Have These Approaches Been?

18.4.1. CCR5 Antagonists for HIV: The Success Story—Maybe

The finding, several years ago, that the chemokine receptors CCR5 and CXCR4 were major coreceptors, along with CD4, for human immunodeficiency virus (HIV-1) invasion resulted in the rapid development of chemokine receptor antagonists by the pharmaceutical industry, and CCR5 antagonists for the treatment of HIV have progressed the fastest through the clinic *(12)*.

HIV-1 resistance exhibited by some exposed but uninfected individuals is due, in part, to a 32-base-pair deletion in the CCR5 gene (CCR5Δ32), which results in a truncated protein that is not expressed on the cell surface *(17)*. About 1% of Caucasians are homozygous for the CCR5Δ32 allele and appear to be healthy with no untoward signs of disease *(17)*. In fact, recent findings suggest that homozygosity for the CCR5Δ32 alleles confers other selective advantages to these individuals, rendering them less susceptible to rheumatoid arthritis *(18)* and asthma *(19)* and prolonging survival of transplanted solid organs *(20)*.

The clinical relevance of CCR5 in solid-organ transplantation was assessed by looking at the effect of the CCR5Δ32 alleles in renal transplant survival. This study examined a total of 1227 renal transplant recipients, of which 21 were homozygous for CCR5Δ32. Analysis of the data demonstrated that individuals who were homozygous for CCRΔ32 had a survival advantage over individuals homozygous for wild-type CCR5. In fact, only one of the 21 CCR5Δ32 patients lost transplant function during follow-up, compared with 78 of the 555 patients with a CCR5 wild-type or heterozygous CCRΔ32 genotype. This study shows that CCR5 can play an important role in enhancing long-term allograft survival and, together with the studies referenced above, underscores the fact that CCR5 antagonists could be therapeutically useful in a variety of clinical situations including organ transplantation, asthma, rheumatoid arthritis, and HIV-1 infection.

Many pharmaceutical companies including Pfizer, Schering-Plough, and Novartis have programs targeting CCR5 *(13)*. Pfizer has the most advanced program and is currently conducting a phase III study for the treatment of HIV with a small-molecule CCR5 antagonist, maraviroc/UK-427,857 (Fig. 4), to which the U.S. Food and Drug Administration (FDA) granted fast-track designation, and a new drug application (NDA) is expected to be filed in 2006. If approved, this would be the first success story for a chemokine receptor antagonist.

The development of maraviroc *(21)*, much like other chemokine receptor antagonists, started with a high-throughput screen employing a competition binding assay and led to the hit compound UK-107,543. Chemical optimization of this compound led to the development candidate UK-427,857; during this optimization phase, a parallel characterization of the compounds was performed

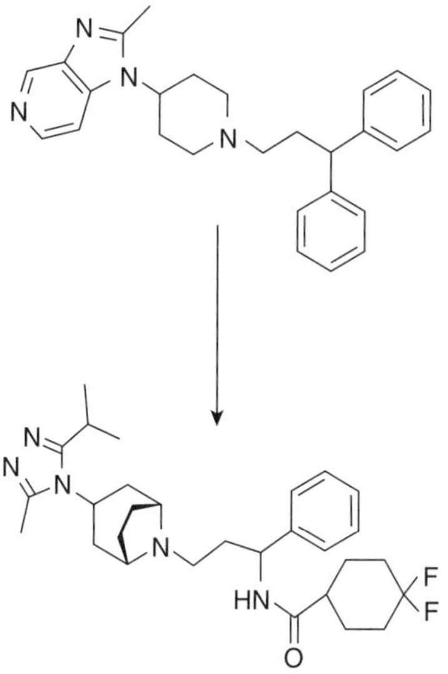

UK-107,543

Maraviroc
(UK-427,857)

Fig. 4. Structure of UK-427,857 (maraviroc), a CCR5-specific antagonist from Pfizer currently in phase III studies for HIV treatment. The original hit from the high-throughput screening, compound UK107,543, is also shown.

involving binding affinity, antiviral activity, absorption, pharmacokinetics, and selectivity assays for human ether-a-go-go related gene (hERG) channel. These particular assays reflect the need for developing early-on in the screening process a series of disease-oriented higher throughput assays, in this case the determination of antiviral activity using two different methods: (i) competition assay that measures inhibition of soluble recombinant HIV-1 glycoprotein 120 (gp120) (Ba-L strain) binding to CCR5 and (ii) inhibition of HIV-1 gp160-CCR5–mediated cell-cell fusion. The first method has a fluorescent readout and uses europium-labeled anti-gp120 antibody as a measure of total binding of the virus to CD4 cells in the presence or absence of a CCR5 antagonist. The inhibition of cell-cell fusion assay is a β-galactosidase reporter gene assay that takes advantage of Tat-induced expression and subsequent catalytic activity of HIV-1 long-terminal-repeat-regulated β-galactosidase.

Independently of the therapeutic approach, the basic pharmacology of the development candidate always includes measurement of calcium flux, GTPγS binding, receptor internalization (membrane localization is determined, e.g., by Fluorescence Activated Cell Sorting (FACS) studies with a fluorescent anti-CCR5 antibody in the presence or absence of an antagonist and natural ligand), and chemotaxis. It is also worthwhile to mention that the same assays that were performed to test the activity of the compounds on human CCR5 were used for testing its selectivity against other chemokine receptors or activity in other species *(22)*.

18.4.2. Clinical Failures: CCR1 and CXCR3

In contrast with the successful clinical trials of chemokine receptor antagonists for the treatment of AIDS, clinical trials of chemokine receptor antagonists for autoimmune diseases have been disappointingly unsuccessful. Three trials have been reported *(23)*. The Pfizer compound CP-481,715 entered phase II trials in February 2004, but based on data suggesting that it did not exhibit efficacy in patients with rheumatoid arthritis after 6 weeks of treatment, the clinical trial was terminated. The Berlex/Schering AG oral CCR1 antagonist BX 471 entered phase II clinical trials for multiple sclerosis in early 2004. Although the drug was well tolerated and showed no safety concerns, its development was stopped after the clinical phase II study failed to show a reduction in the number of new inflammatory CNS lesions (detected by magnetic resonance imaging). A CXCR3 antagonist from Amgen, T487, recently entered a phase II trial for psoriasis. Unfortunately, the inhibitor failed to demonstrate any signs of efficacy, and the trial was terminated. Figure 5 shows the structures of these development candidates.

The discovery process for all of these development candidates was similar. Antagonists exhibiting low nanomolar affinities were discovered through competition binding assays, and the antagonists were able to inhibit receptor func-

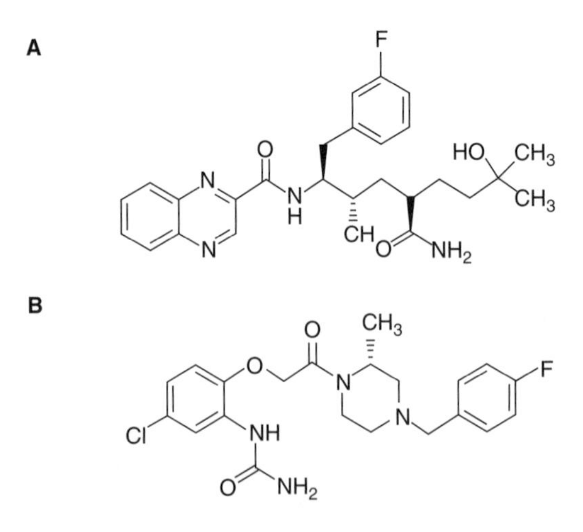

Fig. 5. Structures of (A) CP-481,715, (B) BX 471, and (C) T487. CP-481,715 and BX 471 are specific CCR1 antagonists from Pfizer and Berlex/Schering AG, and both failed to exhibit efficacy for rheumatoid arthritis and multiple sclerosis, respectively, in phase II trials. T487 is a specific CXCR3 antagonist from Amgen/Turalik that failed in phase IIa psoriasis clinical trials due to lack of efficacy.

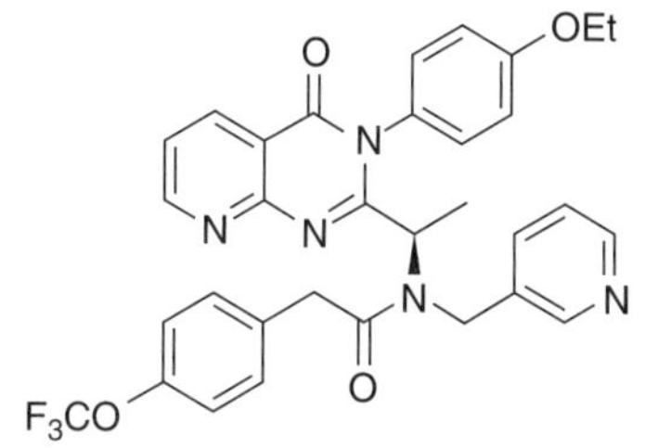

Fig. 5. (*Continued*)

tion as monitored by GTPγS binding, calcium flux, and chemotaxis assays *(24,25)*. The striking difference between this panel of assays and the ones described above, for the only success story with a chemokine receptor antagonist, is the lack of any disease-related high-throughput assay—contrary to AIDS, the complexity of autoimmune diseases cannot be simulated with any known in vitro assays with successful outcomes. Chemotaxis, although an important component of these diseases and an indicator of the ability of these compounds to inhibit migration of peripheral blood mononuclear cells (from patients or healthy individuals) toward a specific chemokine, certainly cannot predict the activity of these compounds in vivo. Limited cross-species activity has also been a factor limiting the use of in vivo models for prediction of outcome in clinical trials, although considerable effort has been made to engineer, for example, human CCR1 transgenic mice *(26)*. Biomarker assays for these diseases has also been challenging. CD11b upregulation and matrix metalloproteinase 9 (MMP-9) generation assays for assessment of both CCR1 antagonists was performed and indicated that both CP-481,715 and BX 471 inhibited CD11b upregulation, although both groups report high variability in these types of assays *(24,25)*.

18.5. Chemokine Receptors Are Targetable but Are They Good Targets?

Because the clinical data from these three failed trials have not been reported in any detail, it is hard to determine whether their lack of efficacy is attributable to the bioavailability of the molecules themselves or whether it could be more directly related to the role of the receptors CCR1 and CXCR3 in the pathogenesis of the diseases targeted. More likely, the heterogeneity of autoimmune diseases like rheumatoid arthritis and multiple sclerosis coupled with the promiscuity of chemokines and their receptors, described above, played some role in the failure of these therapeutic approaches. It has been demonstrated, for example, that CCR1, CCR2, CCR5, and CXCR3 are all implicated in the pathophysiology of multiple sclerosis. Thus, depending on the expression and activation of these receptors in the disease, more than one chemokine receptor may have to be inhibited to show efficacy. As we have discussed above, the regulation and expression of these receptors can be influenced by a variety of factors, and because specific biomarkers to predict the stage of the disease and receptor expression during disease progression are not available, it might be more effective to target multiple chemokine receptors to obtain efficacy. A recent report demonstrates the effectiveness of targeting multiple chemokines and receptors *(27)*. In this study, a model of intimal hyperplasia in response to femoral arterial injury was examined in transgenic mice that were induced to express M3, a herpesvirus protein that binds and inhibits multiple chemokines of all classes. Induction of M3 expression resulted in a 67% reduction in intimal area and a 68% reduction in intimal:medial ratio after femoral artery injury. In contrast with these data, targeted deletion of CCL2 alone resulted in only a 29% reduction in intimal hyperplasia *(28)* suggesting that multiple chemokines and receptors are involved in the pathophysiology.

In a mouse dextran sodium sulfate (DSS) model of colitis, TAK-779, a dual antagonist of CCR2 and CCR5, inhibited the recruitment of monocytes/macrophages into the colonic mucosa and delayed the onset of disease *(29)*. The expression of CCR2, CCR5, and CXCR3 mRNAs was inhibited in the TAK-779–treated mice. Consistent with these results, infiltration of monocytes/macrophages into the lamina propria was almost completely inhibited, and the blockade of CCR2, CCR5, and CXCR3 prevented murine experimental colitis by inhibiting the recruitment of inflammatory cells into the mucosa.

It follows from the discussion above that the design of promiscuous chemokine receptor antagonists that target several receptors could be quite advantageous in a number of situations. For example, joint CXCR4 and CCR5 inhibitors would be very useful therapeutically as viral fusion inhibitors to treat AIDS patients. Whether such an approach is feasible obviously depends on the similarity of the binding pockets of the receptors of interest. A positive example of this type of approach is provided by the angiotensin II receptor AT_1 and the endothelin receptor ET_A. Dual antagonists of these two receptors should be of greater benefit in the treatment of pulmonary hypertension, congestive heart failure, and arteriosclerosis. Although the overall sequence identity of AT_1 and ET_A is low (only 19%), it has proved to be possible to design antagonists that bind both receptors with high affinity in the low nanomolar range *(30)*.

18.6. How Relevant Are Heterodimers as Disease Targets?

It is now well accepted that many GPCRs can exist as homo- and heterodimers. However, the physiologic relevance of receptor dimerization is still largely unknown. It is clear from some studies that GPCR dimerization can alter ligand function; for example, a number of anti-parkinsonian agents have been reported to have a higher affinity with dopamine D_3/D_2 heterodimers than with the equivalent homodimers *(31)*. Another example is provided by the receptors CB_1 (cannabinoid) and orexin. When these are coexpressed, CB_1 enhances the ability of orexin A to activate the mitogen-activated protein (MAP) kinase pathway *(32)*. This effect requires a functional CB_1 receptor and is blocked by the specific antagonist rimonabant. Further studies have shown that the orexin receptor and the CB_1 receptor form heterodimers and that the appetite suppression effect of the CB_1 antagonist rimonabant targets this complex (G. Milligan, personal communication). This is a classic example of GPCR heterodimers playing a role in physiology. Drug discovery will obviously have to pay very careful attention to heterodimeric complexes of GPCRs.

An intriguing report by McGraw et al. demonstrates that even quite unrelated receptors can form heterodimers *(33)*. The authors reasoned that the successful response of asthmatics to β_2-adrenergic receptor agonists might involve GPCR interactions. They followed up on this idea by looking at the signaling of the prostanoid EP_1 receptor, because its endogenous agonist PGE_2 is abundant in the airway, but its functional implications are poorly defined. Although activation of EP_1 failed to elicit airway smooth muscle contraction by itself, it did significantly reduce the bronchodilatory function of a β_2-adrenergic receptor agonist. Using bioluminescence resonance energy transfer (BRET), the authors demonstrated that EP_1 and β_2-adrenergic receptor formed heterodimers that were responsive to an EP_1 agonist.

There is already a large literature establishing that chemokine receptors can form hetero- and homodimers. For example, CCR5 has been shown to form homo- and heterodimers with CCR2, a closely related receptor *(34)*. This dimerization appears to be induced by the CCR2 ligand CCL2. A recent study reveals that CCR5 and CCR2 heterodimerize with the same efficiency as they homodimerize *(35)*. Interestingly CCL4, a CCR5-specific ligand that was unable to compete for CCL2 binding on cells expressing CCR2 alone, efficiently inhibited CCL2 binding when CCR5 and CCR2 were coexpressed *(35)*. These findings suggest that CCR5 and CCR2 can form both homo- and heterodimers with similar efficiencies and that a receptor dimer only binds a single chemokine. Finally, another report suggests that a specific CCR5 antibody blocks HIV-1 replication in vitro as well as in vivo *(36)*. The antibody does not induce receptor downregulation or compete with chemokine binding and signaling on the receptor or interfere with the R5 JRFL viral strain gp120 binding to CCR5. It appears that the anti-CCR5 antibody efficiently prevents HIV-1 infection by inducing receptor dimerization. These findings suggest that it might be possible to develop new and interesting therapeutics that target the ability of chemokine receptors to dimerize without having any of the undesired proinflammatory side effects of the chemokines themselves.

Not all of the reports of chemokine dimerization are consistent, however, and whereas some studies have reported that CXCL12 induces dimerization of CXCR4 that is almost undetectable in the absence of the ligand *(37)*, others have suggested that CXCR4 is a constitutive dimer not affected by the CXCL12 *(38)*. Finally, even highly related chemokine receptors such as CXCR1 and CXCR2 that both bind CXCL8 with high affinity appear to show very different patterns of dimerization. For example, studies that clearly revealed the ligand-independent dimerization of CXCR2 also reported that CXCR1 did not dimerize *(39)*. However, these findings have recently been challenged by a study that used a combination of coimmunoprecipitation, saturation BRET, and a novel endoplasmic reticulum–trapping strategy *(40)*. In this study, the authors were able to demonstrate that CXCR1 is able to form both homo- and heterodimers with CXCR2.

Although we do not currently understand the physiologic consequences of receptor homo- and heterodimerization, the examples discussed above suggest that such receptor interactions may have important functional consequences. These receptor interactions can clearly affect surface expression of receptors, rates of receptor desensitization, and the pharmacology of ligands for the receptor. The clinical significance of these findings is just now being appreciated. For example, it is clear from the discussion above that the ability of chemokine receptors to act as coreceptors for the HIV virus is directly influenced by receptor heterodimerization. In line with this idea, a polymorphism (V64I) in the chemokine receptor CCR2 has been found to correlate with a markedly decreased rate of AIDS progression *(41)*. The V64I CCR2 polymorphism has also been shown to enhance heterodimerization between CCR2/CCR5 and CCR2/CXCR4 *(42)*. These findings suggest that heterodimerization of chemokine receptors is a key determinant in the ability of HIV to use these receptors to gain entry into cells, and it might suggest a new avenue of therapeutic intervention in the future.

18.7. Conclusions

The success of chemokine receptor antagonists as therapeutics will depend on a number of factors. It is clear that CCR5 antagonists will probably pave the way as the first approved chemokine drugs to treat disease in man. Here the situation is quite clear. The AIDS virus, HIV, utilizes CCR5 as a coreceptor to gain entry into the cell, and interfering with this process is likely to be of benefit in infected individuals. In autoimmune diseases, the utility of chemokine receptor antagonists is currently less clear. Why have several approaches from pharmaceutical industries targeting different chemokine receptors failed? Were the wrong receptors targeted? Did poorly correlative animal models of human disease mislead them? Did they fail because they are complex diseases in which several chemokine receptors, acting in concert, drive the pathophysiology? The list of possible reasons for the failure of these approaches could by itself fill an entire review, and it is apparent that we will need to consider these and many other questions and obtain answers to them if we are to make any real progress in bringing additional chemokine receptor antagonists to market.

Despite the failures discussed above, we remain cautiously optimistic that chemokine drugs can be of benefit in man. Their ultimate success, however, will depend on our ability to more clearly understand the role of chemokine receptors in driving the pathophysiology of complex autoimmune diseases than we currently do. This coupled with a parallel understanding of the animal models used as predictors of the human disease will also need to be more appreciated. Finally, much better clinical markers of the disease process in man will also be required not only to set up clinical trials more intelligently but also ultimately to monitor their progress. If we can make real progress in coming to grips with the issues discussed above, we might finally realize the promise of chemokine receptor antagonists as registered drugs.

References

1. Metchnikoff, II. Uber eine Sprosspilzkrankheit der Daphnien; Beitrag zur Lehre über den Kampf der Phagozyten gegen Krankheitserreger. Archiv fur pathologische Anatomie und Physiologie und fur klinische Medicin 1884;96:177–195.
2. Deuel TF, Keim PS, Farmer M, Heinrikson RL. Amino acid sequence of human platelet factor 4. Proc Natl Acad Sci U S A 1977;74:2256–2258.
3. Walz A, Peveri P, Aschauer H, Baggiolini M. Purification and amino acid sequencing of NAF, a novel neutrophil-activating factor produced by monocytes. Biochem Biophys Res Commun 1987;149:755–761.
4. Yoshimura T, Matsushima K, Tanaka S, et al. Purification of a human monocyte-derived neutrophil chemotactic factor that has peptide sequence similarity to other host defense cytokines. Proc Natl Acad Sci U S A 1987;84:9233–9237.
5. Baggiolini M. Chemokines and leukocyte traffic. Nature 1998;392:565–568.
6. Fredriksson R, Schioth HB. The repertoire of G-protein-coupled receptors in fully sequenced genomes. Mol Pharmacol 2005;67:1414–1425.
7. Horuk R, Chitnis CE, Darbonne WC, et al. A receptor for the malarial parasite Plasmodium vivax: the erythrocyte chemokine receptor. Science 1993;261:1182–1184.

8. van Berkel V, Barrett J, Tiffany HL, et al. Identification of a gammaherpesvirus selective chemokine binding protein that inhibits chemokine action. J Virol 2000;74:6741–6747.
9. Loetscher P, Seitz M, Baggiolini M, Moser B. Interleukin-2 regulates CC chemokine receptor expression and chemotactic responsiveness in T lymphocytes. J Exp Med 1996;184:569–577.
10. Struyf S, Proost P, Schols D, et al. CD26/dipeptidyl-peptidase IV down-regulates the eosinophil chemotactic potency, but not the anti-HIV activity of human eotaxin by affecting its interaction with CC chemokine receptor 3. J Immunol 1999;162: 4903–4909.
11. Petkovic V, Moghini C, Paoletti S, Uguccioni M, Gerber B. I-TAC/CXCL11 is a natural antagonist for CCR5. J Leukoc Biol 2004;76:701–708.
12. Horuk R. Chemokine receptors. Growth Factor Reviews 2001;12:313–335.
13. Ribeiro S, Horuk R. The clinical potential of chemokine receptor antagonists. Pharmacol Ther 2005;107:44–58.
14. Ribeiro S, Horuk R. Chemokine receptor antagonists from the bench to the clinic. In: Hannan A, Engelhardt B, eds. Leukocyte Trafficking-Molecular mechanism, therapeutic targets, and methods. Weinheim: Wiley-VCH; 2005:371–402.
15. Edwards BS, Oprea T, Prossnitz ER, Sklar LA. Flow cytometry for high-throughput, high-content screening. Curr Opin Chem Biol 2004;8:392–398.
16. Taub DD, Proost P, Murphy WJ, et al. Monocyte chemotactic protein-1 (MCP-1), -2, and -3 are chemotactic for human T lymphocytes. J Clin Invest 1995;95:1370–1376.
17. Liu R, Paxton WA, Choe S, et al. Homozygous defect in HIV-1 coreceptor accounts for resistance of some multiply-exposed individuals to HIV-1 infection. Cell 1996;86:367–377.
18. Ditzel HJ, Rosenkilde MM, Garred P, et al. The CCR5 receptor acts as an alloantigen in CCR5Delta32 homozygous individuals: identification of chemokine and HIV-1-blocking human antibodies. Proc Natl Acad Sci U S A 1998;95:5241–5245.
19. Ahmed RK, Nilsson C, Biberfeld G, Thorstensson R. Role of $CD8^+$ cell-produced anti-viral factors in protective immunity in HIV-2-exposed but seronegative macaques resistant to intrarectal SIVsm challenge. Scand J Immunol 2001;53:245–253.
20. Fischereder M, Luckow B, Hocher B, et al. CC chemokine receptor 5 and renal-transplant survival. Lancet 2001;357:1758–1761.
21. Dorr P, Westby M, Dobbs S, et al. Maraviroc (UK-427,857), a potent, orally bioavailable, and selective small-molecule inhibitor of chemokine receptor CCR5 with broad-spectrum anti-human immunodeficiency virus type 1 activity. Antimicrob Agents Chemother 2005;49:4721–4732.
22. Napier C, Sale H, Mosley M, et al. Molecular cloning and radioligand binding characterization of the chemokine receptor CCR5 from rhesus macaque and human. Biochem Pharmacol 2005;71:163–172.
23. Pease JE, Horuk R. CCR1 antagonists in clinical development. Expert Opin Invest Drugs 2005;14:785–796.

24. Liang M, Mallari C, Rosser M, et al. Identification and characterization of a potent, selective, and orally active antagonist of the CC chemokine receptor-1. J Biol Chem 2000;275:19000–19008.
25. Gladue RP, Tylaska LA, Brissette WH, et al. CP-481,715: A potent and selective CCR1 antagonist with potential therapeutic implications for inflammatory diseases. J Biol Chem 2003;278:40473–40480.
26. Gladue RP, Cole SH, Roach ML, et al. The human specific CCR1 antagonist CP-481,715 inhibits cell infiltration and inflammatory responses in human CCR1 transgenic mice. J Immunol 2006;176:3141–3148.
27. Pyo R, Jensen KK, Wiekowski MT, et al. Inhibition of intimal hyperplasia in transgenic mice conditionally expressing the chemokine-binding protein M3. Am J Pathol 2004;164:2289–2297.
28. Kim WJ, Chereshnev I, Gazdoiu M, Fallon JT, Rollins BJ, Taubman MB. MCP-1 deficiency is associated with reduced intimal hyperplasia after arterial injury. Biochem Biophys Res Commun 2003;310:936–942.
29. Tokuyama H, Ueha S, Kurachi M, et al. The simultaneous blockade of chemokine receptors CCR2, CCR5 and CXCR3 by a non-peptide chemokine receptor antagonist protects mice from dextran sodium sulfate-mediated colitis. Int Immunol 2005;17:1023–1034.
30. Murugesan N, Gu Z, Fadnis L, et al. Dual angiotensin II and endothelin A receptor antagonists: synthesis of 2'-substituted N-3-isoxazolyl biphenylsulfonamides with improved potency and pharmacokinetics. J Med Chem 2005;48:171–179.
31. Maggio R, Scarselli M, Novi F, Millan MJ, Corsini GU. Potent activation of dopamine D3/D2 heterodimers by the antiparkinsonian agents, S32504, pramipexole and ropinirole. J Neurochem 2003;87:631–641.
32. Hilairet S, Bouaboula M, Carriere D, Le Fur G, Casellas P. Hypersensitization of the orexin 1 receptor by the CB1 receptor: evidence for cross-talk blocked by the specific CB1 antagonist, SR141716. J Biol Chem 2003;278:23731–23737.
33. McGraw DW, Mihlbachler KA, Schwarb MR, et al. Airway smooth muscle prostaglandin-EP1 receptors directly modulate beta2-adrenergic receptors within a unique heterodimeric complex. J Clin Invest 2006;116:1400–1409.
34. Rodriguez-Frade JM, Vila-Coro AJ, de Ana AM, Albar JP, Martinez AC, Mellado M. The chemokine monocyte chemoattractant protein-1 induces functional responses through dimerization of its receptor CCR2. Proc Natl Acad Sci U S A 1999;96:3628–3633.
35. El-Asmar L, Springael JY, Ballet S, Urizar Andrieu E, Vassart G, Parmentier M. Evidence for negative binding cooperativity within CCR5-CCR2b heterodimers. Mol Pharmacol 2005;67(2):460–469.
36. Vila-Coro AJ, Mellado M, Martin de Ana A, et al. HIV-1 infection through the CCR5 receptor is blocked by receptor dimerization. Proc Natl Acad Sci U S A 2000;97:3388–3393.
37. Vila-Coro AJ, Rodriguez-Frade JM, Martin De Ana A, Moreno-Ortiz MC, Martinez AC, Mellado M. The chemokine SDF-1alpha triggers CXCR4 receptor dimerization and activates the JAK/STAT pathway. FASEB J 1999;13:1699–1710.

38. Babcock GJ, Farzan M, Sodroski J. Ligand-independent dimerization of CXCR4, a principal HIV-1 coreceptor. J Biol Chem 2003;278:3378–3385.
39. Trettel F, Di Bartolomeo S, Lauro C, Catalano M, Ciotti MT, Limatola C. Ligand-independent CXCR2 dimerization. J Biol Chem 2003;278:40980–40988.
40. Milligan G, Wilson S, Lopez-Gimenez JF. The specificity and molecular basis of alpha1-adrenoceptor and CXCR chemokine receptor dimerization. J Mol Neurosci 2005;26:161–168.
41. Smith MW, Dean M, Carrington M, et al. Contrasting genetic influence of CCR2 and CCR5 variants on HIV-1 infection and disease progression. Science 1997;277:959–965.
42. Mellado M, Rodriguez-Frade JM, Vila-Coro AJ, de Ana AM, Martinez AC. Chemokine control of HIV-1 infection. Nature 1999;400:723–724.

Index

T